Fundamentals of Geomorphology

Fundamentals of Geomorphology   **R J Rice**

LONGMAN
London and New York

W e are grateful to the following for permission to re-produce copyright material:

American Association for the Advancement of Science for figures from an article 'The Brunhes Epoch: Isotopic Paleotemperatures and Geochronology' by C. Emiliani and N. J. Shackleton in *Science* Vol. 183. Copyright 1974 by the American Association for the Advancement of Science; The American Association of Petroleum Geologists for figures from *Bulletin of American Association of Petroleum Geologists* Nos. 3, 1964 and 54, 1970; American Geophysical Union and the authors for figures by Isacks and Opdyke from *Journal of Geographical Research* 73, 1968; The American Journal of Science and the authors for figures by Prof. S. A. Schumm and Prof. R. J. Chorley from *American Journal of Science* 262, 1964; Edward Arnold Publishers Ltd. for a figure from *Beaches and Coasts* by Cuchlaine King; The author for a figure from his book *The Relief and Drainage of Wales* by Prof. E. H. Brown; Cambridge University Press for data from *Hillslope Form and Process* by Carson & Kirkby; Constable Publishers for data from *Figures from Geographical Essays* by W. M. Davis 1954; Dover Publications Inc. for a figure from *The Origin of Continents and Oceans* by Wegener; Elsevier Publishing Co. for data from *Marine Geology* No. 9, 1970 by L. W. Wright; The Geological Society of America and the authors for a diagram from a paper prepared by Dr. D. B. Prior and Dr. A. J. Eardley from *Bulletin* 72, 1961. Reprinted by courtesy, The Geological Society of America Inc.; The authors for data from material by Herron, Pitman and Larson Lamont-Doherty entitled 'The Age of The Ocean Floors'; The Institute of British Geographers for a figure by A. Goudie et al from *Area* 1 (4) 1970, a figure after G. de Boer from *Trans IBG* 34, 1964 and two figures after J. T. Andrews from *IBG Special Publication* No. 2 1970; The author for figures by André Journaux from Centre de Geomorphology, Caen; The author for a figure from *Morphology of the Earth* by Prof. Lester King; Macmillan Publishers Ltd. for a figure from *Morphological Analysis of Landforms* by W. Penck. Reprinted by permission of Macmillan London and Basingstoke; Scottish Geographical Society and the author for figures by Moisley and Linton from the *Scottish Geographical Magazine* 76, 1960; National Science Foundation for figures from *Initial Reports of the Deep Sea Drilling Project* Vol. 12, 1972; Nature Magazine and the author for this figures giving schematic diagram by Dr. F. J. Vine and Matthews from *Nature* 199; Pergamon Press Ltd. for figures from *The Physics of Glaciers* by Paterson, 1969 and *Physics and Chemistry of the Earth* Vol. 4; Princeton University Press for a figure from 'Quaternary Paleohydrology' by S. A. Schumm in *The Quaternary of the United States* ed. by H. E. Wright, Jr. and David G. Frey (c) 1965 by Princeton University Press; The Royal Society and the author for figures from an article 'Creep of Polycrystallic Ice' by J. W. Glen from *Proc.* 1955, A.228; Seismological Society of America for figures from 'World Seismicity Maps Compiled from ESSA, Coast and Geodetic Survey, Epicenter Data, 1961–1967' by Barazangi and Dorman in *Bulletin* SSA 59: 1; The author for his figures from *Geografiska* Annaler 38 by Prof. Sundborg; University of California Press for a figure from *The Incomparable Valley: A Geologic Interpretation of the Yosemite* by Francois E. Matthes. Copyright (c) 1950 by The Regents of the University of California; John Wiley & Sons Inc. for a figure from 'The Quaternary of Sweden' by J. Lundquist in *The Geologic Systems* ed. by K. Rankama; The authors for figures from *Structure, Surface and Drainage* by S. W. Wooldridge and D. L. Linton.

We are also grateful to the following for permission to use copyright material: The United States Geological Survey for Fig. 4.5 and Fig. 14.5(A); Aerofilms Ltd. for Fig. 9.13 and Fig. 17.3; Cambridge University Collection, copyright reserved, for Fig. 10.4; A. V. Rice for Fig. 14.3; Professor A. V. Morgan and the Royal Society for Fig. 14.5(B) from *Phil. Trans.*

Whilst every effort has been made to trace the owners of copyright, in a few cases this has proved impossible and we take this opportunity to offer our apologies to any authors whose rights may have been unwittingly infringed.

This book stems from a course of lectures given to first-year undergraduates in the department of Geography at the University of Leicester. The aim of the course — and therefore of the book — has been to build on the foundations normally provided by sixth-form studies at school. At the same time a conscious effort has been made to shift the student's thinking from some of the more sterile aspects of school work in geomorphology towards paths that seem, in the author's views potentially much more rewarding. The term 'fundamentals' in the title is intended to imply an introductory survey from which the reader can progress smoothly to more advanced studies in future years. Limitations of space have precluded full coverage of the multifarious aspects of modern geomorphology; instead the aim has been to provide a conspectus of selected trends within the subject. Similar considerations of space have prevented the full development of a number of more difficult topics that have nevertheless been introduced in outline; the justification resides in the view that exposure to the underlying ideas will indicate to the reader the direction in which future work is likely to lead and, with luck, whet his appetite for these later studies.

In order to facilitate the transition from school to university work a conventional arrangement of the material has been adopted. The chapters are divided into four major groups: Chapters 1 to 5 review those aspects of physical geology that are especially pertinent to landform studies; Chapters 6 to 10 outline the processes whereby 'normal erosion converts the deformed surface of the continents into the familiar pattern of hills and valleys; Chapters 11 to 14 cover the fields of glacial and periglacial geomorphology; finally, Chapters 15 to 17 examine the factors controlling the evolution of coastal landforms. By contrast with several recent texts on geomorphology, a relatively full treatment is accorded to relevant aspects of physical geology. Many reasons might be adduced in support of this return to an older tradition but the foremost is the general philosophy that the surface of the globe is best viewed as the interface between two energy systems, one fuelled by the sun and the other by internal processes within the earth. The overriding importance of the former cannot be denied, but equally it would seem foolish to ignore the role of 'endogenetic' forces at the very time when geophysicists are emphasizing the general mobility of the crust.

Throughout the book the approach may be described as quantitative but non-mathematical. Experience has taught that even able undergraduates may be deterred by a too rigorously mathematical introductory course; on the other hand it is vital that such students should come to appreciate, often for the first time, the necessity for quantitative measurement and analysis. A primary objective of the book has therefore been to instil such an appreciation without ever assuming more than a most rudimentary knowledge of mathematics.

Authorship of a book that spans so many aspects of geomorphology means that I have inevitably had to rely upon persons and writings too numerous to mention individually. At the end of each chapter I have included a list of articles and books which proved particularly useful in the initial drafting; the lists are by no means exhaustive and to any individual who finds his original ideas incorporated without due acknowledgement I tender sincere apologies. I should particularly like to thank my mother, Mrs A. E. Rice, for her speedy and painstaking typing on my behalf. Last but by no means least, I should like to express my gratitude to my immediate family for their forebearance during the period of the writing.

R. J. RICE *Leicester* *January 1976*

The aim of this first chapter is to review the way in which knowledge of the shape and constitution of the earth has been gathered since the initial attempts to measure the size of the sphere over 2 000 years ago. It is concerned with the study and measurement of the earth as it exists at the present day; consideration of the way the sphere may have evolved through time forms the essential theme of Chapters 3 to 5. Even the seemingly limited objective of describing the shape and relief of the globe faces formidable problems. Until very recently the proportion of the solid surface that could be directly observed was severely restricted. Water obscures the exterior relief over more than two-thirds of the globe with the result that, for many centuries, scientific investigation was confined to the continents. Bathymetric sounding did not begin in a systematic fashion until the second half of the nineteenth century, and only since the Second World War has it proceeded at a rapid rate. Even more recent is the innovation of using submersibles to examine the nature of the ocean floors directly; as yet very few areas have been studied by this method.

Visual examination of the materials composing the earth is extraordinarily restricted. The depth to which boreholes and mines have penetrated is a minute fraction of the total radius of the sphere. Even at the present day the deepest mines go down little more than 3 km and the longest boreholes penetrate no more than 10 km. This is less than one six-hundredth part of the distance from the surface to the centre of the earth. Moreover, the distribution of boreholes is extremely erratic. In particular, study of the materials of the ocean floor has lagged far behind that of the continents because of the immense problems of boring in deep water. On some nineteenth-century oceanographic expeditions specially designed tubes enabled recovery of samples from the top 30 cm of sediment, but it was regarded as a great advance when, in 1925, the German research ship 'Meteor' recovered cores more than 1 m long. As late as 1960 very few cores exceeding 20 m in length had been obtained from deep-water sites, the normal instrument at this time still being the piston corer which uses hydrostatic pressure to force the sampling tube into the sediments. It was only in the late 1960s that rotary drilling was adopted, notably on board the American research ship 'Glomar Challenger'. Launched in 1968 with a capacity for drilling 770 m into the sea-floor beneath water up to 6 000 m deep, this vessel in the first 5 years of the DSDP (deep sea drilling project) bored more than 300 holes in the floors of the Atlantic, Indian and Pacific Oceans.

In the following account attention will first be directed to the basic shape and gross relief features of the earth. Thereafter consideration will be given

1|1

Gross morphology and structure of the earth

to the two major lines of evidence bearing upon the internal constitution of the earth ; namely, variations in the strength of the earth's gravity field and the way in which vibratory motions are transmitted through the globe.

Shape and dimensions of the Earth

The spherical form of the earth had already been conjectured by the Greeks in the sixth century B.C., and the first important attempt to measure its size is commonly credited to Eratosthenes during the third century B.C. He employed the simple observation that, during the summer solstice, the sun is almost directly overhead at Aswan whereas it is $7°12'$ from the zenith at Alexandria (Fig. 1.1). Assessing the distance between Alexandria and Aswan as 5 000 stadia, he calculated the overall circumference of the earth as 250 000 stadia. The measurements were necessarily crude but if, as is believed, the stadium was 185 m the computed dimensions of the earth were only about 15 per cent too large. Later Greek and Arab astronomers made many similar assessments using the same basic principles, but interest in the whole subject was dramatically revived in the

Fig. 1.1. The simple geometrical principles by which Eratosthenes calculated the overall size of the globe.

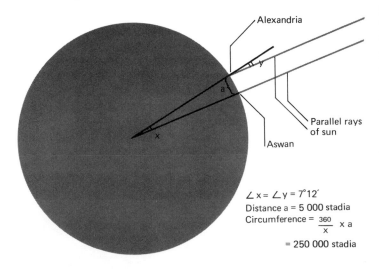

Alexandria

Parallel rays of sun

Aswan

$\angle x = \angle y = 7°12'$
Distance a = 5 000 stadia
Circumference = $\dfrac{360}{x}$ x a

= 250 000 stadia

seventeenth century with the prediction by Sir Isaac Newton that the earth would prove to be a spheroid flattened at the poles. Early observational support for this contention came from the fact that an accurate pendulum clock, regulated to keep time in Paris, lost almost $2\frac{1}{2}$ minutes a day when transferred to equatorial latitudes in South America. This suggested that gravitational attraction, and therefore distance from the centre of the earth, might vary with latitude. On the other hand, survey work in France failed to reveal the expected poleward increase in the length of a degree of latitude and in order to resolve the problem the French Academy of Sciences in the 1730s despatched scientific expeditions to what are now Ecuador and Finland to measure the length of a meridian degree at contrasting latitudes. The results clearly confirmed Newton's original prediction.

The investigation sponsored by the Academy of Sciences heralded the birth of the modern science of geodesy. At first this depended almost entirely on painstaking measurement of huge arcs across the surface of the globe, but since the 1950s increasing use has been made of data from artificial satellites. Two concepts are basic to most geodetic investigations. The first is the spheroid, an imaginary surface which may be visualized as the shape the earth would take if, whilst still retaining its present mass and motion, its materials were redistributed to give uniform concentric shells, the outermost of which would be a continuous ocean about 2 400 m deep. The combined effect of gravitational and centrifugal forces would produce an ellipsoid with a semi-major axis length of 6 378·16 km, a semi-minor axis length of 6 356·18 km and a flattening of 1 298·3. The second concept is the geoid which may be visualized as the surface described by mean sea-level in oceanic areas and by the level to which sea water would rise in hypothetical interconnected tunnels in the continental areas. The geoid lacks the geometrical simplicity of the spheroid since it reflects the gravitational attraction of the irregularly distributed features of the Earth's outer shell. Comparison of the form of the spheroid and the geoid has disclosed a number of significant features. For reasons that will become apparent below, the two surfaces show much less discrepancy than might be anticipated from such obvious topographic features as ocean basins, continental masses and mountain ranges. On the other hand, precise measurements of the orbits of artificial satellites, and in particular the way these deviate from Kepler's laws of planetary motion, have disclosed undulations in the geoid not related in any obvious way to surface features. The cause of these undulations remains uncertain, but they presumably reflect deep-seated variations within the earth and may ultimately provide an important clue to what is going on at depth.

Distribution on relief of the earth's surface

Compared with only a few decades ago, the form of the solid outer surface of the globe is now very thoroughly surveyed. Over the continental areas the pace of detailed mapping has been much accelerated by the use of aerial photography and application of the general principles of stereoscopy. Over the

Fig 1.2 The principle of the continuous recording echo-sounder. Reflections are regularly received not only from the sea floor itself but also from suitable surfaces within the submarine sediments. Consequently much valuable information about submarine structure is normally obtained in the course of echo-sounding.

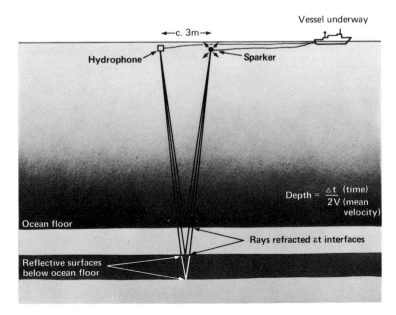

oceans instruments and techniques originally developed for submarine warfare have been employed since 1945 for bathymetric purposes. Particularly important in the latter context has been the continuous recording echo-sounder capable of producing a complete profile of the submarine topography beneath a moving vessel. Simple echo-sounding equipment can easily be attached to an ordinary ship. However, for more accurate oceanographic work a source of sound pulses, commonly a 'sparker' generating the pulses by means of an electrical discharge, is towed behind the ship just below the surface, together with a receiving hydrophone (Fig. 1.2). Although the speed of sound in water varies with pressure, temperature and salinity, a precision of ± 2 m may be attained in water that is 5 000 m deep. The result is that, with very few exceptions, the basic form of the ocean floors is now well known.

One of the most formidable problems facing any surveyor interested in the bedrock relief is a perennial cover of ice. This applies both to continental areas such as Antarctica and Greenland with their massive ice-sheets, and to the Arctic Ocean with its persistent cover of sea ice. However, for each of these unfavourable environments methods have now been devised that permit accurate and rapid charting of the rock surface. In the Arctic this has involved the use of submarines in place of surface vessels, while over the major ice-sheets planes carrying special radar equipment have, during the last few years, added rapidly to our knowledge of the bedrock topography (see p. 230). As a consequence further exploratory work is unlikely to reveal basic lineaments not already discovered, but will be concerned primarily with elaborating topographic detail.

The most striking aspect of the form of the solid surface is the dominance of two distinct levels. From a hypsographic curve (Fig. 1.3) it can be seen that approximately 30 per cent of the earth's surface lies between 2 000 m and −200 m, and a further 50 per cent between −3 000 m and −6 000 m. The former comprises what is termed the continental level, the latter the oceanic level. Of very much smaller extent are the intermediate slopes, the high mountain chains, and the oceanic deeps. Complicating this simple division of global relief is the fact that not only the oceanic level but also part of the conintental level is currently submerged. As will be shown below,

4 in terms of the deep-seated structure of the earth the most fundamental distinction is that between continents and ocean basins, the former to include the drowned fringes of the continental level. On the other hand, erosional and depositional processes beneath the sea are so different from those on land that it is natural to seek a distinction between subaerial and submarine environments. If these two principles are to be accommodated, a threefold division of global relief must be employed, namely, the continents, the submerged continental margins, and the ocean basins.

Relief of the continents The importance and variety of relief patterns on the continents is acknowledged in the everyday use of such expressions as mountain range, hill country or coastal plain. They indicate an intuitive recognition of an association of features repeated at intervals over the globe. At first sight it might seem that greater precision could readily be accorded these descriptive terms and that it would then be a simple task to map their distribution over the land surfaces of the world.

However, such is the complexity of relief forms that their scientific classification has proved a most formidable problem. Moreover, geomorphologists have usually tended to adopt genetic classifications in which as much attention is paid to structure and presumed geological evolution as to pure morphology. This method is well illustrated by the classic division of the United States into physiographic provinces by Fenneman. He wrote

'Unity or similarity of physiographic history' is a formula which *almost* designates the basis here in mind for the delineation of provinces. It implies that the topography throughout the province is all to be explained and described in the same story. It does not imply that it is superficially the same throughout the area. . . . Hence in basing provinces on topography it is understood that the features throughout a province are *essentially related* rather than superficially alike.

Most workers seeking a purely morphological classification acknowledge that a minimum of four variables must be considered: relief amplitude, slope angle frequency, drainage

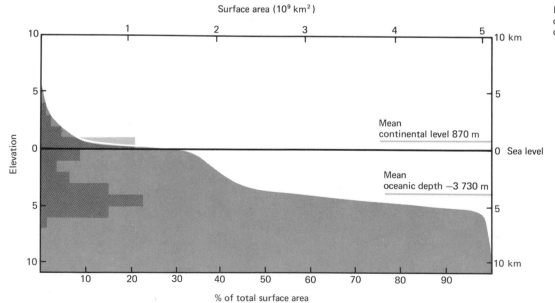

Fig 1.3. The distribution of global relief, depicted by means of both a hypsographic curve and a bar graph.

texture and characteristic profile form. The need to introduce the first two is obvious. The third, although closely related to relief and slope, is necessary to distinguish between a landscape of ravines and one with more widely spaced valleys. The last is essential to distinguish, for example, between dissected tablelands and plains with upstanding hills since both could conceivably have identical values for relief and slope. Having identified four critical parameters, there is still the problem of specifying the method by which actual measurement and analysis should be made. No general agreement on this point has ever been reached, and rigorous application of various quantitative measures of form has so far achieved limited success. One of the most persistent advocates of landform classification based on morphological criteria has been the American geographer, Hammond. He devised a seven-category classification based upon three major characteristics : relative amount of gently sloping land, local relief and generalized slope profiles. The way in which the principal terrain types are related to each other, together with their world distribution, is shown in Fig. 1.4.

The most striking characteristic of the world map is the spotty distribution. Unlike the ocean basins to be described below, there is no gross patterning common to all continents and very few systematic arrangements can be discerned. The one-terrain type well represented on all land masses is the plain, over one third of the total continental area falling into that category. High mountains, on the other hand, are totally absent from the Australian mainland and occupy a very small proportion of Africa. Many disparities of this sort can be related to recent geological history. For instance, the relatively large proportion of Eurasia and the Americas occupied by high mountains appears to be related to the participation by those continents in recent orogenic activity. Conversely the great extent of rolling and irregular plains in Africa and Australia may be associated with their lack of recent large-scale deformation. However, comments of this nature are prompted by a knowledge of tectonic history ; viewed solely in terms of a geometrical pattern the most abiding impression is of a haphazard arrangement of terrain types.

Relief of the continental margins By convention the **5** continental margins are divided into two separate units, the continental shelf and the continental slope, well exemplified in the submarine topography off the coast of the north-eastern United States (Fig. 1.5).

(a) *The continental shelf.* The shelf comprises the underwater extension of the continental coastal plain. Its width averages about 60 km, but ranges from zero around parts of the Pacific Ocean to 900 km around parts of the Arctic Ocean. Its seaward edge is marked by a rapid increase in gradient known as the shelf break. This boundary normally lies at a depth of about 200 m but considerable variations occur in different parts of the globe. It is often particularly deep in high-latitude areas, being found at depths of 350–400 m around the Arctic Ocean and parts of the North Atlantic, and at 300 m along the fringes of the Antarctic continent. If sea-level were to fall so as to expose the shelf, the land area of the globe would be increased from 29 to almost 33 per cent of the total surface.

Where adjacent to economically advanced nations the shelf has often been surveyed in great detail and even examined visually by means of submersibles. It has been possible to recognize topographic forms directly comparable to those found on the adjacent landmass. This is true, for instance, in the apparent seaward extension of river valleys around many coasts of the world, and also in submerged glacial landforms traceable off the coasts of both Europe and North America. Yet such comparisons should not be allowed to obscure the greater overall smoothness of the submarine topography, especially along the outer parts of the shelf, which results primarily from a cover of unconsolidated sediments. Furthermore, the mean seaward gradient is less than quarter of one degree.

(b) *The continental slope.* Lying immediately below the shelf break, this is one of the most distinctive relief features of the whole globe. It is almost always present to mark the true edge of the continent and its total area is actually greater than that of the shelf. In detail its form is quite variable. In some localities gradients exceeding 25° have been recorded, but a more representative value would be 2–5°. To those familiar with

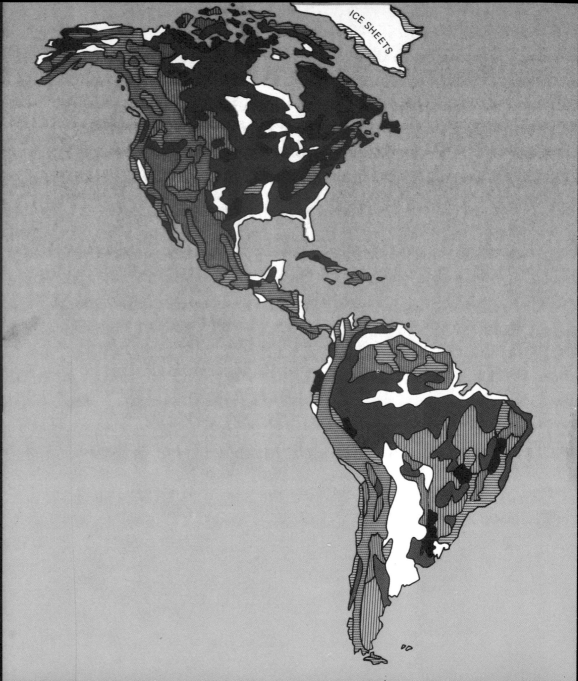

Fig 1.4 The relief of the continents classified into seven terrain-types defined purely on the basis of their morphology. The key shows diagrammatically the relationship between the different terrain-types and the criteria employed in their definition (simplified from an original map in Trewartha, Robinson and Hammond 'Fundamentals of physical geography'.

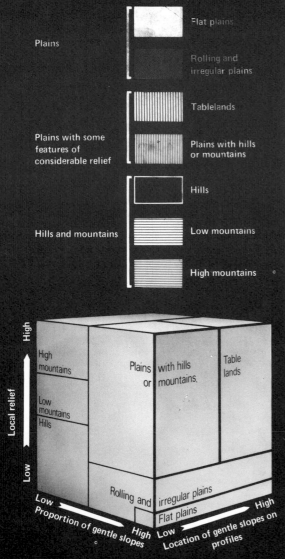

Plains
- Flat plains
- Rolling and irregular plains

Plains with some features of considerable relief
- Tablelands
- Plains with hills or mountains

Hills and mountains
- Hills
- Low mountains
- High mountains

ICE SHEETS

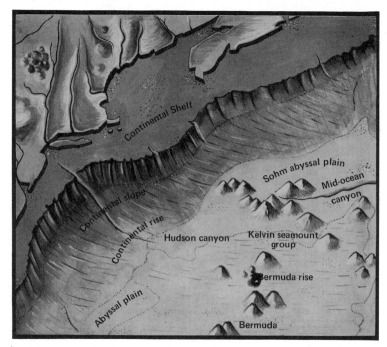

continental relief these angles may appear quite low but it must be remembered that they can be maintained over distances of 50 km or more and so represent changes in elevation of several thousand metres. Although the slope may sometimes pass directly into deep oceanic trenches, it is normally bounded by the continental rise which, with gentle gradients of less than quarter of a degree, leads down to the true ocean floor. Although the slope is conventionally regarded as terminating at the 2 000 m isobath, its actual foot varies considerably in elevation, ranging between 1 500 and 4 000 m around the Atlantic. It is not unusual to find step-like features along the face of the slope and these are occasionally so wide as to constitute extensive underwater plateaus; the most thoroughly studied is the Blake Plateau off the Atlantic coast of Florida which is nearly 300 km wide at depths between 700 and 1 000 m and is bounded on its seaward side by an exceptionally steep lower segment of the continental slope. Further common diversifying features are submarine canyons. These vary in size and form from deep sinuous gorges to broad flat-floored troughs. Some appear to be confined to the face of the slope, whereas others extend back across the shelf to link with valleys on to the continental surface. The Hudson Canyon illustrated in Fig. 1.5 is one of the best explored and may be divided into an Upper Gorge incised into the slope and a Lower Gorge incised into the rise. The largest submarine canyons attain a size comparable to that of the Grand Canyon in Arizona.

Relief of the ocean basins Unlike the continents, the ocean basins all display a common relief pattern. This consists of a broad central ridge flanked on both sides by low, extremely flat plainlands. In addition certain basins exhibit deep marginal trenches. Recognition of this gross patterning permits description of the ocean basins under three separate heads:

(a) *Mid-ocean ridges.* One of the most significant developments in earth science since the Second World War has been the recognition of a virtually continuous submerged ridge system traceable over a total distance of about 60 000 km (Fig. 1.6). The term mid-ocean ridge is not entirely appropriate as the feature is frequently positioned asymmetrically within the ocean basin, and its enormous dimensions are scarcely conveyed by the word ridge. A better idea of its size comes from an appreciation of the fact that it occupies about one third of the total width of the Atlantic and that its crest at 2 000 m is commonly only half as deep as the fringing plainlands. The existence of shoaling in the mid-Atlantic was recorded over 100 years ago, but it took many further decades of bathymetric survey to establish its continuity with similar shallows in the other oceans. With improved instruments allowing a much finer resolution of the relief, it became apparent that ridge topography often displays a marked symmetry about an axial line. This led earlier workers to subdivide each flank of a ridge into a descending sequence of steps. Although the procedure was arbitrary and has not been pursued further, it foreshadowed the later discovery of a geomagnetic symmetry which has been of major importance in the interpretation of the ridges (see p. 40).

The most distinctive topography is to be found near the ridge crest. Here the relief is at its most rugged with abrupt falls of 1 000 m or more from the culminating peaks to the highest of the flanking plateaus. The culminating peaks themselves are often cleft by a deep central fossa which may be as much as 2 000 m deep and 25 to 50 km wide. Such an axial trench is usually well developed in the Atlantic and Indian oceans, but much less obvious in the Pacific where ridge relief is more subdued. As early as the 1950s it was noted from detailed bathymetric charts that the mid-Atlantic ridge between West Africa and Brazil is abruptly offset, with individual displacements of the central fossa locally exceeding 100 km. Subsequent work has shown that such lateral shifts are quite commonplace and indeed appear to be a characteristic feature of all mid-oceanic ridges.

(b) *Ocean-basin floors.* The floors of all the major ocean basins lie at approximately the same depth, generally between 4 500 and 5 500 m. Their most characteristic feature is the abyssal plain which has been defined as an area within which gradients do not exceed 1 in 1 000. Although indicative of general form this definition has not been universally adopted since some workers have stressed the arbitrariness of separating the plains from the very gentle lower slopes of the continental rise into

Fig 1.6. The world distribution of the mid-oceanic ridge system and the major ocean trenches.

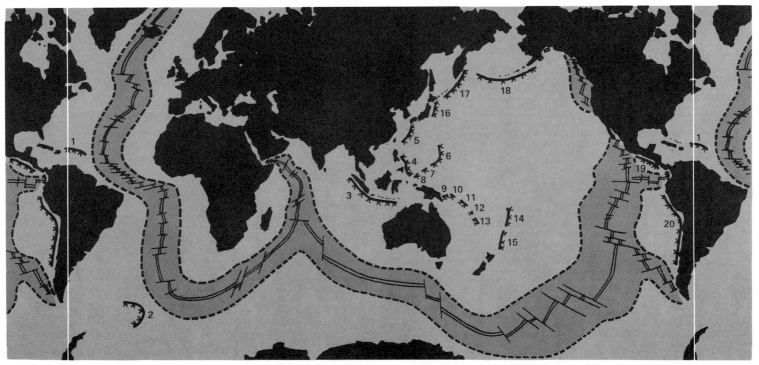

1. Puerto Rico 2. South Sandwich 3. Java 4. Philippine 5. Ryukyu 6. Marianas 7. Yap 8. Palau 9. New Britain 10. N. Solomons 11. S. Solomons
12. New Hebrides (N) 13. New Hebrides (S) 14. Tonga 15. Kermadec 16 Idzu-Bonin 17. Kuril 18 Aleutian 19. Middle America 20. Peru-Chile

which they often seem to grade. Abyssal plains are found in each of the major ocean basins although they are often restricted to a series of individual depressions rather than occupying a single very large unit. The sills between the depressions tend to be rougher in surface texture but may not rise greatly above the level of the plains. Where narrow they are often crossed by channel systems which lead from the higher plain to the lower through 'abyssal gaps'. On a much grander scale are long steep-sided troughs rather misleadingly called mid-ocean canyons since they do not normally occur near the centre of the oceans. The best known has been traced over 2 000 km along the foot of the continental rise off eastern Canada where it is typically 5–8 km wide and 20–200 m deep. Comparable but smaller canyons have been recorded from several of the other major ocean basins.

A number of distinctive forms diversify the abyssal plains. These include abyssal hills which range up to a few hundred metres high and often occur in large groups. They are particularly numerous on the Pacific floor and also occur as linear clusters along the flanks of the mid-ocean ridge in the Atlantic. More spectacular are the so-called seamounts. By definition these are submerged conical peaks rising at least 1 000 m from the adjacent sea floor. It is estimated that there are between 10 000 and 20 000 scattered through all the oceans of the world. They are not confined to the ocean-basin floors but also occur on the continental slopes and mid-ocean ridges. Although many appear to be scattered in almost random fashion, others fall into obvious linear patterns that may be traceable for 1 000 km or more. A small proportion of the seamounts are distinguished by having prominent flat tops and are known as guyots. Despite being found in each of the ocean basins their total number does not exceed a few hundred and although their summits mostly lie at depths of 1 000 to 2 000 m there is little consistency in elevation. It is on the summits of seamounts and guyots that many of the coral reefs in the Pacific are to be found.

A final noteworthy feature of the ocean-basin floors is the presence of linear zones of relatively rugged relief. These were first detected off the Pacific coast of North America before the Second World War, but their full extent was not appreciated until over a decade later when it was shown that at least five zones of greater surface roughness can be traced some 2 000 km through the abyssal plains and hills of the Pacific floor. Each zone consists of narrow but immensely long troughs and ridges separated by scarps several hundred metres high and sloping at angles of 5–10°.

(c) *Oceanic trenches.* It was recognized more than 100 years ago that the deepest parts of the ocean are not centrally placed, but lie relatively close to the continental margins where they form narrow elongated basins. The 'Challenger' voyage of the 1870s recorded its greatest depth of 8 000 m in what is now called the Marianas trench. Later soundings by other survey vessels penetrated even deeper in the Kuril and Kermadec trenches, but in 1950 the new HMS 'Challenger' returned to the Marianas trench and recorded a depth of 10 863 m. This stood as a record for a number of years but soundings in the same area later exceeded 10 900 m. Some uncertainty has attended a number of these very deep soundings since 'false' echoes reflected from the steep trench walls have posed difficult problems of interpretation. However, when the bathyscaphe 'Trieste' was piloted by Piccard and Walsh into the deepest part of the Marianas trench in 1960, a depth of 10 910 m was recorded. Since this measurement was made by a pressure gauge it provided powerful confirmation of the previous echo-sounding results.

Twenty separate trenches are now recognized, the majority being grouped around the edges of the Pacific basin (Fig. 1.6). They display a number of features in common. Characteristically they descend on one side far below the level of the adjacent ocean-basin floor, and on the other are flanked by a steep continental slope culminating in either a major mountain chain or in a line of oceanic islands. Normally about 100 km wide and 1 000–4 000 km long, they are all more or less arcuate in plan. The slopes tend to be steeper on the landward side where they may average 10 to 15° compared with 5 to 10° on the seaward side. Several detailed surveys have disclosed crenulate trench walls with the gradients, if anything, seeming to increase with depth before abruptly giving way to a narrow flat floor that is typically 1–3 km wide.

The earth's gravitational attraction

In outlining the shape and gross relief of the earth emphasis has inevitably been laid upon certain contrasts between the continents and ocean basins. The question that naturally follows is the extent to which these contrasts are a reflection of deep-seated structural variations lying beyond the limits of direct observation. One field of study shedding much light on this problem is investigation of the earth's gravitational attraction. Basic to gravimetry, or the science of gravity measurement, is Newton's law that two bodies attract one another with a force proportional to the product of their masses and inversely proportional to the square of the distance between them.

Fig 1.7 A schematic illustration of the apparatus used by Cavendish to measure the gravitational constant G. In the Newtonian expression $F = G\frac{MM^1}{d^2}$ the gravitational constant is the only unknown quantity. A fine fibre of known torsion (yielding a value for **F**) is used to suspend a bar from which hang two lead balls (of mass **M**). Two larger lead balls (of mass **M¹**) can be rotated so as to lie alternately on either side of the small ones (at distance **d**). Once **G** is known the same formula can be applied to the attraction of the earth for one of the lead balls, the only unknown in that case being the mass of the earth.

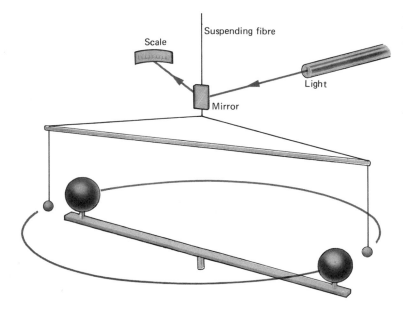

Mathematically this may be written as $F = G\frac{M\,M'}{d^2}$, where F is the force between two bodies of mass M and M', d is the distance between them, and G is the gravitational constant, one of the fundamental constants of the universe. By manipulation of this apparently simple law, major discoveries about the nature of the earth's interior have been made. One of the most important relates to the mean density of the materials composing the globe; a second relates to the distribution of different materials within the sphere.

The mean density of the earth When Newton published his gravitational law in 1687, he recognized that it offered the basis for measuring directly the total mass and thereby the mean density of the earth. However, realization of this objective was delayed for over a century by the extremely delicate experimental work that was required. The most celebrated early attempt to 'weigh the earth' was made by Cavendish in his home at Clapham Common in 1798. He endeavoured to measure the deflective force when large lead weights were swung close to two lead balls suspended from a very sensitive torsion balance (Fig. 1.7). In this remarkably elegant and carefully controlled experiment, he calculated the ratio of the pull exerted by one of the weights to the pull exerted by the earth on the same ball. In this way he was able to compute the total mass of the earth and thereby estimate its mean density as '5·48 times greater than that of water'. A slightly different method was employed by Poynting in 1878. He used a large beam balance from which were hung two lead balls each weighing about 20 kg. Beneath the balance was a turntable bearing a 150 kg lead weight so arranged that it could be moved beneath each ball in turn. The resulting deflection of the beam proved to be over one second of arc, or the equivalent to a change in weight of about 0·4 mg. From this Poynting calculated the mean density of the earth as 5·49 g cm⁻³.

These two classic experiments illustrate how application of Newton's law of gravitation permits evaluation of the earth's mass and density. They have been repeated with increasing refinements on many subsequent occasions, but since the launching of artificial satellites in the 1950s a totally new method of computing the mean density of the earth has become available. From precisely measured satellite trajectories, it is now estimated that the total mass of the earth is 5.98×10^{27} g, and its mean density 5.517 g cm^{-3}. Since the corresponding figure for rocks exposed at the surface of the globe ranges between 2.6 and 3.3 g cm^{-3}, there must clearly be a concentration of much denser material towards the centre of the earth; several lines of argument conduce to the view that the density here approaches a value of 11 g cm^{-3}.

The earth's gravity field On a non-rotating, smooth and perfectly spherical globe with all the constituent material arranged in homogeneous concentric shells, gravitational attraction would obviously be uniform over the entire surface. As indicated on p. 2, workers in the eighteenth century had already deduced that gravity varies from one region to another, and since that time much effort has been devoted to the accurate assessment of these variations. For that purpose it was first essential to establish a suitable unit of measurement. Consideration of Newton's law shows that the gravitational force varies with the mass not only of the earth but also of the other body involved. This is clearly unsatisfactory as a measure of the earth's gravitational attraction and in consequence the latter is normally expressed in terms of an acceleration, conventionally designated g. Being the quotient of force and mass, g is independent of the second body involved. The acceleration of gravity describes the rate at which any body falling in a vacuum at the earth's surface will increase its velocity per unit of time. The unit of acceleration is the gal (named after Galileo) and is equal to 1 cm sec^{-1} sec^{-1}. Numerous observations have shown that the acceleration of gravity is usually about 980 cm sec^{-1} sec^{-1}. Regional deviations from this figure are so small that it is essential in practice to work with a unit equal to only one-thousandth part of the gal, known as a milligal.

A second necessity before the regional variations in gravity could be accurately measured was development of suitable instruments. The most obvious technique, that of actually measuring the acceleration of a falling body, was not widely employed until quite recently for a variety of technical reasons. Instead the most precise measurements were made with a specially designed instrument known as the reversible pendulum. This permitted determination of absolute gravity values at a number of base stations located at intervals round the globe. These base stations were later used for calibrating simpler portable instruments that allow local gravity variations to be much more rapidly surveyed. Two types are currently in widespread use. One depends upon the period of swing of a pendulum, the other and more common upon the extension of a delicate spring. Both require extraordinarily sensitive measuring devices; the period of the pendulum is usually about 1 second and needs to be timed to an accuracy of one part in 10^8, while a 100 mm spring will extend by only 1×10^{-5} mm for each milligal increase. Despite severe technical problems arising from the motion of the transporting vehicle, these instruments have even been adapted for use in ships and aircraft. In consequence a relatively comprehensive network of gravity measurements is now available for many parts of the world.

Surveys have revealed significant regional variations. Some of this variation is readily explicable. As pointed out earlier, on an idealized globe that is non-rotating, spherical, smooth and composed of homogeneous concentric shells there would be no regional deviations. If each constraint is relaxed in turn so that the idealized globe conforms more and more closely to the actual earth, the cause of some of the regional variation is immediately apparent. If it is first assumed that the globe rotates, a 'centrifugal force' is introduced which diminishes from a maximum at the equator to zero at the poles. From the time taken for the earth to rotate on its axis, it is a simple matter to calculate the acceleration due to rotation and at the equator this proves to be about 1/300 of the acceleration due to gravity; since the centrifugal effect acts in the opposite direction to gravitational acceleration it means that an object will weigh less at the equator than at the poles. Lines of equal value of g, known as isogals, would run parallel to lines of latitude on such a rotating sphere. As seen earlier in this chapter, the angular velocity of the earth is sufficient to distort the globe into an oblate spheroid so that perfect sphericity is the second constraint that must be relaxed. The values of gravity on a rotating ellipsoid can be calculated without undue difficulty, and again the isogals would run along the parallels of latitude. Although the situation now being envisaged is still idealized, it does begin to approximate conditions over large areas of the ocean basins.

For much of the earth's surface, however, a third constraint must be relaxed, namely that relating to smoothness. It is clear that ascent of a mountain involves movement away from the centre of the earth and that g is consequently going to decrease. If the material above sea-level is assumed to have zero density, the decrease amounts to 0·3086 milligals for each metre of ascent. For purposes of standardization observed gravity values are normally referred to the sea-level datum by adding an amount for elevation known as the free-air correction. However, the actual gravity at sea level would be affected by the overlying rock whose mass can be estimated by employing an appropriate density; the acceleration due to this mass is known as the Bouguer correction factor. Since it acts in the opposite direction to g it has to be subtracted from the observed gravity value. In practice small additional corrections have to be made for local topography and the gravitational pull of the sun and moon. Once all these adjustments are complete, it might be anticipated that the resultant value would correspond to that for the appropriate latitude on the ellipsoid. In fact, however, there are often residual discrepancies which are known as Bouguer anomalies.

If the foregoing arguments are sound, Bouguer anomalies must result from the one constraint that has not so far been relaxed, namely, the arrangement of materials of different density within the earth. As early as the first half of the eighteenth century one of the expeditions despatched by the French Academy of Sciences to measure the length of a meridian degree noted that the Andes exerted less influence on a plumb-bob than would be predicted by a simple assumption of concentric shells. The unexpectedly slight attraction of mountain masses was confirmed during later surveys along the Himalayan foothills where the pull due to topography was found to be only one-third of that anticipated. Two famous hypotheses were offered in explanation. In 1854 Pratt surmised that the density of elevated mountains might be in inverse relation to their height, thus theorizing that beneath the Himalayas would be found the least dense part of the earth's crust. A year later Airy attributed the weak gravitational attraction to a root of crustal

material extending down into a denser substratum. Both hypotheses predicate a relatively shallow level within the earth's interior above which the mass per unit area is approximately equalized; this notion of hydrostatic equilibrium was termed 'isostasy' by Dutton in 1889.

This brief review of a highly complex subject serves to show how detailed measurement has demonstrated a clear relationship between surface topography and deep-seated structure. However, several possible models of density distribution within the earth fit the observed gravity values so that gravimetry by itself does not define the structures. Highly significant supplementary information has come from a second line of investigation, the study of earthquake shock-waves.

The transmission of seismic waves through the earth

Study of vibratory motions of the earth is the province of science known as seismology. Vibratory motions may be initiated in a vast number of ways, ranging from the pounding of surf along the coast and the passage of traffic on roads, to underground nuclear explosions, earthquakes and even the gravitational pull of the moon. Most information about the earth's interior has come from the study of earthquake shock waves originating in displacements along faults. The waves emanating from the grinding together of the two sides of the fault move outwards from the focus in all directions and may then be recorded on instruments known as seismographs located in observatories throughout the world. The primary function of such equipment is to time and analyse earthquake shock waves so as to provide evidence regarding the structure and composition of the earth's interior.

The nature of earthquake waves Two major types of wave may be distinguished, surface waves and body waves. The former travel across the surface of the earth, spreading out from the epicentre rather like ripples on a pond. Such is the sensitivity of modern instruments that, after a major earthquake, surface waves can be detected making several complete circuits of the globe, reappearing in a more subdued form on the seismograph record at intervals of $2\frac{1}{2}$ to $2\frac{3}{4}$ hours. More significant to present

14 purposes, however, are the body waves which travel through the earth's interior. Two types of body wave may be distinguished, primary or P-waves and secondary or S-waves. The P-waves are compressional with the transmitting medium vibrating backwards and forwards parallel to the direction of propagation. The S-waves are shear waves in which the transmitting medium oscillates in a direction perpendicular to that of propagation. Invariably in the seismograph record of a single earthquake the P-waves arrive first, the S-waves shortly there-after and the last to arrive are the surface waves. Since all may be regarded as originating instantaneously their rates of travel must differ by appreciable amounts.

By comparing the records from three widely spaced seismographs, the time and focus of an earthquake can be fixed within quite precise limits, and the rate of travel of the various seismic waves thereby computed. Even in the early days of seismology it was realized that the time taken for a body wave to reach the recording instrument is not simply proportional to the distance travelled, but that those waves which penetrate moderately deeply into the interior move at a rather greater speed. One consequence is that the waves do not follow straight-line paths but paths that are slightly convex towards the centre of the earth. Waves encountering abrupt changes in physical properties may be either reflected or refracted depending upon their angle of incidence.

Evidence relating to the internal structure of the earth
The single most significant conclusion to emerge from the study of body waves is the existence of a number of abrupt discontinuities within the earth's interior. In 1913 Gutenberg demonstrated the presence of a clearly defined core with markedly different properties from those of the rest of the interior. Its effect is seen most obviously in the 'shadow zone' that extends from 103° to 142° from the epicentre and in which no direct body waves are recorded (Fig. 1.8). Beyond 142° no normal S-waves are recorded but the P-waves reappear; their transmission times, however, are much retarded when compared with those received at 103°. In the failure to propagate S-waves and in the slow transmission of P-waves the core is exhibiting typical properties of a liquid; its diameter is about 6 945 km. Lehmann showed in 1936 that hitherto puzzling wave refractions and accelerated movements through the core could be explained by the existence of an inner solid core with a diameter of about 2 800 km. The composition of the core remains uncertain although it has generally been thought to consist largely of nickel and iron which would give values roughly consistent with its calculated density.

A second major discontinuity was discovered in 1909 by the eminent seismologist Mohorovicic. While examining records of

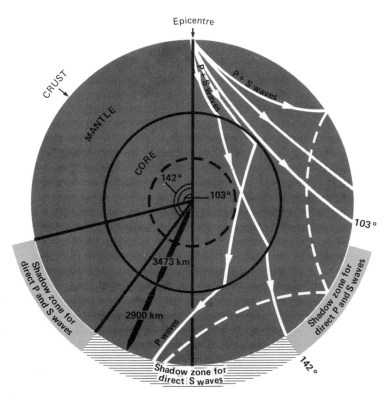

Fig 1.8 The internal structure of the earth as revealed by the propagation of seismic waves. The actual pattern of travel from a single large earthquake may be extraordinarily complex; in addition to the small sample of direct waves depicted below there are numerous reflected waves of which just one example is shown by the dashed line.

the Kulpa valley earthquake in Yugoslavia he realized that, at distances between 200 and 720 km from the epicentre, both the P- and S-waves displayed two distinct bursts of movement. He inferred from this observation that the separate phases of each wave must have followed different routes and concluded that the only satisfactory explanation lay in the presence of a shallow discontinuity within the earth's interior. One phase was assumed to have travelled through the overlying rock layer, the other to have descended to greater depths and to have been refracted. Arrival times indicated more rapid travel beneath the discontinuity, and Mohorovicic estimated the depth of this sudden acceleration to be about 60 km. Subsequent investigators have honoured Mohorovicic by naming the discontinuity after him (unfortunately often contracted to Moho), and have shown that it is virtually a world-wide feature occurring at a depth of 10 km beneath the surface of the oceans and at a depth varying between 20 and 65 km beneath the continents. The greatest figures are recorded beneath the continental mountain chains with the result that the form of the discontinuity resembles a greatly exaggerated mirror image of the earth's surface. It undoubtedly constitutes one of the major features in the overall structure of the earth and by convention is regarded as separating the crust from the underlying mantle.

In 1923 Conrad detected a shallower and less distinct discontinuity at a depth of 10 to 25 km beneath the continents. This is generally regarded as dividing the crust into an upper part restricted to the continents and a lower part which not only extends beneath the continents but also floors the major

ocean basins. The discoveries of both Mohorovicic and Conrad were of the utmost importance in confirming that the division of the surface relief into continents and ocean basins is not a purely superficial phenomenon but reflects fundamental contrasts within the outer layers of the earth.

The Conrad and Mohorovicic discontinuities were detected because the transmission speeds of seismic waves change abruptly at certain depths. In the upper crust the velocity of P-waves is about $6 \cdot 1$ km sec^{-1}, increasing to $6 \cdot 9$ km sec^{-1} in the lower crust. These velocities are consistent with a granitic composition above the Conrad discontinuity, and a basaltic composition below it; the materials of these two layers are often referred to as sial and sima indicating the relative significance of silicon, aluminium and magnesium in their make-up. It should be noted, however, that recent investigations suggest that the deep crust beneath the continents may differ slightly from that beneath the oceans. Below the Mohorovicic discontinuity there is an abrupt leap in the speed of P-wave transmission to over 8 km sec^{-1}, and the physical properties of this upper part of the mantle are generally consistent with a composition resembling that of the olivine-rich rock known as peridotite.

The general tendency for the velocity of seismic waves to increase with depth is temporarily reversed in a layer of the mantle between 100 and 200 km below the surface. Known as the 'low velocity layer', this was first detected in the 1920s when Gutenberg recognized that waves emanating from earthquake foci at depths of much over 50 km travel more slowly than

Fig 1.9 Schematic sections to illustrate the characteristic structure of the outer layers of the globe.

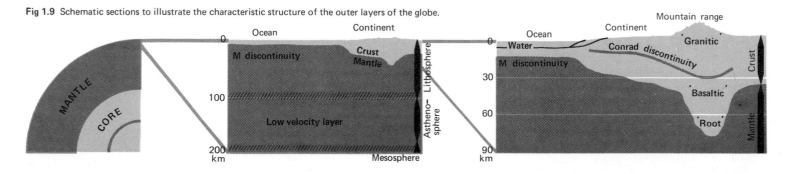

16 might be anticipated on theoretical grounds. Recent investigations have confirmed this slower transmission with velocities of about 7·8 km sec^{-1} characteristic of a thick layer which extends rather deeper beneath the continents than beneath the oceans. The cause of this low-velocity zone remains uncertain, but it is possibly due to partial fusion of the solid rock at the high temperatures and pressures that exist at this depth. Some workers have argued that in the dynamics of the earth's interior this low velocity layer may be of more significance than the better known Mohorovicic discontinuity; it has therefore become desirable to have a new nomenclature recognizing this possibility, and the terms commonly employed are asthenosphere for the low-velocity layer itself, and lithosphere for the overlying mantle and crust. Major structural features of the earth's outer shells are illustrated in Fig. 1.9.

FENNEMAN, N. M., 1916. 'Physiographic divisions of the United States', *Ann. Ass. Am. Geogr.,* 6, 19–98.

HAMMOND, E. H., 1954. 'Small-scale continental landform maps', *Ann. Ass. Am. Geogr.* 44, 33–42.

A valuable standard text on the topography of the ocean basins is provided by the series entitled *The Sea* (Wiley) ; especially useful are vols 3 (*The Earth Beneath the Sea* – Ed M. N. Hill) and 4 (*New Concepts of Sea-floor Evolution* – Ed A. E. Maxwell). A modern account of the overall structure of the globe is provided by M. H. P. Bott *The Interior of the Earth* Arnold, 1971.

1|2

Concepts of time

In Chapter 1 the earth was viewed as an unchanging planet on which measurements could be made to determine its form and structure Much of the rest of this book is concerned with modifications the global surface undergoes, whether as a result of deep-seated tectonic movements or due to the activity of erosional agents. This clearly introduces the dimension of time which must be examined in some detail before proceeding to the dynamics of surface change.

Much of classical geology was concerned with establishing the relative ages of the rocks exposed at the surface. A few fundamental principles sufficed in the execution of this work. Outstanding was that of superposition, by which the upper beds in an uninverted succession were dated as younger than the lower ones. Of almost equal importance was palaeontological dating by which distinctive fossil assemblages were recognized as characteristic of certain time periods throughout the world ; this method was employed even before Darwin provided the philosophical basis through his concept of evolution. These principles were applied with such skill that nineteenth-century workers were able to establish a stratigraphic column which is still used world-wide today with only minor modifications. The standard nomenclature is shown in Table 2.1.

The geological time-scale

The nineteenth-century investigations could only provide the basis for relative dating, and their reliance on palaeontological evidence meant that correlations between rocks older than the origin of life were wellnigh impossible. Varied arguments were advanced in estimates of the length of geological time. These ranged from the biblical calculations of the intellectual descendants of Bishop Ussher, who had placed the date of the Creation in the year 4004 BC, to the more scientific estimates that it would have required at least 20 or 30 Ma (million years) for the total recorded thicknesses of sediment to have accumulated. Of a similar order of magnitude was the estimate by Lord Kelvin that if the earth had started as a molten body its present temperature indicated a cooling history lasting between 20 and 40 Ma. The crude assumptions underlying all such estimates made the results very suspect, and gave added impetus to the search for an independent and more precise method of dating. As Sollas wrote in 1900 'How immeasurable would be the advance of our science could we but bring the chief events which it records into some relation with a standard of time.' In practice the basis for such an advance had been established several years earlier when Madame Curie had identified the radioactive elements uranium and thorium. By 1899 Rutherford had

Table 2.1 Geological time scale　　　　　　　　　　　　　　　**19**

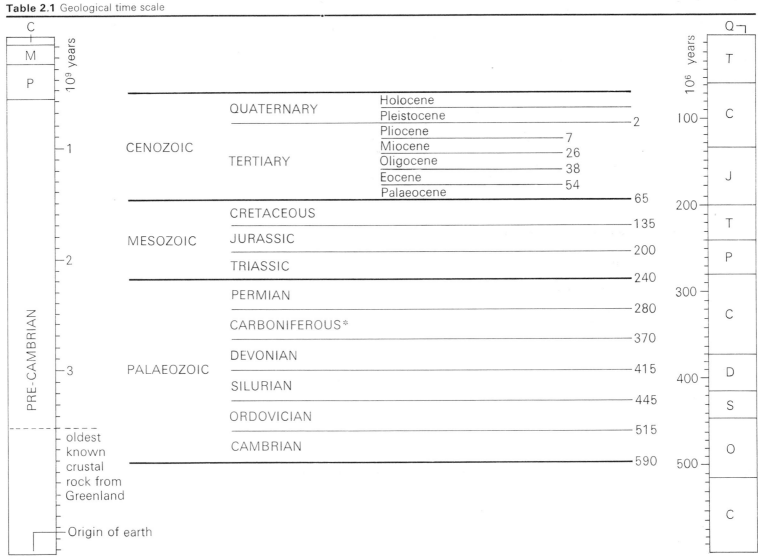

		Holocene	
	QUATERNARY	Pleistocene	2
CENOZOIC		Pliocene	7
		Miocene	26
	TERTIARY	Oligocene	38
		Eocene	54
		Palaeocene	65
MESOZOIC	CRETACEOUS		135
	JURASSIC		200
	TRIASSIC		240
	PERMIAN		280
	CARBONIFEROUS*		370
PALAEOZOIC	DEVONIAN		415
	SILURIAN		445
	ORDOVICIAN		515
	CAMBRIAN		590

Left bar: C / M / P — 10⁹ years — 1, 2, 3 — PRE-CAMBRIAN — oldest known crustal rock from Greenland — Origin of earth

Right bar: Q / T / C / J / T / P / C / D / S / O / C — 10⁶ years — 100, 200, 300, 400, 500

*Known as Pennsylvanian and Mississippian in N. America.
The divided bars to the left and right show the relative duration of the different divisions of geological time.

Fig 2.1. Uranium and thorium dating. The decay series for U^{238} and U^{235} and Th^{232} are shown, each horizontal transformation representing emission of an alpha particle, each diagonal transformation representing emission of a beta particle. The half-lives of nuclides in common use for dating purposes are also indicated.

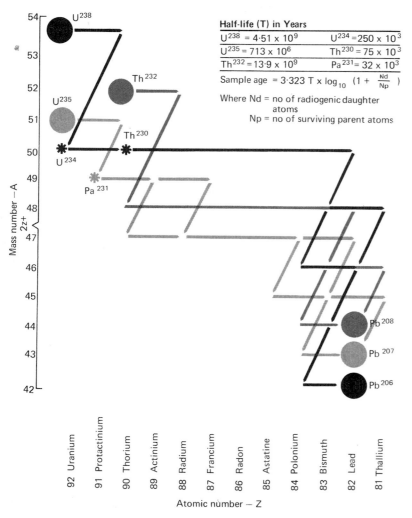

Half-life (T) in Years	
U^{238} = 4·51 x 10^9	U^{234} = 250 x 10^3
U^{235} = 713 x 10^6	Th^{230} = 75 x 10^3
Th^{232} = 13·9 x 10^9	Pa^{231} = 32 x 10^3

Sample age = $3 \cdot 323\ T \times \log_{10}\ (1 + \frac{Nd}{Np})$

Where Nd = no of radiogenic daughter atoms

Np = no of surviving parent atoms

demonstrated the association of radioactivity with atomic disintegration and had postulated the emission of two different types of radiation which he named alpha and beta. In 1903 he showed that alpha radiation consists of streams of positively-charged particles similar to the nuclei of helium atoms (2 protons +2 neutrons), whereas beta radiation consists of negatively-charged electrons. He maintained that the emission of this radiation leads to a change in the radioactive element, and in a classic paper with Soddy argued that 'the proportional amount of radioactive matter that changes in unit time is a constant . . . for each type of active matter a fixed and characteristic value'. The tools were thus at hand for the great advance Sollas had desired, and within 2 or 3 years several workers were attempting to date rocks experimentally.

The principle employed was simple. Radioactive disinte-gration is a random process dependent solely on the number of atoms of the unstable element present at any particular instant of time. A set proportion of these atoms, represented by the value known as the decay constant, will disintegrate in the next unit of time; expressed in an alternative way, half the atoms will disintegrate in a period known as the half-life of the element concerned. Daughter elements will be produced and if the quantity of parent nuclide and daughter element can be accurately measured the period during which radioactive decay has been going on can be calculated. By 1910 measurements on uranium-bearing minerals were already suggesting a much longer span of geological time than had previously been supposed. This was the start of a protracted controversy centring around the accuracy of the analyses employed and the possibility of different forms of contamination.

The original determinations were based upon the transfor-mation in igneous rocks of either uranium or thorium into lead (Fig. 2.1). A limitation of the early U/Pb and Th/Pb techniques was the shortage of suitable minerals containing these nuclides, and the development of methods based upon other elements was important in both enlarging the potentialities for dating and affording opportunities of testing one method against another. A vital technical innovation was the development of mass spectroscopy permitting detailed isotopic analyses of a mineral. Nowadays much dating is based either upon the decay of the

Rb[87] isotope of rubidium to the Sr[87] isotope of strontium by the emission of beta particles, or upon the transformation of the K[40] isotope of potassium into the Ar[40] isotope of argon by the capture of an electron. Both the Rb/Sr and K/Ar methods are subject to a number of potential errors, but these can be minimized by modern technical refinements.

The immense half-life of each of the above nuclides means that radiometric dating is particularly well adapted for dealing with the older rocks of the earth's crust. In addition to indicating the age of the previously established divisions of the geological column (Table 2.1, p. 19) radiometric techniques have also permitted the likely age of the crust itself to be assessed. The relative abundance in crustal rocks of the various isotopes of uranium, thorium and lead will obviously change with time; one guide to the present composition is provided by analyses of lead derived from erosion of the continents and now incorporated in recent marine sediments. By this method the age of the crustal material has been assessed at about 4.5×10^9 years. The same age has also been recorded when K/Ar and Rb/Sr dating has been applied to stony meteorites; although the history of meteorites remains obscure, there is reason to believe that they came into existence at about the same time as the crust of the earth. Finally, it is worth noting that the ages of lunar rock samples recovered during manned space flights range from 3 to 4.5×10^9 years. There is growing speculation that the earth is one of a whole group of solar planets formed approximately 4,600 Ma ago.

Problems of late Cenozoic chronology

The radiometric techniques discussed in the preceding section have provided a dating framework for the whole of the stratigraphic column. However, for the geomorphologist it is the chronology and time-scale of the late Cenozoic era that is particularly important. It has been claimed by some workers that the vast majority of landforms are Quaternary in age and the rest no older than late Tertiary. This has been contested by others and there is an obvious problem of defining exactly what is meant by the age of a landform that undergoes slight but continuous modification over an extended period of time. Nevertheless, it can hardly be denied that the last few million years are of

critical importance to the geomorphologist and require much closer examination than the preceding aeons of geological time.

General considerations There are at least four major problems confronting any attempt to formulate a satisfactory chronology for the late Cenozoic era. Most of the earlier part of the stratigraphic column was constructed on the basis of marine sediments and faunas raised above modern sea-level. Continuous marine successions of late Cenozoic age are rare on the continents at the present day. The contemporaneous terrestrial beds tend to be very fragmentary and to span only a short interval at any one locality. In consequence it is necessary to devise methods of correlating one section with another and building up a composite picture from many diverse sources. A second problem arises from the relatively brief duration of the late Cenozoic era. This means that biological evolution during that period was so restricted that one of the basic principles on which the rest of the stratigraphic column was founded is of severely limited application. A third difficulty results from the rapidity of environmental change during the late Cenozoic era. This is attested by an immense amount of both biological and physical evidence that points unequivocally to rapid fluctuations of climate. The exact number of major oscillations remains uncertain although it is unlikely to be less than seven or eight and may be many more. With relatively little to distinguish one oscillation from another it has proved extremely difficult to build up long chronological sequences from the highly fragmented evidence that is normally available. A final problem is the absence of any single method of radiometric dating that can span the whole of the period under discussion. The usefulness of methods based upon the production of radiogenic lead declines rapidly for rocks younger than 100 Ma, while the Rb/Sr technique is difficult to apply to materials younger than 20 Ma. Although K/Ar dating has been used on rocks less than 30 000 years old, the amount of radiogenic argon in such cases is so slight that considerable doubt surrounds the accuracy of the results. Alternative methods are available for application to younger materials (see below, p. 29) but most suffer from severe restrictions in the age range for which they are appropriate.

22 The Pliocene-Pleistocene boundary Despite much discussion there is still no agreement regarding the criteria to be adopted in fixing the boundary between Pliocene and Pleistocene times, and obviously until this is done there is no way of determining the age of the boundary. Some workers have wanted to employ the evidence of fossils so as to maintain consistency with earlier geological eras subdivided on the basis of palaeontology. However, as already pointed out, evolutionary changes in the late Cenozoic tend to be over-shadowed by the effects of rapid climatic fluctuations. Other workers have argued that the most distinctive attribute of the Pleistocene epoch is the development of large continental ice sheets over Europe and North America, and that consequently the onset of the Pleistocene should be equated with the first dramatic fall in temperature. This idea was implicitly endorsed by the International Geological Congress in 1948 which recommended that the boundary should be fixed by reference to sections in southern Italy which show a sudden influx of 'cold species' at the end of a long succession of marine sediments known to be of Pliocene age. Even if adopted this recommendation leaves unanswered many problems of correlation with other parts of the world.

If evidence of glaciation were to be adopted as the basic criterion, it would push the Pliocene–Pleistocene boundary back to several million years ago at least. In both California and Alaska glacial deposits interstratified with dated igneous rocks must be more than 2·7 Ma old. If the existence of an ice sheet in Antarctica were to be admitted as proof of glaciation, a wholly illogical situation would be created since it now seems probable that the southernmost continent was covered with ice at least 15 Ma ago during Miocene times. This illustrates very clearly the potential pitfalls of an unwise change in the criteria for subdividing geological time.

Several workers have suggested that, if the characteristic feature of the Pleistocene epoch is large-scale temperature fluctuations, the onset of the earliest of these should be held to mark the Pliocene–Pleistocene boundary. This raises problems of definition since there is no reason to believe that climate was unchanging during earlier geological periods. Yet there does seem to be some justification for the view that the amplitude and rapidity of climatic fluctuations was especially great during the Pleistocene epoch. Workers in several different regions have concluded that, during early and middle Cenozoic times, mean annual temperatures showed a slow decline on which were superimposed minor oscillations with an amplitude of perhaps 2°–3°C. This compares with commonly estimated values of about 5°C for the Pleistocene oscillations. The precision of these estimates is open to question and the approach scarcely provides a satisfactory basis for defining the boundary between two geological epochs. One must therefore conclude that no adequate basis for demarcating the Pleistocene has yet been proposed, although the dominant characteristics of the period are well established. It appears that most if not all of the globe has experienced sharp fluctuations of climate for a minimum of 2 to 3 Ma.

The Pleistocene–Holocene boundary Almost as much confusion and uncertainty attends the fixing of this boundary as that at the beginning of the Pleistocene. So far as we know, there is nothing to distinguish the amelioration of climate at the end of the last cold period from the many others that must have preceded it earlier in the Pleistocene. This has led several eminent workers to argue that it would be better to abandon the term Holocene and to recognize that we live during one of the climatic fluctuations of the Pleistocene. On the other hand, in middle latitudes it is obviously desirable to distinguish between glacial and post-glacial events, and the term Holocene was originally intended as a designation for the post-glacial period. However, the melting of an ice-sheet is transgressive of time so that what is literally post-glacial in one locality is contemporaneous with what is glacial in another.

In such circumstances it is probably best to fix an arbitrary boundary, even if it appears to make little physical sense in some parts of the globe. The most widely acknowledged boundary is that originally proposed by Scandinavian workers. By means of pollen analysis they identified a period of rapid warming that led to an accelerated withdrawal of the ice margin across the Baltic region. The start of this warming phase has subsequently been fixed at about 10 000 years ago, a date now adopted by convention as the beginning of the Holocene. If the term Holocene were dropped, the recommended Pleistocene stage name is the Flandrian. This has the disadvantage that the term Flandrian was originally applied to the rise in sea-level that accompanied the melting of the last major ice-sheets, and by 10 000 **23** years ago a substantial proportion of this rise had already taken place. Throughout this book the term Holocene will be retained, even though the logic of the argument that we live within the Pleistocene epoch is overwhelming.

Nature of the evidence bearing on late Cenozoic chronology

The evidence of deep-sea cores It has already been stressed that on the continents late Cenozoic deposits are sporadic in occurrence and rarely span a long time interval at a single locality. With the growing capacity to recover cores from the ocean floor, it has become possible to sample materials that appear to have accumulated as a relatively continuous rain of very fine sediments. The sediment itself is divisible into lutite or terrigenous debris and the tests and hard parts of such minute marine organisms as Foraminifera and diatoms. These microfossils occur in vast numbers and, because the organisms have different environmental requirements, are capable of yielding important evidence ragarding the changing temperature of the oceanic water. Most significance attaches to the surface-water planktonic Foraminifera which include, among the temperature-sensitive species, *Globorotalia menardii*, indicative of warm tropical conditions, and *Globigerina pachyderma* found today even at the North Pole. The relative frequency of these species varies at different levels within a single core (Fig. 2.2) and is believed to indicate fluctuations in water temperature and therefore in climate. Careful analysis of different horizons within deep-sea cores has revealed other systematic changes. The proportion of calcium carbonate in the sediment varies widely and is believed to reflect the number of foraminiferal shells falling to the sea floor. The productivity of the surface waters almost certainly varies with temperature so that carbonate fluctuations may be regarded as indices of climatic change. In some cores as many as 20 fluctuations have been found in sediments regarded as no more than 2 Ma old. Another fascinating aspect of the cores is the periodic alternation in the coiling direction of certain foraminiferal species, the best known example being *Globorotalia truncatulinoides*. The proportion of

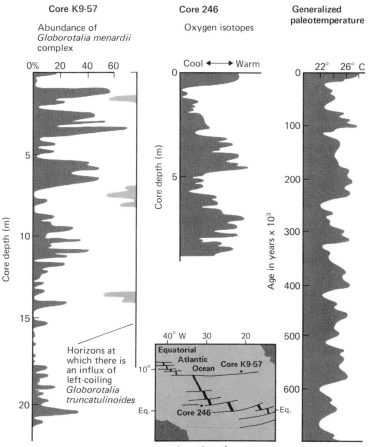

Fig. 2.2. Pleistocene environmental changes as evidenced by low-latitude oceanic cores. Left and centre, two cores from the equatorial Atlantic analysed on the basis of their fauna and oxygen isotope ratios.
Right, a generalized paleotemperature curve for the last 700 000 years (after Emiliani and Shackelton, 1974).

right- and left-coiling specimens alters radically over a few centimetres of sediment. The reason for the changes is unknown, but at the present day some areas of the Atlantic are known to have a predominance of right-coiling and other areas of left-coiling individuals. If there is an environmental control of this distribution, radical changes have presumably taken place on many occasions in the past. The faunas of the deep-sea cores therefore underline the frequency of Pleistocene climatic oscillations.

The calcareous foraminiferal tests have also been subject to oxygen isotope analysis. The ratio of the isotopes O^{16} and O^{18} theoretically depends upon two major factors, the isotopic composition and temperature of the sea water at the time the shells were forming. If one assumes for the moment that the composition remained constant, the O^{18}/O^{16} ratio may simply be regarded as a function of temperature. Shells formed in cold water are found to be relatively richer in O^{18} than those formed in warm water. The rate of enrichment is extremely small, but when a mass-spectrometer is used for analysing the tests from oceanic cores, the O^{18}/O^{16} ratio is found to vary significantly at different horizons (Fig. 2.2). At first sight the method appears ideal for determining the changes in sea-water temperature throughout the late Cenozoic, but in practice interpretation of the O^{18}/O^{16} values has raised many controversial issues. These have centred around the other major factor controlling the constitution of pelagic shells, namely, the isotopic composition of the water.

Owing to their greater vibrational energy, water molecules with O^{16} atoms are preferentially evaporated to leave the oceans temporarily depleted in that isotope. If the water were returned immediately in the form of local precipitation the effect would be minimal. However, areas of evaporation and precipitation do not coincide and the mass transfer of moisture in its gaseous state leads to latitudinal variations in the composition of modern sea water. It has consequently been necessary to define a Standard Mean Ocean Water (SMOW) to which other measurements may be referred. The most severe problem, however, arises from possible long-term changes in the isotopic composition of the water. The outstanding control is likely to be the growth of ice-sheets which will store water of different make-up from that in the sea, with consequent repercussions on the residue left in the ocean basins. In order to evaluate this effect it is necessary to calculate not only the likely O^{18}/O^{16} ratios in Pleistocene ice-sheets but also the total volume of those ice masses. Neither can be assessed with any great precision. The isotopic composition of both the Greenland and Antarctic ice-sheets has been examined, although the samples obtained from great depths are extremely limited. As might be expected from basic physical principles, the O^{18}/O^{16} values vary with the temperatures prevailing at the time of precipitation. This interesting property, discussed in greater detail below, complicates the calculations to be made in assessing the effect on oceanic water. Expressed as a deviation from SMOW, estimates of the average composition of the European and North American ice-sheets during a major glacial period vary from $-30‰$ to $-9‰$. If the former figure is adopted and applied to the best estimates of ice volume, it means that virtually all the isotopic variation in the ocean cores can be attributed to changes in water composition. At the other extreme, about one-third of the observed variation is attributable to that cause and the remaining two-thirds to temperature oscillations. On the latter assumption Emiliani suggested temperature fluctuations of about 5°C in the equatorial Atlantic and of about 3°C in the equatorial Pacific (Fig. 2.2). There is still much controversy surrounding these figures. The shells of benthonic organisms provide a potential means of checking the actual isotopic composition of the water since the temperature of the deep oceans can have varied little between glacial and interglacial periods. Such materials generally support the view that the temperature fluctuations were rather less than Emiliani estimated.

Whether or not oxygen isotope ratios form a good basis for assessing temperature changes, they almost certainly reflect environmental fluctuations of global significance. It has been contended, for example, that the curves depicting variations in the O^{18}/O^{16} ratio, if they do not refer to temperature, must refer to the extent of the contemporaneous ice-sheets. In that respect they still make a major contribution to the understanding of Pleistocene chronology.

The evidence of shallow marine and estuarine sediments
In a number of localities scattered around the world there are important sequences of shallow-water sediments spanning much of the Pleistocene epoch. Correlation from one site to another poses many difficulties, but significant contributions to Pleistocene chronology have been made by careful faunal and floral analysis of the sediments. A good example is afforded by the deposits around the southern margin of the North Sea. These have been closely studied in both the Netherlands and eastern England. In East Anglia they are known as crag deposits and consist of shelly marine sands and estuarine silts and clays. They are particularly interesting since they underlie what are believed to be among the earliest true glacial deposits anywhere in Britain. As the lowest members of the crag are classified on faunal grounds as Pliocene, it implies that the succession almost certainly spans some part of the local pre-glacial Pleistocene. The sequence has been studied by means of a core from a borehole at Ludham in Norfolk (Table 2.2). Both Foraminifera and pollen bear witness to major climatic oscillations with three cold phases separated by intervals that were probably no cooler than today. The basal sediments of the Ludham core yield a

Table 2.2 The succession of glacial and interglacial episodes as identified from the Ludham borehole in East Anglia. (Note that only the three most recent glaciations are represented by tills, the earlier ones being inferred from biological evidence in East Anglia.)

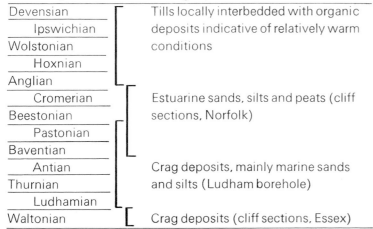

Devensian	Tills locally interbedded with organic deposits indicative of relatively warm conditions
Ipswichian	
Wolstonian	
Hoxnian	
Anglian	
Cromerian	Estuarine sands, silts and peats (cliff sections, Norfolk)
Beestonian	
Pastonian	
Baventian	
Antian	Crag deposits, mainly marine sands and silts (Ludham borehole)
Thurnian	
Ludhamian	
Waltonian	Crag deposits (cliff sections, Essex)

temperate fauna and flora, but it is unlikely that they are of Pliocene age. More probably, the borehole is sited where the lowest Pleistocene crags are absent, and by analogy with the succession in the Netherlands at least one further cold phase is unrepresented.

It is clear that around the southern margins of the North Sea a minimum of four cold intervals preceded the first major ice invasion. This evidence tends to confirm the multiplicity of cold phases indicated by oceanic cores, although there is little sign of agreement on the exact number of climatic cycles within a specified time interval.

The evidence of terrestrial deposits It would be quite impossible within the compass of a few pages to survey the multifarious evidence on which the Pleistocene chronology of the continental land surfaces has been built up during the last few decades. Interpretation of the evidence is often highly controversial and remarkably few generalizations can be made that would meet with universal assent. It is essential in the first place to distinguish between areas known to have been covered by the Pleistocene ice-sheets and those which lay beyond the edge of the ice-sheets. In glaciated areas attention for a long time was concentrated on the interpretation of till sheets laid down one on top of the other and believed to bear witness to multiple glacial advances. With increasing use of pollen analysis equal attention has been devoted to biogenic materials interleaved with the tills. Accompanying this emphasis on pollen-bearing sediments has been more thorough study of such fossil remains as non-marine molluscs and beetles. Most fossiliferous sediments are of very limited extent so that correlation from one site to another must depend either on the use of more continuous stratigraphic horizons such as till sheets and spreads of glacifluvial gravel, or on some intrinsic quality of the fossil record itself. It has been claimed, for example, that *Abies* pollen is abundant in the penultimate interglacial of north-western Europe but rare in the last interglacial and that *Corylus* pollen becomes frequent at a much earlier stage in the last than in the penultimate interglacial. These and similar diagnostic features, if substantiated, would permit correlation over relatively long distances without the necessity of further

Fig 2.3. The prominent shoreline that marks the former level of Lake Bonneville near Salt Lake City, Utah. The modern Great Salt Lake lies some 300m lower in the centre of the basin to the left of the photograph.

Fig. 2.4. Pleistocene environmental changes attested by relativity full continental sequences. In no instance is the dating certain, but the evidence from each points unequivocally to numerous rapid fluctuations.
(A) Cyclical fluctuations between soil formation and lacustrine sedimentation in the Great Basin, Utah, USA (after Eardley, et al., 1973).
(B) Fluctuations between soil formation and renewed loess accumulation in in central Europe (after Eardley, et al., 1973).
(C) Variations in O^{18}/O^{16} ratio in the Camp Century ice core from Greenland (after Dansgaard, et al., 1971).

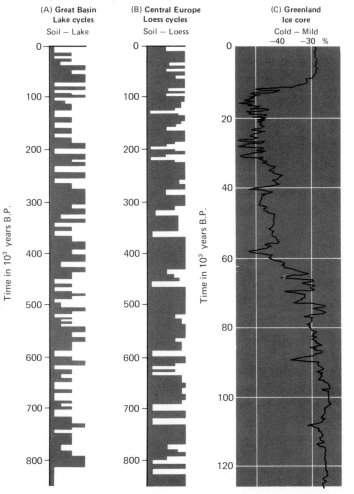

marker horizons. It should be emphasized, however, that the record of terrestrial sediments in glaciated regions is usually short and confined to the later part of the Pleistocene epoch. It sheds little light on such questions as the total number of climatic oscillations since Pliocene times. This is a point that needs stressing since for a long time there was a tendency to equate the Pleistocene with the period represented by identifiable glacial deposits.

In unglaciated regions the nature of the evidence relating to late Cenozoic climatic fluctuations varies widely. Pollen analysis has again made a significant contribution, the most important limitation being the shortage of sites in which biogenic sediments could accumulate continuously for a long period. A number of cores have been recovered from basins subject to sustained tectonic depression. Most basins of this type contain sediments with occasional fossiliferous horizons, but a continuous floral or faunal record is more rare. Since 1960 several long pollen-bearing cores have been obtained in southern Europe, in some cases apparently extending back to Pliocene times. They imply numerous climatic fluctuations comparable to those recorded, for instance, in north-western Europe. Another major source of information in extraglacial regions is the sequence of sediments in the larger lake basins. This is particularly valuable in desert areas where temporary lakes existed during wetter climatic phases but dried up during more arid periods. An excellent example is afforded by the Great Salt Lake basin in Utah. Here abandoned strandlines denote a former water level some 330 m higher than at present (Fig. 4.7). This is believed to date from a period broadly contemporaneous with the last glaciation in both north-western Europe and North America. A 307 m core of clastic and evaporite sediments from the floor of the old lake bears witness to numerous earlier alternations between deep water and desiccated salt flats. It has been estimated that the bottom of the core is some 3 Ma old and confirms that throughout this period the region has been subject to sharp climatic oscillations. The uppermost 110 m of the core has been analysed in detail and 25 phases of desiccation identified during an interval believed to cover the last 700 000 years (Fig. 2.4). A similarly complex sequence of environmental changes has been inferred from study of the loess deposits in central Europe.

Many other lines of evidence might be quoted to illustrate the methods by which Pleistocene chronologies have been built up in various parts of the world. Least information is available for tropical regions but even here there are clear indications of significant environmental changes during at least the later part of Pleistocene time. For each major glaciated region there is now a recognized nomenclature covering the period during which large continental ice-sheets developed. The names in common usage for the Alps, for the north European plain, for Britain and for North America east of the Rockies are shown in Table 2.3. There are problems of correlation between these four regions, but the relationships shown are those which command widest support at the present time. More uncertainty attends the earlier glacial and interglacial periods than the later ones and it would be foolish to deny that further research might lead to a considerable reorganization of the lower part of the table.

The evidence of modern ice-sheets The shortcomings of sedimentary successions on the continents have already been stressed, but there exist in Antarctica and Greenland long stratigraphic columns of ice which, if they can be properly deciphered, may yield information about conditions at the time of accumulation. Two problems have to be overcome. The first concerns determination of the age of the ice at different levels within the column, the second the nature of the evidence that can be derived from physical examination of the ice. Several radiometric methods have been proposed for dating young ice, but none can yet be applied to the deeper parts of an ice-sheet. Instead, the age of ice at different levels has had to be estimated by reference to annual rates of accumulation and models of flow in an ice sheet. This is the procedure adopted by Dansgaard in an analysis of a 1 390 m core from the Greenland ice-sheet (Fig. 2.4). He divided the core into 7 500 layers which were analysed for their O^{18} content. Owing to preferential evaporation from the oceans, all glacier ice is enriched in O^{16}. The deficit in O^{18} varies with the temperature at the point of condensation, being greatest when the temperature is lowest. Snow currently accumulating on the ice-sheet near the point where the core was taken has O^{18} values that deviate from SMOW by about $-28‰$. This figure is maintained with only minor variations to a depth of more than 1 100 m. Below that level there is a sudden fall to values that are consistently below $-35‰$ and occasionally sink to less than $-40‰$. Figures of this order extend to a depth of 1 330 m, beyond which there is a rise to values at least as high as those found in snow falling today. It is believed that the core between depths of 1 100 and 1 330 m consists of ice that accumulated during the last glacial period, while the basal layers are composed of ice dating from the preceding interglacial.

An important aspect of this continuous stratigraphic column is the opportunity it affords of studying relatively minor climatic changes superimposed upon the larger fluctuations. The potentiality in this direction is indicated by trends in the upper part of the core which can be equated with known historical periods of warm or cold climate. For example, it is possible to

Table 2.3 The terminology of glacial and interglacial periods in four areas of the northern hemisphere

North-western Europe	British Isles	Alps	North America
Weichselian	Devensian	Würm	Wisconsin
Eemian	Ipswichian	Riss/Würm	Sangamon
Saalian	Wolstonian	Riss	Illinoian
Holsteinian	Hoxnian	Mindel/Riss	Yarmouth
Elsterian	Anglian	Mindel	Kansan
Cromerian	Cromerian	Günz/Mindel	Aftonian
		Günz	Nebraskan

detect relatively warm periods around 1930, 1750 and 1550, together with colder intervals around 1820 and 1690. The latter coincide with well-authenticated glacier advances and periods of unusually extensive pack ice in the North Atlantic. It may be reading too much into the evidence, but there are some indications that minor fluctuations in the O^{18}/O^{16} ratio earlier in the core correspond to known interstadials, or intervals of milder climate, during the last glacial period. Analysis of a core from Antarctica has shown broadly comparable features although much more work is required to establish whether there is true synchroneity of relatively minor events in the two hemispheres. The potentiality of the method is clear; its major limitation is the small fraction of Pleistocene time still represented by ice in the modern ice-sheets.

Radiometric dating of the Pleistocene and Holocene epochs

Radiocarbon dating Of several techniques introduced during the last few decades to provide reliable dates for the Pleistocene and Holocene epochs by far the most important has proved to be that based upon the C^{14} isotope of carbon. The method depends on the fact that the atmosphere and hydrosphere represent reservoirs of radioactive carbon which are tapped by living organisms, both animal and plant, to build their various structures and tissues. The source of the radioactive carbon lies in cosmic ray bombardment which converts a minute proportion of atmospheric nitrogen into C^{14}. This in turn is rapidly oxidized to CO_2 and widely dispersed, not only through the atmosphere, but also through all surface waters. There are three isotopes of carbon, C^{12}, C^{13} and C^{14}, the last being by far the most common with little more than one atom in a million million being of the C^{14} variety. Of the three, C^{14} is the only unstable isotope and has a half-life of 5 730 years. It appears that, prior to interference by Man, an approximate long-term equilibrium had been established with the quantity of new carbon arising from cosmic radiation balanced by that lost through disintegration; the abundance of background radiocarbon may therefore be regarded as approximately constant. Since the radioactive isotope is evenly distributed throughout the lower atmosphere, and since living organisms absorb CO_2 with little or no discrimination between the various isotopes, each organism during its lifetime incorporates a constant proportion of unstable carbon. After death replenishment with C^{14} ceases and there follows a continuous decline in the C^{14} content as radioactive decay proceeds. The ratio of radioactive to stable carbon is therefore a measure of the age of such diverse organic materials as bones, tusks, shells, hides, peat and wood.

Analysis is made by counting the number of radioactive disintegrations per minute per gram of carbon. Since the number of disintegrations of modern carbon is only 13·8, and this value is halved for every 5 730-year increase in the age of the sample, the count soon becomes extremely low and difficult to differentiate from background radiation even within a massive lead shield. This limitation restricts radiocarbon dating to organic materials that are less than 50 000 years old. There is an uncertainty in the age determination since repeated counting produces a Gaussian distribution of the number of disintegrations per unit time. It is normal practice to quote the mean value as the age of the sample, and the standard deviation as the ± value. In other words, if the assumptions behind the method are sound, there is a 68 per cent probability that the true age lies within one standard deviation of the mean, and a 95 per cent probability that it lies within two standard deviations.

Precautions need to be taken when collecting samples for radiocarbon dating. Contamination by either younger or older material may seriously affect the accuracy of the results. Modern rootlets penetrating a bed of ancient peat are an obvious hazard, making the peat appear much younger than its true age. On the other hand, carbonates dissolved from ancient rocks may be absorbed in photosynthesis by submerged aquatic plants, which will then appear older than their true age. In general contamination by modern carbon is the more serious problem since it has a much greater effect on the calculated age (Fig. 2.5). Techniques have been evolved to detect contamination, but it is a hazard which needs to be constantly borne in mind.

The principle of radiocarbon dating rests upon one fundamental assumption that deserves rather closer examination. This is the assumption that the concentration of atmospheric

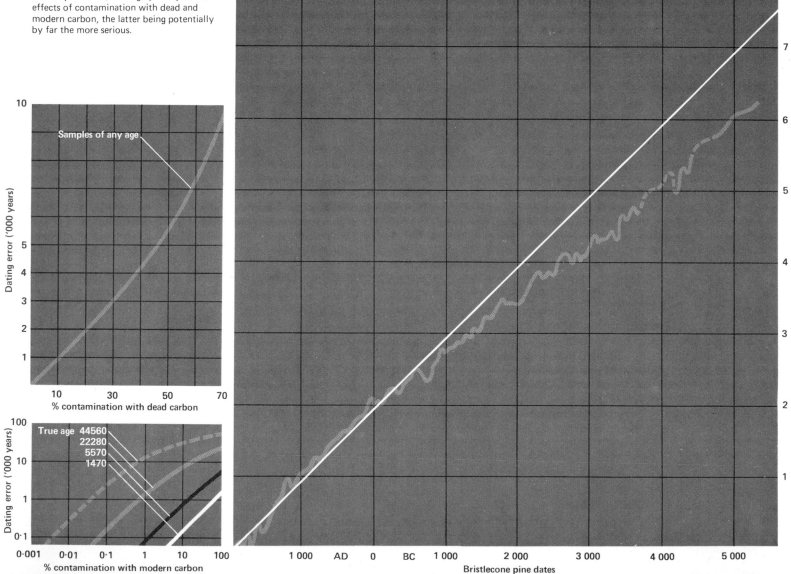

30 Fig 2.5. Calibration of C^{14} dates against bristlecone pine dates for the last 7 000 years (the straight line indicates the position of exact equivalence). The graphs depict the effects of contamination with dead and modern carbon, the latter being potentially by far the more serious.

Samples of any age

Dating error ('000 years)

% contamination with dead carbon

True age 44560
22280
5570
1470

Dating error ('000 years)

% contamination with modern carbon

C^{14} dates ('000 years B.P.)

Bristlecone pine dates

radiocarbon has remained constant through time. It is a proposition that is not difficult to test since the known age of selected sample materials can be compared with their 'radiocarbon age'. Research of this nature has revealed systematic deviations that are most easily explained as due to temporal changes in the amount of atmospheric C^{14}. For instance, in the period between A.D. 1500 and 1700 there seems to have been an abnormally high concentration of C^{14}, with the result that organic materials formed in 1700 appear on radiocarbon dating to be only about 100 years old. By analysing annual rings from the bristlecone pine (*Pinus aristarta*) in California, the world's oldest living tree, it has recently been possible to assess C^{14} variations during the last 7 000 years (Fig. 2.5). It is now believed that the excess of C^{14} between 4000 and 3000 B.C. was so great that radiocarbon dates from this period may be at least 800 years too young. Such discrepancies are of particular importance to the archaeologist and have already been the cause of controversial new hypotheses about the evolution and spread of cultures; but they also need to be remembered by the geomorphologist as he endeavours to measure with increasing precision the rates of erosional and depositional processes.

The reason for the changes in atmospheric C^{14} remains obscure; among suggested causes are variations in the strength of the earth's magnetic field and in the frequency of sunspot activity. It is obviously important to test for discrepancies between 'true' and 'radiocarbon ages' prior to 7 000 years ago, but unfortunately the older the material the more difficult it becomes to apply independent checks. Attempts have been made to compare radiocarbon dates with those derived from counting annual layers of varves in lake-floor sediments, but the evidence currently available is very conflicting. However, it is worth recording that there is no indication that the discrepancies increase systematically with age, nor that they are directly related to climatic changes. It is reasonable to conclude that radiocarbon dating provides a good guide to age over the range of time for which it can be employed, although the apparent precision of the figures must not lead to their being treated uncritically.

The great contribution of radiocarbon dating has been to provide a time scale for the final stages of the Pleistocene epoch and the whole of the Holocene. It has enabled vital correlations to be established between widely separated regions and vastly different environments. Deep-ocean cores have been calibrated, glacial advances and retreats dated, and the histories of the two hemispheres compared. Broad comparability in the timing of environmental change during the last 50 000 years in North America and Europe has been demonstrated. The maximum ice advance on the two continents appears to have taken place about 18 000 years ago and to have been the culmination of a long period of cold climate with only brief milder interludes. However, the limit of radiocarbon dating does not extend as far back as the Eemian/Sangamon interglacial. Subdivision of Holocene time in Europe was originally based upon the sequence of floral changes as revealed by pollen analysis. This led to recognition of a number of distinctive pollen zones which have more recently been accorded radiocarbon dates (Table 2.4).

Uranium-series dating The long half-life involved in the production of lead from either uranium or thorium militates against the use of this natural decay process for Pleistocene dating. However, the transformation takes place through a series of intermediate stages in which many relatively short-lived daughter products are formed (Fig. 2.1). These intermediate members usually have half-lives too short for dating purposes, but three lie between 1 000 years and 1 Ma. These are U^{234}, Th^{230} and Pa^{231} with half-lives of 250 000, 75 000 and 32 000 years respectively. Of the many attempts to use these three isotopes for dating, few have proved really successful. One method depends on the discovery that marine coral may form with several parts per million uranium but virtually none of the unstable thorium isotope, Th^{230} (also known as Ionium). The uranium will consist of both U^{238} and U^{234}, and with the passage of time the latter will yield Th^{230}. In theory, therefore, the ratio of Th^{230} to uranium is a measure of the age of the coral. The method is capable of dating marine carbonates up to about 200 000 years old, but is based on the assumption that the material forms a closed system with no addition or exchange of either uranium or thorium. In some instances, at least, there appears to have been late enrichment with uranium

Table 2.4 Subdivision of the Holocene, together with an indication of the associated vegetational changes in England and Wales. (after West, *Pleistocene Geology and Biology*, 1968)

Time based on C[14] dating	Blytt & Sernander periods	Pollen zone	Zone characteristics for England and Wales	
		VIII modern	Afforestation	
1000	Sub-Atlantic	VIII	Alnus–Quercus Betula(–Fagus–Carpinus)	Deforestation
A.D.				
B.C.				
1000	Sub-Boreal	VIIb	Alnus–Quercus–Tilia	
2000				
3000			Ulmus decline	
4000	Atlantic	VIIa	Alnus–Quercus–Ulmus–Tilia	
5000				
6000	Boreal	VI	(c) Quercus–Ulmus–Tilia Pinus–Corylus (b) Quercus–Ulmus (a) Ulmus–Corylus	
7000		V·	Corylus–Betula–Pinus	
8000	Pre-Boreal	IV	Betula–Pinus	

making the calculated ages too low.

A second technique based upon intermediate stages in the uranium series is that which employs the ratio of Pa^{231} to Th^{230} to assess the age of deep-sea sediments. It is founded on the fact that uranium in sea water, derived from erosion of the continents, consists of two major isotopes, U^{238} and U^{235}, in the proportions of approximately 138 to 1. By radioactive decay these isotopes yield Th^{230} and Pa^{231} which become attached to clay minerals settling to the ocean floor. The thorium and protactinium then disintegrate at rates corresponding to their respective half-lives. If the amount of uranium in sea water could be assumed constant, it would be relatively easy to calculate the age of sediment recovered in oceanic cores by measuring the amount of either Th^{230} or Pa^{231}. In practice such an assumption would be difficult to justify and an alternative approach must be adopted. This involves making the much more reasonable assumption that the proportion of U^{238} to U^{235} has remained constant, and concentrating therefore on the Pa^{231}/Th^{230} ratio in the cores. In theory this ratio should be $10 \cdot 8 : 1$ at the time of deposition and gradually increase with age; with the diminishing amounts of the isotopes available for analysis the method is capable of dating sediments up to about 200 000 years old. There has been considerable controversy regarding some of the technicalities involved, but the method, if sound, might contribute significantly to the correlation of oceanic cores. One advantage of a dating technique extending back more than 100 000 years is that it permits the rate of sedimentation in both glacial and interglacial periods to be assessed. Using the relevant figures the earlier parts of the cores can then be time-calibrated and in this way approximate dating extended back beyond the practical limits of the Pa^{231}/Th^{230} method. On this basis several workers have concluded that the cores indicate ten or more glacial periods during the last million years.

A final point worthy of note is that periodic reversals of the earth's magnetic field, more fully discussed in Chapter 3, are proving an increasingly important means of correlation between individual oceanic cores; moreover, they can occasionally be employed for establishing relationships with long stratigraphic successions on the continents. The uses made of changes in the geomagnetic field will almost certainly grow rapidly in the future.

DANSGAARD, W., et al., 1971. 'Climatic record revealed by the Camp Century ice core' in *Late-Cenozoic Glacial Ages* (Ed. K. K. Turekian) Yale Univ. Press. (see also N-A Mörner, *Geol. Mag.*, 1974.)

EARDLEY, A. J., et al., 1973. 'Lake cycles in the Bonneville basin, Utah', *Bull. geol. Soc. Am.*, 84, 211–16

EMILIANI, C. and SHACKLETON, N. J., 1974. 'The Brunhes epoch : isotopic paleotemperatures and geochronology', *Science. N.Y.*, 183, 511–14.

A valuable work on dating techniques and time scales is, *The Phanerozoic Time Scale – A Supplement*, Spec. Pap. geol. Soc. Lond., 5, 123–356, 1971. This volume includes a special section by P. Evans entitled 'Towards a Pleistocene time scale'. A full discussion of C^{14} dating is provided by I. U. Olsson (Ed.) 'Radiocarbon variations and absolute chronology', *Nobel Symposium 12*, Almqvist and Wiksell, 1970.

1|3

Global tectonics

With an established time scale, it is now feasible to examine the evolution of the gross relief forms described in Chapter 1. Until about 1960 it was conventional wisdom that the pattern of continents and ocean basins is unchanging, although for many years a number of individuals had voiced the contrary opinion that large-scale translocation of the continents might have occurred. The most passionate and influential advocate of this latter idea was Alfred Wegener through his famous book *Die Entstehung der Kontinents und Ozeane*. Publication of this work in 1915 was followed by decades of bitter controversy since so much of the evidence adduced by Wegener and his disciples was capable of alternative interpretations. Although a meteorologist by profession, Wegener was first drawn to the idea of continental drift by the congruence of the coastlines on either side of the Atlantic ocean. Initially rejecting the hypothesis as improbable, he was led to re-examine it by reading literature describing palaeontological and geological similarities on the bordering continents. It was both the strength and weakness of his position that it impinged upon so many scientific fields; on the one hand an enormous wealth of corroborative evidence could be accumulated, but on the other numerous workers in very diverse specialisms had to be convinced by his arguments. So widespread were the repercussions of the whole concept that by 1928 Wegener admitted that he found himself unable to keep abreast of the rapidly growing literature.

It would obviously be impractical to attempt a summary of all the conflicting arguments advanced in this early period. Instead it is easier to review the research that provoked such a profound change of attitude on the part of the scientific community. Without doubt the most influential research was that concerned with rock magnetism and the reconstruction of the ancient geomagnetic field; but once the likelihood of continental movement had been established, a wide range of further investigations added strong supporting evidence. Two of the more important, heat-flow and seismic studies, will be briefly examined before attention is finally turned to the modern unifying concept of plate tectonics.

Palaeomagnetism

The geomagnetic field Modern scientific study of the earth's magnetic field is usually said to begin with the experiments of Sir William Gilbert, physician to Queen Elizabeth I. He showed that the field resembles that which would result from a giant bar magnet located near the centre of the earth and aligned approximately along the axis of rotation. Extensions of the long axis of the magnet intersect the surface of the globe at points known as the north and south magnetic poles. This dipole field at the surface of the earth may be described in terms of the magnitude and direction of the magnetic force. The magnitude attains its maximum value close to the magnetic poles and its minimum value around the magnetic equator. The direction is usually specified in terms of inclination and declination (Fig. 3.1). The inclination is the angle which a freely suspended magnetized needle makes with the horizontal. In the case of a perfectly regular dipole field such a needle would stand vertically at the magnetic poles and horizontally at the magnetic equator ; between these two extremes the angle would vary as a function of the latitudinal distance from the magnetic pole. The declination is the angular difference between the geographic meridian and the horizontal component of the earth's magnetic field. Its value reaches the extremes of zero and 180° along the great circle passing through both the geographic and magnetic poles.

The observed geomagnetic field differs in a number of **35** important respects from the simple dipole field assumed so far. Measurements of magnetic intensity show the expected general decline from poles to equator, but superimposed on this are longitudinal variations such that the intensity over Australia, for instance, differs considerably from that over South America in the same latitude. Similarly inclination and declination show marked divergences from the regular dipole pattern. It should be emphasized that these divergences are not due to shallow crustal peculiarities such as local iron-ore bodies, but are part of a world-wide pattern which appears to be independent of surface form and structure and to have its origin much deeper within the earth. In addition to these spatial complications there are also important temporal changes. Continuous recording stations widely distributed over the globe reveal small daily fluctuations in intensity, declination and inclination. These same properties also vary on a longer, secular time scale. Ever since the seventeenth century it has been known that the positions of the

Fig 3.1. The geomagnetic field (A) Inclination (B) Declination.

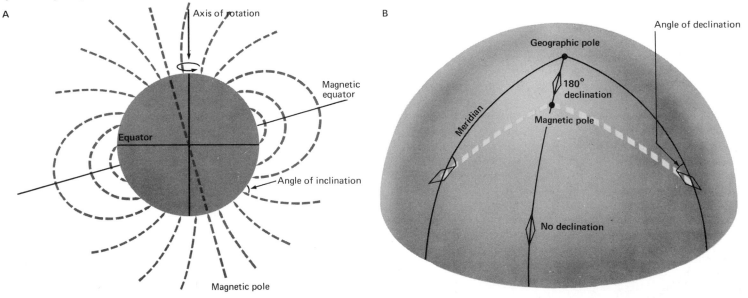

magnetic poles slowly move. When first recorded the declination at London was about 10°E; by 1800 it had reached over 20°W but is now decreasing and in 1975 was about 7°W. In the same period the inclination has varied between 75° and 67° and the intensity has gradually weakened. Viewed on a global scale many of the deviations from the simple dipole field migrate slowly westwards, moving at a rate of 0·18° of longitude each year, that is, making one revolution in about 2 000 years. On an even longer time scale the most striking change is undoubtedly reversal of the earth's polarity so that what is now the north pole becomes the south pole and vice versa.

It is from detailed studies of the modern magnetic field that the most acceptable hypotheses of origin arise. These envisage that the outer core of the earth, being liquid and an electrical conductor, acts as a 'self-exciting dynamo'. The precise mechanism is still far from clear, but it seems that some form of energy in the earth's interior is ultimately converted into electric currents which encircle the core and produce the dipole magnetic field. Most significantly, all the widely supported hypotheses demand a permanently dipolar field with an alignment approximately parallel to the axis of rotation.

Remanent magnetism When an igneous rock containing iron-rich minerals cools from the molten state it becomes magnetized in accordance with the prevailing geomagnetic field. Laboratory investigations show that most of the magnetization is acquired quite abruptly during cooling, the critical temperature being designated the Curie point of the rock. So long as the rock is not heated again to near its Curie point, which commonly lies between 400° and 600°C, the acquired magnetization is remarkably stable and may be retained for hundreds of millions of years. In other words, when an iron-rich lava cools through the Curie temperature a permanent record is frozen into the material. The record is contained in the remanent magnetism of the rock, its strength varying according to the mineralogical composition of the rock and the intensity of the geomagnetic field at the time of cooling. At best the remanent magnetism is extremely weak, a fact which undoubtedly delayed the widespread application of palaeomagnetic principles. A major advance came with the development of the astatic magnet-

ometer by the distinguished British physicist, Blackett. With this highly sensitive instrument it became possible to measure not only the remanent magnetism of igneous rocks but also the even weaker magnetism of certain sedimentary rocks. As detrital particles settle through water, those fragments which have already been magnetized tend to align themselves in conformity with the ambient magnetic field. In this way iron-bearing grains assume a preferred orientation that confers a weak remanent magnetism on the sediment and ultimately on the resulting rock. This is sometimes destroyed by chemical changes during and after consolidation so that a sedimentary rock with a weaker and less stable magnetization presents greater problems of interpretation than an igneous rock; nevertheless, with adequate precautions in sampling and measurement it may still yield much valuable information.

The basis of palaeomagnetic reconstruction When a rock sample is selected for palaeomagnetic study, its orientation is first carefully measured and any necessary allowance made to compensate for tectonic disturbance. Subsequent analysis of the sample in a magnetometer theoretically specifies the magnitude, declination and inclination of the local force at the time the remanent magnetism was acquired. At this stage a major assumption needs to be made, namely that the total geomagnetic field was dipolar in form. On that basis it is possible to reconstruct the contemporaneous positions of the magnetic poles. Of course there are several sources of potential error. The allowance for tectonic disturbance is difficult to assess and the final estimate may often be accurate to no more than $\pm 5°$. Secular fluctuations distorting the pure dipolar field place an additional constraint on the precision of the method. In practice, the remanence of many samples of the same age is measured, calculations made for each individual sample and then statistical procedures employed to define confidence limits for the positions of the poles.

Most of the early palaeomagnetic investigations were concerned with Cenozoic lavas in such countries as Japan, Italy and France. Then in the 1950s came the dramatic studies initiated by Blackett on the Triassic sandstones of the British Isles. These revealed polar positions some 200 Ma ago widely differing from those of the present day. One obvious explanation, known as the polar wandering hypothesis, was that the magnetic poles had slowly migrated over the earth's surface. This led to attempts at drawing what are termed polar wandering curves. These involve plotting the apparent polar positions at different geological periods and joining the points by a smooth curve. When this is done for a single continent a consistent pattern emerges and seems to support the polar wandering hypothesis. However, when a polar wandering curve is calculated for a second continent, it is found to differ from the first. It was this vital discovery which led to rejection of the polar wandering hypothesis and revival for continental drift as the only satisfactory explanation for the palaeomagnetic findings. If the continents have shifted their relative positions, all contemporaneous rocks from a single continent should yield a unique polar position, whereas those from different continents should yield different polar positions. In essence this is what palaeomagnetic investigators claim to have found.

At first it might seem that an assumed identity between magnetic and geographic poles, justified on theoretical grounds, would permit detailed reconstruction of continental movement. However, brief reflection will show that this is not so. Whereas inclination defines palaeolatitude, declination does not define palaeolongitude ; former longitudinal positions with respect to some arbitrary datum such as the Greenwich meridian must remain indeterminate. Nevertheless, relative displacements

Fig. 3.2. The 'super-continent' Pangaea as reconstructed by Wegener for Upper Carboniferous times.

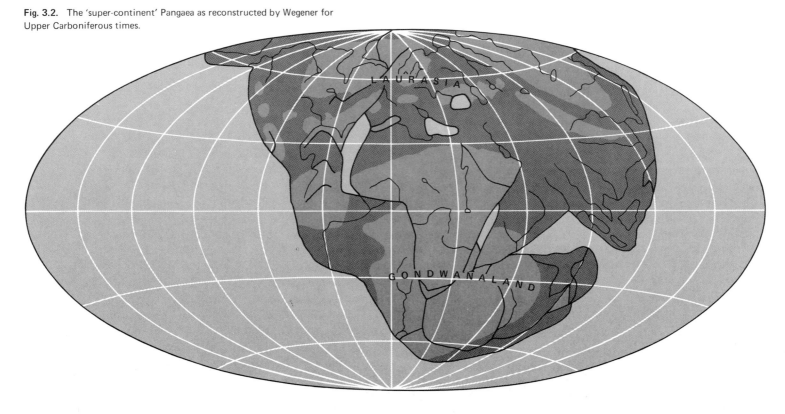

of the continents can be evaluated by drawing polar wandering curves. In theory, as long as two continental blocks are fused together, or not moving with respect to each other, they should yield a curve of identical form. Once they start drifting apart their individual curves should begin to deviate. Employing this principle it is feasible to check observed palaeomagnetic values against the predictions of any particular model of continental drift. For instance, Wegener suggested that the major continental blocks once formed a supercontinent which he termed Pangaea (Fig. 3.2). He argued that this was in existence until at least Permian times since evidence for Permo-Carboniferous glaciation in South America, Africa, India and Australia required contiguity of these continents if the global ice cover were to be kept to acceptable proportions. The opening of the Atlantic Ocean as the Americas separated from the rest of Pangaea was believed to have begun in the south in Jurassic times and gradually spread northwards until Europe and North America finally separated in late Cenozoic times. The drifting apart of Africa, Asia, Australia and Antarctica was less closely dated but was believed to have begun in early Jurassic times. If Wegener's reconstruction of Pangaea is correct, all continents when fitted into their respective positions should indicate a single palaeomagnetic pole during late Palaeozoic times. In broad outline this proves to be the case and constitutes important support for Wegener's original thesis. The polar wandering curves for individual continents begin to split in late Triassic times. This is clearly seen in a comparison of the curves for Africa, South America and North America. Australia and Antarctica appear to have moved away as a separate unit at about the same period. Palaeomagnetic data suggest that the final splitting of India from Africa and of Australia from Antarctica occurred rather later, the latter possibly as recently as early Cenozoic times.

Reversals of polarity It was not long after the first systematic study of remanent magnetism that instances of reversed polarity were recorded. One of the early discoveries was made by the French worker Brunhes in 1906, but for a long time there was uncertainty about the correct interpretation of such findings. At one stage it was believed that there might be a 'self-reversal' mechanism by which a rock could become magnetized in the opposite sense to the ambient field. Laboratory experiments in Japan showed that this could indeed happen, but that it required such exceptional circumstances that it could not be expected to be a common occurrence. Therefore when reversed remanence was found to be almost as common as normal remanence it was obvious that the geomagnetic field must have changed. This was confirmed when dating was undertaken and it was shown that the field reversals were synchronous all over the globe.

Further research has disclosed well over 100 polarity reversals during the Mesozoic and Cenozoic eras (Fig. 3.3). Detailed analysis of the transition phases has shown that, prior to a reversal, geomagnetic intensity declines rapidly to about one-quarter of its normal value. This is accompanied by growing instability of the poles until, quite abruptly, they move along a great circle route to approximately antipodal positions. Thereafter the field gradually regains its normal strength. The whole process appears to take about 10 000 years and the actual movement of the poles is completed in less than 5 000 years. At the present time the intensity of the field is declining quite rapidly, but as fluctuations are a normal feature of the field there is no proof that we are heading for a polar reversal; equally there is no proof that we are not!

The upper Cenozoic era has been divided into intervals of either reversed or normal polarity, each lasting about 1 Ma. These periods have been designated 'epochs' and accorded the names of famous research workers in the field of geomagnetism (Fig. 3.3). During each epoch there were much shorter 'events' when the field was temporarily changed; there may even have been very brief 'flips' if evidence accumulating for the end of the Brunhes epoch is finally substantiated. For reversals prior to the Gilbert epoch a numbering system has now been adopted. Owing to the frequency of reversals, isolated rock samples can rarely be dated on the basis of their remanent magnetism. More interest attaches to continuous sedimentary accumulations in which long sequences of polarity changes might be expected. Deep-sea cores fulfil these requirements and analysis of many oceanic cores has demonstrated a succession of reversals directly comparable to those inferred from the study of continental rocks (Fig. 3.3). This not only provides a valuable way

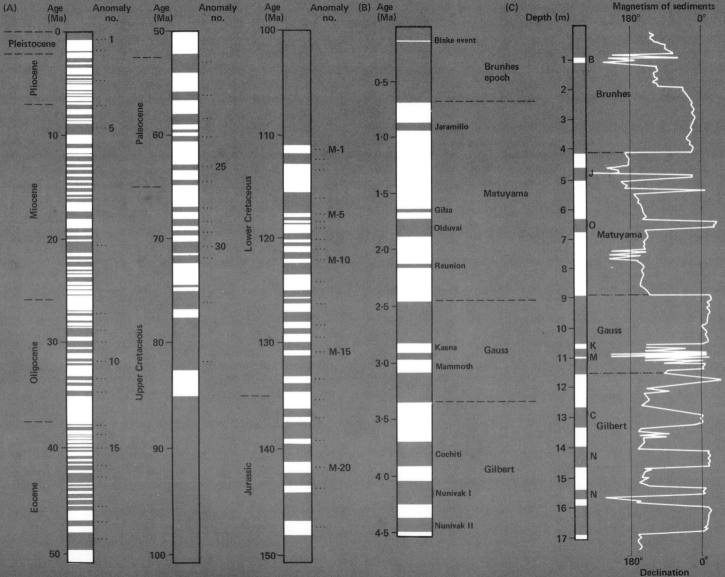

Fig. 3.3. The chronology of polarity reversals (A) Reversals during the last 150 Ma together with an indication of distinctive anomalies used for dating the ocean floors (Fig 3.6). (B) Detail of the reversals during the last 4·5 Ma. (C) An illustrative ocean core showing the nature of the evidence that may be derived from this source.

of correlating different cores, but also permits dating by means of the time scale that has been worked out for the polarity changes.

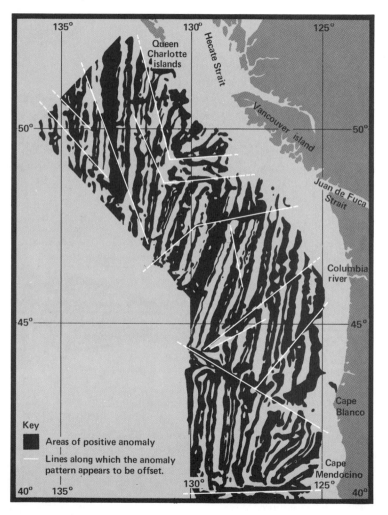

Fig. 3.4. The linear pattern of magnetic anomalies as first detected over the north–eastern Pacific by Raff and Mason.

Key

■ Areas of positive anomaly

⎯ Lines along which the anomaly pattern appears to be offset.

Variations in magnetic intensity over the oceans The preceding paragraph alludes to one sphere in which polarity reversals have affected oceanographic studies. Their major impact, however, was to be felt in a rather different sphere. In the late 1950s a magnetic survey of the north-eastern Pacific was carried out by a research team from the Scripps Institute at La Jolla, California. This disclosed a remarkable pattern of linear zones, each trending approximately north–south and averaging 30–50 km in width, in which the intensity of the geomagnetic field is alternately above and below average (Fig. 3.4). Such an orderly arrangement of magnetic anomalies is absent from the continents and at first defied all explanation. The interest that was aroused stimulated research off the Atlantic seaboard of North America where a similar magnetic striping was discovered.

In a famous paper in 1963 Vine and Matthews linked this linear patterning with two separate ideas. The first was an earlier notion of Hess that the ocean floors might be the sites of crustal generation with new crust being formed at the mid-oceanic ridges and gradually carried away towards the continental margins. The second idea was the rapidly growing belief that the earth had experienced numerous reversals of magnetic polarity. Vine and Matthews suggested that, if the hypothesis of sea-floor spreading were correct, hot new crust might be imprinted with the contemporaneous magnetic field as it cooled through the Curie point at the mid-oceanic ridge (Fig. 3.5). The linear pattern could then be explained as the result of periodic changes in polarity, since the crust that had cooled during an epoch of normal polarity and subsequently moved away from the ridge would tend to reinforce the present geomagnetic field and so produce a zone of positive anomalies. On the other hand, reversely magnetized crust would produce a zone of negative anomalies. According to this hypothesis, the anomalies on either side of a mid-oceanic ridge should form a mirror image of each other. The whole process has been likened to a giant tape-recorder in which the tapes move away from the central ridge and, as they do so, pick up messages about the nature of the earth's magnetic field. The continental rocks provide an independent record of the messages and their timing, so that the movement of the 'tapes' can be deduced.

Much current research is directed to correlating the magnetic anomalies with the polarity record of the continents. In this way isochrons, or lines joining points of equal age, may be drawn on maps of the ocean floors (Fig. 3.6).

If the foregoing hypothesis is correct, two important inferences may be drawn from the pattern of anomalies. Firstly, the age at which sea-floor spreading began in any particular area can be assessed. As an illustration, the oldest positive anomaly in the North Atlantic between Norway and Greenland is believed to be number 24 which indicates that that section of ocean began to open up about 60 Ma ago. Secondly, rates of sea-floor spreading can be computed. This was first attempted after an aeromagnetic survey in 1963 of the Reykjanes ridge south-west of Iceland where each flank of the ridge was estimated to have moved at about 10 mm yr^{-1} for the last 4 Ma. This is known as the half-spreading rate since it is half the rate at which the ocean floor has widened. Since the study of the Reykjanes ridge all the mid-oceanic ridges of the world have been magnetically surveyed, and by matching magnetic profiles against the record of polarity reversals half-spreading rates have

been calculated. The greatest rate has been found along the East Pacific ridge between the equator and 30°S where the computed movement is as high as 60 to 90 mm yr^{-1} for the last 9 Ma. Over the same period the half-spreading rate in the South Atlantic has averaged 20 mm yr^{-1} and in the Indian Ocean between 15 and 30 mm yr^{-1}.

Further tests of sea-floor spreading A corollary of sea-floor spreading is that oceanic crust should be youngest near the mid-oceanic ridge and become progressively older as distance from the ridge increases. Two simple tests of this proposition have been devised, the first relating to the age of oceanic islands and the second to the age of the oldest marine sediments. If oceanic islands, mainly composed of basaltic outpourings, are formed close to the axis of a mid-oceanic ridge, they should be slowly transported away from their point of origin by the moving crust. In 1963 Wilson analysed the age of the oldest known rock on each island in the Atlantic; if this is assumed to indicate the age of the island itself, it can be shown that in general the youngest lie closest to the central ridge and

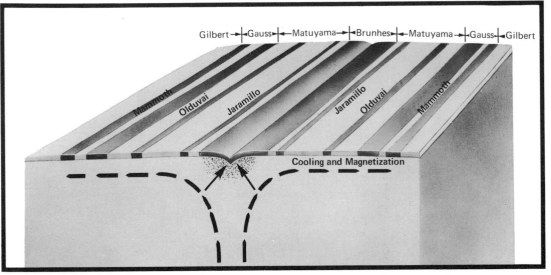

Fig. 3.5. Schematic diagram of the explanation offered by Vine and Matthews for the linear magnetic anomalies over the oceans.

Gilbert → | ← Gauss → | ← Matuyama → | ← Brunhes → | ← Matuyama → | ← Gauss → | ← Gilbert

Mammoth   Olduvai   Jaramillo   Jaramillo   Olduvai   Mammoth

Cooling and Magnetization

Fig 3.6. The age of the Atlantic floor as deduced from the pattern of magnetic anomalies.

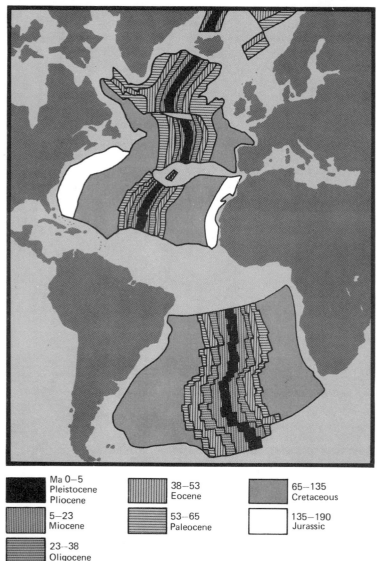

■ Ma 0–5 Pleistocene Pliocene	▦ 38–53 Eocene	▓ 65–135 Cretaceous
▦ 5–23 Miocene	▤ 53–65 Paleocene	□ 135–190 Jurassic
▦ 23–38 Oligocene		

the oldest furthest from it (Fig. 3.7).

Comparable changes would be expected in the age of the basal sediments on the ocean floor. One of the aims of the Deep Sea Drilling Project has been to determine whether this is the case or not. In general the results have accorded remarkably well with the ages inferred from the study of magnetic anomalies. The oldest sediments recovered are Jurassic in age and are confined to one or two very marginal sites. Over the vastly greater part of the ocean floor the basal sediments are either Cretaceous or Cenozoic in age, with the older material usually restricted to localities well away from the mid-oceanic ridges. As might be anticipated, the sedimentary cover is much thinner near the central ridges, although this may not be due entirely to the shorter period of accumulation; the thicker cover along the ocean margins is in part attributable to the abundance of sediments supplied by erosion of the nearby continents. On the other hand, even in the west Pacific where the abyssal floor is separated from the Asiatic continent by deep trenches a similar relationship between sediment thickness and distance from the mid-oceanic ridge still seems to hold.

Heat flow measurements

It has been known for a very long time that the temperature of rocks in the continental crust increases with depth. This was obvious from measurements in early mines where it was found that the temperature commonly increases at a rate of 30°C for every kilometre of descent. Assessment of actual heat flow requires measurement not only of the thermal gradient but also of the thermal conductivity of the rocks. The geothermal flux is normally calculated in units of calories $\times 10^{-6}$ cm^{-2} sec^{-1},[*] and it is found that the average value for the continents is about 1·46 such units, with a range from less than 0·5 to over 10. The major source of heat is believed to be radioactive minerals in the continental crust. This view led to the supposition that heat flow beneath the oceans would prove to be considerably smaller, since the oceanic crust is thought to contain a much lower proportion of radioactive elements. However, in the early 1950s

---

[*] $4 \cdot 2 \times 10^{-6}$ J cm^{-2} sec^{-1}. See Appendix A.

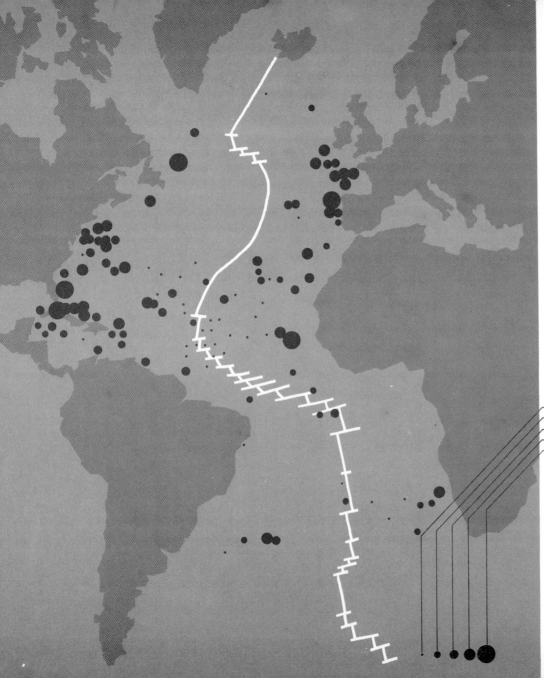

Fig 3.7. The age of basal sediments recovered from deep-sea cores. The graph shows the age of the Atlantic islands expressed as a function of distance from the mid-oceanic ridge. Also shown is the age of the basal sediments recovered on a single leg of the deep-sea drilling project. The dotted line corresponds to a half-spreading rate of 20 mm yr^{-1}.

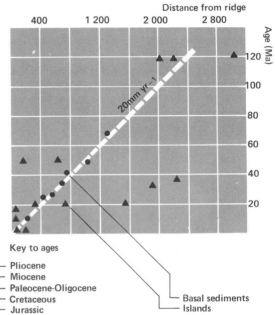

Key to ages

— Pliocene
— Miocene
— Paleocene-Oligocene
— Cretaceous
— Jurassic

— Basal sediments
— Islands

44 Fig. 3.8. Areas of current seismicity (after Baranzagi and Dorman, *Bull. seism. Soc. Am.,* 1969).

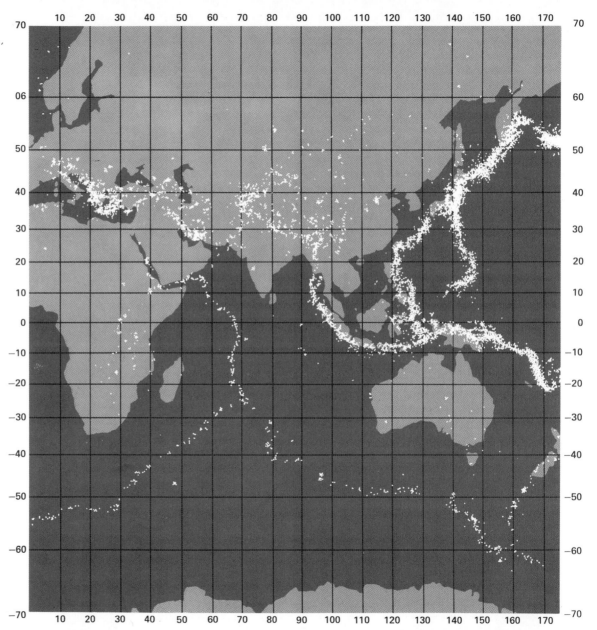

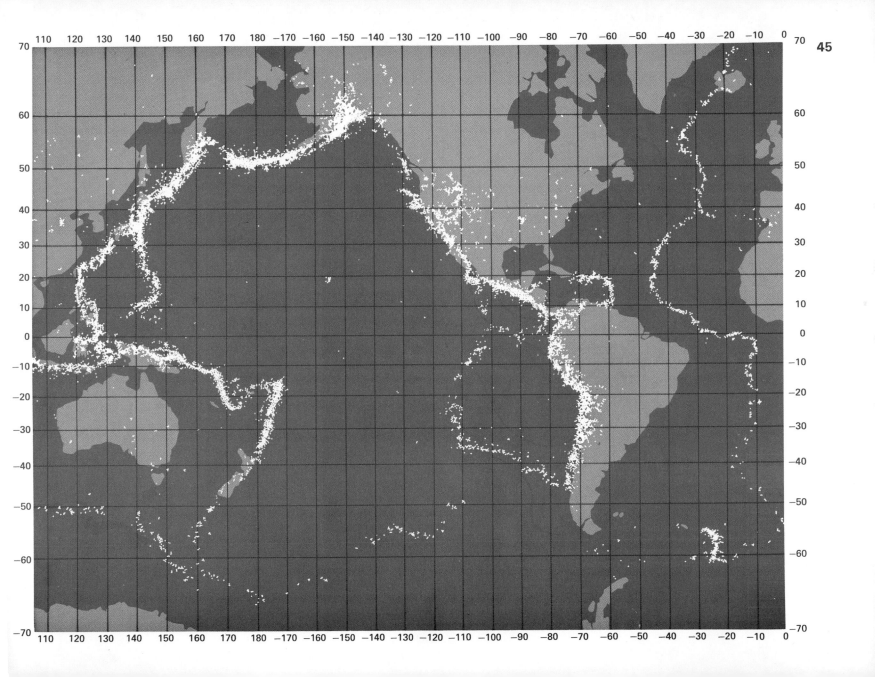

a technique was devised for measuring directly the heat flow through the ocean floors. A probe inserted several metres into the sea-floor sediments enabled the temperature gradient to be determined by means of a thermocouple or thermistors; a core from the same site provided a sample for assessment of thermal conductivity. The results were startling for they showed that, contrary to expectations, the mean heat flow through the ocean floors is rather greater than that from the continents. Thousands of further measurements have now been made and important regional variations have become discernible. These include particularly high values, occasionally exceeding 8, over the axes of the mid-oceanic ridges. On the other hand, the ridge flanks tend to be characterized by unusually low values so that there are sharp variations within relatively short distances. Other areas that display rapid lateral changes are the trench and island-arc systems. High values are normal on the continental side of the arc, whereas very low values are typical of the trench and the adjacent abyssal plain.

Many workers have seen in these results support for a very old idea, that of thermal convection currents within the earth's mantle. Numerous variants of this basic hypothesis have been proposed, but the geothermal measurements provide the most direct evidence for the positioning of the currents. The majority of recent authorities have envisaged hot currents rising beneath the mid-oceanic ridges, with the trenches sited along zones of convergence where material in the convective cells begins to descend. Friction along the upper side of the descending limb has been suggested as the source of heat on the continental side of each island arc. Such a model of movement within the mantle and crust has commanded widespread assent. Controversy among its proponents has tended to centre round the source of the necessary energy, whether there is a close geographical relationship between surface relief and the pattern of convective cells, and the depth within the mantle at which convection occurs. Some workers have argued that convection may affect the total thickness of the mantle, while others have contended that the cells are much shallower and do not extend below the asthenosphere.

Seismic studies

The analysis of seismic waves has made a major contribution to the study not only of the internal constitution of the earth (p. 14), but also of the fracture zones within the crust and upper mantle. By plotting the sites of all earthquake epicentres recorded over a period of several years, it is possible to identify the main areas of current fault activity (Fig. 3.8). Seismicity is a highly localized phenomenon. Most epicentres are located in a circum-Pacific belt, with a secondary concentration extending from Indonesia through south-central Asia to the Mediterranean regions of Europe. Within the oceanic areas there is a striking localization along the lines of the mid-oceanic ridges. However, in addition to indicating the distribution of active faults, careful analysis of modern seismograph traces has enabled the sense of motion along individual faults to be determined. The method depends on the fact that the first displacements recorded for a single earthquake are at some stations compressional, and at others dilatational. It is found that these two types of first-motion are arranged in quadrants separated by two mutually perpendicular planes known as nodal planes (Fig. 3.9). It can be shown that one of the latter must correspond to the fault plane, and it is usually easy to choose which is the more likely on geological grounds. Once this has been done the sense of relative displacement can be inferred; it should be parallel to the fault plane, normal to the intersection of the nodal planes, and towards the compressional quadrant. First-motion studies of this type have been particularly significant in elucidating the nature of displacements associated with two groups of features, namely transform faults and Benioff zones.

Transform faults The presence of major linear offsets affecting the trend of mid-oceanic ridges was first noted in the course of bathymetric charting, but became particularly obvious as a result of detailed magnetic surveys. Because the zones of magnetic anomaly vary in width and strength, the pattern on either side of the offsets can be matched and the amount of misalignment estimated. The outcome was the discovery that apparent displacements frequently exceed 10 km and not uncommonly surpass 100 km. As they affect magnetic patterns formed in the very recent geological past, the dislocations must be regarded as presently active. In 1965 they were termed transform faults by Wilson who argued that they should possess properties quite different from those of ordinary transcurrent faults such as are illustrated in Fig. 4.1. In the first place, there is no reason to believe that the separate sections of a ridge were ever in continuous alignment and have later been parted; instead, the close matching of anomaly patterns is due to almost identical sequences being generated from each spreading centre. It follows that shearing should normally be confined to that part of the dislocation lying between the ridges. A further consequence is that the sense of displacement should be the reverse of that forecast by conventional analysis as a transcurrent fault. In 1967 Sykes was able to confirm by seismic studies many of the properties predicted by Wilson. He showed that there are two kinds of earthquake initiated along the mid-oceanic ridges. The first occurs along the axis of each ridge and is of the type associated with dip-slip faulting. The second is restricted to the inter-ridge segments of the offsetting faults and is of the type associated with predominant strike-slip move-movement; first-motion studies also demonstrated that the seismic waves are consistent with transform but not transcurrent faulting.

Benioff zones It is evident that on a sphere of constant size, creation of new crust along the mid-oceanic ridges must be matched by destruction elsewhere (Fig. 3.10). There are many reasons for believing that the oceanic trenches mark such zones of crustal destruction, but probably the most convincing evidence comes from plots of earthquake foci in the vicinity of the trenches. It is found that the foci define steeply sloping zones of earthquake activity which plunge at an angle of approximately 45° from the floors of the trenches beneath the adjacent island arcs or continental margins. These seismic zones have been named Benioff zones in honour of the eminent seismologist who first described them in detail. They are distinguished by being the location of the deepest earthquakes recorded anywhere on the globe, many having been detected at depths in excess of 500 km. First-motion studies indicate compression parallel to the dip of the Benioff zone, apart from a very shallow section near the trench where there appears to be tension. These results seem consistent with the idea that a rigid crustal slab moving away from the mid-oceanic ridge is first buckled downwards with consequent tension in its upper layers, and then thrust much deeper into the mantle with resulting compression. The presence of lighter crustal material forced down to considerable depths

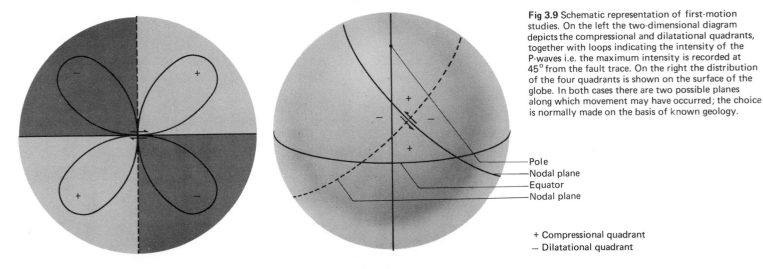

Fig 3.9 Schematic representation of first-motion studies. On the left the two-dimensional diagram depicts the compressional and dilatational quadrants, together with loops indicating the intensity of the P-waves i.e. the maximum intensity is recorded at 45° from the fault trace. On the right the distribution of the four quadrants is shown on the surface of the globe. In both cases there are two possible planes along which movement may have occurred; the choice is normally made on the basis of known geology.

Pole
Nodal plane
Equator
Nodal plane

+ Compressional quadrant
− Dilatational quadrant

48 explains the negative gravity anomalies recorded over most island-arc systems. Similarly, disappearance of the pattern of magnetic anomalies at the trenches may be explained by the heating of the descending slab to above the Curie point with consequent erasure of the magnetic record. The rate of under-thrusting has been computed for various arc-trench systems and has been shown to vary from about 10 mm yr^{-1} for the Aleutians to 90 mm yr^{-1} for the Marianas.

Plate tectonics

Basic concepts In the preceding sections of this chapter reference has been made to two wide-ranging concepts that have exercised the minds of earth scientists for many years. The first is the idea of continental drift, and the second that of sea-floor spreading. One arose from prolonged comparative study of the continents, the other from more recent investigations into the nature of the ocean floors. They are related but not identical ideas, and it is largely a fusion of the basic tenets of each that has given birth to the new concept of lithospheric plates.

It is a fundamental postulate of much recent geophysical theory that the lithosphere is composed of a series of internally rigid plates (Fig. 3.11). Originally six major plates were identified, but it is now recognized that there must be at least a further 20 smaller plates that may move independently of their larger counterparts. Each plate may carry both oceanic and continental crust and there is no simple relationship between the pattern of the plates and the distribution of the continents. The Pacific plate, for instance, is almost entirely 'oceanic' and the Eurasian plate very largely 'continental'. Investigation of the nature and

Fig. 3.10. A schematic sketch showing the generation of new lithosphere at so-called spreading centres and its destruction at steeply dipping Benioff zones (after Isacks, et al., 1968).

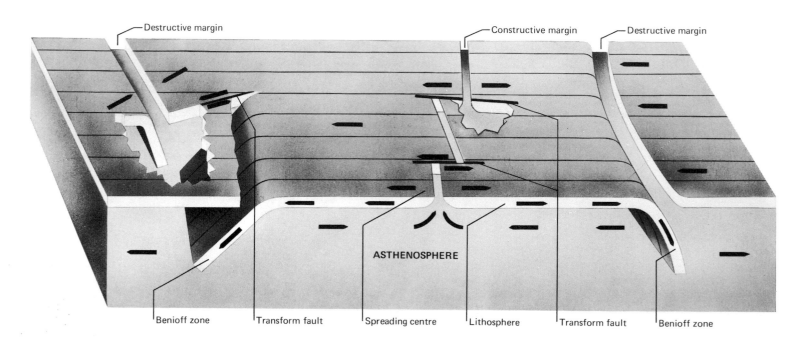

Destructive margin — Constructive margin — Destructive margin

ASTHENOSPHERE

Benioff zone — Transform fault — Spreading centre — Lithosphere — Transform fault — Benioff zone

movement of the plates is the field of study known as plate tectonics. Much research is devoted to studying the margins of the plates for it is here that most seismic, volcanic and tectonic activity seems to be concentrated. Three contrasting types of plate margin are recognized, constructive, destructive and conservative. At a constructive margin new material is generated and as already seen this takes place along the mid-oceanic ridges. At a destructive margin plate material is consumed by subduction into the underlying mantle, normally along a Benioff zone. At a conservative margin two plates slip **49** past each other without either experiencing a change in surface area; a transform fault constitutes the obvious example of a conservative margin.

At most destructive margins only one plate is actively consumed, which means that the margin itself must migrate in order to accommodate the advance of the second plate. Equally it seems that the positions of the constructive margins must move relative to each other. This is probably best illustrated by the

Fig. 3.11. The major lithospheric plates into which the surface of the globe is divided. Many of the boundaries are still tentative, especially in Asia, and there are almost certainly numerous small plates not yet adequately defined. Compare with Fig. 3.8.

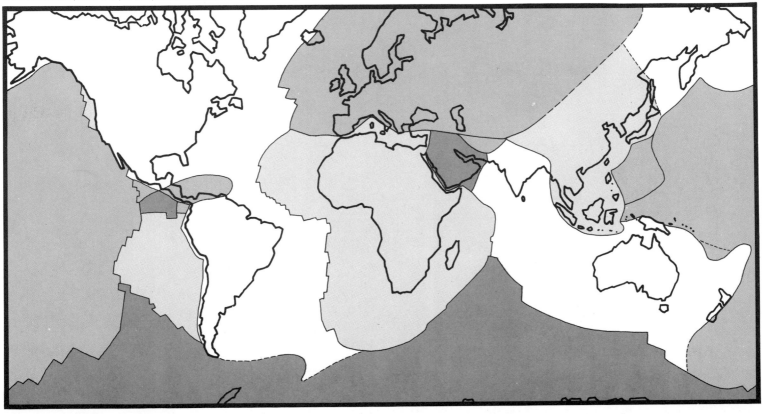

50 African plate which is flanked to both east and west by mid-oceanic ridges believed to be centres of active spreading. The absence of any known destructive margin beneath Africa can be explained if the two mid-oceanic ridges are themselves moving apart at a rate equal to their combined spreading rates. In other words, not only do the plates move but the pattern they form also changes with time.

Plate geometry There are certain geometrical controls which must govern the movement of the plates. Morgan has pointed out that, in conformity with Euler's geometrical theorem, the movement of a rigid plate from one position to another can always be described in terms of a simple rotation about an axis passing through the centre of the sphere. Each point on the plate then describes a small circle path about the pole position of that rotation axis. It follows that in the case of a conservative margin, where material is neither created nor destroyed, the motion of the plates must be along small circle routes round the pole of rotation. A further consequence of the geometrical properties of the sphere is that the rate of relative movement at

Fig. 3.12. The opening of the North Atlantic based upon interpretation of magnetic anomalies (after Laughton, 1972). Spreading centres operative at each period are shown and it is notable how these have periodically changed their location.

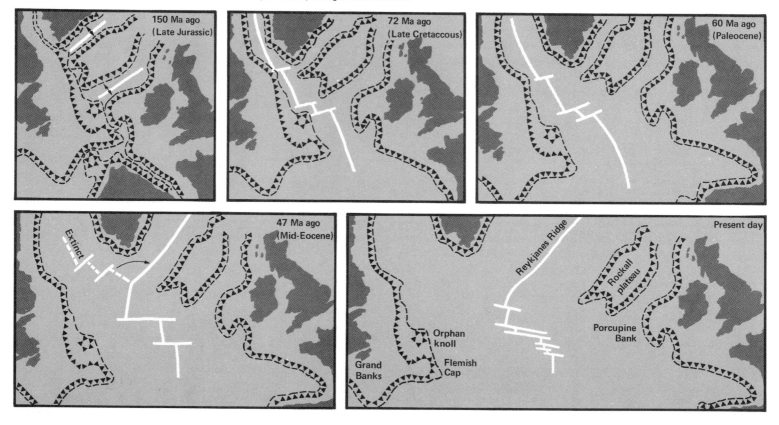

constructive and destructive margins must vary as the sine of the angular distance from the rotation pole. This is exactly analogous to the variations in velocity at different geographical latitudes on the rotating earth. Morgan was able to test the concept of rigid plates moving over the earth's surface by comparing these consequences with observational data from the mid-Atlantic ridge. He first defined the pole of rotation by drawing great circles normal to the known transform faults. He then examined whether the observed spreading rates vary consistently with change in angular distance from the rotation pole. Given the uncertainties involved, all the observational data accorded remarkably well with the predictions of the model he was testing. Poles of rotation have now been determined for all the major lithospheric plates as they appear to be moving at the present time.

It is relatively easy to calculate the rotations that would be necessary to move the American, African and Eurasian continental blocks from their respective pre-drift locations in Pangaea to their modern positions. This clearly does not define the actual routes taken since the continents could, for instance, have followed zig-zag paths with many successive rotations over short angular distances. With increasing information about the age of the ocean floor it is possible to determine in much greater detail the likely movements of the plates (Fig. 3.12). It has been claimed that a minimum of six poles of rotation is required to simulate the movements of the North American and Eurasian continental blocks since late Mesozoic times. Differing angular velocities about these poles produced variations in the rate of sea-floor spreading and continental separation. Whether the rate also varied during the operation of a single pole of rotation remains very uncertain. At one time it was believed that both the Atlantic and Pacific floors showed abrupt changes in spreading rates during the last 10 Ma, but more recent work has cast doubt on this interpretation.

Dissentient views The concept of lithospheric plates, incorating many aspects of the earlier continental drift and sea-floor spreading hypotheses, appears to provide a convincing if partial explanation of the gross relief features of the earth. As the preceding section illustrates, it offers the potentiality for reconstructing the evolution of continents and ocean basins in very considerable detail. However, it would be both misleading and unwise to omit reference to important dissentient views. These range from extreme 'fixists' who deny the reality of continental drift at all, to 'drifters' who doubt the validity of certain aspects of the plate hypothesis. Critics generally point to a long list of assumptions upon which many recent interpretations have been based; if just a few of those assumptions were shown to be false it would shake to the foundations much of the elaborate superstructure that has been constructed. As in the early days of the continental drift hypothesis there are accusations of data selection to fit the favoured theory, but now the accusers tend to be in the minority, not the majority. In many instances it is a question of emphasis. This is well illustrated by the divergent views that have been expressed regarding the fit of the continents bordering the Atlantic ocean. A computer has been used to ascertain the mathematically best solution at the level of the continental slope. In general the deviations are remarkably small, and it seems difficult to envisage such close similarity of form being produced purely by chance. Nevertheless, there are difficulties in completing the jig-saw fit of Africa and Europe against North and South America. The Iberian peninsula in particular poses a problem since its present position precludes a neat closure of the Atlantic. This can easily be resolved if a counter-clockwise rotation of Spain and Portugal has taken place since the splitting away of the Americas. There is some independent evidence for such a rotation, although it must then be admitted that the picture of a passive continental mass being transported on a totally rigid plate needs some modification. In this case as in a number of others, some form of decoupling of part of the continent from the rest of the plate has been invoked, but exactly how or why such decoupling occurs remains uncertain. In such circumstances, does one stress the overall fit of the continents, or the necessity to rotate Iberia by an unknown mechanism in order to achieve closure of the Atlantic basin? Does one emphasize the structural similarities between the conjoined continents, or the differences they exhibit? In a modern guise these are the same arguments as those that surrounded Wegener more than half a century ago.

A disagreement about possible driving mechanisms also echoes earlier disputes. It is now widely held that, if the plate model is correct, movement within the mantle is required. Yet no one has ever recovered rock from the mantle so that its composition and physical properties can only be deduced indirectly. A number of workers have maintained that the inferred properties preclude any form of convective cell; others have argued that the lower viscosity of the asthenosphere may provide suitable conditions. It is certainly true that no single model of convective cells yet commands general assent, and alternative mechanisms have been proposed involving such processes as gravitational sliding away from the elevated mid-oceanic ridges, or the forcing apart of lithospheric plates by igneous intrusion along the ridge axes.

As already pointed out, it was palaeomagnetism that revitalized the drift hypothesis and led to its widespread acceptance. Yet there are problems associated with the interpretation of palaeomagnetic data. Initially the technique may appear remarkably precise, yielding a unique position for the magnetic pole from simple measurements of inclination and declination. However, the results need to be treated with caution. Any disturbance a sample has suffered after the main magnetization can be very difficult to assess. Experiments show that sediments accumulating on a sloping surface often yield misleading measurements of inclination. Many rocks have undergone a secondary magnetization which needs to be removed in the laboratory to disclose the most stable remanent magnetism.

This most stable direction is then compared with other observations and, if consistent, is normally deemed primary in origin. Critics contend that this leads to circularity of argument, since only those results conforming with the initial hypothesis are regarded as satisfactory. They also point out that published pole positions from a single continent sometimes deviate from each other by as much as 30 or 40° angular distance, and maintain that this casts doubt either on the assumed dipolarity or on the rigidity of the lithospheric plates; at best it suggests that palaeomagnetism is a rather less precise tool for tracing continental movement than is sometimes claimed.

Concern has also been expressed about the interpretation of ocean-floor magnetic anomalies. It has been argued that we have too little understanding of the basic processes involved to justify elaborate hypotheses about rates and directions of sea-floor spreading. Admittedly the early DSDP results appear consistent with the inferred age of the ocean floor, but there remain certain ambiguities in that the basal sediments are occasionally baked and therefore older than the underlying igneous rock. In some instances it is possible that the drilling has terminated in a sill rather than in the true basement rock which consequently has not been dated.

As a final comment it is worth recording that Russian scientists who tend to be great proponents of vertical movements in the crust do not generally support the concept of lithospheric plates and have always been rather sceptical of the arguments adduced in favour of drift.

CREER, K. M., 1970. 'A review of palaeomagnetism', *Earth Sci. Rev.*, 6, 369–466.

ISACKS, B. L., et al., 1968. 'Seismology and the new global tectonics', *J. geophys. Res.*, 73, 5855–99.

LAUGHTON, A. S., 1972. 'The southern Labrador Sea – a key to the Mesozoic and early Tertiary evolution of the North Atlantic' in *Initial Reports of the Deep Sea Drilling Project*, vol. 12, Nat. Sci. Found.

MASON, R. G. and A. D. RAFF, 1961. 'Magnetic survey off the west coast of North America, 32 N latitude to 42 N latitude', *Bull. geol. Soc. Am.*, 72, 1259–66.

McELHINNY, M. W. and B. J. J. EMBLETON, 1974, 'Australian paleomagnetism and the Phanerozoic plate tectonics of eastern Gondwanaland', *Tectonophysics*, 22, 1–29.

MORGAN, W. J., 1968. 'Rises, trenches, great faults and crustal blocks', *J. geophys. Res.*, 73, 1959–82.

PHILLIPS, J. D. and D. FORSYTH, 1972. 'Plate tectonics, paleomagnetism and the opening of the Atlantic', *Bull. geol. Soc. Am.*, 83, 1579–600.

RAFF, A. D. and R. G. MASON, 1961. 'Magnetic survey off the west coast of North America, 40°N latitude to 52°N latitude', *Bull. geol. Soc. Am.*, 72, 1267–70.

SYKES, L. R., 1967. 'Mechanism of earthquakes and nature of faulting on the mid-oceanic ridges', *J. geophys. Res.* 72,, 2131–53.

VINE, F. J. and D. H. MATTHEWS, 1963. 'Magnetic anomalies over oceanic ridges', *Nature, Lond.*, 199, 947–49.

WEGENER, A., 1967. *The Origin of Continents and Oceans* (trans. J. Biram from 4th Ed.), Methuen.

WESSON, P. S., 1972. 'Objections to continental drift and plate tectonics', *J. Geol.*, 80, 185–97.

WILSON, J. T., 1963. 'Evidence from islands on the spreading of ocean floors', *Nature, Lond.*, 197, 536–8.

WILSON, J. T., 1965. 'A new class of faults and their bearing on continental drift', *Nature, Lond.*, 207, 343–7.

The recent literature on global tectonics is vast. Useful summaries and bibliographies are to be found in : I. G. Gass, et al., *Understanding the Earth*, Open Univ. Press, 1972 ; T. Kasbeer, 'Bibliography of continental drift and plate tectonics', *Spec. Pap. geol. Soc. Am.*, 142, 1972 and 164, 1975 ; X. LePichon, et al., *Plate Tectonics*, Elsevier, 1973 ; E. R. Oxburgh, 'The plain man's guide to plate tectonics', *Proc. Geol. Ass*, 85, 1974, pp. 299–358 ; P. J. Wyllie, *The Dynamic Earth: Textbook in Geosciences*, Wiley, 1971. A collection of dissentient views is to be found in 'Plate tectonics – assessments and reassessments', *Am. Ass. Petrol. Geol. Mem.* 23, 1974.

1|4

Localized movements of the continental crust

While being displaced by the large-scale movements described in Chapter 3, the continental blocks can also be subject to important deforming stresses. Some of these are directly attributable to the positioning of the blocks at the margin of a lithospheric plate, but others appear to be largely independent of plate tectonics. It is to these more localized movements of the continental crust that the present chapter is devoted. Two fundamental causes of the movements may be distinguished. The first is the loading and unloading that results from such events as the growth and decay of an ice-sheet, or the impounding and drainage of a deep lake. During loading the crust is depressed in the form of a large downwarped basin, while unloading is followed by recovery in the form of a broad domed uplift. Deformation of this type is clearly predicted by the concept of isostasy. The second and much more common cause of crustal movement is tectonic in origin. It is due to deep-seated changes taking place either near the base of the crust or in the mantle. The nature of the changes cannot be observed directly but must be inferred from the effects seen at the surface. The most obvious are rock structures such as folds and faults which have long been the subject of study and classification by geologists (Fig. 4.1).

Classification of tectonic activity has generally stressed the distinction between the radial and tangential components of earth movement. Gilbert, for instance, emphasized the contrast between what he termed epeirogenic and orogenic activity. In the former, displacement was held to be primarily radial, allowing extensive areas of the crust to be uplifted or depressed with little internal deformation. In the latter the tangential component was held to be dominant, resulting in folding and dislocation of strata and ultimately, as the name implies, in the building of major mountain chains. Beloussov, a Russian geophysicist, has argued that this simple two-fold classification is unsatisfactory in a number of respects. In particular he underlines the importance of what he calls oscillatory movement. This undulatory motion involves alternating elevation and depression without any permanent record of the displacements being imprinted on the local rocks. Working in Africa, King has argued that yet another type of movement needs to be distinguished, namely cymatogeny. This consists of a broad arching of the continental crust, accompanied by the simultaneous generation of gneissose rock at depth. Uplift at the crest may amount to 1 000 m or more, while the overall width of the structure extends to hundreds of kilometres. King regards cymatogeny as extremely common and has gone so far as to regard it as the characteristic tectonic activity of late-Cenozoic times.

The ideas of Gilbert, Beloussov and King illustrate the considerable diversity of views that exist regarding the classification of earth movements.

They also exemplify the various lines of evidence that may be employed in elucidating the nature of the movements. Gilbert laid stress on the evidence of geological structure. Epeirogeny is attested by horizontal strata elevated high above the level at which they must originally have accumulated. Orogeny is indicated by contorted and crumpled rocks that have clearly been subject to powerful deforming stresses. Later workers have endeavoured to supplement these purely geological arguments by reference to other lines of evidence. Beloussov, for instance, emphasizes the need to investigate contemporary crustal mobility by modern survey techniques, while King stresses the importance of correctly interpreting surface morphology which he believes can provide one of the fullest guides to recent tectonic history. Geological structure provides the best indication of the broad pattern of tectonic activity; yet it is essential for the geomorphologist to recognize that it relates specifically to the rocks and not to the surface of the globe. It is extremely difficult to translate the evidence of folded strata that were once deeply buried into, say, simple vectors of surface displacement. Modern survey techniques and geomorphic studies both have the advantage of dealing directly with surface form, but are severely limited in other respects. Although current instrumental surveys can attain remarkable standards of accuracy, the results often have to be compared with much older records of doubtful reliability. Moreover, the period that can be covered by this approach is so short that the question inevitably arises as to how representative it is of the vastly longer span of geological time. Geomorphic evidence falls into an intermediate category. It lacks the precision of modern survey methods, but covers a much longer time interval.

In the following account each of the major lines of evidence outlined in the preceding paragraph will be examined in more detail. In practice this will involve considering successively longer periods of time; for contemporary movement a representative figure is up to 10^3 years, for geomorphic evidence it is 10^3–10^6 years, and for geological evidence it is over 10^6 years.

Evidence for contemporary crustal movement

Tidal gauge records Detection of short-term changes in the relative level of land and sea requires accurate instrumentation, normally provided nowadays by continuous-recording tidal

Fig. 4.1. Conventional classification of structures resulting from folding and faulting.

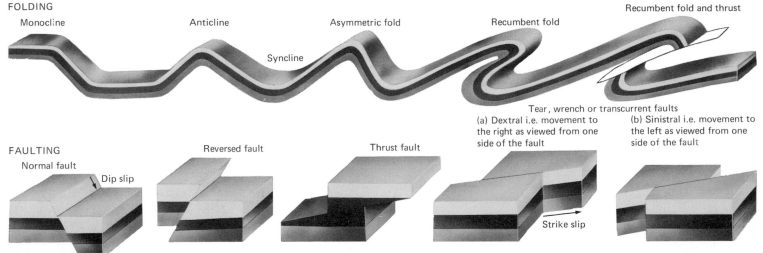

FOLDING

Monocline   Anticline   Asymmetric fold   Recumbent fold   Recumbent fold and thrust

Syncline

Tear, wrench or transcurrent faults
(a) Dextral i.e. movement to the right as viewed from one side of the fault
(b) Sinistral i.e. movement to the left as viewed from one side of the fault

FAULTING
Normal fault   Reversed fault   Thrust fault

Dip slip

Strike slip

gauges. Some ports have very long histories of measuring tidal heights, but there are often difficulties in comparing old and modern records owing to changes in the type of instrument used. Even where acceptable modern records are available, interpretation demands considerable care. For example, collation of tidal data from many sites scattered around the globe suggests that, during the first half of the present century, there was a slow rise in sea-level of the order of 1 mm yr^{-1}. Indications are that this rise may now have ceased, but at least for the earlier period some correction factor needs to be applied if tidal gauge data are to be used for assessing local land movements. It must also be recognized that sea-level is affected by a variety of factors that are neither obvious nor easy to evaluate. It is known, for instance, that sea-level fluctuates seasonally on many coasts owing to changes in wind and current direction. Given this sensitivity it is obviously hazardous to compute a global rate of sea-level change and to assume without further thought that any local residual is entirely due to crustal deformation.

Despite these limitations, careful analysis of tidal data has revealed that many coastal areas are currently suffering displacement. The highest rates tend to be concentrated in those areas of the globe that are still rebounding after removal of the weight of a massive ice sheet. For example, the northern shores of the Gulf of Bothnia are rising at just over 9 mm yr^{-1}, parts of the Alaskan coastline at over 10 mm yr^{-1}, and areas of northern Britain at about 3 mm yr^{-1}. Such results are consistent with long-term trends that have been recognized for many years. More surprising is the rate of movement implied by gauges sited on coasts beyond the limits reached by the Pleistocene ice-sheets. The most detailed information comes from the coasts of the United States. On the Atlantic seaboard south of New York much of the coastline appears to be sinking at a rate of at least 1 mm yr^{-1} with some segments exceeding a rate of 2 mm yr^{-1}. On the Pacific seaboard the tidal gauge at San Francisco is slowly sinking, while that at Crescent City in northern California is rising. Taken at face value the records from the two sites imply a differential movement since 1940 averaging about 3 mm yr^{-1}. Viewed in the context of a human lifetime all these rates may seem of relatively small consequence, but if sustained for a prolonged geological period each is capable of producing major topographic changes.

Archaeological evidence Coastal settlements often provide indications of sea-level change over a rather longer time interval than that so far considered. A harbour built at Torne on the Gulf of Bothnia in 1620 had to be abandoned within a century because of the shoaling of the approaches. Along the coasts of the Aegean several settlements have been totally submerged by subsidence of the land. However, the best known example of a town that affords evidence for rapid changes in level is Puteoli on the shores of the Bay of Naples. Much of the original building is believed to have taken place during the first century A.D. when the site clearly stood above sea-level. Markings left on marble columns by rock-boring marine molluscs indicate a submergence prior to 1500 of over 6 m. Yet it is known that by the mid-sixteenth century the locality was undergoing uplift and it is possible that total emergence followed. Since the early nineteenth century subsidence has again been the dominant movement; in 1828 the base of the columns was about 0·3 m below sea level, but by 1950 this figure had increased to over 2 m. The history of Puteoli is particularly noteworthy as an indication of the way in which uplift and subsidence may follow each other in quick succession.

It has often been claimed that south-eastern England has subsided 4 or more metres during the last two millennia, largely on the basis of Roman remains now found well below high-water level in the London area. However, such evidence needs to be treated with caution since factors other than true crustal subsidence might explain many of the observed relationships. Akeroyd has plotted data from some 28 sites in eastern England between the north Kent coast and the Fenlands (Fig. 4.2); the information suggests that this region has been subject to only modest subsidence during the last few millennia, probably averaging little more than 1 mm yr^{-1}.

Geodetic or high-precision levelling There is no single definition of what constitutes a high-precision standard of levelling, and as instruments and techniques have improved so more stringent requirements have been laid down. Some surveys

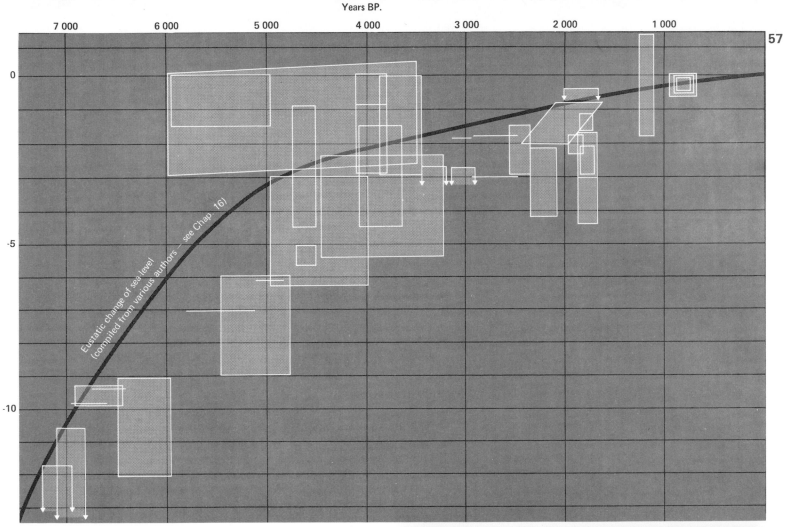

Fig. 4.2. A plot of archaeological and geomorphological sites in south-eastern England yielding estimates of sea-level during the last 7 000 years (after Akeroyd). Also shown is a cure depicting world-wide sea-level changes in the same period. The concentration of symbols below the eustatic curve suggests that the whole region has probably been subject to slight subsidence.

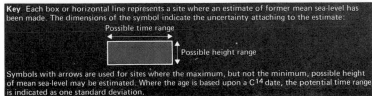

Key Each box or horizontal line represents a site where an estimate of former mean sea-level has been made. The dimensions of the symbol indicate the uncertainty attaching to the estimate:

Possible time range

Possible height range

Symbols with arrows are used for sites where the maximum, but not the minimum, possible height of mean sea-level may be estimated. Where the age is based upon a C14 date, the potential time range is indicated as one standard deviation.

Years BP.

Elevation (m)

Eustatic change of sea-level (compiled from various authors – see Chap. 16)

have specified that closing errors on traverses should not exceed $4\sqrt{K}$ mm where K is the length of the traverse in kilometres. This would require the closure on a 100 km run to be 40 mm or less to avoid the need for a re-survey. In practice a much better standard is often attained. In the 1960s, for example, a survey of the Great Lakes region in the United States had a prescribed accuracy of $1.5\sqrt{K}$ mm, but the finished traverse showed actual closures of less than $0.6\sqrt{K}$ mm. This means that after running a line of levels for a distance of 10 km the deviation from true height would amount to less than 2 mm. By repeating surveys of this precision at intervals of several years it is clearly possible to detect quite small changes in surface elevation.

Once such displacements have been measured, there remains the problem of interpreting the results. It certainly cannot be presumed that all changes in surface elevation are due to normal crustal movements. It is well known, for instance, that mining subsidence can easily exceed 1 m in only a few decades. It is less obvious, but equally well-established, that extraction of oil, gas or water from porous strata can lead to subsidence. In California pumping of irrigation water in the Central Valley lowered the surface by as much as 7 m between 1943 and 1964, while further south around Los Angeles harbour abstraction of oil and gas has caused a basin-shaped depression about 7 km across and over 8 m deep at its centre. The previously rapid subsidence at Los Angeles now seems to have been halted by pumping over a million gallons of sea water per day back into the rock to replace the oil and gas being removed. Very detailed studies in this locality have shown that another factor determining the surface level close to the coast is the state of the tide. The land surface rises and falls a perceptible amount in sympathy with the tides, presumably because of changes in the pressure of water trapped in the pore spaces of the rocks. In Europe the potentially disastrous sinking of Venice is believed to be partly due to removal of groundwater from porous underlying strata; in this case further contributory factors may be compaction of the geologically young sediments around the northern margin of the Adriatic Sea, together with true tectonic downwarping. The weight of water impounded behind large dams, and even the weight of closely spaced high-rise city buildings, has been shown to cause local depression of the earth's surface.

Even after allowance is made for these anthropogenic factors there remains abundant evidence for crustal movement attributable to deep-seated tectonic causes. The idea of using accurate survey techniques to monitor natural changes in surface level is an old one, but it was first employed on a large scale at the end of the nineteenth century. Following a disastrous earthquake in 1891 the Japanese instituted a programme of systematic levelling aimed at establishing the nature and rate of crustal movement. In 1923 Tsuboi reported on the evidence that had accumulated in central Japan. He found that in this region uplift was taking place at an average rate of 5mm yr^{-1}, with values ranging locally from 80 mm to 0.75 mm. More recent investigations have not only confirmed the general magnitude of these figures, but have also demonstrated a highly complex mosaic of small blocks all rising at different and irregular rates. Two illustrative examples may be briefly quoted. For over 50 years parts of the Niigata region on the west coast of Honshu rose at a rate of about 2 mm yr^{-1}, but in 1955 the uplift gradually accelerated to 10 mm yr^{-1}. After 1959 the upheaval slowed down and even turned into a slight fall. Following a major earthquake in 1964 a subsidence of about 100 mm was recorded. The second example concerns a remarkable series of tremors that affected Matsushiro in central Japan during 1965 and 1966. In the course of many thousands of tremors, several different types of physical change were noted. One of the most prominent was erratic uplift and tilting of the land surface. By careful analysis of the changes that were taking place it was possible for a period of several months to offer public forecasts of likely earthquake intensity. Particularly violent phases in April and August 1966 were successfully foretold in this way. The importance attached to the evidence of levelling in Japan is reflected in plans to resurvey traverses totalling 20 000 km at regular 5-year intervals.

Following the pioneer work in Japan, many surveys have now been undertaken in other parts of the world, most frequently in regions of active seismicity where research into crustal movement has obvious practical value. In Russia, for instance, repeated surveys of part of the Caucasus have shown marked differential displacement to be taking place. Certain blocks are being elevated at about 4 mm yr⁻¹ while a number of marginal basins seem to be subsiding at about the same rate. The pattern is very reminiscent of that recorded in Japan, but the rates of movement are on the whole rather lower.

Some of the most striking results of geodetic levelling come from the Pacific margin of the Americas. Over 80 000 km of precise levels have been run within the State of California alone, and these have defined quite clearly regions of both elevation and depression. Repeated surveys in the Los Angeles area have disclosed considerable differences in the rates at which local hill ranges are being uplifted; among the fastest is the Tehachapi (for location see Fig. 4.6) which is currently rising at 15 mm yr⁻¹. Re-surveys following earthquakes often disclose very significant changes. After an earthquake centred near San

Fernando in 1971 it was found that parts of the San Gabriel Mountains had suddenly been elevated by as much as 2·3 m. Although other displacements lack any obvious association with major tremors, there is increasing evidence that they too may be spasmodic in occurrence. Such is the precision of modern instruments that the period between surveys can be reduced to less than 1 year and reliable results still obtained. In 1964–5 fifteen separate traverses across southern California detected vertical displacements ranging from zero to 26 mm. The exceptionally high figures at one or two sites suggest movement is pulsatory and that decadal averages probably conceal a very episodic motion.

In South America traverses of sufficient accuracy to assess crustal movement are very much rarer. It was a fortunate coincidence that a section of the Inter-American Highway near

Fig 4.3. Provisional maps of contemporary crustal movement in the United States and European Russia as revealed by geodetic surveys.

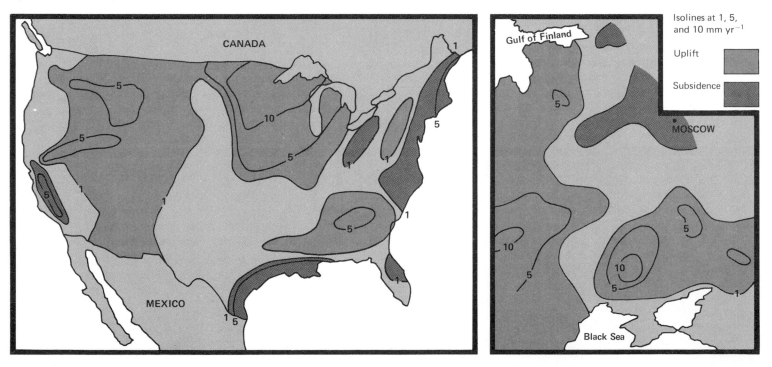

60 **Fig 4.4.** A tentative estimate of the rate of vertical surface displacement in Great Britain during the last few centuries.

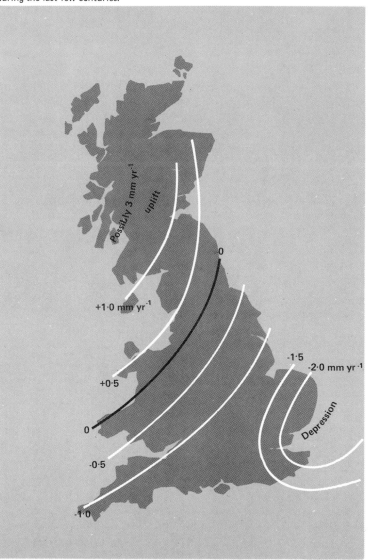

Valdivia in Chile had been surveyed shortly before the whole area was shaken by a major earthquake in 1960. A new survey in 1964 showed that part of the route had been depressed by 2·3 m compared with 7 years earlier. Supplementary evidence came from a study of such displaced shoreline features as the upper limit of mussel and seaweed growth. This confirmed the area of subsidence around Valdivia and demonstrated a zone of uplift immediately to the west locally reaching an amplitude of 5·7 m. The problem in this instance is to estimate the frequency with which similar earthquakes can be expected to recur. Historical records since the early sixteenth century suggest destructive earthquakes are experienced at intervals of less than 100 years, although whether they have all been comparable to the 1960 occurrence is much less certain. Differential movements along the coast at a mean rate of 10 mm yr^{-1} are by no means improbable.

Examples so far have all been drawn from areas of known Cenozoic mountain-building. It should not be assumed, however, that crustal movement of tectonic origin is confined to such regions. In the United States (Fig. 4.3) detailed surveys have disclosed a broad area of subsidence currently depressing New York city by about 4 mm yr^{-1} and parts of Delaware and coastal Virginia by over 2 mm yr^{-1}. By contrast, parts of Georgia and Alabama are being elevated at rates in excess of 5 mm yr^{-1}. In European Russia a similarly intricate pattern of uplift and subsidence has been demonstrated. Recorded movements commonly lie between +3 and −3 mm yr^{-1}. There are in addition a number of local centres of very strong uplift, one dome-shaped area in the Ukraine currently rising at about 10 mm yr^{-1}. Where levelling along the same traverse has been repeated several times during the last few decades, the rate of movement has often proved far from constant. In several instances the direction of displacement has actually been reversed so that regions which were formerly rising are now subsiding and vice versa. This has led Russian workers to propound the idea that the crust may experience small-scale undulatory movements that make total stability the exception rather than the rule. In Great Britain the Ordnance Survey has adopted a conservative attitude, attributing differences in successive surveys to human frailty or instrumental error rather than to crustal movement. There is little doubt that workers in some other countries have gone to the opposite extreme and pay too little attention to the precision of the data they are using. It may be said that much of the levelling evidence in Britain is consistent with movements inferred from other evidence, but does not by itself constitute proof. For example, Ordnance Survey records may be interpreted as indicating uplift in south-east Scotland at about 0.5 mm yr^{-1}, conforming with tidal-gauge analyses that seem to indicate a tilting of the British Isles with elevation in the north and subsidence in the south and east (Fig. 4.4).

There can be no doubt that vertical movements of the crust are taking place in many areas of the world at the present day, and that they are sufficiently rapid to be measured by modern survey techniques. Much interest will attach to the extension of these techniques to other parts of the world since at the moment the global coverage is very limited.

Measurement of horizontal displacements So far attention has been concentrated exclusively on vertical movements of the crust. Relatively few areas have been intensively studied with a view to assessing contemporary horizontal displacements. One major exception is the area of California traversed by the San Andreas fault system (Fig. 4.5), and this region will be used to illustrate some of the principles involved. It has been known for well over a century that the dominant movement on the San Andreas system is lateral rather than vertical (Fig. 4.6). In 1857 the Fort Tejon earthquake caused visible surface ruptures over a distance of 300 km and in places horizontal displacements of up to 9 m. The 1906 San Francisco earthquake was accompanied by lateral movements of 6.5 m, while the Imperial Valley earthquake of 1940 translocated man-made structures more than 5 m. In each of these cases the movement was dextral with the coastal margin of California moving northwards relative to the rest of the state.

Following the 1906 San Francisco shock Reid proposed the 'elastic rebound theory' of earthquake activity which has since formed a cornerstone of modern seismology. It envisages accumulation of elastic strain energy in a zone on either side of a fault; it may be likened to the accumulation of such energy in a bow that is flexed before the shooting of an arrow. Eventually

Fig 4.5. The scar of the San Andreas fault as viewed from the air over southern California.

62 the stress along the fault will exceed the frictional resistance and slip is then initiated. This will occur first at a single point on the fault, but release at that point will throw increased stress on adjacent sections with the result that movement will be propagated along the fracture at an estimated velocity of 3–4 km sec^{-1}. On major dislocations the length of the rupture during a single earthquake may reach as much as 1 000 km. Movement ceases at the point where the stress no longer exceeds the frictional resistance and the fault remains locked in position. The longer a section of fault remains locked in position the greater the accumulated strain energy and the more devastating the shock is likely to be when it finally comes. This is one reason for concern about the interval that has now elapsed since the last major movement along the northern branch of the San Andreas fault around San Francisco.

One implication of the elastic rebound theory is that, with increased knowledge of fault dynamics, it may be possible to foretell when a major earthquake is imminent. The two variables that would need to be measured are the amount of accumulated strain energy and the frictional resistance of the fault. Such considerations have elicited much support for attempts to monitor the gradual deformation of the crust and so to get an approximate idea of the build-up of strain energy. By means of repeated triangulation surveys, some of which date back to the 1880s, the angular changes between fixed points on the ground have been measured. Extremely accurate instruments such as the geodimeter have recently been used for monitoring variations in the distance between points situated on opposite sides of the fault trace. Special strain gauges and lasers have been installed at critical sites, and with this modern instrumentation changes of less than a millimetre in a measured kilometre can be detected. The work has established as a fact that it is possible to measure surface strain near the San Andreas fault, but the pattern which emerges is so complex that many uncertainties remain. Along some branches of the fault system south of San Francisco continuous creep has been observed; the most famous instance is at Hollister where the creep is gradually splitting apart the buildings of a winery at a rate of about 12 mm yr^{-1}. This creep is only accompanied by very small tremors at the present day, apparently indicating that lateral slip can occur without inter-mittent locking. On the other hand, the creep is slower than the average rate of movement along certain other sections of the fault system, and triangulation surveys and geodimeter measurements suggest that strain is still accumulating in the crust for 20 km on both sides of Hollister. If this is the case the stored energy may well be released by intermittent earthquakes of greater magnitude. Further south where the fault trace for a short distance trends nearly east-west the crustal strain is indicative of compression almost normal to the fracture. This is in an area that has not experienced a major earthquake during the last 100 years. It seems possible that the east–west segment of the fracture is subject to exceptionally powerful locking with infrequent but very intense earthquakes. Still further south in the Imperial Valley systematic triangulation and geodimeter measurements have established what appears to be the simplest sequence of strain accumulation and release anywhere along the fault system. Surveys in 1935, 1941, 1954 and 1967 disclosed deformation affecting a zone that extends 60 km on both sides of the Imperial fault. Since the major Imperial Valley earthquake of 1940 strain has been accumulating at a rate equivalent to slip along the fault of 85 mm yr^{-1}. Actual movement along the fault in the same period has averaged only 13 mm yr^{-1} or about 15 per cent of the total displacement. The residue is presumably released during the occasional major earthquake. As the mean displacement along the fault in 1940 was about 1·7 m, an earthquake of similar magnitude might be expected to occur once every 25 years or so.

Geomorphic evidence for crustal movement

Deformed strandlines Of all types of geomorphic evidence, the most convincing is that presented by deformed shorelines. This is because the initial horizontal form is beyond dispute, whereas the original shape of a river terrace or erosion surface may be a matter of controversy. Although instances of disturbance from deep-seated tectonic causes are known, most attention has focused on the strandline deformation that arises from crustal unloading. There are two such circumstances in particular that lead to isostatic uplift and warping of strand-lines. The simpler, although less frequent, is the desiccation of a

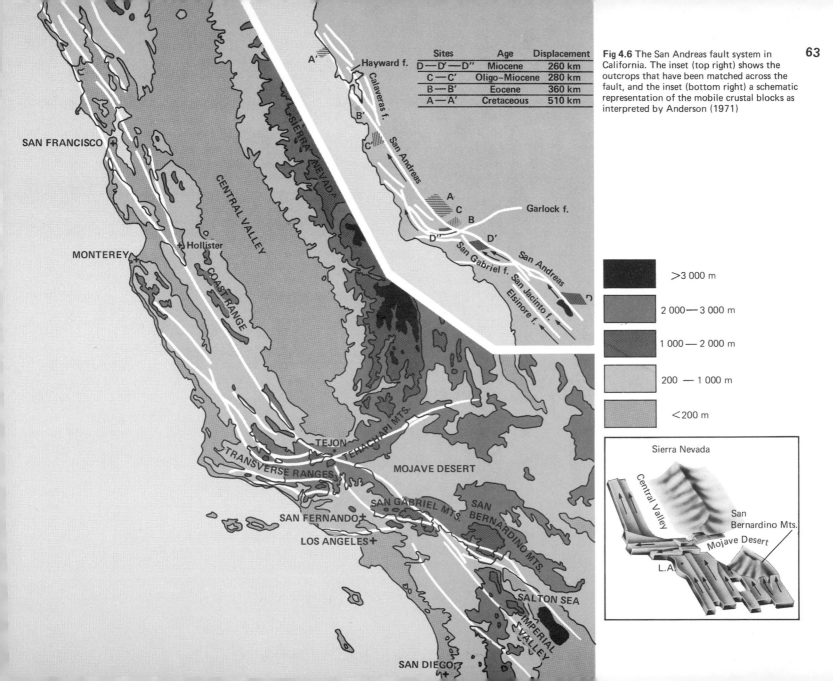

Fig 4.6 The San Andreas fault system in California. The inset (top right) shows the outcrops that have been matched across the fault, and the inset (bottom right) a schematic representation of the mobile crustal blocks as interpreted by Anderson (1971)

Sites	Age	Displacement
D — D' — D''	Miocene	260 km
C — C'	Oligo–Miocene	280 km
B — B'	Eocene	360 km
A — A'	Cretaceous	510 km

> 3 000 m

2 000 — 3 000 m

1 000 — 2 000 m

200 — 1 000 m

< 200 m

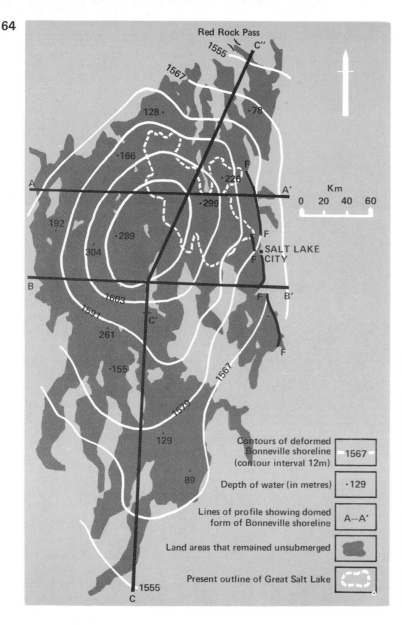

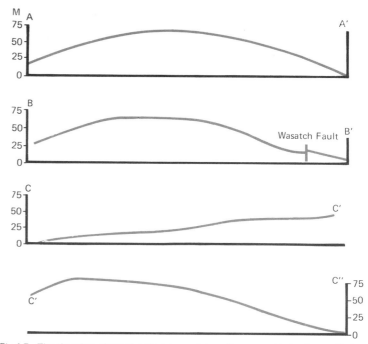

Fig 4.7. The elevation of the highest deformed strandline around Lake Bonneville in Utah, USA (data from Crittenden, 1963).

lake that was sufficiently large to have caused local depression of the crust. A good illustration is provided by Lake Bonneville in Utah (Fig. 4.7). This forerunner of the Great Salt Lake has left striking testimony to its old levels in prominent strandlines that are found not only around the margins of the basin but also fringing a number of central islands. It formed during a cooler, wetter climatic phase and at its maximum occupied an area of 50 000 km² and reached a depth of 330 m. The total mass of the water was 9×10^{12} tonnes, a large proportion of which was abruptly evacuated when a natural dam at Red Rock Pass was breached in a catastrophic event, probably occurring about 16 000 years ago. Mapping of the highest shoreline shows that it has undergone broad domed uplift, with the greatest displacement of 64 m sited where the water was deepest. Over a large central area the water averaged 290 m deep and if the density of the displaced sub-crustal material is assumed to be 3·3 g cm⁻³ a total recovery of over 85 m might be expected. Such calculations suggest that uplift may still be incomplete, and this accords with levelling in 1911 and 1958 that indicates doming to be continuing at a rate of about 1 mm yr⁻¹. Because the forces acting to restore the basin to isostatic equilibrium decrease as uplift takes place, the rate of movement itself declines. Just as with radiometric dating, the passage of a certain unit of time sees the rate halved, and it can be shown that in the case of Lake Bonneville the appropriate unit of time is about 2 800 years. This implies that if there had been instantaneous removal of the whole weight of Lake Bonneville, some 42 m of uplift would have taken place in the first 2 800 years, that is, deformation of the crust would have proceeded at an average rate of about 15 mm yr⁻¹. In practice the unloading was not instantaneous and the uplift was correspondingly slower; nevertheless, the data provide an indication of the viscosity of the sub-crustal rock and the rate at which movement can theoretically take place. The viscosity is much lower than is appropriate to the lithosphere and is believed to underline the important role played by the asthenosphere in surface deformation.

The more common cause of deformed strandlines is the melting of an ice-sheet. Large lakes may develop around the decaying ice-mass, or the sea may invade areas vacated by the retreating ice front. As melting continues the load on the crust is reduced and the land differentially uplifted. Any beaches that have developed are elevated and warped. By analysing the height and distribution of the raised beaches it is possible to piece together the history of isostatic recovery. However, two intrinsic problems that do not arise in the case of a desiccated lake are estimating the original ice thickness and calculating the amount of uplift that may have preceded local deglaciation. Moreover, the bearing of raised marine shorelines on the subject of isostatic rebound can only be fully appreciated in the context of world-wide changes in sea-level. For this reason the main discussion of post-glacial recoil is deferred until Chapter 16 in

Fig 4.8. A 2m-high fault scarp cutting across the floodplain of a small stream in Montana, U.S.A. The scarp was initiated at the time of the Hebgen earthquake in 1959; it produced a knickpoint in the stream long profile that at first migrated up the valley but later degenerated into a series of minor rapids.

66 which the Pleistocene oscillations of sea-level are discussed in detail. Here it will suffice to note some of the main conclusions arising from the study of elevated strandlines in Scandinavia and North America. In the former area, the oldest and highest shorelines attest to a recoil of at least 500 m during the last 12 000 years. Yet a significant part of the full isostatic recovery is still to come to judge from the negative gravity anomalies that still characterize the Gulf of Bothnia. In North America the highest recorded marine shorelines are found around Hudson Bay at an elevation of about 300 m in an area that became ice-free only 8 000 years ago. The amount of uplift still to come has been variously estimated as between 100 and 250 m. These figures derived from strandlines in recently deglaciated regions confirm the general mobility of the crust when subject to

vertical stress. Rates of movement can obviously average over 40 mm yr^{-1} for extended periods. As in the case of Lake Bonneville, the rapid deformation appears to result from much smaller variations in stress than the physical properties of the crustal rocks seem to demand; once again the explanation almost certainly lies in the low viscosity of the asthenosphere.

Aggradational features Recent aggradational features can display the results of crustal warping and faulting in a variety of ways. The most direct is by simple disruption of the completed surface form (Fig. 4 8). River terraces, glacial moraines and alluvial fans are among the many features known to have been modified in this way. On the other hand, if accumulation and tectonic disturbance occur simultaneously, patterns of sedi-

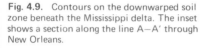

Fig. 4.9. Contours on the downwarped soil zone beneath the Mississippi delta. The inset shows a section along the line A—A′ through New Orleans.

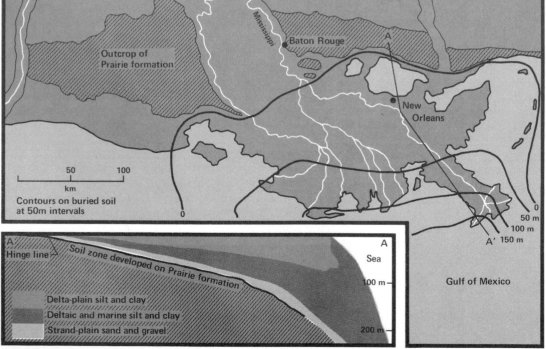

mentation are likely to be affected and the clearest evidence for the earth movements will then lie in the stratigraphy of the deposits.

Large deltas frequently bear witness to continual down-warping. Immense thicknesses of shallow-water debris imply that local subsidence has occurred pari passu with the influx of new sediment. In part this may be ascribed to isostasy which demands that each new increment of load be compensated by further depression of the underlying crust. However, the density of deltaic material is so small compared with that of the mantle that this cannot be the full explanation; there must be

further internal processes involved. The Mississippi delta (Fig. 4.9) is the most thoroughly studied in the world, and in its upper layers a soil horizon has been identified which is believed to date from the mid- or even late-Wisconsinan period. Today this marker level near the extremity of the modern delta lies at a depth of almost 300 m. Of this figure approximately 100 m can be attributed to the fall in sea-level that occurred during the last glaciation. The residue is presumably due to down-warping which has thus, for the last 20 000 years, averaged about 10 mm yr^{-1}.

The effects of faulting, unlike warping, are most commonly

Fig. 4.10 The faulted terraces of the Branch River in South Island, New Zealand. The large block diagram represents the current position, the smaller insets the sequence of erosion and displacement phases by which it has evolved (based on data in Lensen, 1968).

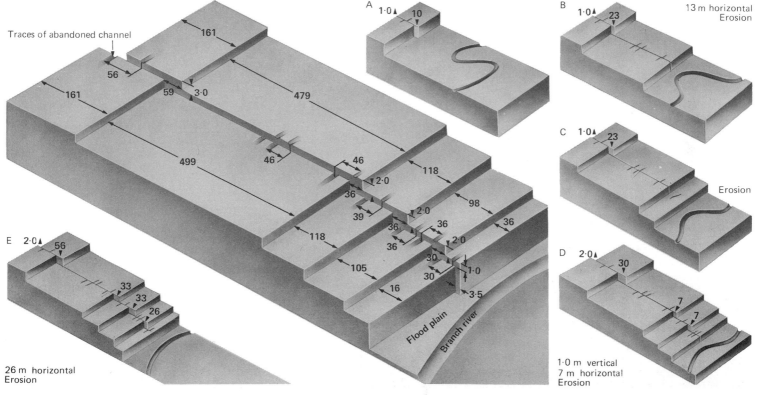

68 seen in relatively small aggradational forms. Alluvial fans accumulating against the face of an active fault scarp may bear witness to a number of discrete phases of tectonic movement. Along the western face of the Sierra Nevada in southern Spain spasmodic uplift has clearly initiated several separate cycles of fan development. Some of the oldest examples have been disturbed during more recent movements along the fault zone. In several instances the apex of a fan has been elevated as part of the Sierra Nevada massif and is now separated from the main section of the accumulation by a prominent fault scarp.

Although transcurrent faulting is most frequently attested by offset valleys producing a characteristic zigzag pattern, the fullest record of movement may be contained in the displacement of river terraces. Where a fault is active during the fashioning of a suite of terraces the earlier members of the succession will be displaced more than the later since the effects are cumulative. If the terraces can be dated, the history of tectonic activity should be decipherable in some detail. After careful survey of the Branch River terraces in New Zealand (Fig. 4.10) Lensen has concluded that, in the last 20 000 years, lateral movement along the Wairau fault has averaged at least 3·4 mm yr^{-1}.

Uplifted and warped erosion surfaces It has long been held that the morphology of the continents bears the imprint of intermittent tectonic uplift. The basic concept requires a prolonged period of stability during which the landscape evolves towards a surface of low relief related to the contemporaneous sea-level. If uplift of the land ensues, the subdued surface is first elevated and then progressively consumed as a new landscape evolves at a lower level. Should the sequence be repeated several times, the result is a staircase of gently sloping platforms separated by steeper bluffs. Within this basic conceptual framework, several models of landscape evolution have been proposed. These are discussed at length in Chapter 10, but the details are of little consequence in the present context.

There can be very few upland areas in the world where this idea of alternating planation and uplift has not been applied at some stage in the history of geomorphological investigation. Contrasting examples are afforded by the Welsh uplands on the one hand and the Rocky Mountains on the other. In a painstaking study of Wales, Brown has identified three major plateau surfaces at levels of approximately 525–600 m, 375–475 m and 250–300 m (Fig. 4.11). He regards each as the product of sub-aerial erosion acting during an early phase in the history of the present drainage. There is little evidence of warping and the form of the three features implies either intermittent uplift of the Welsh massif as a structural unit or spasmodic falls in sea-level. Both explanations pose problems. If the latter were true, similar features at the same altitude would be expected in other upland areas on the oceanic margins. Although analogous erosion surfaces have been identified, there is little consistency in altitude. Moreover, world-wide sea-level changes of the requisite amplitude would have immense implications for the volume of sea water or the capacity of the ocean basins. It therefore seems more likely that the Welsh massif has been uplifted as a block. Yet such a block must presumably be limited in size and bounded by either faults or monoclinal flexures. Neither dislocated nor warped erosion surfaces have yet been identified around the massif, but with the crudity of analytical tools available to the geomorphologist this is scarcely surprising.

Many authors have noted that the crests of the Rocky Mountains in parts of Colorado consist not of jagged peaks, but of undulating areas which they have interpreted as the residuum of an uplifted erosion surface now undergoing dissection. Most writers agree that the surface was warped during uplift, and many contend that it can be traced south-westwards to a point where it passes beneath a thick cover of Tertiary lavas. Extrusion of these lavas is said to have preserved the feature relatively intact from erosion, but not from tectonic warping. By mapping the base of the lavas it should be possible to determine the form of the surface and to assess the amount of deformation it has suffered.

These two examples illustrate both the value and limitations of erosion surfaces as guides to crustal movement. Their great advantage lies in coverage of regions and periods for which alternative evidence is almost totally lacking. They occur in upland areas where, prima facie, crustal elevation has occurred at some time during the Cenozoic era. Yet the very process of uplift has led to destruction of those sedimentary sequences from which geological history is normally inferred. Unfortunately mor-

Fig 4.11. Simplified diagram of the erosion
surfaces of Wales as mapped by Brown.

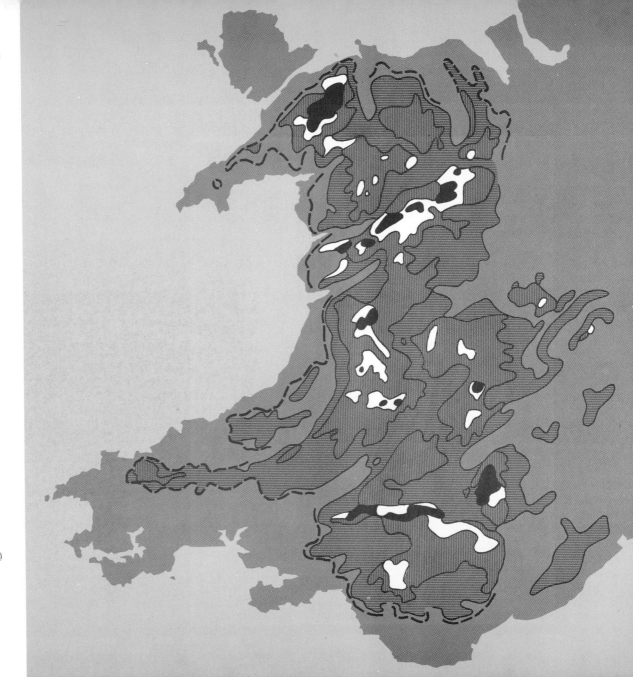

Key

 Upstanding residuals

 High Plateau (525–600 m)

 Middle Peneplain (375–425 m)

 Low Peneplain (250–300 m)

 Approximate run of a former
coastline, now at 200 m
(? of Calabrian age)

phology alone is often capable of more than one interpretation. Jones, for instance, has maintained that the gross form of the Welsh upland is due to stripping away of a former cover of Triassic sediments and exhumation of a much older landscape. Recent studies of the Tertiary lavas in Colorado have cast doubt on the existence of an underlying erosion surface that is everywhere of the same age. This in turn must raise doubts about the origin of the undulating summits at the crest of the Rockies. An increasing group of workers has expressed scepticism about the traditional way of interpreting accordant summit levels, arguing that in many cases they are not remnants of earlier landscapes and therefore have little historical significance. In Chapter 10 it is argued that this viewpoint has had a beneficial effect in forcing geomorphologists to examine much more critically the concepts and evidence on which they have formulated their denudation chronologies, but that it has not entirely undermined traditional

Fig. 4.12. Contours of the sub-Eocene surface in south-eastern England (after Wooldridge and Linton). The map illustrates the minimum amount of tectonic deformation during Cenozoic times. The inset exemplifies an alternative way of depicting tectonic structures. In both cases, however, it needs to be remembered that the 'structural relief' may have evolved slowly over a long period and have borne little relationship to surface topography.

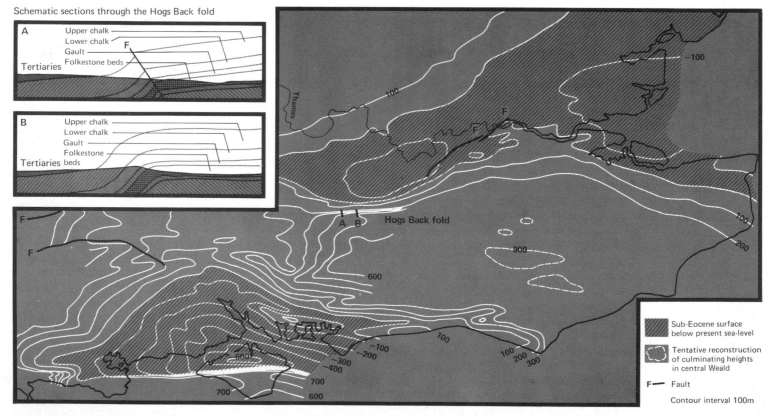

Schematic sections through the Hogs Back fold

interpretations. There is little doubt that the morphology of many upland areas reflects local tectonic history, and that movements recorded in this way have often been pulsatory.

Geological evidence for crustal movement

General principles It is evident from the writings of classical Greece that relative movements of land and sea were envisaged to explain the presence of fossil marine organisms in rocks now situated far above sea-level. When William Smith, the eighteenth-century founder of stratigraphy, demonstrated that a single sheet of rock was of the same age at points many kilometres apart and at levels differing by hundreds of metres, it became apparent that differential movement of the land must be involved. With the further development of simple stratigraphic principles, it became possible to unravel the structures and to assess the nature and age of the earth movements that produced them.

The present disposition of the thick stratum of Chalk in south-eastern England provides an illustrative example of the methods of analysis that may be employed. Originally laid down as a continuous horizontal sheet of marine sediments, the Chalk now remains as a dissected remnant underlying the London and Hampshire basins but removed by erosion from the crest of the major upfolds in the Weald and Isle of Wight. The deformation which the Chalk has suffered may be depicted by means of structure contours (Fig. 4.12). These are simply contours drawn on a chosen datum plane such as that separating the Chalk from the overlying Tertiary beds, and they show the form and amplitude of the folding which that plane has suffered. Where denudation has removed the Chalk the elevation of the datum plane can be estimated by assuming appropriate thicknesses for the intervening strata and projecting upwards from contours on a lower datum surface. In this way the structural relief of south-eastern England can be shown to exceed 1 500 m. This is a measure of the differential movement to which the region has been subject during Cenozoic time. It is, however, possible to identify with much greater precision the stages by which the present structural relief evolved. This is achieved by considering the evidence of the more recent strata that overlie

the Chalk. In both the London and Hampshire basins an angular unconformity between the Chalk and the oldest Cenozoic strata indicates a major phase of tectonic movement and associated erosion during Paleocene times. The numerous cycles of marine transgression and withdrawal which characterize the Eocene and Oligocene beds attest to continuing movement, and although the amplitude of each folding episode was small the cumulative effect was considerable. Nevertheless, the major deformation seems to have occurred during a relatively brief episode in late Oligocene and early Miocene times, a tectonic phase also responsible for fashioning the basic outline of the North Sea basin. Subsequent erosion produced a surface of relatively low relief in south-eastern England across which the sea transgressed in late Pliocene times. The resulting marine sediments are sporadically preserved in East Anglia and on the dip-slope of the North Downs in Kent, the contrast in elevation between the various sites implying a differential movement of about 180 m. Early Pleistocene deposits seem to be similarly tilted, thus confirming the considerable amplitude of warping in south-eastern England during Quaternary times.

It cannot be emphasized too strongly that the contours shown on Fig. 4.10 depict structures without any direct reference to surface topography; tectonic displacements are cumulative and have been going on spasmodically since earliest Cenozoic times. Throughout this period the crustal movements must have been accompanied by erosion, and it must not be assumed, for instance, that at any stage the Wealden area reached an elevation of 900 m. Similarly it needs to be remembered that the dotted lines commonly shown above geological cross-sections depict structural relationships, not former topography. In the case of the Weald it seems that the cover of Chalk had already been breached in some areas by mid-Eocene times; it appears likely, although difficult to prove, that for prolonged periods in the Cenozoic era there was little disparity in the mean rates of crustal uplift and subaerial denudation.

The concept of orogenic cycles The pattern of events outlined for south-eastern England is characteristic of many other continental regions. Long periods of relative quiescence are broken by shorter paroxysms of intense deformation. It is during these relatively brief interludes that many of the major cordilleran ranges of the world seem to have formed. The actual uplift of the mountains may constitute the final stage of a much longer cycle of events beginning with accumulation of a thick sequence of sedimentary rock. In a classic study of the Appalachians Hall and Dana drew attention to the fact that the materials forming those mountains must have been laid down in a continually subsiding basin to which they gave the name geosyncline. This idea was adopted and elaborated by workers on both sides of the Atlantic with general agreement that a complete orogenic or mountain-building cycle consists of three distinct phases: firstly a prolonged period of geosynclinal sedimentation, secondly a briefer episode of tectogenesis in which the strata are intensely crumpled and intruded by plutonic rock, and finally a rapid uplift of the deformed rocks to produce the actual mountain chain. Subdivision of these phases has often proved possible. The tectogenic phase commonly involves an early development period in which material known as flysch is eroded from growing structural highs and deposited in nearby subsiding basins; the later folding episode affects not only the true geosynclinal sediments but also the flysch. The close of the tectogenic phase and early part of the uplift is often accompanied by deposition of coarse debris in marginal depressions. This material is known as molasse and may remain undisturbed by the intense folding and thrusting affecting all the earlier sediments. By dating both the flysch and molasse, particularly in areas of recent mountain-building, it is possible to assess the duration of the tectogenic and uplift phases. In general they appear to be comparatively short intervals with the implication that the associated tectonic movement is extremely rapid.

Recent studies have shown that this simple model of mountain-building is not universally applicable. In many instances it is difficult to discern the archetypal geosyncline, while in others several separate tectogenic phases may be distinguished. In the western Alps, for example, it is now doubtful whether the sediments accumulated in a conventional geosyncline, and although the major tectonic movement was concentrated in middle Miocene times intense deformation also occurred during both the late Eocene and late Oligocene. In its simplest form the traditional view of orogenesis is one of the casualties of the revolution in geology and geophysics during the 1960s. Most mountain-building is now believed to be associated with deformation along the leading edge of a continental block at a destructive plate margin. However, no single model has yet been proposed that entirely replaces the traditional one, and it appears that cordilleran ranges differ considerably from each other in their evolutionary sequences.

The periodicity of orogenic activity The complications in the idealized orogenic cycle alluded to in the previous paragraph do not nullify the generalization that tectonic movement in any one region appears to be spasmodic. It was recognition of this fact that enabled nineteenth-century geologists in Europe to identify three Phanerozoic mountain-building periods:

(a) The Caledonian during which folding was predominantly along north-east–south-west axes in an area extending from northern Norway through Scotland to central Ireland. Prolonged sedimentation appears to have characterized this region from late pre-Cambrian times until the culminating tectogenic activity at the close of Silurian times. Recent radiometric dating, however, has permitted identification of several further tectogenic phases spanning a total period of 100 Ma or more.

(b) The Hercynian during which fold axes ranged from east–west in Ireland to north–south in parts of the Massif Central and Spanish Meseta. The main phase of movement appears to have been concentrated in late Carboniferous times, although in some areas a secondary phase was experienced in early Permian times.

(c) The Alpine during which the structures in the mountain chains that today characterize much of central and southern Europe were formed. Although there is great uncertainty about the exact sequence of events during the Alpine orogeny, the maximum phase of tectonic activity is traditionally placed in early to middle Miocene times.

This dating of orogenies was originally worked out in Europe and the question naturally arises whether the same sequence is applicable to other continents. If it is, then clearly there must be global cycles during each of which a world-wide cordilleran system is formed, gradually worn down by erosion, and then replaced by a new system. It would follow that there must also be cyclical changes in the form of the hypsographic curve, the extent of the high ground gradually diminishing during each long period between orogenies. As less than 25 Ma has elapsed since the Alpine orogeny and the usual interval between mountain-building episodes appears to exceed 100 Ma it might be assumed that the current relief of the continents is abnormally great. On the other hand, if orogenic activity is not synchronous throughout the world, it is possible to envisage a steady-state situation in which the hypsographic curve would remain relatively constant.

Views have differed on the question of global synchroneity. European geologists, as they moved into other continents, tended to interpret local tectonic history in terms of European orogenic cycles. This is reflected in the way the terms Caledonian, Hercynian and Alpine have often been applied on a world-wide scale. The exact time equivalence of orogenic activity needs to be examined without any preconceptions, and there is no doubt that similarities have often been emphasized at the expense of differences. A close parallel is to be found in the way Scottish raised beaches were formerly assigned to the '100-foot', '50-foot' and '25-foot' levels, despite the fact that field observations indicated significant deviations from those values (see p. 346) Table 4.1 is an attempt to summarize the ages of the major orogenies in selected regions of the world according to various local authorities. It is evident that Europe possesses neither a full nor even a particularly representative suite of orogenic cycles. The table also suggests that during virtually the whole of Phanerozoic time mountain-building was in progress in one area or another. Nevertheless, tectonic movement does seem to have been particularly intense during at least four periods, the mid-Palaeozoic, the late Palaeozoic, the late Mesozoic and the mid-Cenozoic. Neither of the two models discussed above therefore appears entirely appropriate. It does not seem that great cordilleras are produced in a global paroxysm, and then slowly reduced by the processes of erosion until the next upheaval. Yet neither is the relief of the continents as represented by the hypsographic curve constant through time; instead it probably shows a subdued cyclical pattern.

Contemporary orogenic activity If it is true that there is always orogenic activity taking place, it should be possible to find examples where it is occurring at the present day. Since one cannot directly observe the growth of geological structures, the main source of evidence must be the measurement of contemporary surface uplift as outlined earlier in this chapter. As early as 1835, Charles Darwin during the voyage of the 'Beagle' suggested that the displacement of the Andes, vividly attested by the frequent earthquakes which he experienced, would be sufficient to account for their elevation; he seems to have had no doubt that he was witnessing mountain-building in progress. The problem, however, is that the time-scale over which direct measurements may be made is almost infinitesimal compared with the time scale of an orogeny, and there is a justifiable

Table 4.1. A schematic portrayal, continent by continent, of the age of the world's major orogenies since the end of pre-Cambrian times.

Period	W. Europe (traditional)	European Russia	N. America	Japan	Australia	South America	Africa
T	Alpine	Alpine	Laramide	Oyashima			
C							
J		Cimmeridian	Nevadan	Sakawa			
T				Akiyoshi			
P			Appalachian				
C	Hercynian	Hercynian		? ?			
D			Acadian				
S	Caledonian	Caledonian					
O			Taconic				
C							

hesitation in multiplying contemporary measurements by a factor as large as 10^5 or 10^6. Nevertheless, it is instructive to consider the potential effects if presentday movements were to continue for a prolonged period. In many areas of current seismicity, levelling has disclosed uplift exceeding 5 mm yr^{-1}. Were this rate to be sustained for a million years it would elevate the surface by 5 000 m and thereby engender a major mountain chain. Although such a calculation ignores any concurrent erosion, it still demonstrates that there is no inherent difficulty in the view that orogenic uplift is taking place at the present day.

Large-scale horizontal displacements Just as geologists for many years resisted the idea of continental drift, so they also tended to minimize the effects of transcurrent faulting. In many instances it is not possible to define precisely the nature of displacement along a fault, since either dip-slip or strike-slip can explain the observed relationships. In such cases dip-slip was often assumed and the possibility of significant strike-slip either ignored or discounted. However, as the importance of lateral movements on a global scale became increasingly clear, re-interpretation of a number of faults originally regarded as either normal or reverse was undertaken.

An excellent example of changing attitudes is afforded by studies of the Great Glen fault in Scotland. Prior to the Second World War this was usually regarded as a normal fault with a vertical throw of hundreds or even thousands of metres. Following detailed investigations completed in 1939 Kennedy argued that the dislocation is a transcurrent fault with a horizontal displacement of approximately 100 km. He likened certain features of the fault, especially its remarkably straight trace and abnormally wide shatter belt, to those of the San Andreas fault where observed movements at the present time are predominantly horizontal. He found masses of granite on opposite sides of the fault so matched in structural details that they are best explained as one original stock split into two by translocation along the Great Glen fault. The relative movement was thus fixed as 100 km and was dated on geological grounds as Devonian in age.

The comparison with the San Andreas fault was a bold pronouncement by Kennedy since at the time it was far from universally agreed that there had been large-scale transcurrent movements in California. As early as 1925 a few writers were arguing that strike-slip on the San Andreas system might amount to between 10 and 40 km, but it was not until 1953 that Hill and Dibblee assembled data which they claimed to demonstrate offsets totalling at least 500 km since Cretaceous times. Their basic idea was that markers could be identified on both sides of the fault showing diminishing displacement with diminishing age (Fig. 4.6, p. 63). According to their reconstruction Cretaceous beds are offset 510 km, Eocene beds 360 km, Miocene beds 280 km and Pleistocene beds 15 km. Not all workers have accepted these detailed findings, but there has undoubtedly been a much greater willingness to accept the validity of large-scale lateral displacement. In a more recent study of southern California, Crowell has argued for a displacement of 260 km since early Miocene times but is doubtful about the earlier history of the fault. Increasing uncertainty about the more distant period is to be expected on theoretical grounds. The older the materials involved the wider their displacement will be and the greater their liability to independent metamorphism after separation along the fault. If the mean figure of 270 km of movement since Miocene times is accepted, the rate of displacement has averaged just over 10 mm yr^{-1} for the last 25 Ma. This is appreciably lower than many estimates of current movement and suggests either that displacement is accelerating or that we live in a period of unusually intense activity.

AKEROYD, A., 1972. 'Archaeological and historical evidence for subsidence in southern Britain', *Phil. Trans. R. Soc. A 272,*, 151–69.

ANDERSON, D. L., 1971. 'The San Andreas fault', *Scient. Am.*, 225, 53–68.

BELOUSSOV, V. V., 1962. *Basic Problems in Geotectonics*, McGraw-Hill.

BOLLINGER, G. A., 1973. 'Seismicity and crustal uplift in the southeastern United States', *Am. J. Sci. 273A*, 396–408.

BROWN, E. H., 1960. *The Relief and Drainage of Wales*, University of Wales Press.

CRITTENDEN, M. D., 1963. New data on the isostatic deformation of Lake Bonneville, *Prof. Pap. U.S. geol. Surv.*, 454-E.

CROWELL, J. C., 1962. 'Displacement along the San Andreas fault, California', *Spec. Pap. geol. Soc. Am.*, 71.

HILL, M. L. and T. W. DIBBLEE, 1953. 'San Andreas, Garlock and Big Pine faults, California', *Bull. geol. Soc. Am.*, 64, 443–58.

JONES, O. T., 1951. 'The drainage system of Wales and adjacent regions', *Q. Jl. geol. Soc. Lond.*, 107, 201–25.

KENNEDY, W. Q., 1946. 'The Great Glen fault', *Q. Jl. geol. Soc. Lond.*, 102, 41–76 (see also K. M. Storetvedt, *Geol. Mag.*, 111, 1974).

KING, L., 1962. *Morphology Of The Earth*, Oliver and Boyd.

LENSEN, G. J., 1968. 'Analysis of progressive fault displacement during downcutting at the Branch River terraces, South Island, New Zealand', *Bull. geol. Scc. Am.*, 79, 545–55.

RONA, P. A., 1974. 'Subsidence of Atlantic continental margins', *Tectonophysics,* 22, 283–99.

SAVAGE, J. C. and R. O. BURFORD, 1970. 'Accumulation of strain in California', *Bull. seism. Soc. Am.,* 60, 1877–96.

SAVAGE, J. C. and R. O. BURFORD, 1973. 'Geodetic determination of relative plate motion in central California', *J. geophys, Res.*, 78, 832–45.

WOOLDRIDGE, S. W. and D. L. LINTON, 1955. *Structure, Surface and Drainage of Southeastern England*, Philip.

The field discussed in this chapter is very unevenly covered in the literature at the present time. Interesting discussions of specific aspects are to be found in : T. Rikitake, 'Earthquake prediction', *Earth Sci. Rev.*, 4, 1968, pp. 245–82 ; 'Symposium on recent crustal movements', *Can. Jl. Earth Sci.*, 7, 1970 ; 'Recent crustal movements', *Bull. R. Soc. N.Z.*, 9, 1971 ; 'A discussion on problems associated with the subsidence of southeastern England', *Phil. Trans. R. Soc.*, A 272, 1972.

1|5

Patterns of global relief and crustal mobility

As the previous two chapters demonstrate, the trend of recent investigations has undoubtedly been towards recognizing both the rapidity and pervasive nature of crustal movement. Despite the scepticism expressed in certain quarters, the concept of mobile lithospheric plates commands wide support among geophysicists at the present time. As currently formulated, the plate model has obvious implications for the geomorphologist. These do not arise merely from the explanation it affords of the global distribution of the continents and ocean basins. It is probably in defining areas of relative vertical stability and zones of intense deformation that the concept of lithospheric plates is most important. In the former areas one might anticipate the continental surfaces to be dominated by low relief produced as a result of sustained denudation uninterrupted by uplift; similarly the older parts of the ocean floors away from the spreading centres should be dominated by the smooth surface of accumulated sediments. The plate margins, on the other hand, should be characterized by topography reflecting active tectonism, and comparison of Figs 1.4 and 3.11 confirms a general accordance between plate edges and mountainous relief; there are, however, noteworthy exceptions as, for instance, in southern Africa and the central Sahara.

In the following account the nature of topographic relief at constructive and destructive plate margins will first be examined in some detail. Thereafter the possibility of vertical movements affecting the central regions of the plates will be discussed. Finally attention will be turned to two additional topics that can appropriately be treated within the framework of plate tectonics, volcanic landforms and eustatic changes of sea-level.

Constructive plate margins

It is axiomatic that nearly all constructive margins should occupy oceanic positions. The characteristic topographic forms of the mid-oceanic ridges have already been described and will not be discussed further. However, great interest attaches to those places where oceanic spreading centres abut against the continental blocks since it is here that it may be possible to discern what happens during the early phases of continental separation. Two such locations have attracted particular attention during recent years, the Gulf of Aden and the Gulf of California. The former is particularly instructive because it has been claimed to represent the initial stages of a split between Africa and Arabia.

The Gulf of Aden and Red Sea region The Gulf of Aden and Red Sea (Fig. 5.1) together constitute a marine inlet over 3 000 km long but nowhere

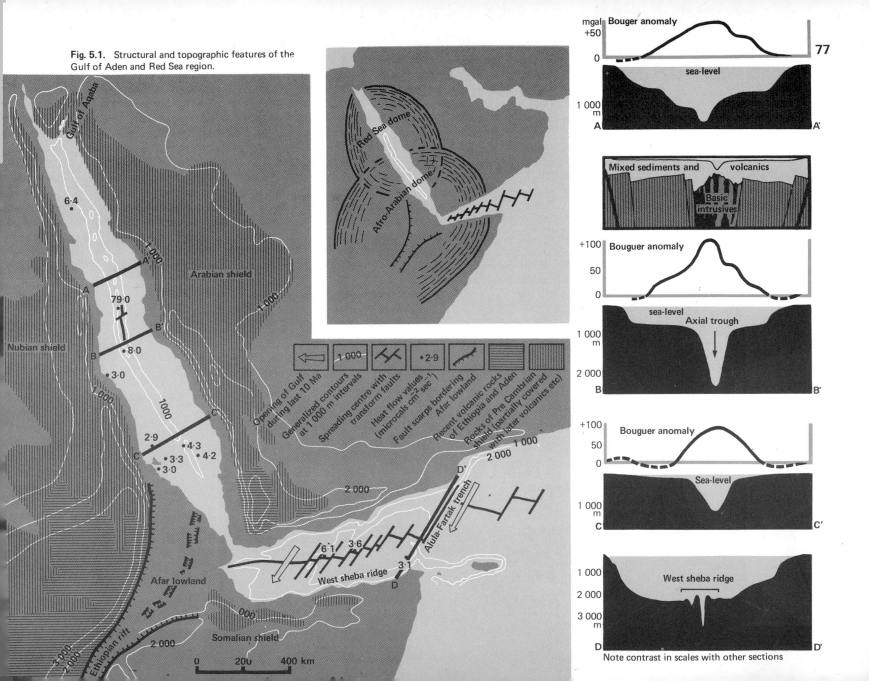

Fig. 5.1. Structural and topographic features of the Gulf of Aden and Red Sea region.

Gulf of Aqaba

6·4

1 000

A'

A

79·0

Arabian shield

B

· 8·0

B

· 3·0

1 000

Nubian shield

1000

C

2·9

· 4·3

C

· 3·3   · 4·2

· 3·0

2 000

2 000   1 000

D'

Alula-Fartak trench

6·1   3·6

3·1

West sheba ridge

D

Afar lowland

Ethiopian rift

3 000

2 000

Somalian shield

0   200   400 km

Red Sea dome

Afro-Arabian dome

Opening of Gulf during last 10 Ma

Generalized contours at 1 000 m intervals

Spreading centre with transform faults

Heat flow values (microcals cm⁻² sec⁻¹)

Fault scarps bordering Afar lowland

Recent volcanic rocks of Ethiopia and Aden

Rocks of Pre Cambrian shield (partially covered with later volcanics etc)

1 000   · 2·9

77

mgal   Bouger anomaly
+50
0

sea-level

1 000
m
A   A'

Mixed sediments and   volcanics

Basic intrusives

+100   Bouguer anomaly
50
0

sea-level   Axial trough

1 000
m

2 000

B   B'

+100   Bouguer anomaly
50
0

Sea-level

1 000
m
C   C'

1 000

West sheba ridge

2 000

3 000
m
D   D'

Note contrast in scales with other sections

more than 350 km wide; from the head of the Red Sea to the Mediterranean is a mere 120 km. The inlet transects a dome-shaped area of Precambrian rock which reaches elevations of 2 000 m on the African side and 3 000 m on the Arabian side. Within 150 km of these crests the dome slopes outwards to 1 000 m in Africa and 1 500 m in Arabia. Bathymetric surveys have charted depths exceeding 2 200 m in both the Sea and the Gulf, but the forms of the two areas differ sharply. A transverse profile of the central Red Sea shows marginal shelves less than 200 m deep, steep slopes leading down to the main trough at 600 to 1 000 m, and then further steep slopes descending to what has been termed the inner or axial trough. The elements in this profile do not extend along the full length of the Sea, the axial trough being confined to the central part and the main trough to the central and northern parts. By contrast the Gulf of Aden, except at its western end, has a much broader central deep flanked by steep slopes that begin within a few kilometres of the coast. Along the middle of the Gulf the floor is diversified by a topographically rough zone known as the West Sheba Ridge in which there are prominent north-east—south-west lineaments. By far the most striking is the 250 km Alula-Fartak trench descending to a depth of 5 360 m.

The major relief features are associated with distinctive geophysical properties. The dome-shaped shield is generally characterized by low gravity values, with the Bouguer anomaly reaching a value of −270 mgal beneath part of the Ethiopian plateau. By contrast the whole of the Red Sea south of the Gulf of Aqaba is characterized by high gravity values, with particularly large Bouguer anomalies over the axial trough where they locally surpass 150 mgal. Over the Gulf of Aden gravimetric surveys have revealed only very small Bouguer anomalies. These variations clearly indicate that the surface forms are accompanied by deep-seated crustal disturbances.

Almost all heat-flow measurements on the floor of both the Gulf and the Sea are well above the world mean. The highest values tend to be concentrated along the West Sheba Ridge and the axial trough, and it is in the latter location that the highest submarine figures anywhere in the world have been recorded. Considerable excitement was aroused by the discovery that hot spots in the axial trough are accompanied by pools of acidic brine and sea-floor sediments remarkably rich in metals; it seems likely that the metals have been precipitated from colloidal suspension after migrating from the upper mantle by way of faults.

Surveys in the Gulf of Aden have revealed a very strong lineation of magnetic anomalies, approximately parallel to the West Sheba Ridge. The pattern is typical of that over the mid-oceanic ridges and has permitted the plotting of isochrons. The earliest recognizable isochron is that for 10 Ma ago when the Gulf would have been a narrow inlet ending just east of Aden. In the last 10 Ma Arabia and Somalia seem to have been drifting apart at a rate of approximately 20 mm yr^{-1}. Over the Red Sea pronounced linear anomalies are confined to the axial trough. Although very restricted in area the pattern is undoubtedly more characteristic of oceanic than continental regions and has been interpreted as the product of sea-floor spreading operating for only 3 to 4 Ma. The narrowness of the magnetic pattern makes interpretation difficult, but it is likely that during this period Arabia and Africa have been separating at a rate of about 20 mm yr^{-1}. Seismic activity along both the West Sheba Ridge and the axial trough supports the view that these are sites of crustal spreading and transform faulting.

There is an unresolved conflict between the evidence from the Red Sea and the Gulf of Aden on the exact motion of the two continental blocks since no single rotational pole can accommodate all the inferred displacements. As a consequence several workers have argued that the region is a triple junction, that is, that three plates and not merely two are involved. On this view the African side is composed of independent Nubian and Somalian plates divided along the line of the Ethiopian rift; in order to allow for opening up of the Gulf of Aden a clockwise rotation of the Somalian plate has been postulated but whether this is consistent with observed movement on the Ethiopian rift needs more thorough investigation.

The principal events in the recent evolution of the Aden and Red Sea region appear to have been doming followed by the development of linear collapse structures. The initial stages of collapse seem to have involved both flexuring and faulting, with tension along the crest of the dome inducing 'necking' or thinning of the continental crust. This was soon succeeded by

Fig 5.2. The East African rift system. The areas of domed uplift are outlined by shading, with the amplitude indicated by contours drawn on the surface of tilted erosion surfaces of probable mid-Cenozoic age. Both dating and details of form have been the subject of much controversy but the broad pattern of movement seems clear. The inset shows schematically the nature of the uplift and its effect on local drainage development approximately along the line of the equator. Again details of dating are uncertain but (A) probably represents conditions in early Cenozoic times (D) conditions since mid-Pleistocene times.

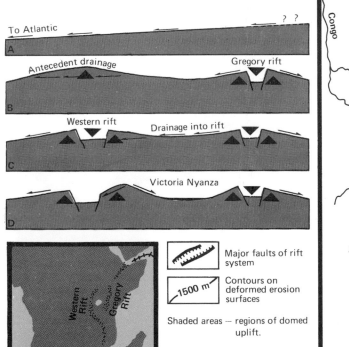

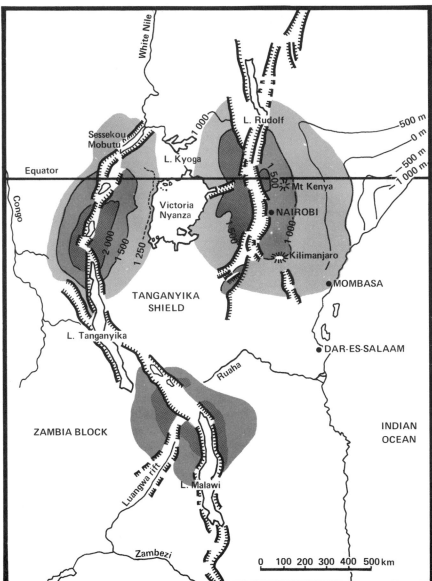

Major faults of rift system

Contours on deformed erosion surfaces

Shaded areas — regions of domed uplift.

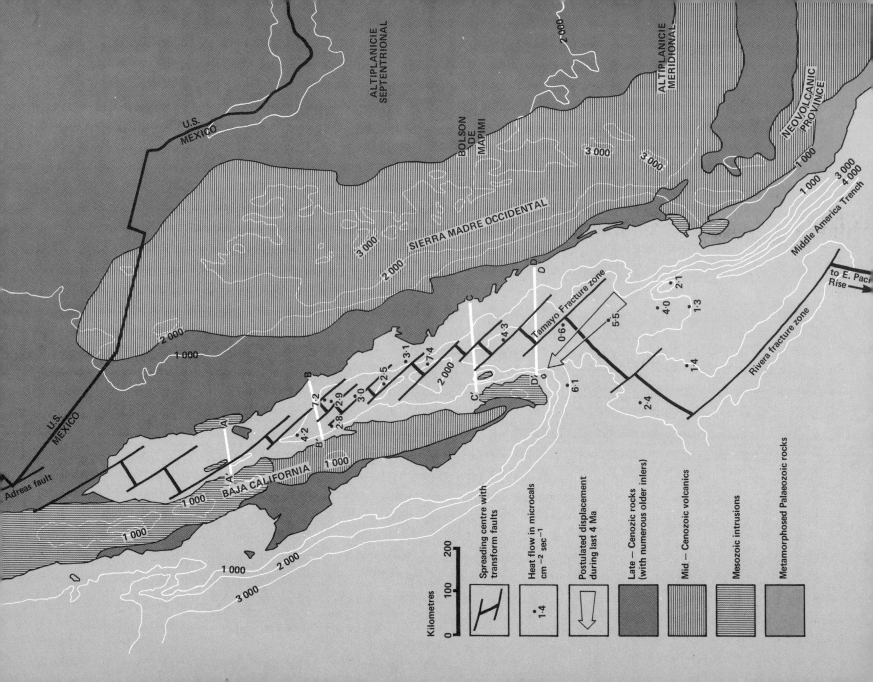

ALTIPLANICIE SEPTENTRIONAL

ALTIPLANICIE MERIDIONAL

NEOVOLCANIC PROVINCE

U.S.
MEXICO

BOLSON DE MAPIMI

3 000

3 000

2 000

Middle America Trench

3 000

SIERRA MADRE OCCIDENTAL

Tamayo Fracture zone

to E. Paci
Rise

2 000

3 000

2·1

1 000

1·3

4·0

2 000

1 000

7·4

4·3

5·5

Rivera fracture zone

U.S.
MEXICO

•3·1

0·6

1·4

2·5

2·0

6·1

2 000

·7·2

·2·9

3·0

D'

2·4

A

·2·8

B'

4·2

BAJA CALIFORNIA

1 000

Adreas fault

1 000

A'

1 000

1 000

1 000

2 000

1 000

2 000

3 000

Kilometres

200

100

0

Spreading centre with transform faults

Heat flow in microcals cm^{-2} sec^{-1}

•1·4

Postulated displacement during last 4 Ma

Late – Cenozoic rocks (with numerous older inliers)

Mid – Cenozoic volcanics

Mesozoic intrusions

Metamorphosed Palaeozoic rocks

actual rupture and the formation of new oceanic crust along the line of the suture. The doming is believed to have elevated a relatively featureless erosion surface which received its final shaping in early Cenozoic times. The present radial drainage pattern reflects this domed uplift, with only short streams flowing into the rifted depressions. The asymmetry of the resultant divides has led to many instances of stream piracy, and it may even be possible to reconstruct drainage lines directly dis-

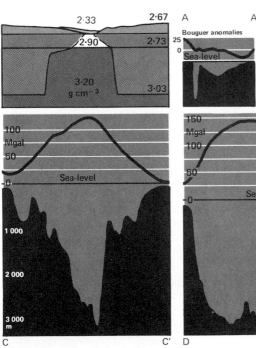

Fig. 5.3. Structural and morphological features of the Gulf of California.

Each of the four insets shows a topographic profile across the gulf with, above it, the measured Bouguer gravity anomalies. For profiles B— B' and C— C' possible structural interpretations of the anomalies are shown.(based upon Harrison and Mathur, *Am. Ass. Pet. Geol. Mem. 3*, 1964)

membered by the opening of the Gulf of Aden although the evidence on this latter point remains inconclusive. One reason is the growing recognition that, even where the continental separation is restricted to a few hundred kilometres and has occurred during the last 10 Ma, the fit is not like that of a well-cut jigsaw. It seems that crustal thinning and disruption prior to actual separation are so severe as to preclude exact fitting. The best reconstruction of the Gulf of Aden 10 Ma ago leaves a 75 km gap between the continental slopes, while closure of the Red Sea involves overlap between the 'nose' of Arabia and the Afar lowland. The rocks of this remote region have been closely studied to determine whether they could have evolved after the detachment of Arabia; in this way the problem of overlap might be solved. Although many of the rocks are basaltic flows compatible with recent tensional rifting and extrusion of oceanic crust, there are also several large horsts composed of continental crust which seem to preclude a simple and exact jigsaw fit.

For many years it was believed that a simple rift system extended from the Dead Sea in the north, via the Red Sea and East Africa, to the Zambezi valley in the south (Fig. 5.2). The central section of this system has been shown above to be an area of recent continental separation, and at first sight it is tempting to regard the East African rifts as incipient centres of the same process. However, among other things, the age of the rifts makes this very unlikely. The troughs in the vicinity of the lower Zambezi appear to date back at least to early Mesozoic times, and further north many of the tectonic movements are certainly no younger than those round the Red Sea and Gulf of Aden. Nevertheless, there are some notable similarities between the two regions. The rifting is concentrated close to the crest of a series of strongly arched structures and is often accompanied by sharp monoclinal flexuring. Warped erosion surfaces characterize the flanks of the domes, while the later faulting has induced striking instances of stream piracy and drainage reversal. The sequence of events is clearly exemplified in the region around Victoria Nyanza. This area originally drained westwards to the Congo, and during a period of crustal stability extending from Mesozoic into early Cenozoic times was reduced to an extensive erosion surface of very subdued relief. In mid-Cenozoic times upwarping along the line of the Western Rift required erosion by the west-flowing rivers to maintain their antecedent courses. This upwarping was later accompanied by a central rift which deepened so rapidly in Miocene times that the connection with the Congo was finally severed. The streams of southern Uganda continued to flow into the rift until accelerated Pleistocene movements completed the disruption of the drainage; part was reversed to flow in to the newly formed Victoria Nyanza and part drained down very steep gradients to the floor of the rift. From the elevation of erosion surfaces along the rift margins differential warping of about 850 m since late Oligocene times has been postulated for south-west Uganda. The depression of the rift valley has been even greater since over 2 600 m of infill has been proved in a borehole near Lake Sessekou Mobutu (formerly Lake Albert). Comparable movements seem to have characterized the Gregory Rift in Kenya. Domed Cenozoic uplift has here been estimated at more than 1 500 m, and again there is clear evidence of a linear downwarp along the crest of the arch prior to the actual rifting.

As already mentioned, there is no reason for supposing that the rift system of East Africa represents a line of current continental separation. Yet it does seem to have evolved in a way very reminiscent of the early development of the Red Sea, and it has even been suggested that it represents a site of lithospheric spreading which ceased to function before actual dismemberment took place.

The Gulf of California Oceanographic research within the Gulf of California (Fig. 5.3) has revealed many indications of an offset spreading centre that appears to be a northward extension of the East Pacific Rise. The more significant findings include a distinctive pattern of deep rhomboid basins, active fault scarps, above-average heat flow values and Bouguer gravity anomalies that suggest an intrusion of relatively high-density material along the axis of the gulf. Although magnetic surveys have failed to disclose linear anomaly patterns, there is probably a single spreading centre offset by frequent transform faults. The latter are aligned approximately parallel to the San Andreas fault which may simply be the largest of their number. Supporting this interpretation is the presence of further spreading centres at the northern end of the San Andreas fault off the coast of Oregon, Washington and British Columbia. This would clearly imply that the continuation of the East Pacific Rise does not lie beneath the continent. Nevertheless, there are indications of tectonic movements around the Gulf of California reminiscent of those already described from Africa. In north-west Mexico the Sierra Madre Occidental has been uplifted along its western edge by as much as 2 000 m, while to the north in the United States the Colorado plateau province has been updomed by an equal or even greater amount. It was this rapid elevation that led to erosion of the Grand Canyon of the Colorado (p. 202) and the almost equally spectacular but less well-known gorges of Mexico. The peninsula of Baja California has undoubtedly suffered recent tectonic deformation, but the details remain obscure; along its northern segment differential uplift and tilting has amounted to 1 500 m, while further south alternating phases of both elevation and depression have been detected. These findings leave no doubt regarding the tectonic instability of the region since mid-Cenozoic times, and it is even possible to see in them evidence of a collapsed dome. An elongated trough had already developed along the line of the gulf by Miocene times, but the creation of new oceanic crust and the relative north-westerly drift of Baja California seems to have been delayed until little more than 4 Ma ago. Although analogies with the sequence of events in the Afro-Arabian region may be discerned, it should also be noted that the evolution of the plate margins off the Mexican coast has been particularly complex. At present a destructive margin represented by the Middle America trench ends immediately south of the entrance to the gulf but is believed to have extended much further north along the coast in mid-Cenozoic times. Within the relatively recent geological past, therefore, this section of the continental edge was probably experiencing subduction rather than lateral extension.

Characteristic features of continental separation The foregoing review of constructive margins in Africa and North America suggests that continental separation is likely to be preceded by massive upwarping and central collapse before the actual drifting starts. In many ways this is a predictable sequence in that the oceanic spreading zones are invariably marked by broad ridges which, if they developed beneath a continental block, would lead to linear arching of the surface. Various reasons have been advanced for the topographic highs along the centres of spreading. Seismic and gravity studies have shown anomalous structural conditions beneath the mid-oceanic ridges. The Mohorovicic discontinuity is either absent or modified, and the transmission of seismic waves is abnormally slow. Bouguer anomalies can be interpreted as indicating mantle material of unusually low density. These properties have been attributed to thermal expansion of the hot upper mantle, a condition that is also implied by the high heat-flow measurements. An alternative hypothesis that may explain localized elevation of the crust is the concept of phase change in the upper mantle. The underlying principle is simply that the phase in which the mantle material occurs is both temperature- and pressure-dependent. Phase in this context implies not only the liquid and solid states, but also the structural changes which solid materials undergo as temperature and pressure vary. The volumetric changes arising from rearrangement in the packing of the atoms may amount to 10 per cent or more. The attraction of this hypothesis is that it offers an explanation for variations in surface elevation without requiring lateral movement of material at depth; on the other hand, the concept may be linked with mantle convection currents to explain combined lateral and vertical movements. The simplest model envisages a constructive margin as being located above the upward-moving limb of a pair of convective cells. The crust is raised between 1 000 and 3 000 m and then splits symmetrically to drag apart any overlying continental mass. During updoming the sialic layer becomes so attenuated that later movement away from the spreading centre inevitably causes large-scale subsidence.

Older examples of continental separation So far attention has been confined to continental areas where separation by sea-floor spreading has been active for 10 Ma or less. In these cases the present surface topography seems to reflect deep-seated tectonic processes initiated no earlier than mid-Cenozoic times. There are many continental margins which, on the hypothesis of rigid lithospheric plates, must have originated rather earlier by separation along oceanic spreading centres. These have been termed Atlantic-type margins, but the extent to which they display common topographic forms directly attributable to their early tectonic history is still uncertain. In most instances continental separation took place at such a distant period that morphological expression of domed uplift might be expected to have been destroyed by erosion. More likely to be preserved is the attenuated margin of the continent, subject to subsidence as it moved away from the spreading centre.

Both the North American and European borders of the Atlantic appear to fit this evolutionary model reasonably well. The continental shelves bear thick sedimentary accumulations of Cretaceous and Cenozoic age resting directly on a Palaeozoic basement (Fig. 5.4). The beds generally thicken seawards and at the continental slope may attain an aggregate thickness of many kilometres. In the United States it has been estimated that the total volume of Mesozoic and later sediments between the Appalachians and a line 1 000 km seaward of the coast amounts to 8·7 million km^3; this is equivalent to a mean thickness over the whole area of about 2 500 m. The Cenozoic contribution to this total is 2·8 million km^3, or a mean thickness of 800 m. Two implications of these vast quantities of sediment deserve mention. The first is the very considerable amount of continental erosion that must have taken place to supply the necessary detritus. Although in the south a high proportion of the material consists of biogenic carbonates, further north it is dominantly terrigenous in origin. Even at the most conservative estimate the adjacent Appalachian region seems to have been denuded of more than 2 500 m of rock during Cretaceous and Cenozoic times. The second implication is that the shelf has subsided to accommodate the thickness of overlying material. Most of the material is shallow-water marine in origin and requires that the Palaeozoic basement has sunk pari passu with sedimentation. It has been calculated that the mean rate of subsidence must have been about 0·01 mm yr^{-1} throughout Cenozoic times. In part this can be ascribed to isostatic depression beneath the growing weight of new strata but there must also be an element of tectonic movement. One way of assessing the subsidence that may be attributable to sea-floor spreading is to consider the relationship between age and depth of the Atlantic ocean. The modern mid-oceanic ridge rises on average to within about 2 500 m of sea-level whereas the basement beneath the oldest part of the ocean floor lies at over 5 500 m. In other words, as oceanic crust moves away from the ridge its level may be expected to fall by some 3 000 m in 150 Ma. This yields a mean rate of 0·02 mm yr^{-1}, rather larger than the subsidence calculated for the shelf. However, plots of elevation against age for the ocean floor suggest that as new crust moves away from the spreading centre it descends relatively rapidly at first, and then more slowly. This may partly account for the low figure on the shelf during Cenozoic times, and there seems to be no gross discrepancy between the two independent methods of calculation.

Of the gross morphological features along an Atlantic-type margin, it is the continental shelf that seems most readily ascribed to sea-floor spreading. However, a few workers have also sought evidence for the domed uplift that it believed to precede actual rupturing. It has been suggested, for instance, that the basic form of the Atlantic seaboard in Canada may be attributable to events that occurred when North America split away from Europe and Africa. Broad arching took place in Permo-Triassic times and was immediately succeeded by deep rifting to produce the numerous graben structures that are still a prominent feature of the tectonics along the Atlantic coast. As the land masses separated the trailing edge of the continent subsided to form the continental shelf. Marine transgression ensued, but the more elevated remnants of the original arch were never submerged. This would mean that they have been the site of continuous subaerial erosion since early Mesozoic times. It seems clear that the modern relief cannot incorporate remnants of the original Mesozoic landscape because the amount of erosion has been too great. Rather the present landforms are the product of prolonged erosion acting on a region of resistant rocks last subject to major tectonic uplift at the time of continental separation; this would not preclude the minor spasmodic elevation that might be required to maintain isostatic equilibrium. It is worth recalling that, despite much argument about the evolution of the river pattern in the northern Ap-

Fig. 5.4. Section through the continental margin of Cape Hatteras, North Carolina. Note that a ridge (? reef complex) appears to have served as a dam until the end of Cretaceous times (after Emery, 1970).

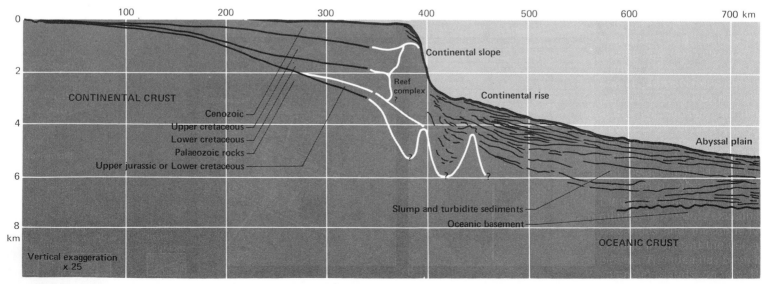

palachians, many workers have contended that the original drainage flowed westwards from an axis of uplift near the continental edge. They have ascribed the present pattern to reversal by piratical streams draining to the Atlantic coast.

If a comparison is made with the Atlantic margin of Europe and northern Africa, certain similarities are immediately discernible. Deep graben filled with Triassic sediments separate many of the upland areas of Britain, and analogous structures have now been located on parts of the continental shelf to the west. The main cover on the shelf is of Cretaceous and Cenozoic age, and although less thick than on the other side of the Atlantic still bears witness to long-term subsidence. In the British Isles attempts to trace the evolution of the present drainage pattern have normally envisaged an original system of rivers flowing eastwards away from the continental edge. Yet there are equally striking contrasts. The Cretaceous transgression in north-west Europe was extremely widespread. It was responsible for a bed of chalk which acts as a very useful datum in deciphering later earth movements. During Cenozoic times the chalk was sharply folded and within the intra-continental North Sea basin was downwarped to permit accumulation of some 3 000 m of later sediments. The main flexuring was contemporaneous with intense orogenic activity in the Alps and has been referred to as the 'outer ripples of the Alpine storm'. These major tectonic movements, subsequent to those associated with the rupture from North America, have greatly complicated the structural evolution of the Atlantic fringe of Europe. This is evident also in the contortion of strata on the continental shelf, contrasting with the relatively undisturbed disposition of the same beds off the North American coast.

Attention has been concentrated on the continental margins around the North Atlantic since these are among the most thoroughly studied in the world. They have the disadvantage that continental separation is believed to have begun in the south at least 200 Ma ago so that there has been an immensely long period of subsequent erosion. In the north the opening between Greenland and Scandinavia is more recent, having been dated to about 60 Ma ago. The gross relief of the Scandinavian peninsula appears to correspond reasonably well to the model of continental rupture by sea-floor spreading. An elongated marginal block, faulted along its western edge, is tilted eastwards so as to direct most of the drainage to the Baltic rather than the Atlantic. However, there are at least two complicating factors. As noted above, north-western Europe suffered major tectonic dislocation during the Alpine orogeny, and many workers have ascribed uplift of the Norwegian mountains to this episode. Secondly, part of the basin-shaped form of Scandinavia is due to depression beneath the Pleistocene ice-sheets, and the potential effects of glaciation need to be borne in mind in evaluating the present relief pattern.

Destructive plate margins

The lack of identity between the pattern of plates and continents results in three different types of destructive margin, ocean-to-ocean, ocean-to-continent, and continent-to-continent. The associated structural and relief features are so contrasted as to require separate consideration.

Ocean-to-ocean type The destructive margin at the junction of two oceanic plates is characterized by trenches and island arcs. The essential topographic features of the trenches have already been described (p. 10) and need no further elaboration here. The associated lines of islands are commonly divided into two groups. The first comprises the single arc system in which the trench is bordered at some distance by a long line of volcanic islands. The second comprises the double arc system in which a parallel festoon of islands, mainly composed of sedimentary strata, is interposed next to the trench. The single arc system is well exemplified by the Kuril Islands, the double arc by Indonesia. In relating surface form to inferred structure, it is presumed that the descending lithospheric plate generates sufficient heat to produce a regular line of volcanoes at a distance of about 200 km from the trench. Later erosion of the volcanic peaks may be sufficient to build a thick wedge of sediment extending to the edge of the trench; it is this material which is folded and contorted to form the second line of islands. Although destructive margins of this type are a very active part of the earth's crust, they do not necessarily evolve rapidly into major topographic features. Much depends upon the supply of sediment. Without a nearby continent to yield abundant material, the trench-and-arc system may persist little altered for long geological periods. The arcs were at one time interpreted as mountain chains in statu nascendi, but several examples are now known to be older than the analogous mountain ranges into which they were supposed to develop. They must therefore be regarded as enduring features, but if the plate which is undergoing subduction carries a continental 'passenger', the destructive margin will ultimately be converted into the continent-to-ocean type with potentially dramatic results.

Continent-to-ocean type The mountains along the western border of North and South America provide the best example of forms and structures believed to result from an oceanic plate being thrust beneath the edge of a continental block. In North America the relationships are complicated by the presence of a zone of sea-floor spreading in the Gulf of California, but the Andean area seems to provide a relatively simple illustration of the continent-to-ocean type of plate margin. The Peru–Chile trench extends as a structural feature some 4 500 km from Panama to Tierra del Fuego, although near its northern and southern ends it becomes so filled with debris that it almost disappears as a topographic feature. The middle reach of the trench descends to depths of over 7 000 m while many ranges in the Andes rise to over 5 000 m. The maximum amplitude of relief is about 15 000 m which is the greatest found anywhere on the globe within a horizontal distance of 400 km. Seismic activity defines a well-marked Benioff zone descending eastwards beneath the continent. The Mohorovicic discontinuity declines from 11 km beneath the Pacific to over 70 km below the Andes and then rises steeply again beneath the eastern edge of the mountains. The central section of the Andes is divisible into three physiographic and structural units. On the east a mountain system rising to well over 5 000 m consists mainly of metamorphosed Palaeozoic sediments. On the west a second mountain system of comparable height is composed of abundant Mesozoic and Cenozoic volcanics, interlayered with shallow-water marine sediments; in addition there are numerous batholithic intrusions varying in age from Jurassic to late Cenozoic but tending to become younger in an eastward direction. Between these two elevated units is a region known as the altiplano which has been a basin of accumulation for continental clastic debris since at least Cretaceous times; the combined thickness of sediments and volcanic materials may here exceed 10 000 m.

Proponents of lithospheric plates have hypothesized the following sequence of development. The Andean chain began to form in early Mesozoic times with underthrusting of the Pacific plate beneath a continental margin laden with a great thickness of Palaeozoic marine sediments. A volcanic arc developed above the descending plate and the accumulated Palaeozoic rocks suffered their first major deformation. Although this earliest episode seems to have preceded the first opening of the Atlantic, later developments were almost certainly linked with westward drift of the South American continent. Activity quickened in late Mesozoic and early Cenozoic time which is one of the traditional periods of Andean orogeny. The volcanic and associated marine sediments in the west were sharply folded and intruded by granite plutons, while the earlier rocks of the east suffered additional compression and uplift. This pattern continued into later Cenozoic times with plutonic activity slowly migrating eastwards. The migration of plutonism away from the trench is puzzling but has been attributed to changing thermal conditions along the upper surface of the descending plate; as a plate descends it cools the adjacent mantle and in consequence progressively deeper penetration is required before conditions suitable for magma generation are attained.

Although phases of slightly more intense activity may be distinguished, the central Andes appear to have suffered almost continuous tectonic deformation since early Mesozoic times. The volume of clastic material testifies to concomitant erosion on a prodigious scale. It seems highly unlikely that in the environment of a growing mountain range topographic forms of any great antiquity can be preserved. Nevertheless, many workers have discerned in the relief of the Andes indications of uplifted erosion surfaces, explicable only if elevation has been spasmodic but extremely rapid. It has been suggested that the pulses of greater orogenic activity are related to periodic acceleration in the rate of sea-floor spreading, but no precise correlation has yet been established.

The concept of plate subduction along a Benioff zone seems to place a natural limit of a few hundred kilometres on the width of a mountain chain. Where a cordilleran system exceeds that width, a more complicated evolutionary history must be envisaged. This applies to the western cordillera of the United States extending 1 500 km from Colorado to the Californian coast. It is presumed that much of the orogenesis is due to westward drift of the continent, but the width of the mountain zone must result from more complex tectonics than those of simple subduction. The exact pattern of plate movement has yet to be determined, but some of the results are clear. Whereas undisturbed Cretaceous and Cenozoic sediments on the Atlantic continental shelf have an aggregate volume of 8·7 million km³, on the Pacific coast the corresponding offshore sediments total no more than 2 million km³ of which at least half is pyroclastic in origin. However, an immense amount of Mesozoic and Cenozoic material has been returned to the continent in the form of uplifted and contorted sediments comprising the various coastal mountain ranges ; the total volume has been estimated at more than 5 million km³. As North America has drifted westwards, the detritus from erosion of the continent has apparently been subject to continual orogenic recycling.

Continent-to-continent type Where approaching plates carry continental 'passengers' the most complex orogenic conditions are found. The situation is illustrated by the Alps and Himalayas which lie between the Eurasian continental block on the north, and Africa and India on the south. The precise sequence of events has varied greatly along different segments of the mountain system. The Alps consist in the main of materials laid down in the broad Tethyan sea between Europe and Africa. The exact dimensions of this sea are unknown, but parts of it reached true oceanic depths. There were also broad shallow shelves on which thick sedimentary sequences accumulated. The closure of the Tethyan sea began in Mesozoic times and was accompanied by one or more subduction zones. Eventually in Cenozoic times the accumulated sediments were squeezed between the converging continents. In the Himalayas, on the other hand, there is appreciably less Tethyan sediment, the main constituents of the mountain chain being gneiss and crystalline rock similar to that which underlies the majority of the Indian sub-continent. Compared with the Alps there was apparently

much more direct contact between the ancient continental blocks, with India being thrust northwards under Asia. It has been estimated that the amount of crustal shortening totalled some 500 km. Although there has been little detailed research, the great Tibetan plateau may testify to intense uplift analogous to that of the central Andes. No directly comparable feature is to be found in the Alps, although several workers have identified what they claim to be severely dissected erosion surfaces.

Structural conditions are so varied that it is extremely difficult to generalize about the evolution of mountains where two continental blocks collide. The Alpine system is the most thoroughly studied but is also unusually complex, as is obvious from the enclosure of the deep Mediterranean sea within the mountain framework. The concept of destructive plate margins appears to offer a feasible explanation for the location and origin of all the major mountain chains resulting from recent orogenic activity. The traditional idea of geosynclinal subsidence as an essential prelude to mountain building seems to require modification, but there is still much debate whether the idea of prolonged marine sedimentation along a continental edge is an adequate substitute.

Intra-plate movements

Emphasis in the concept of lithospheric plates is on horizontal rather than vertical displacements. Major vertical movements are only envisaged at plate margins where new crust is being either created or destroyed. Yet, as indicated in Chapter 4, there is abundant evidence for epeirogenic movement in areas far removed from accepted plate edges. Indeed geodetic levelling in some continental regions appears to indicate that a plate surface can undergo almost constant vertical adjustment. It is clearly possible to exaggerate the rigidity of plates and the very shape of the geoid requires that, as they move across the surface, they suffer some deformation. It should also be recalled that many areas nowhere near a plate margin experience the occasional major earthquake. For example, although it is California that is renowned for its seismicity, two of the largest earthquakes in the United States have been centred on the Atlantic seaboard near Charleston and in the Mississippi valley near New Madrid.

In the oceanic regions Menard has drawn attention to features that he terms midplate rises. Typically these are broad gentle swells, up to 500 000 km² in area, and with a relief amplitude that may exceed 2 000 m. Examples are found in the Shatsky Rise of the north-western Pacific, the Bermuda Rise in the western North Atlantic and the Rio Grande Rise off southern Brazil. Structures in the sediments across the crest of these rises indicate that there has been both uplift and faulting. Low lying regions that were formerly fed by turbidity currents have been elevated and broken by tensional fractures. This took place while the whole area remained part of an actively spreading plate and seems to indicate the presence of small transient bulges in the mantle.

There appears no reason why midplate rises should be confined to oceanic areas, and it was noted earlier that many geomorphologists have postulated domed uplift to explain continental relief patterns. In some cases uplift may be due to initiation of a fresh spreading centre, but in others it seems to be unrelated to plate margins. Africa probably provides the clearest examples of epeirogenic uplift and depression, but most continental blocks have suffered some degree of deformation in this way. There remains much uncertainty about the mechanism involved. Several workers have invoked convective cells in the mantle of much smaller dimensions than those responsible for driving the major plates themselves. Others have stressed the potential effects of phase changes in the mantle, although this leaves unanswered the question of the trigger mechanism for the phase changes. Yet others have pointed out the need to supplement deep-seated movements with shallower transfers of rock to maintain isostasy and account for variations in the depth of the Mohorovicic discontinuity.

The hypothesis of lithospheric plates undoubtedly provides one of the most satisfactory explanations yet offered for global relief patterns. Yet many features still defy full explanation, and it is clear that plate surfaces cannot be assumed to be totally inert regions. It seems very likely that deep-seated phase changes are capable of producing significant localized uplift. The evidence from north-western Europe also suggests that orogenesis resulting from continental collision can cause appreciable deformation 1 000 km or more from the actual plate margin.

Volcanic landforms

Just as plate margins are the site of the most intense tectonic deformation, so they are also the location of much of the most spectacular volcanic activity. A distinction may be drawn between constructive margins where, once continental separation is complete, nearly all the extruded material is basaltic in character, and destructive margins where the range of chemical composition is much wider. Since there is a close relationship between magma composition and the resulting surface morphology, a contrast may be drawn between the volcanic landforms characteristic of mid-oceanic ridges and those found in association with island arcs and orogenic zones. In addition to these two groups, volcanic features situated in a number of other locations require brief examination.

Mid-ocean ridges Vulcanicity along the mid-oceanic ridges consists mainly of quiet fissure eruptions by which lava is extruded in vast quantities over the sea floor. The material is almost entirely a form of basalt relatively rich in silica and known as tholeiite. Eruptions from more centralized vents can build up immense cones rising several thousand metres from the sea floor. In these the predominant material tends to be alkali basalt. Among examples of oceanic islands formed in this way close to the axis of a mid-oceanic ridge are Tristan da Cunha and Gough Island in the Atlantic, Prince Edward and St Paul Islands in the Indian Ocean, and Easter Island in the Pacific.

The only large island built up by extrusive activity astride a mid-oceanic ridge is Iceland (Fig. 5.5). Some 100 000 km² in area, this landmass is divisible into three parts, a median trough-faulted zone generally between 100 and 200 km wide, and two flanking plateaus mainly constructed of innumerable basaltic flows. The range of volcanic forms is very wide. Large areas have been covered by fissure eruptions, with the basic lava sometimes of such low viscosity that it has flowed more than 100 km before solidifying. Gently sloping shield volcanoes have been constructed by protracted emission of lava with only small quantities of pyroclastic debris; Skjalbreidur, for example, is 600 m high, 10 km in diameter and has slopes that average about 7°. Complex strato-volcanoes with abundant pyroclastic material are less well developed in Iceland than in many continental areas, but Hekla and Oraefajokull are more appropriately described as strato-volcanoes than simple shield volcanoes. In one or two localities a viscous acidic magma has been extruded to form steep-sided domes. The occurrence of acidic rock in this mid-oceanic situation has been the subject of considerable speculation. One possible explanation is the contamination of upward-moving basaltic magma by a remnant of continental crust preserved beneath the island. A more widely supported idea invokes fractionation of the original

magma prior to eruption. The heaviest of the early formed crystals are held to sink to the base of the magma chamber, thereby ensuring that the material reaching the surface is relatively rich in silica.

Lying astride an active spreading centre, Iceland might be expected to suffer gradual stretching with the emplacement of new volcanic material along the median zone. There is much evidence that lateral extension of this type is currently taking place. The oldest rocks were formed 20 Ma ago and are generally

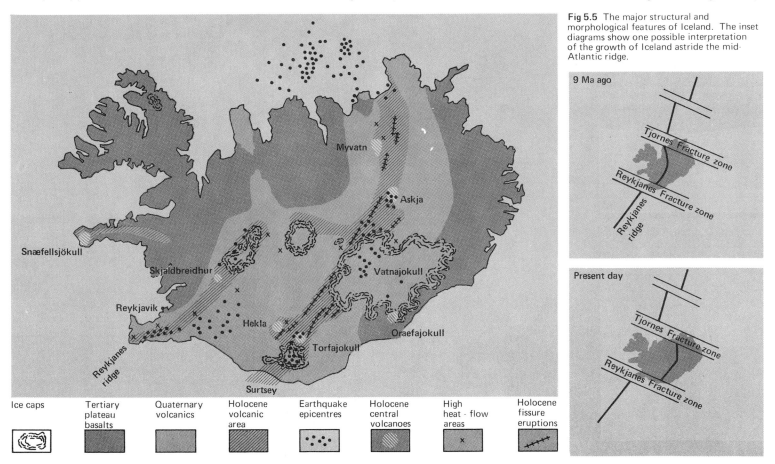

Fig 5.5 The major structural and morphological features of Iceland. The inset diagrams show one possible interpretation of the growth of Iceland astride the mid-Atlantic ridge.

9 Ma ago

Tjornes Fracture zone

Reykjanes Fracture zone

Reykjanes ridge

Present day

Tjornes Fracture zone

Reykjanes Fracture zone

Myvatn

Askja

Snæfellsjökull

Skjaldbreidhur

Vatnajokull

Reykjavik

Hekla

Oraefajokull

Torfajokull

Reykjanes ridge

Surtsey

Ice caps	Tertiary plateau basalts	Quaternary volcanics	Holocene volcanic area	Earthquake epicentres	Holocene central volcanoes	High heat-flow areas	Holocene fissure eruptions

found at the eastern and western extremities of the island ; if they are assumed to have originated near the median zone they imply a half-spreading rate of about 10 mm yr^{-1}. A striking characteristic of the trough-faulted median zone is the profusion of fissures. Along selected traverses these have been shown to occupy some 6 per cent of the total distance, and from an assessment of their age have been held to represent a lateral extension amounting to between 10 and 20 mm yr^{-1}. Repeated measurements across the median graben have also indicated extension of the crust ; a specially surveyed line 25 km long was found to have lengthened by 65 mm in 3 years.

Many of the seamounts and guyots on the ocean floors may have originated by volcanic activity close to a spreading centre, and been carried to their present positions by plate movement. It should be possible to determine the proportions that were initiated in this way when their ages are better known and it can be seen whether they correspond to the date of the nearby seafloor. It is believed that many if not all guyots owe their flat tops to wave action. The later subsidence that this implies may be ascribed to lowering of the plate surface as it moves away from the mid-oceanic ridge. The great thickness of coral on many Pacific islands, built on a volcanic cone but extending to depths far below those at which coral will grow today, might be explained by the same mechanism.

Island arcs and orogenic zones Subduction along a Benioff zone appears to engender large-scale production of magma which rises to the surface along conduits through the overlying crust. This is seen most clearly in the development of volcanic island arcs, but is also responsible for extrusive activity in many orogenic regions. The characteristic lava of island arcs is andesite rather than basalt. At one time andesitic lava was ascribed to the mixing of basaltic magma with continental crust, but more recent laboratory experiments suggest that it may be due to special conditions of high temperature and pressure experienced along a Benioff zone. It now seems likely that andesitic magma is generated directly in the mantle, with local variations in composition due to assimilation of other rocks. Even more acidic rocks, such as rhyolite, may be produced either by assimilation or by direct melting of the lower continental crust.

The variety of erupted materials results in a wide range of volcanic landforms. In general the greater the acidity of the magma, the higher the proportion of pyroclastic debris and the more explosive the average eruption (Table 5.1). Using the percentage of fragmentary materials as a measure of explosive violence, Rittman has estimated that areas in the central Pacific would have an explosion index of 3, whereas the Indonesian island arc would have an index of 99. The characteristic volcanic forms along Benioff zones include many features built either wholly or in large measure of pyroclastic fragments. Cinder or scoria cones can be constructed very rapidly around the pipe from which the material is being ejected. Of much greater extent are the plains built by ash flows in which fine solid particles are suspended in volcanic gases. When first erupted these flows move rapidly with many of the properties of a fluid, but after settling and cooling they form the rock type known as ignimbrite or welded tuff. Total thicknesses of over 100 m have been recorded and in Sumatra one area of ignimbrite covers more than 25 000 km^2.

Table 5.1 Regional contrasts in the nature of volcanic activity (after Rittman)

Location	Explosion Index
Island arcs	95
Cordilleras (Andes)	97
Grecian Islands	83
Southern Italy	41
African rift system (continental)	40
Iceland	39
Atlantic and Indian Oceans	16
Pacific Ocean	3

Explosion index – fragmentary material expressed as percentage of all material emitted.

It is, however, the composite strato-volcano that must be regarded as the most characteristic form associated with an active Benioff zone. The eruptions that build volcanoes composed of both lava and detrital fragments are commonly classified as either Strombolian, Vulcanian or Pelean depending on their nature and degree of violence. Strombolian eruptions are either continuous or have a cycle of activity that is repeated after only a short period of dormancy. The magma is often basaltic in composition and the proportion of pyroclastic material low. Vulcanian eruptions tend to be more explosive and are associated with increasingly viscous lava that tends to solidify quickly. The period between eruptions is greater and activity is often renewed by ejection of vast amounts of ash as any blockage in the vent is first forced apart. In a Pelean eruption the solid summit of the volcano is first elevated and then disintegrated as one or more 'glowing avalanches' or clouds of ash descends the slopes of the cone. The Pelean eruption is usually associated with the most acidic type of magma. Occasionally where the magma is extremely viscous and the pressure from beneath is limited, a volcanic dome may be built over the vent in preference to the more common strato-volcano.

Composite cones possess a simple outward form but an intricate structure. The external slopes are typically either straight or gently concave, with angles between 10° and 35°. Although the main conduit rises vertically through the volcano, subsidiary dikes often feed small lateral cones; it is not uncommon for the top of the main cone to be composed primarily of pyroclastic material while the base contains proportionately more lava extruded from fissures on the lower slopes. Nearly all composite volcanoes have a central crater, but its size may vary considerably. After an eruption has taken place and the top of the conduit has been emptied by a fall in magma level, the unsupported walls of the crater tend to collapse. The morphology depends in large part on the composition of the walls. If the material is mainly pyroclastic, loose screes develop, whereas if it is mainly lava, steep rock faces can be preserved until the next eruption takes place. Many composite cones after a long and complex history finally collapse to form the large depressions known as calderas. These may be 10 km or more in diameter. Famous examples include Mt Mazama, the crest of which disappeared some 6 000 years ago to form Crater Lake in Oregon, Monte Somma which collapsed in A.D. 79 overwhelmthe Roman towns of Pompeii and Herculaneum with ash and mud-flows, and Krakatoa in the Sunda Strait which was destroyed during a violent explosion in 1883.

Although most areas of recent orogenesis include active volcanoes, one notable exception is the chain of mountains running from the Himalayas into south-east Asia. The Himalayas differ from most other mountain ranges in resulting from the direct collision of two continental blocks and it has yet to be established that there are sub-crustal structures directly resembling those of the typical Benioff zone.

Vulcanicity in other regions Although most recent vulcanicity is concentrated along mid-oceanic ridges and destructive plate margins, several other localities have experienced major eruptions during the Mesozoic and Cenozoic eras. One important group of eruptions is that responsible for the outpouring of immense quantities of tholeiitic basalt on to the continental surfaces. Examples covering well over 100 000 km are afforded by the Columbia plateau basalts in the United States, the Deccan traps in India and the Parana basalts in South America. Smaller areas include the Drakensberg region of South Africa and the Antrim plateau in Northern Ireland. The quantities of extruded material are enormous, with the total volume in the Parana basin estimated at some 500 000 km^3. Several of the basaltic accumulations can be related to extrusive activity occurring close to a spreading centre during a period of continental separation. For instance, the Parana basalts are Jurassic in age and may be linked to the initial rupture between South America and Africa; similarly the Karroo lavas in South Africa may be related to an early phase of Gondwanaland fragmentation. A third possible example is afforded by the Antrim plateau basalts and the associated extrusives of western Scotland. Most evidence points to these rocks being no younger than Oligocene and one K/Ar assay has yielded an age as great as 74 Ma. Such ages are entirely consonant with vulcanicity at the time northwestern Europe was actively splitting from Greenland and North America. In each of these three cases the basalt might be construed as an outpouring near a constructive margin that has been carried to its present position by plate movement. The Columbia plateau basalts and Deccan traps, although less readily explained in this way, are generally acknowledged as a manifestation of local tension in the continental crust. The underlying causes of that tension have yet to be determined.

A second group of eruptive centres occurs in association with continental rift systems. The outstanding example is that of East Africa. Here the volcanic rocks tend to be extremely alkaline in composition, some consisting almost entirely of carbonates with little or no silica. The ash from such volcanoes may be so rich in sodium carbonate that after heavy rain it renders local waterholes highly poisonous for both cattle and wildlife. There has been much discussion regarding the origin of 'carbonatites' and most workers feel that they originate from a distinctive magma type. It is interesting that other major rift systems, such as the Rhine valley in Germany and Lake Baikal in Siberia, have experienced highly alkaline vulcanicity, and it seems likely that both structures and igneous activity connote rather unusual conditions within the mantle and lower crust.

A third group of volcanoes not directly related to active plate margins includes those that lie in distinctive linear patterns across the ocean basins. They are particularly characteristic of the Pacific and the best example is provided by the Hawaiian archipelago. This chain of islands and seamounts can be traced some 3 000 km across the central Pacific. At its north-western end it consists of submerged basaltic shields, in the central part of old dissected cones, and at its south-eastern end of active volcanoes. K/Ar dating shows that the age of the extruded lavas generally increases in a north-westerly direction, the seamounts near the extremity of the chain having an estimated age of just over 25 Ma. The oceanic crust on which the archipelago rests is believed to be more than 70 Ma old, and it has been suggested that the whole volcanic chain results from the movement of the Pacific plate over a static hot spot in the mantle. This would explain both the disparity in age between the volcanoes and the underlying crust, and also the progressive change in date of the vulcanicity along the length of the archipelago. The idea has been tentatively extended to other lines of volcanic islands on the Pacific floor, and even to cases of extrusive activity in other regions, but the supporting evidence is not always entirely convincing.

Eustatic changes of sea-level

Mid-oceanic ridges are of sufficient size to provide a mechanism by which eustatic or worldwide changes in sea-level could be induced. The idea of eustatic changes is a recurrent theme in geomorphological studies concerned with the preglacial evolution of uplands in both Europe and North America. However, it is a topic in which definitive evidence is hard to obtain and, until the advent of plate tectonics, it was difficult to envisage on a globe of constant size any process that could significantly change world sea-level other than the growth and decay of ice-sheets. This glacio-eustatic effect is considered in more detail in Chapter 16, and attention here will be confined to the range of sea-level changes that might theoretically be induced by the development of multiple spreading centres.

At the present time the total area of the oceans is 362×10^6 km^2 and the volume of water $1\,350 \times 10^6$ km^3, giving a mean depth of 3 730 m. On a smooth sphere with all the water concentrated in a single uniform ocean, the water would be about 2 440 m deep and its surface measured from the centre of the earth would lie some 240 m above present sea-level. This constitutes a theoretical maximum height for the ocean surface. If more and more of the crust is then visualized as rising above the ocean level, that level must in turn sink. The present hypsographic curve represents one particular state of this relationship, and the fundamental distinction between continents and ocean basins suggests that it is unlikely to alter radically with time. However, sea-level would fall well below its present altitude if the mid-oceanic ridges did not occupy large parts of the ocean basins. The total volume of the ridges has been estimated at 119×10^6 km^3, which is equivalent to a water layer over the oceans about 330 m thick. Given constancy in the size of the globe and the volume of water, it is reasonable to conclude that sea-level has never been more than 240 m above, or more than 330 m below, its present height. It is most unlikely that either of these extremes has ever been approached, let alone reached. The first demands total disappearance of the continents, for which there is no evidence. The second implies an unwarranted destruction of the matter composing the present mid-oceanic ridges. More realistically the latter represent a phase change in the mantle by which material of high density is converted into material of low density; the likely volumetric increase has been calculated at no more than 27×10^6 km³, which is equivalent to a change in sea-level of about 74 m. This value could be reduced still further on the very reasonable assumption that the absence of spreading centres would also involve the disappearance of ocean trenches.

The above figures are very hypothetical and merely demonstrate the capacity of mid-oceanic ridges to affect global sea-level. Whether they have actually done so is still a matter of conjecture. The problem is intimately linked with the periodicity of major tectonic processes in general. It is possible to envisage a steady-state situation in which plates are constantly moving but there is very little change in sea-level. On the other hand, if spreading centres periodically become inactive and the ocean floor subsides, long-term eustatic changes might be expected to occur. In theory the pattern of magnetic anomalies over the ocean floor should provide a guide to changes in spreading rates, but current evidence on this point is inconclusive. As mentioned earlier, several workers have claimed to detect variations associated with individual spreading centres, but the global synchroneity of such variations remains very uncertain. Another method of assessing constancy in plate movement is to examine changes in the rate of orogenic activity at the destructive margins. Several writers have considered the evidence from the cordilleran chain along the Pacific coast of the Americas, but have come to different conclusions. Damon has argued in favour of a marked periodicity, whereas Gilluly has collected K/Ar dates from deep-seated plutons that seem to imply virtually continuous orogenic activity since early Mesozoic times. There has even been disagreement whether orogenic movements are associated with fast or slow plate motion. At first sight rapid motion might seem likely to promote intense orogenic activity; on the other hand, some workers have maintained that the onset of mountain-building at a destructive margin might itself cause a reduction in previous spreading rates.

BISHOP, W. W. and A. F. TRENDALL, 1966–7. 'Erosion surfaces, tectonics and volcanic activity in Uganda', Q. Jl. geol. Soc. Lond., 122, 385–420.

DAMON, P., 1971. 'The relationship between late Cenozoic volcanism and tectonism and orogenic-epeirogenic periodicity' in Late-Cenozoic Glacial Ages (Ed. K. K. Turekian), Yale Univ. Press.

DECKER, R. W., et al., 1971. 'Rifting in Iceland – new geodetic data', Science, N.Y., 173, 530–2.

DOORNKAMP, J. C., 1968. 'The nature, correlation and ages of planation surfaces in southern Uganda', Geogr. Annlr., 50-A, 151–61.

EMERY, K. O., et al., 1970. 'Continental rise off eastern North America', Bull. Am. Ass. Petrol. Geol., 54, 44–108.

FLEMMING, N. C. and D. G. ROBERTS, 1973. 'Tectono-eustatic changes in sea level and seafloor spreading', Nature Lond., 243, 19–22.

GANSSER, A., 1973. 'Facts and theories on the Andes', Jl. geol. Soc. Lond., 129, 93–131.

GILLULY, J., 1973. 'Steady plate motion and episodic orogeny and magmatism', Bull. geol. Soc. Am., 84, 499–514.

MENARD, H. W., 1969. 'Elevation and subsidence of the oceanic crust', Earth Planet. Sci. Lett., 6, 275–84.

MENARD, H. W., 1973. 'Depth anomalies and the bobbing motion of drifting islands', J. geophys. Res., 78, 5128–37.

MITCHELL, A. H. and H. G. READING, 1971. 'Evolution of island arcs', J. Geol., 79, 253–84.

RITTMAN, A., 1962. Volcanoes and Their Activity, Wiley.

The structure and relief of the Red Sea region is thoroughly surveyed by a symposium published in the Phil. Trans. R. Soc., A 267, 1970, pp. 1–415. The literature on East Africa is more scattered but a useful bibliography of part of the region is to be found in B. H. Baker, et al., 'Geology of the eastern rift of Africa', Spec. Pap. geol. Soc. Am., 136, 1972. A basic reference, now rather dated, for the Gulf of California is 'Marine geology of the Gulf of California' (Ed. van Andel and Shor), Am. Ass. Petrol. Geol. Mem. 3, 1964. Orogenic activity within a plate tectonic framework is discussed by J. F. Dewey and J. M. Bird in J. geophys. Res., 75, 1970, pp. 2625–47.

The evolution of continental margins, particularly those around the Atlantic, is discussed at length in Reports 70/13–70/16 of the Inst. Geol. Sci., HMSO, 1970. See also The Geology of Continental Margins (Ed. C. A. Burk and C. L. Drake), Springer-Verlag, 1974. Vulcanicity is treated from an essentially geomorphological rather than a petrological viewpoint in C. Ollier, 'Volcanoes', M.I.T. Press, 1969.

Many different approaches have been employed in attempts at the systematic description of fluvially-eroded landscapes. One of the most influential has been the interpretative approach of the eminent American geomorphologist Davis. He used two major criteria, the first being the geomorphic history of a region, and the second the relationship of surface form to geological structure. Often the two were combined as when he wrote: 'The rivers of eastern England are now in the mature stage of the second cycle of subaerial denudation of a great mass of gently dipping sedimentary rocks, and they have in this second cycle extended the adjustments of streams to structures that were already begun in the first cycle.' There can be no doubt regarding the effectiveness of the Davisian descriptive method as a pedagogic tool. It has fired the imagination of generations of students and has been in no small measure responsible for the popularity of geomorphology as a field of study. Yet this should not be allowed to hide the fact that it is severely limiting as a general approach to landform studies. Moreover, its attractive simplicity has an undeniable blinkering effect on many who have been led to the study of geomorphology through the writings of Davis and his disciples. They find it hard to view landscapes in terms other than those proposed by Davis, a difficulty reinforced by the failure of geomorphologists to offer wholly satisfactory alternatives.

It seems axiomatic that a first essential is some method of describing, in reasonably precise and objective terms, those objects it is proposed to study, namely, the landforms. Yet the three-dimensional geometry of fluvially-eroded landscapes has proved bewilderingly complex. Two different strategies may be distinguished. The first is that which essays a quantitative description of the general relief properties of an area. At the global scale the difficulties that this encounters have already been alluded to on p. 4 and it is an approach that will not be pursued further here. The second strategy has been to focus attention on the basic units, the building blocks, of which a fluvially-eroded landscape is composed. The component unit is the drainage basin and analysis of basin form has proved a remarkably fertile approach to the general problem of quantitative description. An outstanding advantage is that the drainage basin also constitutes the natural unit for studying geomorphic systems within fluvially-eroded terrains. However, any attempt to measure and compare the morphological properties of drainage basins requires some system of classification. The key has been found in the procedure known as stream ordering.

The geometry and energetics of fluvially-eroded terrains

Methods of ordering drainage networks Several early workers experimented with the notion of analysing the organization of drainage networks, hoping in this way to devise a rational classificatory system for both streams and the basins that they drain. However, their efforts were completely overshadowed by the genetic approach to drainage patterns adopted by Davis, and it was not until 1945 that an important alternative method of analysis was conceived. The acknowledged father of most recent network studies is the American engineer, Horton. The basis of his method is the procedure known as stream ordering (Fig. 6.1). This involves assigning each stream a number dependant upon its relative position in the drainage network. For example, all headstreams above the first confluence points are designated first-order streams. Where two first-order streams join, they combine to form a second-order stream;

where two second-order streams join, they combine to form a third-order stream, and so on. A high-order stream may receive a low-order tributary without its designation altering; a change only occurs at the confluence of two streams of like order. In the scheme originally proposed by Horton, the initial ordering was followed by a process of re-designation in which the higher-order streams were traced back to what might be regarded as their sources. Although rules for carrying out this operation were suggested, the whole process seems rather arbitrary and lacking in any real justification. In 1952 Strahler suggested that this second part of the Horton procedure should be abandoned, and since that time most work has employed what is now known as the Strahler system of ordering. More recently a number of weaknesses in this system have been discussed, attention focusing on its failure to take any account of those confluences that involve streams of different order. Shreve has suggested an alternative scheme in which each link

Fig. 6.1. Systems of stream ordering.

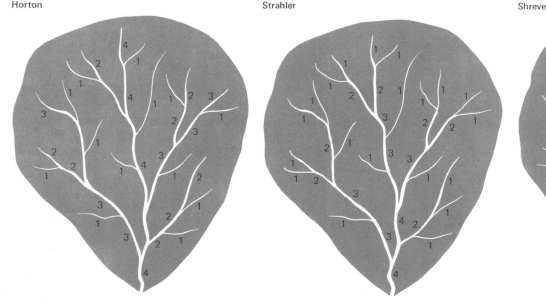

Horton

Strahler

Shreve

of a network is assigned a magnitude equal to the sum of the values for the tributary links immediately upstream. In effect this means that each link has a magnitude equal to the number of undivided headstreams that feed into it. Although the logic of Shreve's approach has much to commend it, the scheme has not yet formed the basis of many morphometric studies.

Once stream ordering is complete, corresponding values may be assigned to the constituent drainage basins. Thus the area drained by a first-order stream is designated a first-order basin, that drained by a second-order stream a second-order basin, and so on. It should be noted that the whole area tributary to a particular stream segment is included in the catchment; a third-order basin therefore includes not only the slopes shedding water directly to the third-order stream, but also the area draining into that stream by way of the first- and second-order tributaries.

The rigour of procedural rules for ordering streams and drainage basins should not be allowed to obscure the very

real practical difficulties that are often encountered. There has been much discussion regarding the best way to map a drainage network. Although the blue lines on ordinary topographic maps have frequently been used, field checking has revealed serious inconsistencies, even on the same map, in the way minor headstreams are depicted. In some instances virtually all stream channels are shown, whereas in others only a selected proportion are portrayed. Consistent mapping of small headwaters is crucial to the proper ordering of a drainage net. In consequence many geomorphologists have prepared their own plans of a drainage system, relying either on field-mapping or on

Fig 6.2. An example of a 'Horton analysis' applied to an individual drainage basin (data from Brush, 1961).

Susquehanna river basin U.S.A. (Horton ordering system)

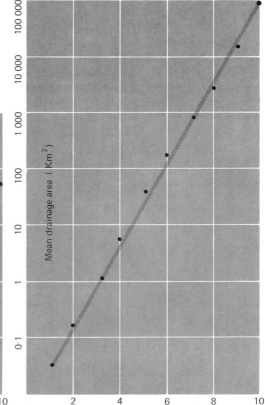
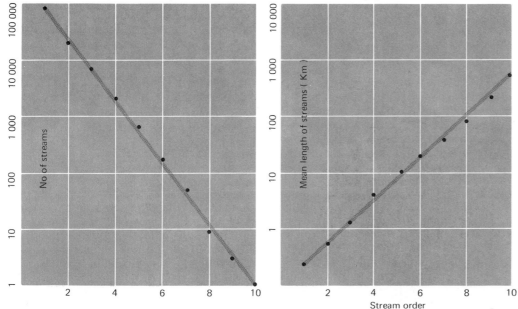

aerial photography. Recourse has occasionally been had to contour crenulations as the best cartographic guide to the positioning of headstreams, but this seems only marginally better than reliance on watercourses shown in blue. Delimitation of drainage basins may be attempted by means of the contours on topographic maps, but for accurate results field survey or careful plotting from air photographs must be undertaken.

Analysis of data Through his introduction of stream ordering Horton equipped the geomorphologist with a logical system for comparing the form of different water-eroded landscapes. The field of inquiry thereby opened up is so immense that it is beyond the scope of a few paragraphs to indicate the full range of possibilities, but it needs little imagination to visualize some of the more significant potentialities. In his original work Horton investigated three relationships which have since come to be known as the component parts of a 'Horton analysis'. The first concerns the number of streams of different orders within a single network. Studies have repeatedly shown that the total number of streams of successively lower orders increases as a simple geometrical progression. This may be demonstrated by plotting stream order on an arithmetic scale against the number of streams on a logarithmic scale (Fig. 6.2). The ratio of the number of streams of one order to that of the next higher order is known as the bifurcation ratio; its value commonly lies between 3 and 5. The second relationship concerns the average length of streams of different orders within a single network. Many studies have revealed that the mean channel length of successively higher orders increases geometrically. Again this is evident from a plot of the relevant data on semi-logarithmic graph paper. The third and final relationship concerns the average size of drainage basins within a single catchment unit. It is found once more that the mean size for successively higher orders increases as a geometrical progression. Following a Horton analysis of two separate regions, it is possible to construct hypothetical drainage nets incorporating the observed values of stream number, channel length and basin area; these show schematically the sharp contrasts that can exist between different drainage systems (Fig. 6.3).

Within the framework offered by an ordering system many other morphological properties can be measured and compared. Of course, operational procedures need to be carefully

Fig. 6.3. Hypothetical fourth-order drainage basins in three contrasting environments (after Selby).

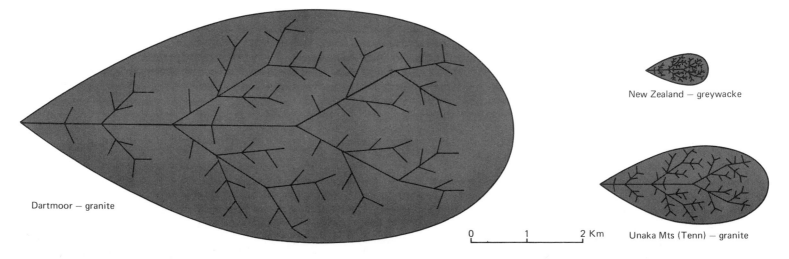

Dartmoor — granite

New Zealand — greywacke

Unaka Mts (Tenn) — granite

0   1   2 Km

specified to ensure true comparability and in many instances there has been prolonged debate about the best techniques to adopt. Reference may first be made to two additional planimetric properties, drainage density and basin shape. Drainage density is normally defined as the total channel length within a basin divided by the basin area. The inverse relationship, known as the constant of channel maintenance, is of interest as a guide to the catchment area required to sustain a unit length of channel. A further derivative is the value known as the distance of overland flow; by convention this is defined as the reciprocal

of twice the drainage density. Basin shape, by contrast, has proved much less readily amenable to simple analysis; its circularity has sometimes been assessed by comparing its area with that of a circle having the same perimeter.

Introduction of the third dimension, altitude, affords many further parameters worthy of investigation. The simplest is relief amplitude, or the difference in elevation between the highest and lowest points within a basin. A second derivative of altitude is stream slope. In many instances this has been shown to vary inversely with stream order; as in the cases discussed

Fig. 6.4. Calculation of the hypsometric integral. The integrals for four contrasting landscapes are shown, although there is a strong tendency for most fluvially dissected areas to yield curves between (b) and (c) with integrals between 40 and 60 per cent.

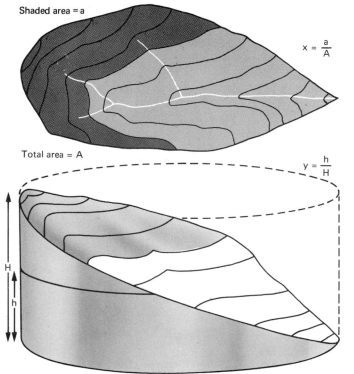

Shaded area = a

$$x = \frac{a}{A}$$

Total area = A

$$y = \frac{h}{H}$$

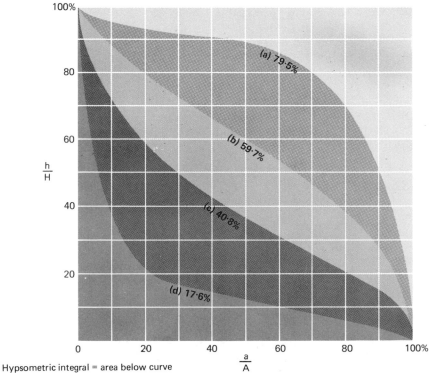

Hypsometric integral = area below curve

(a) Dissected plateau, Maryland
(b) Undulating piemont region, Virginia

(c) Rugged Verdugo Hills, Calif.
(d) Shale lowlands with sandstone residuals, Alabama.

above a law of geometric progression seems to apply. Attempts have been made to investigate the angle of valley-side slopes in the context of basin ordering, but no widely accepted procedure has yet emerged. One technique, however, that has aroused considerable interest is computation of the value known as the hypsometric integral. This involves measuring the area above successively lower contours within a drainage basin. Instead of dealing with absolute figures, both heights and areas are then converted into proportional values of the total basin relief and area (Fig. 6.4). When plotted on a graph these values yield a curve known as the hypsometric curve. The complete area of the graph may be visualized as the gross volume of rock in the catchment before erosion began; the area beneath the curve, known as the hypsometric integral, represents the volume of rock that still remains to be removed. Both rock-type and geomorphic history seem likely to affect the integral; the computation tends to be tedious, but it is one of many values that may appropriately be calculated in comparing drainage-basins.

Energetics of fluvial denudation

Transformation of the warped and dislocated continental surfaces into the intricate geometrical shapes described in the first part of the present chapter is the work of what has aptly been termed the geomorphology machine. Its primary driving force is gravity. Both the rocks raised to form the mountain chains and the water evaporated from the oceans into the atmosphere possess potential energy by virtue of the gravitational attraction of the earth. The energy needed to elevate these materials comes from two sources. In the first instance it is internal terrestrial heat, in the second solar radiation. It is the latter which is by far the more important in the fashioning of fluvially-eroded landscapes; a major goal of geomorphology, it has wisely been claimed, is to understand how the energy of solar radiation is converted into the mechanical work responsible for moulding the landforms.

Solar radiation The radiant energy emitted by the sun is enormous. At a distance of 149 million km, equivalent to the mean radius of the earth's orbit, the energy falling on a surface normal to the sun's rays is just under 2 calories cm^{-2} min^{-1}; this means that every minute over 12×10^{17} calories of radiant energy are intercepted by the earth. However, of this total only a small fraction actually penetrates to the surface of the globe to become potentially available for geomorphic work. The Russian climatologist Budyko has compiled a map (Fig. 6.5A) depicting the total solar radiation received at the earth's surface in a year. The distribution pattern reflects the interaction of two major factors, the geometry of the earth's orbit round the sun and the shielding effect of the atmosphere. Although of interest as a statement of the energy directly available at the earth's surface, Budyko's map has important limitations. Studies of the global heat budget show that in latitudes below about 38°N, and S. there is an excess of incoming solar radiation over outgoing terrestrial radiation, whereas in higher latitudes the relationship is reversed. To compensate this imbalance which would otherwise lead to a steady heating up of the tropics and cooling down of the polar regions, there must be massive poleward transfers of heat energy. This is achieved by the movement of relatively warm air and water from low to high latitudes, and is a fundamental feature of the atmospheric and oceanic circulation of the planet.

The solar energy penetrating to the earth's surface is responsible for a multitude of complex and interrelated processes. For example, part is used in sustaining the biosphere, and it has been estimated that each year photosynthesis by the continental vegetation cover transforms a total of $1 \cdot 6 \times 10^{20}$ calories from radiant into chemical energy. A further part is absorbed by the ground surface with resultant heating and enhanced rates of chemical and biochemical change. A yet further part falling on wet surfaces is responsible for the evaporation of water moisture into the atmosphere. The proportion of the incoming energy devoted to the different processes obviously varies from place to place and also from time to time. In a desert area, for instance, surface heating normally predominates, whereas in a forested region the vegetation cover shades the soil from the direct effect of the radiant energy. Seasonal variations are well illustrated by deciduous forests which in winter not only consume less energy in photosynthesis and transpiration but also provide a less effective shield for the surface of the ground. Great contrasts are also found in the amount of energy devoted to evaporation. The latent heat required to evaporate 1 g of water ranges from 540 calories at 100°C to 600 calories at 0°C, so that a mean annual precipitation over the globe of 857 mm demands on average about 5×10^4 cal cm^{-2} yr^{-1} or some 30 per cent of available energy. Budyko has compiled a map (Fig. 6.5B) portraying the regional variations in the proportion of energy expended in evaporation. As might be expected, the highest percentages are found over the oceans with values commonly exceeding 50 per cent, the lowest over the deserts with values normally below 15 per cent.

The hydrological cycle The energy expended in evaporation

may be viewed in another way, that is, as the primary source of energy for the hydrological cycle. In essence the cycle is a transfer of water between four major reservoirs, the ocean basins, the atmosphere, the continental surfaces and underground storage. Estimates of the water contained in each reservoir vary because of the difficulty in making accurate global measurements, but the amounts quoted in Fig. 6.6 are believed accurate ±10 per cent. They reveal that over 97 per cent of all the water is contained in the oceans, over 2 per cent on the land surfaces, about 0·6 per cent underground, and less than 0·001 per cent in the atmosphere; of the water on the continental surfaces 99 per cent is held in the form of glacier ice, less than 1 per cent constitutes the lakes and rivers, and a minute fraction is

Fig. 6.5. Solar energy at the earth's surface: (A) Total energy received; (B) Amount of energy expended in evaporation (after Budyko, 1958).

A

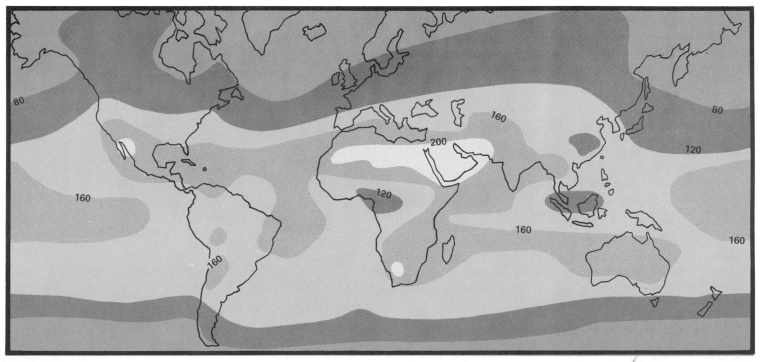

102 stored in the biosphere. Recent estimates of the rate of transfer between the various reservoirs are also incorporated in Fig. 6.6. Combined annual evaporation from the continents and oceans at 423×10^3 km³ is sufficient to replenish the water in the atmosphere once every 11 days. By contrast it would require almost 3 centuries for precipitation to replenish the continental reservoir, the main reason being the very large volumes of ice stored in the major ice-sheets; the water in the lakes and rivers could be replaced in little over 2 years. However, by far the longest residence period for water is in the oceans since it would require almost 4 000 years for evaporation at the current rate to completely dry up the ocean basins.

Defined as weight times head, potential energy is generated on a tremendous scale by the hydrological cycle. The runoff from the continents is 37×10^3 km³ with a total weight of 37×10^{15} kg. If it is assumed that this enormous body of water has descended from the mean continental height of 870 m, the potential energy amounts to over 35×10^{18} kgf m.* It is essential to appreciate, however, that a very high proportion of this energy, once converted into kinetic form as the water moves downhill, is dissipated as frictional heat and so becomes unavailable for the work of erosion. The above figure may also be compared with that for the rock materials elevated above sea-level by tectonic processes. Conversion of this second source of

---

* 34×10^{19} J. See Appendix A.

Fig 6.5 continued

B

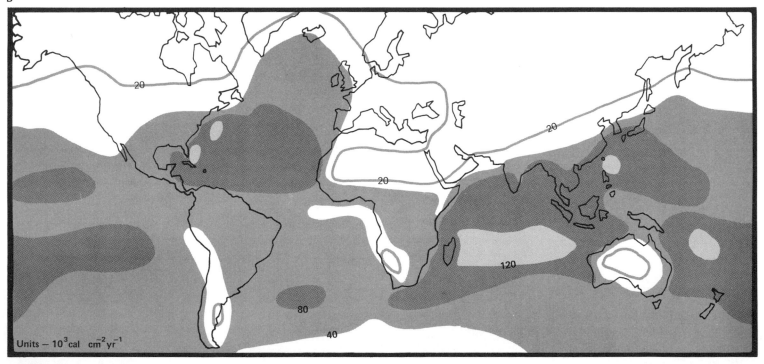

Units – 10^3 cal cm^{-2} yr^{-1}

potential energy into kinetic form is controlled by the rapidity of weathering and disaggregation of material at the earth's surface. It will be shown in Chapter 10 that in areas of moderate relief lowering of the land surface appears to be proceeding at a rate of approximately 0·05 mm yr^{-1}. The total weight of sediment released in this way each year is about 20×10^{12} kg; its potential energy, assuming it to descend from the mean continental height, is a little under 18×10^{15} kgf m, or less than one-thousandth part of the energy available from the hydrological cycle.

The drainage basin as a unit of study The preceding paragraphs have examined some of the energy relationships on a global scale. Although basic to an understanding of the erosional system fashioning the earth's surface, these remain remote from everyday experience. Moreover the variable ways in which solar radiation is converted into other forms of energy means that global averages, even if they could be computed accurately, would have little practical significance for purely local studies. In consequence research has tended to concentrate on small catchments where the geomorphic system can be examined in much closer detail. Such 'catchment studies', as they have come to be known, permit the investigator to select drainage basins having relatively uniform characteristics, thereby reducing the variance of the data to be handled. For example, he may choose as a unit of study a basin underlain by rocks or just one lithology and supporting a homogeneous vegetation cover. Alternatively for purposes of comparison two basins may be chosen which differ radically in only one respect. Thus contiguous catchments having similar geological but contrasting vegetational characteristics might be selected. A third common procedure is to study the temporal changes that occur when one of the controls is radically altered, say, by deforestation.

A great advantage of the drainage basin as a unit of study is

Fig. 6.6 Schematic representation of the hydrological cycle on global scale.

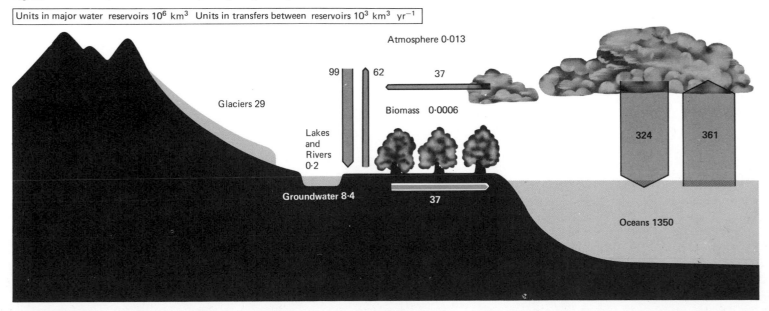

Units in major water reservoirs 10^6 km^3 Units in transfers between reservoirs 10^3 km^3 yr^{-1}

Atmosphere 0·013

Glaciers 29

99   62   37

Biomass 0·0006

Lakes and Rivers 0·2

324   361

Groundwater 8·4   37

Oceans 1350

that it can be analysed as an open system, that is, a system in which there is a constant input and output of both mass and energy. The rest of this chapter is devoted to the techniques by which the two principal inputs and outputs, namely water and rock waste, can be measured and their interrelations thereby elucidated.

The hydrology of individual catchments

A basic requirement in understanding water input and output is the ability to put figures to the hydrological equation:

Precipitation = Runoff + Evapotranspiration ± Storage.

Each of the four components must be separately assessed in order to compile a complete water budget.

Precipitation Normally most of the precipitation input can be measured by means of rain gauges, but where such forms as snow, dew or frost constitute a significant proportion of the total other methods of assessment may be necessary. Two types of rain gauge are in regular use, the non-recording which is usually read by an observer at 24-hour intervals, and the recording which automatically provides a continuous trace on a rotating chart. Recording gauges have the great advantage that they give an indication of rainfall intensity which, for the geomorphologist, may be at least as significant as the rainfall total. A further facility is the easy conversion of the instruments to telemetry and computer tape recording so that even in remote areas data from an array of instruments can be rapidly received and analysed.

However, advances in the electronics used should not be allowed to obscure inherent limitations in accuracy. Precipitation measurement is fundamentally a sampling procedure and two sources of uncertainty are immediately identifiable. The first is instrumental and the second statistical. One of the problems associated with a rain gauge is that it creates a pattern of turbulent air flow which deflects a variable amount of the precipitation away from the orifice. It is difficult to apply any correction factor because the deficiency varies with wind speed and size of raindrop; experimental work suggests that it commonly amounts to 5 per cent and in adverse conditions may exceed 15 per cent. Different designs of shield have been tried,

and gauges have been sunk in pits so that their orifices are level with the ground. The problem in the latter case is to prevent 'insplash' from the surrounding surface but slats styled so as to divert the excess water away from the gauge have proved effective in a number of experimental studies. Where snow constitutes more than, say, 5 per cent of the precipitation it is essential for accurate results to include some form of snow-measuring device. Instruments employed include modified rain gauges where heating elements or antifreeze mixtures are used to melt the snow and special weighing gauges which record the weight of precipitation falling in a pan. In areas of regular heavy snowfall sampling with snow corers may be substituted, but whichever method is adopted, a high proportion of snow undoubtedly reduces the reliability of measured precipitation totals.

The statistical uncertainty is not easily assessed. Various procedures have been adopted for calculating, from the point samples, total precipitation input. They range from the manual drawing of isohyets to geometrical constructions so designed that the value for each gauge is assigned to an area determined solely by the proximity of other gauges. The latter technique is normally used where the gauge network is unevenly distributed and little supplementary information is available to guide the drawing of isohyets. Gauge density can be of critical importance in making accurate overall assessments, especially where a high proportion of the rainfall comes from localized storm centres. The advent of the computer has permitted the use of more elaborate statistical techniques, particularly where data is abundant. It is then possible, for instance, to compute regional trends and by regression to assess the influence of such local factors as altitude, aspect and exposure. This ultimately allows prediction of the precipitation values to be expected at all points within the study area and the drawing of isohyets by computer. Nevertheless, the intrinsic unreliability of the raw data needs to be remembered, and it is doubtful whether total precipitation even in a small catchment can be estimated more accurately than ±5 per cent of the true value.

Run-off Measurement of stream flow is a relatively simple procedure, and with reasonable precautions accuracy should equal or exceed that normally attainable in the assessment of precipitation. Two basic methods are in common use, one relying upon the natural channel and the other involving installation of a short section of artificial channel. Gauging procedures in a natural channel may be divided into three phases. There is firstly measurement of water-level, secondly determination of the corresponding discharge, and thirdly definition of the relationship between the two.

The water-level or stage of a river may be measured either by periodic observation with a manual gauge or by a continuous recording instrument. Various forms of manual gauge have been devised but the simplest is the vertical calibrated staff either set into the bed of the river or attached to a fixture such as a bridge pier. The value of the gauge is obviously dependent upon the frequency with which it is read. The interval between readings is commonly 24 hours, and although adequate for a large river of relatively stable flow, this is quite unsuitable for a small stream with a very 'flashy' regime. If the interest of the investigator lies in flood discharges it may be possible to supplement the regular readings by installation of a peak gauge. This simple device records the maximum water level reached between successive readings. It consists of a tube set vertically into the stream and either coated internally with a water-sensitive dye or containing some form of float which remains at its maximum height even when the water-level falls. The principle is akin to that of estimating a flood-level from the 'tide mark' on buildings or from driftwood trapped in bushes along a river bank. These latter methods are often employed where an exceptionally high flood exceeds the capacity of the installed gauges, or where a drainage basin lacks any instruments at all.

The severe limitations of manual gauges require that, for detailed studies, some form of recording gauge be installed. The commonest type is that in which a float, free to move vertically with changes in water-level, is coupled to a pen that plots the variations on a rotating drum. In order to protect the float and damp the turbulent water movement, a stilling well has to be built and connected to the river by a submerged pipe. The resulting continuous trace of river stage indicates not only the levels of peak and low flow but also the rates of change during the passage of a flood. A second type of continuous recorder is actuated by pressure changes. As the water-level rises or falls, alterations in pressure at the stream bed are measured by the compression of air in a vessel fitted with a flexible diaphragm. A tube transmits the pressure variations to a sensitive aneroid which in turn is coupled to a recording pen. Both this and the float recorder can be readily adapted for telemetry and direct recording on computer tapes, with the advantage that in remote areas the instruments may be left unattended for prolonged periods.

Instantaneous discharge is usually measured by surveying the cross-sectional area of a stream and then determining the velocity of flow perpendicular to that section. The velocity can be measured by means of a current meter fitted with either a propeller or rotating cups like the familiar anemometer. The rate of rotation of propeller or cups is proportional to the speed of water flow and is normally computed by means of an electrically-operated counting device. The greatest problem is the variation in velocity at different points in the cross-section since it is well known that the flow close to the banks and bed is retarded by friction leaving a central thread of faster moving water (Fig. 9.4, p. 167). The standard procedure to overcome this difficulty is to partition the cross-section into ten or more vertical segments depending on the width of the stream. Each segment is then treated as a separate unit within which the water velocity is measured at depths corresponding to 0·2 and 0·8 of the total depth. The mean of these two readings has been found to correspond closely to the mean for the whole segment. More speedy but slightly less reliable results can be obtained by one measurement at a depth corresponding to 0·6 of total depth. The discharge of each segment is then calculated by multiplying its area by its mean velocity, and the total for the whole stream obtained by simple summation.

An alternative means of assessing instantaneous discharge is known as the salt dilution method. It involves feeding into a stream a measured volume of a suitable salt solution for a period of about an hour. Owing to turbulence the solution will be thoroughly mixed with the stream water and by measuring the salt concentration in samples taken a short distance downstream, the discharge may be calculated. The salt dilution technique is particularly useful on small mountain streams where other methods are difficult to apply.

The instantaneous discharge at a gauging station must be measured at a variety of river stages in order to define its relationship to water-level. When measurements have been completed at a number of different stages, the results may be plotted on a graph (Fig. 6.7). A smooth curve drawn through the plotted points is known as a rating curve and may be used in conjunction with a continuous trace of river-level to provide a complete record of discharge. Certain precautions need to be observed in the construction and use of rating curves. The site for gauging should be selected so as to consist of a straight length of channel with a symmetrical cross-section. The banks and bed should preferably be smooth, and must in any case be stable; it is obviously impossible to use a rating curve where a

channel constantly changes it cross-sectional shape, and if necessary the banks and bed must be artificially stabilized.

An alternative procedure avoiding dependence on a suitable reach of natural channel is to install either a weir or a flume (Fig. 6.8). Often impracticable in the case of a large river, these are frequently employed on small streams. They have the advantage that both types of installation can be so constructed and calibrated that discharge is directly calculable from the change in head, that is, the height of the water surface above the crest of the weir or the throat of the flume. It is this head that is normally measured by float recorder. On large rivers measurements of discharge are regularly made as an adjunct of engineering structures built for other purposes. For instance, navigation works commonly involve the construction of locks and weirs

Fig 6.7. Examples of two types of rating curve: (A) the stage-discharge relationship for the Willamette River, Oreg., USA: (B) The sediment-water discharge relationship for the Rio Puerco, New Mex., USA.

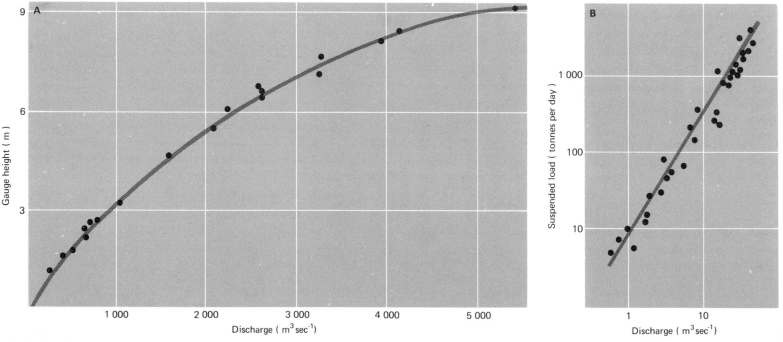

Fig 6.8. A typical premanent gauging station on a small mountain stream. The stilling well and equipment for recording the water level are housed in the structure on the left.

which can be used to provide streamflow records, while the spillways of dams and the turbines of generating plants can both be calibrated to yield data about discharge.

Evapotranspiration Evapotranspiration presents such formidable problems of measurement that it has usually been estimated indirectly. If the hydrological equation is applied to a sufficiently long time-interval, the change in storage becomes proportionately so small that the equation may justifiably be rewritten in the form:

$$\text{Evapotranspiration} = \text{Precipitation} - \text{Runoff}.$$

Evapotranspiration can then be assessed by means of the techniques already described. However, for full understanding of water movement within a catchment direct measurement is essential. The very name evapotranspiration indicates part of the problem, for it is derived from two separate terms – evaporation and transpiration. The former operates alone over large water bodies, but is supplemented over the land surfaces by the transpiration from plant stomata of moisture originally taken in through the roots.

Evaporation from water surfaces is usually assessed by means of pans in which either the fall in water-level is observed or the amount of water required to maintain a constant level is measured. One difficulty is that evaporation from a small pan tends to exceed that from a large water body, partly because of heat exchange with the atmosphere through the pan walls and partly because of the so-called 'oasis effect' in which there is a constant supply of relatively dry air to maintain a high evaporation rate. Because of these limitations the various designs of instrument are assigned a 'pan coefficient' by which the observed values need to be multiplied in order to derive truer values of natural evaporation.

An additional problem when considering land surfaces is the availability of water. Clearly neither evaporation nor transpiration will be the same under natural conditions as it is where water is constantly replenished by artificial means. This difference has led to formulation of the concepts of potential .evaporation and potential transpiration, both much used in climatological and hydrological studies. However, for present purposes it is better to concentrate on the measurement of actual values. Evaporation from an unvegetated soil surface may be assessed by a percolation gauge consisting of a metal cylinder some 40 cm in diameter filled with an undisturbed column of soil. The cylinder is then set in the ground above a receptacle to catch all the downward percolating water. The difference between the measured precipitation and the volume of water caught in the receptacle is the amount lost by evaporation. Such an instrument gives reasonable results over a long time-interval, but for day-to-day changes variations in soil-water storage must be assessed. This is most readily achieved through a mechanism for continually weighing the soil column, as is often done in the more elaborate installations designed to measure the combined effects of evaporation and transpiration. Termed lysimeters, these instruments consist of a much larger soil monolith set in a tank and supporting a vegetation cover. Owing to their greater size they are very expensive to build and maintain, and are not generally suited for use in small catchment studies. Nevertheless, they are of considerable scientific interest since they are one of the few sources of data about a vital phase in the hydrological cycle. The most famous lysimeters are those at Coshocton in Ohio where the tanks containing the monoliths are about 2·5 m deep and have a surface area of 9 m². Despite loads of the order of 50×10^3 kg the weighing mechanism is sensitive to the nearest 3 kg or the equivalent of about 0·3 mm of precipitation. Hydraulic weighing methods in special floating lysimeters have recently increased the sensitivity to the equivalent of $\pm 0·03$ mm of precipitation.

Although the number of lysimeters in operation throughout the world has steadily increased they remain relatively rare instruments, and the shortage of data means that it is still difficult to appraise with any confidence the effects of climate, soil and vegetation type on actual evapotranspiration. This continues to be a serious weakness in attempts to compile full catchment budgets.

Storage Water may follow many different routes in its passage from the point at which it falls as precipitation to its appearance as channelled flow at a stream gauging site. Some may flow over the ground surface, some may pass downslope at shallow depth along various horizons within the soil, and some may sink downwards to become part of the groundwater filling the voids in the zone of saturation. Whereas rain falling on an interfluve may reach a stream channel in a matter of minutes by surface flow, it commonly takes weeks, months or even years to reach the same point by an underground route. Studies by means of radioactive isotopes have shown that some groundwater now in storage was precipitated several thousand years ago. Of course this is not a direct measure of the response time between heavy precipitation and increased flow from springs and seepage lines. Accession of more groundwater tends to steepen the upper surface of the zone of saturation and so to produce a greater head. According to the law formulated by Darcy over 100 years ago but still forming the cornerstone of modern groundwater studies, the velocity of flow through a rock of given permeability is proportional to the head; consequently, addition of water from renewed precipitation soon increases outflow at the point of issue to the ground surface. Conversely, as the zone of saturation subsides during a dry spell, the rate of groundwater discharge tends to decline exponentially. The overall effect of water sinking underground is to reduce the amplitude of the fluctuations in stream discharge.

By convention, storage changes are often deemed to be restricted to variations in groundwater volume. This is because so many attempts to strike a balance in the water budget are concerned with a period ranging from a few months to a year or so, and at this time-scale only groundwater changes are of any real significance. Measurements are normally made by observing fluctuations in the height of the water table. The level of the water surface in suitable wells and boreholes may be monitored either by periodic manual inspection or by the installation of continuous recording gauges. In temperate latitudes these generally reveal a prominent seasonal rhythm, with high winter levels followed by a pronounced decline during the spring and summer. Superimposed on this annual cycle are shorter term changes reflecting periods of wet or dry weather. In addition, where the water table is sufficiently near the surface to be tapped by tree roots, there is often a diurnal rhythm with a falling level during the daylight hours of active transpiration, and a recovery during the night when abstraction declines below the rate of recharge. The irregular spacing of observation points may make volumetric calculations extremely difficult, particularly in areas of complex geological structure, but in view of the crucial importance of water resources in many parts of the world much attention is currently being devoted to this problem.

If attempts are made to balance a water-budget for much shorter periods, additional storage sites need to be taken into account. These include water courses, sites of surface detention, and the soil. The term detention refers to entrapment of precipitation that is either intercepted before reaching ground-level or held temporarily in the numerous shallow depressions that characterize almost any natural surface. Interception is mainly due to vegetation, the effectiveness of which is well recognized in the common practice of sheltering from a storm beneath a tree. The proportion intercepted varies enormously depending upon the nature of both the precipitation and the vegetation. In showery summer weather it may be as high as 90 per cent, and even with heavier storm rainfall values up to 40 per cent have been recorded beneath a Douglas fir cover. These figures underline the desirability of siting gauges so as to measure not only total precipitation but also effective ground-level precipitation. For budgetary purposes the term 'soil water' should include moisture held both in the soil sensu stricto and in the unsaturated rocks above the water table. In practice, however, most measurements are concentrated in the soil layer owing to the difficulty of making observations at greater depth. The simplest way of measuring moisture content is to weigh soil samples before and after drying in a laboratory oven, the difference between the two readings being an indication of water content normally expressed as a percentage of the dry weight of the sample; this however is not a very practical approach to the problem of continuous monitoring. Several methods have been devised that attempt to measure soil moisture content in situ. Various types of porous block have been inserted into the soil, the principle being that as soil moisture fluctuates so will the water content and electrical conductivity of the blocks; this latter property can be measured without removal of the blocks. A

more recent technique involves use of a neutron probe. This instrument consists of two parts, a radioactive source of fast neutrons and a slow neutron detector. When the probe is lowered into the ground the fast neutrons are scattered by the hydrogen in the soil and the resulting slow neutrons are counted by the detector. Since almost all the hydrogen atoms in the soil are in the water, the number of slow neutrons is a measure of the moisture content within about 25 cm of the probe. The advantage of the method is its speed of operation. Once a series of holes has been drilled and lined with aluminium tubing, the actual process of lowering the probe to the required depth and taking a reading can be completed in a matter of minutes. A single observer can monitor a large number of holes at frequent intervals, a facility of great potential value in the study of weathering and mass movement on hill slopes.

The sediment yield of individual catchments

The space devoted above to the hydrological cycle must not be allowed to give the impression that study of the cycle is the primary aim of the geomorphologist. He views a knowledge of the movement of water through the drainage basin as a necessary preliminary to understanding the processes shaping the basin. It is the production and movement of rock debris that constitutes the real core of the geomorphology of water-eroded landscapes. These topics are discussed in detail in Chapters 7–9; here consideration will be restricted to ways of measuring the total sediment yield emanating from a catchment.

The material in transit along a stream channel may be divided into three parts: the suspension load, solution load and bed load. The contribution to the total represented by each of these parts varies greatly, and each poses special problems of assessment.

Suspension load Material is carried in suspension by the turbulent motion of flowing water in which the tendency of the particles to settle is offset by the residual upward momentum of the transporting medium. The water close to the river bed will normally be more heavily laden than that near the surface, so that samples taken from different parts of a stream cross-section will vary in the amount of sediment they contain. Since the water velocity also varies with depth, calculation of the total suspension load passing through a particular cross-section must take account of both variables. In early investigations this meant that many hundreds of samples had to be collected and analysed in order to derive reasonably accurate results. Moreover, the whole procedure needed to be repeated at many different river stages if the amount of suspended material being transported over a period of time was to be correctly assessed.

Design of the instrument known as a depth-integrating sampler came as a vital technical innovation. The equipment consists essentially of a metal body shaped so as to direct the intake end upstream, a projecting nozzle to permit ingress of the water, an internal sampling bottle, and a duct to allow escape of the air as the bottle fills. An electrically controlled valve permits remote opening and closing of the intake and also equalizes pressure in the sample bottle so as to obviate an initial surge. The operating principle is simply that the amount of water entering the nozzle is proportional to the velocity of flow. This means that if the sampler is lowered at a constant rate from the surface to the bed of the stream, each depth in that particular vertical section is represented in the sample by a proportion dependent solely on its relative velocity. In other words, the sample is automatically velocity-depth integrated, and once its sediment concentration has been measured the resulting value need only be multiplied by the appropriate discharge to obtain the total sediment load passing through the vertical section in question. The depth-integrating sampler may conveniently be used in each of the vertical segments used for water-gauging purposes, as outlined on p. 105, and the total suspension load calculated by summation. An alternative procedure, known as equal-transit-rate sampling, is to employ exactly the same rate of lowering and raising the instrument at all verticals; the composite sample then requires no more than a single analysis to yield a sediment concentration figure which can be multiplied by stream discharge to obtain the total suspension load.

The depth-integrating sampler has greatly simplified the work needed to measure the suspension load passing at any one instant through a particular cross-section. There remains, however, the problem of deriving a continuous record to take account of the variations with time. One possible approach is to construct a rating curve that directly relates suspension load to dis-charge. When observed values for certain streams are plotted on logarithmic scales a good correlation is found (Fig. 6.7B), but for others a much wider scatter of points is apparent. This is because the suspension load is not simply a function of dis-charge but is affected by a wide range of other factors. Some of these are seasonal so that the amount of material carried at a specified discharge may differ in summer and winter; others are related to the character of the precipitation, particularly its intensity and duration. Deviation from a consistent suspension load-discharge relationship is well illustrated by the San Juan River (Fig. 6.9) where it has been shown that the amount of material carried on the rising leg of a flood may be ten times that transported by the same discharge on the subsiding leg. Other rivers show the reverse relationship with the peak sediment concentration occurring as the flood-waters recede. In the face of such fluctuations great care must be exercized in the use of sediment-discharge rating curves; for accurate results it is often necessary to pursue a sustained programme of regular sediment sampling.

Attempts are increasingly being made to devise mathematical models that would permit prediction of the total suspension load from a very limited number of point samples; the great prize of this approach would be the ability to rely on data from a few locations at which the sediment concentration was continuously monitored. Techniques such as measuring the scattering of light from a laser beam by the concentration of suspended particles are potential methods of obtaining continuous records. At the present time, however, the mathematical models do not yield total suspension load values very close to the observed values and considerable further refinement appears to be necessary.

Solution load The variations with depth that characterize suspension load do not apply to most dissolved material. The turbulent flow tends to ensure relatively complete mixing and for many purposes a 'bottle on a string' will provide a representative sample. Nevertheless, for a complete chemical analysis, and particularly where the stream is slow-flowing with deep quiet pools, it is preferable to collect a depth-integrated sample. A complete chemical analysis covers a very wide range of

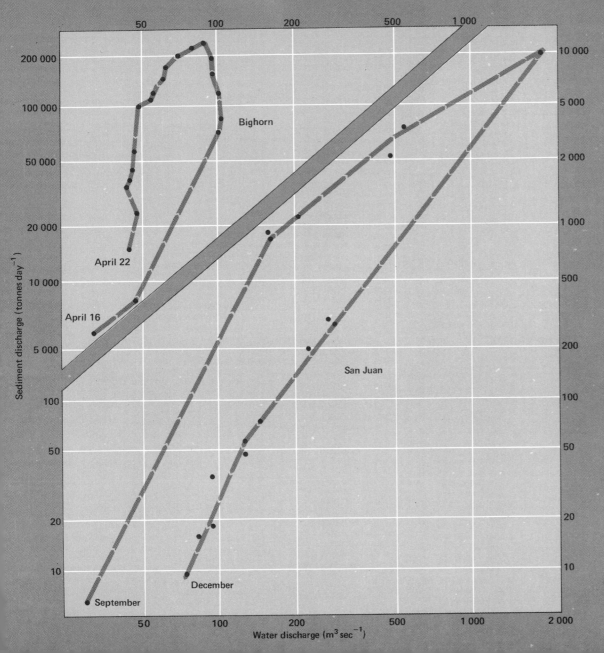

Fig 6.9 Short-term variations in the suspended sediment concentration on the Bighorn and San Juan rivers in the USA. For the same water discharge the 4-month record of the San Juan illustrates a substantially greater sediment load on the rising stage than on the falling stage; the week-long record of the Bighorn illustrates the reverse relationship (after Guy, 1964).

112 dissolved solids and gases, and it is not appropriate to describe here all the complex analytical techniques that may be required. From the geomorphological viewpoint, most interest focuses on those dissolved solids which result from solutional weathering and from which the rate of chemical denudation may be estimated. The simplest and most widely studied example is the calcium carbonate content of streams draining limestone catchments. Approximate determinations of calcium hardness may be made in the field by adding specially prepared tablets to a water sample and observing the ensuing colour changes, but laboratory titration is normally used for the more precise determinations required in a research programme. The results of each analysis are expressed in parts per million but can easily be converted into corresponding values of limestone dissolved. However, in order to ascertain the total amount of limestone being removed from a drainage basin per unit of time it is obviously necessary to monitor both stream discharge and the changing calcium

hardness of the water. This is the problem already encountered in studying the suspension load and again constitutes a major source of error unless frequent sampling is undertaken. It is found that the concentration of specific dissolved solids varies not only with river stage (Fig. 6.10) but also with season owing to the strong biological influence on many chemical processes.

Bed load This third component of the total load, sometimes also called the traction load, is the most difficult to measure. It may be defined as that part of the sediment forming the moving bed of the stream. Displacement of individual particles is by rolling, sliding or occasional saltation into the basal layers of the water. As with many other geomorphological investigations, the difficulty is to make measurements without disturbing the very processes one is attempting to measure. An instrument that has been widely tried is the basket sampler anchored to the bed of a stream. This consists of a box made of mesh screening but open

Fig. 6.10. Two illustrations of the way the solution load of a stream varies with discharge. Note the tendency for the solute concentration to fall with rising discharge although there may sometimes be an initial 'flush' as accumulated salts are washed out of the catchment. Individual basins can show considerable differences in their response to flood events (after Walling, 1975).

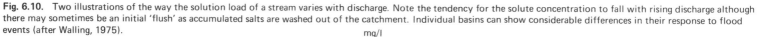

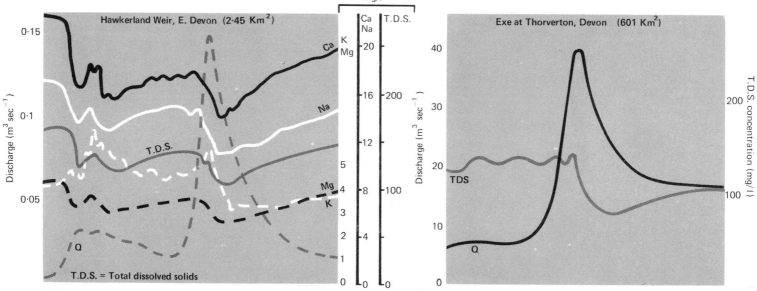

at its upstream end. If the mesh is too large part of the load passes through, whereas if it is too small it is easily clogged and accentuates the intrinsic problem of any basket sampler, the diversion of flow away from the inlet. One way to counter this latter difficulty is to design the sampler so that the walls at the entrance flare outwards in a downstream direction. This creates a pressure drop which can be adjusted to ensure that the inlet velocity approximates that of the surrounding stream. A second instrument that has occasionally been tried is the pit sampler. This involves digging elongated pits across the floor of a stream and installing receptacles which can be removed at intervals for measuring the amount of debris trapped. Although more efficient than the use of basket samplers, this technique demands elaborate preparation and also causes eddying which disturbs the natural flow of the water.

The difficulties in making short-term measurements at a single point in a cross-section are obviously compounded when it comes to assessing total bed-load movement over a relatively long time-interval. Careful work with basket samplers suggests that even at constant discharge the bed load moves in surges with an interval of a few minutes between each peak. As a result of the difficulties in obtaining regular field measurements, recourse has been had to developing mathematical formulae to describe bed-load transport. These must rely on readily measured parameters such as velocity, depth, slope, width and particle size of the bed material. Numerous such formulae have been proposed, often predicting quite different transport rates, but none has proved sufficiently satisfactory to command really widespread support.

When values for total sediment load are quoted, the limitations indicated above need to be borne in mind. The proportions being carried as suspension and traction load appear to vary considerably, although a figure of 5–10 per cent as traction load is often regarded as characteristic. On the other hand, some measurements suggest that it may occasionally exceed 50 per cent where most of the debris supply is of a coarse grade. It should also be remembered that suspension-load samplers do not normally collect material closer to the bed than about 0·15 m. This leaves an important unsampled fraction of the suspension load which may need to be estimated by some form of transportation formula. Increasingly the problem of measuring material in transit at, and close to, the bed of a stream is being recognized by a realistic two-fold division of the detrital load into measured and unmeasured portions, the latter simply being estimated by appropriate formulae.

References

Selected bibliography

BUDYKO, M. I., 1958. *The Heat Balance Of The Earth's surface* (trans. N. Stepanova), US Dept. of Commerce.

GUY, H. P., 1964. 'An analysis of some storm-period variables affecting stream-sediment transport', *Prof. Pap. U.S. geol. Surv.* 462-E.

HORTON, R. E., 1945. 'Erosional development of streams and their drainage basins : hydrophysical approach to quantitative morphology', *Bull. geol. Soc. Am.* 56, 275–370.

SELBY, M. J., 1967. 'Aspects of the geomorphology of the greywacke ranges bordering the lower and middle Waikato basins', *Earth Sci. J.*, 1, 1–22.

SHREVE, R. L., 1967. 'Infinite topologically random channel networks', *J. Geol.* 75, 178–86.

STRAHLER, A. N., 1952. 'Hypsometric analysis of erosional topography', *Bull. geol. Soc. Am.*, 63, 923–38.

WALLING, D. E., 1975. 'Solute variation in small catchment streams : some comments', *Trans. Inst. Br. Geogr.*, 64, 141–7.

A valuable review with extensive bibliography of morphometric techniques is to be found in V. Gardiner, 'Drainage basin morphometry', *Tech. Bull. No. 14*, Brit. Geomorph. Res. Group, 1974. The dynamics of catchment basins are discussed in detail in K. J. Gregory and D. E. Walling, *Drainage Basin Form and Process* Arnold, 1973.

Weathering may be defined as the response of rock materials which were once in equilibrium with conditions in the lithoshere to the changed conditions at the earth's surface. It is in essence a form of metamorphism since the rocks are adjusting to a change in temperature and pressure in the presence of circulating liquids and gases. Ultimately a new equilibrium may be established with little chemical or physical alteration taking place in the weathering products. There are, however, many complex weathering cycles through which material may pass before such a state is achieved, and in the meantime the incompletely altered material has usually migrated downslope under the influence of gravity. This mobility of the surface mantle also has the effect of exposing fresh rock to the weathering processes.

Two basic types of weathering may be distinguished, mechanical and chemical. The former refers to those processes involving disintegration or comminution of the original rock, the latter to those processes involving decomposition or chemical alteration. It must be emphasized that rock material is rarely affected by one process operating alone, but is normally subject to several interactive processes tending to induce concurrent disintegration and decomposition. Although stress is often laid on mechanical weathering in certain environments and chemical weathering in others, this merely indicates their relative importance and does not imply that either is totally absent.

Mechanical weathering

Four major sources of stress leading to fragmentation of the bedrock may be recognized: dilatation; thermal expansion; crystal growth; and organic activity.

Dilatation With a mean density of about $2 \cdot 7$ g cm^{-3} the confining pressures exerted on a deeply buried block of rock by the column of overlying material are enormous. As denudation removes the top of the column these pressures diminish and the block may adjust to this unloading by upward expansion. The clearest manifestation is the development of a closely spaced joint system, and on many quarry faces the rock within a few metres of the ground surface is seen to be divided into innumerable small cuboidal blocks whereas the same material at depth remains massive with only a few widely-spaced joints. Pressure release is a phenomenon well known to quarry operators since it occasionally leads to explosive shattering of freshly opened faces. The actual amount of dilatation has been measured by comparing the dimensions of a block while still in place at the foot of a

Weathering and the weathered mantle

deep quarry wall with the dimensions after it has been removed. Linear expansion of as much as 0·1 per cent has been found in quarried blocks of granite in Georgia. Occasionally pebbles of basalt dredged from the deep ocean floor are found to 'pop' or fracture shortly after being brought to the surface.

The most striking natural dilatation features are found in areas of massive igneous rock where jointing parallel to the surface sometimes produces remarkably regular sheeting. This is particularly spectacular in the Sierra Nevada of California (Fig. 7.1) where concentric layers at the surface may be no more than 0·1 m thick; yet the rock at depth is sufficiently massive to sustain sheer walls along the glaciated Yosemite trough many hundreds of metres high. Pressure release structures are not limited to igneous rock, nor do they necessarily arise from vertical unloading. Distinctive scars along canyon walls in the arid south-western United States have been ascribed to rock falls controlled by joints that developed because of lateral expansion in massive sedimentary strata. It seems reasonable to infer that where layered structures develop parallel to the ground surface they may well tend to perpetuate existing topographic forms. However, the main significance of the joint systems is the opportunity they afford for deep penetration by other weathering agencies.

Thermal expansion The idea of thermally induced expansion and contraction as a mechanism of rock shattering has a long history. Many early explorers in desert regions commented on the number of shattered pebbles they noticed lying on the ground, and some even claimed the actual fracturing to be accompanied by an audible report like a rifle shot. The abundance of broken pebbles was commonly attributed to the extreme temperature ranges experienced in deserts, possibly accentuated by the lack of vegetation and soil cover to provide protection from the direct rays of the sun. Allusion was also made to the poor thermal conductivity of most rock materials; this was held to result in a thin surface rind being heated and cooled over a much wider temperature range than the interior of a pebble. Finally, it was held that the varied coefficients of thermal expansion possessed by different rock minerals might be capable of inducing disintegrative stresses.

As early as 1915 Tarr was casting doubt on the validity of some of the more extravagant claims. Experimenting with granite blocks, he calculated that the stresses due to solar heating are appreciably smaller than the elastic strength of the rock. Later experiments by a variety of investigators have all failed to demonstrate fragmentation due solely to thermal changes. The best known research on thermoclastis is that of Griggs who heated and cooled a cube of granite 89 400 times through a temperature range of 30° to 140°C. If each temperature cycle is equated with a single day, this is equivalent to almost 245 years of thermal weathering. Yet at the end Griggs was unable to detect any significant deterioration in rock structure. Although the general validity of many similar laboratory tests can hardly be doubted, they do not dispose of the field observation that desert surfaces are often littered with fragmented pebbles. Either an alternative process is responsible, or the laboratory tests do not reproduce in some vital particular the natural conditions of the desert floor. Ollier has pointed out that specimens heated in an oven are completely unconfined, whereas the lower halves of many of the pebbles observed in the field are constrained by partial burial; this, it is argued, might induce extra disintegrative stresses. Other workers have maintained that chemical alterations must weaken a rock prior to its final thermal rupturing; they cite additional experiments by Griggs in which he showed that cooling blocks of granite in moist rather than dry air leads to rapid disintegration. This argument is countered by the assertion that some of the fractured desert pebbles are composed of materials such as flint which are not very susceptible to early chemical alteration. The importance of thermoclastis as a weathering mechanism remains unproven, but further study should be able to shed more light on the problem. For example, themocouples may be employed to measure actual rock temperatures in the field. Already surface values of 75°C, diurnal ranges of 40°C and temperature gradients surpassing 1°C per 20 mm have been recorded in this way. Perhaps the most intractable problem is assessing the potential effect of long-term fatigue failure. Two other heat sources that might contribute towards thermal rupturing are fires and lightning, but again any quantitative estimate of their significance is very difficult to obtain.

Fig 7.1 Dilatational sheeting reminiscent of normal bedding planes developed on the surface of igneous rocks in the Sierra Nevada, California. It is instructive to compare this superficial layering with the deep unjointed rock of similar composition that is capable of sustaining the huge vertical faces illustrated from the nearby Yosemite valley in Figs 12.1 and 12.2.

118 Crystal growth The growth of foreign crystalline solids within a rock is known to generate great disruptive stresses. Two types of crystal growth may be distinguished – that due to the freezing of water and that due to the precipitation of solids from solution. Although the former is undoubtedly the more important, the latter has recently been invoked as another potential source of the rupturing stresses encountered in deserts.

Water has the unusual property of reaching its maximum density at $3\cdot98°C$. Above and below this temperature its density falls from the standard value of 1 g cm^{-3}. It reaches $0\cdot9998$ at $0°C$ and declines to $0\cdot917$ on transformation to ice. This phase change from water to ice thus involves an expansion in volume of just over 9 per cent, but the temperature at which it occurs varies according to the confining pressure; for every increase of 100 kgf cm^{-2}*, the temperature at which ice forms declines by about $1°C$. The best-known illustration of the force generated during freezing is the fracture of domestic water pipes. As the air cools below $0°C$ part of the water in the pipes begins to freeze, expanding in volume as it does so. This increases the pressure on the residual water which remains liquid until the temperature has fallen sufficiently to cause freezing at this higher pressure. In practice domestic copper piping succumbs to the stresses that result from sustained temperatures only slightly below $0°C$. Theoretically with a temperature of $-22°C$ a pressure of $2\,100$ kgf cm^{-2} could be exerted on the confining walls of a rock joint; if the pressure were to exceed that value the molecular structure of the ice would change, with denser packing leading to higher densities and diminishing volume. Such extreme pressures are not normal, if for no other reason than that any confining ice seal would first be extruded from the joint. Estimates of the maximum pressures likely to be generated in rock fissures vary considerably, but values of up to 150 kgf cm^{-2} have been claimed.

In practice the situation is undoubtedly more complex than the foregoing analysis indicates. Much of the water in a fine-grained porous rock is held not in wide fissures directly open to the atmosphere, but as thin tightly bonded films that will not change to ice even when the temperature falls significantly below freezing point. At sub-zero temperatures water may still migrate through the finer pore spaces to ice crystals growing in the larger cavities in a rock. These and other factors in frost-riving are capable of systematic study in the laboratory. For example, specially designed freezing chambers have been used to examine the effects on rock samples of different temperature changes. Continental and maritime environments have been simulated by employing so-called Siberian cycles of $-30°$ to $+15°C$ and Icelandic cycles of $-7°$ to $+6°C$. Whereas each Icelandic cycle has normally been timed to occupy 1 day, the Siberian cycle has usually been programmed to take about 4 days. Several workers have begun by testing both wet and dry rock samples but have abandoned the latter when it became obvious that they remained virtually unaffected by the temperature changes. Wiman in Sweden subjected a collection of rocks including slate, mica schist, granite, quartzite and gneiss to 36 Icelandic cycles and 9 Siberian cycles, and found at the end of his experiment that the weight of the weathering products, expressed as a percentage of the original sample weight, ranged from $1\cdot16$ per cent for the slate subject to Icelandic cycles to $0\cdot007$ per cent for quartzite subject to Siberian cycles. For all rock types the Icelandic conditions had more effect than the Siberian, while under identical temperature ranges the rocks of greater porosity tended to shed the larger amounts of detritus. At the laboratories of the Centre de Geomorphologie in Caen elaborate experiments have been made on representative materials from north-western France, as well as on samples from other parts of Europe (Fig. 7.2). Many limestones have been shown to be extremely susceptible to frost action. Chalk in particular was completely fragmented by 100 cycles of either Icelandic or Siberian severity, whereas certain basalts and other fine-grained igneous rocks remained unaltered after 300 such cycles. All these laboratory tests confirm the general effectiveness of freeze–thaw processes. The frequency with which the temperature crosses the freezing point seems to be more significant than the absolute temperature range, although to be fully effective the temperature probably needs to remain below $0°C$ for several hours in each cycle. Lithology influences the character of the weathering products; porous rocks are easily

---

* 10^7 N m^{-2}. See Appendix A.

broken down into numerous small fragments, whereas dense impervious rocks split readily only along joints whose spacing largely determines the size of the resultant debris.

A wide variety of soluble salts may be precipitated in the joints and pore spaces of a rock. The force exerted in this case on the confining walls is unlikely to be due to simple volumetric increases since those that arise directly from crystallization are very small and are usually more than offset by evaporative losses. Yet the importance of salt crystallization has long been recognized by engineers and architects concerned with the durability of building stones, and the deteriorating facades of many historic monuments are now attributed to this form of weathering. It appears that the stress derives from the force of crystallization. Addition of material to a growing crystal can continue

even though it is already pressing against the constraining walls. The major controlling factor seems to be the degree of supersaturation of the surrounding solution. Evans has suggested, for instance, that with 1 per cent supersaturation calcite could crystallize against a pressure of about 10 kgf cm^{-2}.

Many experiments have been made to test the validity of salt crystallization as a disruptive force (Fig. 7.3). As early as 1828 Brard submerged various rocks in a solution of Na_2SO_4 and found that after a few cycles of wetting and drying most lithologies began to show signs of disintegration. More recently Pedro has experimented with granite spheres which were first submerged either in water or in solutions of sulphates, nitrates, carbonates and chlorides. Each sphere was then oven-dried at 80°C and the cycle repeated over 100 times; the detrital material

Fig 7.2. Laboratory experiments simulating freeze-thaw action: (A) Susceptibility of different rock types to 'Siberian' cycles; (B) Relationship between susceptibility and porosity for a range of igneous and metamorphic rocks (data from Centre de Geomorphologie, Caen, France).

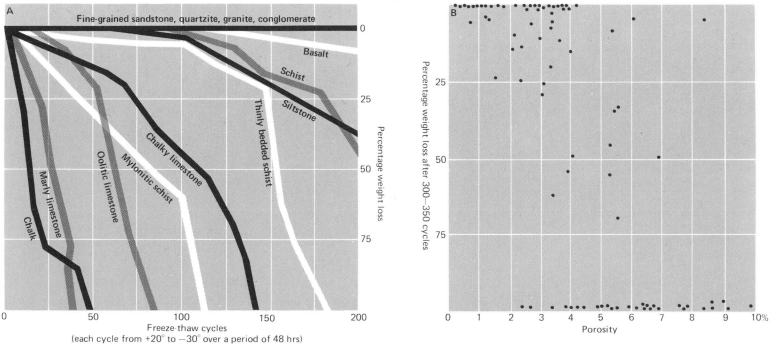

produced by the solutions was always far greater than that produced by the water alone. It now seems clear that a major factor in the weathering of urban building stones is the high concentration of SO_2 in the city atmosphere. This acidulates the rainwater and renders limestone buildings particularly prone to corrosion. Following a complex sequence of chemical reactions, gypsum ($Ca SO_4$) crystallizes just below the surface and thin flakes of rock constantly peel away.

The potential power of salt crystallization is beyond dispute, but the frequency of natural conditions under which this potential is fully realized may be more questionable. The development of major stresses requires high supersaturation implying rapid evaporation and lack of protective soil and vegetation cover. Such conditions are most likely to be en-countered in desert regions, and possibly along coasts. It now seems likely that salt crystallization, at least in arid areas, is a form of mechanical weathering which has been underestimated in the past. Some workers have suggested that the precipitated salts, since they have higher coefficients of thermal expansion than most rocks, may further contribute to rock shattering by expanding under the intense daytime heating of desert areas.

Organic activity It is well known that growing plant roots in favourable situations have the capacity to wedge open bedrock joints. Since the cellulose of cell walls is stronger than many metals, and root systems extend to depths of tens of metres, it might be supposed that this is a major means of physical dis-integration. In practice, however, the actual force exerted by

Fig 7.3. Laboratory experiments on salt crystallization: (A) Tests on the relative effectiveness of different solutions; (B) Tests on the susceptibility of different rock types (after Goudie and Cooke, 1970).

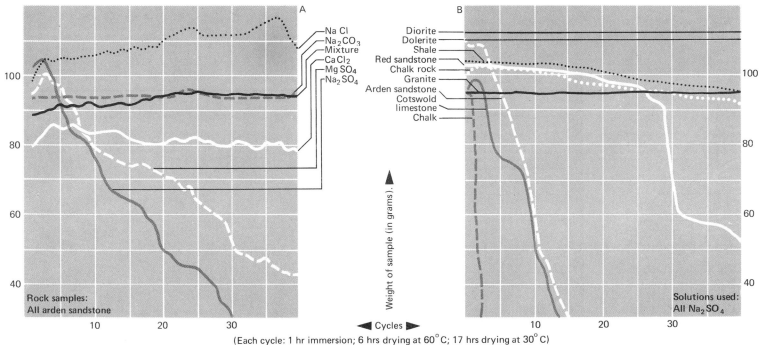

Na Cl
Na₂CO₃
Mixture
CaCl₂
MgSO₄
Na₂SO₄

Diorite
Dolerite
Shale
Red sandstone
Chalk rock
Granite
Arden sandstone
Cotswold limestone
Chalk

Rock samples: All arden sandstone

Solutions used: All Na₂SO₄

Weight of sample (in grams).

◀ Cycles ▶

(Each cycle: 1 hr immersion; 6 hrs drying at 60°C; 17 hrs drying at 30°C)

roots is difficult to ascertain since they very often follow openings already prised apart by some other agency. In assessing the overall role of roots, two contrasting activities need to be remembered. On the one hand, they act as a stabilizing agent by binding weathered materials together and thereby retarding the exposure of fresh rock. On the other, they occasionally have a disruptive effect, as when large trees are blown over.

Animals contribute little to the direct fragmentation of rocks, but have an important role in disturbing the detritus already produced and thereby enhancing the efficacy of other weathering processes.

In addition to the forms of mechanical weathering already discussed, a number of others have been suggested but without proof of their effectiveness in natural environments. One of the most interesting is colloidal plucking since it is known that gelatine drying in a tumbler is capable of detaching small flakes of glass. As an important fraction of soil organic material exists in colloidal form, this type of plucking appears to be a potential means of rock disintegration. However, its significance in the field has not been demonstrated and it is possible that, whilst effective on concave materials, it does not operate on the convex form of the typical rock grain.

Chemical weathering

General background In considering mechanical weathering it was possible to identify four major processes and to relate these to differing climatic environments and such rock properties as porosity and jointing. The web of chemical processes is much more complex and it is impossible to convey in a few pages the enormous wealth of chemical interactions revealed by almost any field investigation. Attention must therefore be concentrated on a few of the more basic processes. Before discussing these it will be helpful to review, firstly, some of the properties of rocks influencing their liability to chemical alteration and, secondly, the nature of the major reactant, rainwater.

Eight elements make up more than 98 per cent of the earth's crust. By far the most abundant is oxygen comprising by weight almost half the total, followed by silicon comprising a little over one-quarter. The six other elements in descending order of

abundance are aluminium, iron, calcium, sodium, potassium and magnesium. On the basis of volume, nearly all the crust is composed of oxygen anions bonded to metal cations in the form of oxides. By far the most important oxide occurring in mineral form is silica (SiO_2) best known as the rock material quartz; other mineral oxides include alumina (Al_2O_3) and haematite (Fe_2O_3). A very high proportion of the two most common elements is bonded together in the form of silicate tetrahedra (SiO_4) (Fig. 7.10). These constitute the basic building blocks not only of most minerals found in igneous rocks, but also of that apparently very dissimilar group, the clay minerals. The immense variety of silicate minerals derives from contrasting linkages between the tetrahedra, from different elements bonding the tetrahedra together, and from replacement of some of the silicon ions by aluminium. All these factors influence the susceptibility of the minerals to chemical alteration since the major weathering processes involve disruption of the tetrahedral lattice and removal of the bonding cations.

Both laboratory and field observations have been employed in attempts to tabulate the relative resistance of the primary rock-forming minerals (Table 7.1). In general minerals crystallizing at the highest temperatures during cooling of a magma are the least resistant to chemical weathering, those that develop last are the most resistant. In detail many variations from this basic pattern may be observed since other factors, such as the com-

Table 7.1 The susceptibility of various common minerals to chemical weathering (after Goldich, J. Geol. 1938)

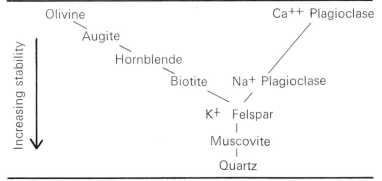

position of the soil water, exert an important influence. Moreover, in considering global weathering patterns, it is vital for the geomorphologist to remember that although sedimentary rocks comprise less than 5 per cent of the total crustal volume, they outcrop over about 75 per cent of the earth's land surface. On the basis of this last figure, it has been estimated that four major mineral groups constitute 85 per cent of the material exposed to weathering: felspars, quartz, clay minerals and micas, calcite and dolomite. Yet the apparent simplicity implied by this statement is rather deceptive. There are many different types of felspar and clay mineral, all altering chemically in different ways. Moreover, the susceptibility of a rock to decomposition is not solely dependent on its mineralogical make-up. Since the major reactant is percolating rainwater, accessibility for water is a highly significant property. One measure of accessibility is porosity which ranges from less than 1 per cent in most igneous rocks to over 15 per cent in some sandstones. Porosity also provides a rough guide to the surface area within a rock over which chemical activity can take place. Volume for volume, finer-grained rocks generally offer a greater surface area than coarse-grained so that they tend to weather more rapidly; for instance, felspars will tend to be more speedily altered in a fine arkosic sandstone than in a coarse-textured granite. An important consequential effect is that the elements released become available to participate in further chemical reactions at a much earlier stage. A second measure of water accessibility is permeability. The importance of routeways followed by percolating water is clearly seen in the development of karstic landforms in limestone regions. It is also a major factor in spheroidal weathering where chemical decomposition is concentrated at the intersections of joint planes.

Rainwater is far from the pure water that is often visualized. It obviously acts as a carrier of dissolved atmospheric oxygen and carbon dioxide, and on analysis is found to have a surprisingly complex chemical make-up. Salts derived from the sea, nitrous oxide produced by lightning and dissolved to form nitric acid, sulphur dioxide emitted by the combustion of sulphurous fuel and dissolved to form sulphuric acid, these and many other common constituents confer on rainwater the capacity to attack rock in almost endless ways. Once the water has percolated beneath the surface, its composition is further modified by reaction with both the mineral and organic fractions of the soil.

Pure water percolating through a rock is capable of inducing three main chemical processes: solution, hydration, and hydrolysis. By virtue of dissolved carbon dioxide and oxygen, rainwater can regularly induce two further processes, carbonation and oxidation. Biological influences within the soil layer are so important as to require discussion as a sixth topic. Although each of these subjects will be treated separately, it must be emphasized that it is extremely rare for one process to operate alone; their overall effectiveness is vary largely the result of interactions between them.

Solution Rocks exposed at the surface are not generally soluble in water, although one locally significant exception is halite or rocksalt. The vital importance of solution lies in its role of transporting the products of other weathering processes, since if this material is not removed it can slow down or radically alter the whole sequence of rock decomposition. Many solutions are dependent on external factors for their maintenance. For instance, the solubility of both ferrous iron and manganese rises rapidly under slightly acidic conditions, but it requires a pH of 4 or less to render alumina soluble. This helps to explain why, in different localities, the weathering residue from the same rock can vary widely. Equally important, it helps to account for re-precipitation as the properties of the solvent change during migration through the soil and weathered mantle.

Hydration Hydration has been described as the forerunner of all the more profound chemical alterations. Many different minerals can incorporate water in their molecular structure. This includes not only those commonly found in igneous rocks but also many constituents of sedimentary rocks. For instance, haematite may be converted to limonite, while many of the silicate clay minerals swell greatly on becoming hydrated. This property of swelling when entering into combination with water is one of the most important aspects of hydration. It is believed to be a major cause of the crumbling of coarse-grained igneous rocks which are disrupted by the progressive expansion of their hydrated minerals.

Hydrolysis The classic experiments in the weathering of the silicate minerals, and particularly of the felspars, were carried out by the Frenchman Daubree in the nineteenth century. He placed 3 kg of orthoclase felspar in water and agitated the mixture for 200 hours. At the end of that time he found the water had gone milky in appearance and the surface of the felspars was decomposed. Later analysis showed the water to contain 2·52 g of K_2O in solution, together with a fine suspension of clay minerals. The chemistry of the process is extremely complex but may be approximately represented by the following equation:

$$2KAlSi_3O_8 + 2H_2O \rightarrow Al_2Si_2O_5(OH)_4 + K_2O + 4SiO_2$$

| Orthoclase | + | Water | | Kaolinite | + | Soluble potassium oxide | + | Soluble silica or 'silicic acid' |

It can be seen that the reaction involves a virtually complete disruption of the original silicate lattice together with removal of the bonding potassium ion. The rearranged silicon and aluminium ions can accommodate more water and the final product is the hydrous clay mineral known as kaolinite. A further part of the silica may be removed in solution. This decomposition of orthoclase illustrates a chemical reaction with water that affects many other silicate minerals. The sodium- and calcium-rich felspars are decomposed in a very similar fashion, as are such less closely related minerals as olivine, augite and hornblende. However, in all these cases chemical changes are accelerated under natural conditions by the impurity of the water, and in particular by the presence of dissolved CO_2.

Carbonation Dissolved atmospheric CO_2 turns rainfall into a very weak carbonic acid with an average pH of about 6. When the water percolates underground more CO_2 is rapidly dissolved from the soil air which is much richer in that gas than the free atmosphere. Experiments by Daubree and numerous later workers have shown orthoclase decomposition to be greatly accelerated when the mineral is churned in water enriched with CO_2. The enhanced reaction appears to arise from the ready ionization of weak carbonic acid into hydrogen and bicarbonate ions. The weathering of orthoclase in such circumstances can be approximately represented by the following formula:

$$2KAlSi_3O_8 + 2H_2O + CO_2 \rightarrow Al_2Si_2O_5(OH)_4 + K_2CO_3 + 4SiO_2$$

| Orthoclase | + | Water | + | Carbon dioxide | | Kaolinite | + | Potassium carbonate | + | Soluble silica |

In essentials the process clearly resembles that already outlined for pure water. The silicate structure of the orthoclase is broken down with the release of the potassium ion, and the end product is a clay mineral together with soluble residues. Weathering of the other main felspar groups also results in the development of clay minerals, but with the release of Na_2CO_3 from the soda-rich minerals and $Ca(HCO_3)_2$ from the calcium-rich minerals. Whereas the sodium and calcium is readily removed to become part of the solution load of streams, more of the potassium tends to be absorbed on to the colloidal clays for ultimate use by plants.

Water enriched in CO_2 attacks many other silicate minerals. Olivine, for example, can be almost entirely dissolved in a sequence of reactions represented in a much simplified fashion by the following formula:

$$Mg_2SiO_4 + 2H_2O + 4CO_2 \rightarrow 2Mg(HCO_3)_2 + SiO_2$$

| Olivine | + | Water | + | Carbon dioxide | | Magnesium bicarbonate | + | Soluble silica |

However, the best-known example of chemical alteration by carbonated water is the dissolution of limestone. Calcium carbonate is only slightly soluble in pure water but undergoes the following reaction in the presence of weak carbonic acid:

$$CaCO_3 + H_2CO_3 \rightarrow Ca^{++} + 2(HCO_3)^-$$

The calcium and bicarbonate ions may then be removed in solution, leaving behind impurities to accumulate as a distinctive superficial residuum; in many areas this contains a considerable proportion of iron which oxidizes to form the bright red material known as terra rossa. The volume of dissolved calcium carbonate that may be transported by the water is very sensitive to variations in the amount of dissolved CO_2. Equilibration with the low CO_2 pressure of cave air, particularly by comparison with soil air, is believed to be a primary cause of re-precipitation in the form of stalactites and stalagmites. Similar reactions at ground level may result in the formation of tufa incrustations.

124 Oxidation To the modern chemist the term oxidation has assumed a rather wider meaning than the original one of chemical combination with oxygen, but it is in this earlier sense that the term is often still applied in the discussion of weathering. Combination with oxygen dissolved in water is one of the most frequently observed weathering phenomena, and is particularly obvious in the case of that almost ubiquitous element, iron. Iron is a constituent of many of the common rock-forming minerals, such as biotite, augite and hornblende, and when released by one of the other chemical processes is rapidly oxidized to the ferric state in the form of haematite or its hydrated equivalent, limonite. It is in the ferric form that iron tinges so many soils with characteristic red, brown or yellow hues.

Biological influences Biological agencies affect chemical weathering both by influencing the rates of the processes already discussed, and by producing reactions that are specifically biochemical in nature. Into the first category come the various controls exercised by vegetation on the quantity and quality of percolating water. By interception the vegetation cover will regulate the amount of precipitation reaching ground level, by the production of humus it will influence the rate at which water moves through the soil horizons. The effects on water quality are of even greater significance. The most fundamental is in the supply of CO_2 which has already been mentioned in the earlier account of carbonation. Oxidative bacteria decomposing organic residues, together with the respiration from plant roots, regularly raise the CO_2 concentration in soil air to between 0·2 and 2·0 per cent, with the figure occasionally rising as high as 10 per cent. Even the lowest of these values represents a six-fold increase by comparison with the mean atmospheric concentration. The actual amount of CO_2 in soil air varies in response to a number of factors. One is temperature since bacterial action declines rapidly as the soil temperature falls below 10°C; another is water content since organic activity falls significantly when the moisture content sinks below about 10 per cent. Further factors include soil aeration and vegetation cover so that the overall effect is a complex pattern of seasonal and regional variations in dissolved CO_2. This in turn has important consequences for the solubility of such weathering products as the iron and aluminium compounds.

One of the most dynamic agents in changing both the quantity and quality of percolating water has been Man. By felling natural vegetation, covering large urban areas with impermeable surfaces and tile-draining agricultural land he has modified water movement through the top few feet of vast areas of the earth's surface. At the same time water composition has been altered by an almost endless list of activities. Emission of substances into the atmosphere, ploughing and harvesting of crops, application of fertilizers, all have affected the character of percolating water. Although Man's activities rarely have the dramatic impact of soil erosion in the United States during the 1930s, this must not be allowed to hide his more subtle yet extremely pervasive influence.

Two important weathering reactions that appear to be primarily biochemical in nature are reduction and chelation. Many anaerobic bacteria obtain part of their oxygen by reducing iron from the ferric to the ferrous form, and in more extreme circumstances from the ferrous form to the metal itself. One consequence is to produce ferrous compounds which are significantly more soluble in water than the original ferric ones; this is a major way in which iron can be mobilized and removed from the soil. Chelation involves the union of metallic cations with the hydrocarbon molecule and is a fundamental process in sustaining plant life. The roots are surrounded by a concentration of hydrogen ions which can exchange with the cations in adjacent minerals; the metallic cations are then absorbed into the plant by chelation. The resulting co-ordination compound, as it is called, is soluble in organic solvents but not in water. Lichens may act in a similar fashion, deriving essential trace elements from the rocks on which they grow until the surface layer is thoroughly decomposed.

Climatic controls

In the preceding pages, reference has frequently been made to the effect of both climate and vegetation on the nature of the weathering processes. Since the influence of the plant cover is often an indirect expression of climatic controls, several workers have been led to seek a closer definition of the relationship between climate and weathering. The best-known scheme is that of Peltier who began by making a number of simplifying assumptions. The first is that mechanical weathering is almost entirely due to freeze–thaw activity, the second that chemical weathering is so dependent on the presence of water that its intensity should bear a fairly simple relationship to precipitation. Arguing that regions could be both too hot and too cold for freeze–thaw to be fully effective, he attempted to identify those climatic regimes where the temperature most frequently crosses the freezing point. Further contending that chemical weathering is accelerated by high temperatures and dense vegetation cover, he defined regimes where he believed such conditions would prevail. By these arguments he was able to delimit climatic regimes characterized by distinctive combinations of mechanical and chemical weathering (Fig. 7.4). Most workers would probably agree that the relative importance of different weathering processes varies from region to region and that Peltier's diagram provides at least a rough guide to these variations.

In a rather different analysis, the Russian worker Strakhov has more recently emphasized the interaction of climate and relief controls. He believes that chemical weathering normally far outstrips mechanical weathering. It is particularly effective under conditions of plentiful moisture, high temperature and abundant vegetation, and so reaches its acme in the tropical zone. An area of secondary maximum is found in moist temperate latitudes, although the rate here is less than one-twentieth of that in the tropics. Only in deserts and extreme northern latitudes do climatic controls encourage relatively vigorous mechanical weathering. However, an additional factor stressed by Strakhov is tectonic uplift. He argues that, with increasing relief amplitude, mechanical denudation in the form of surface wash becomes so intense that it finally suppresses chemical weathering altogether. However, to achieve this state in the humid tropics demands exceptionally rapid uplift; it can be achieved much more readily in temperate latitudes.

Weathering forms

Bedrock may be fashioned into distinctive surface forms owing their shape primarily to weathering processes. Two groups can

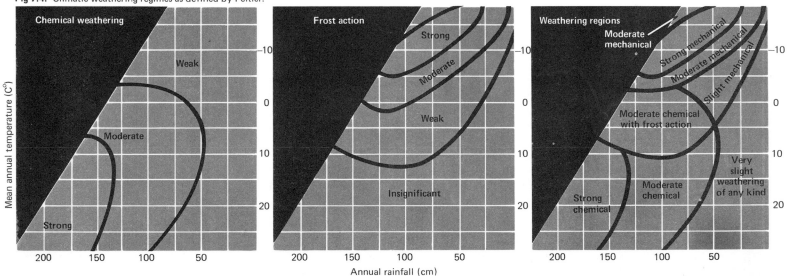

Fig 7.4. Climatic weathering regimes as defined by Peltier.

be recognized, those that originate at ground level and those that initially develop underground but suffer later exhumation. It is often difficult to distinguish between the two, and investigation of many of the features has involved a history of conflict between surface and exhumation hypotheses.

Weathering pits Small-scale depressions known as weathering pits are very widespread. They may be observed on horizontal bedrock surfaces in a variety of climatic environments ranging from arctic Canada to tropical Africa. They develop on many different rock types, but tend to be particularly prominent on granite and similar acid igneous rocks. Commonly a metre or so in diameter they may range in depth from 0·1 to over 0·75 m. Although pitted surfaces are often revealed when a soil cover is stripped back, there seems no doubt that the forms can also develop in a purely subaerial situation. An initial shallow depression becomes the site of a puddle around and beneath which the rock is regularly wetted. Hydration and hydrolysis ensue, with the soluble products removed by percolation. It is characteristic of virtually all weathering pits that they occasionally dry out and these periods of desiccation may be essential to permit removal of the insoluble residue by wind action.

Karren The rocky surface of a limestone outcrop may exhibit shallow depressions very similar to the weathering pits described above. Undoubtedly due to solution, these constitute one minor type of a whole group of limestone features best described by the German term 'Karren'. This word embraces forms ranging from runnels only a few millimetres in depth to prominent clefts locally several metres deep. All are due to solutional weathering, but some appear to form at the surface and others beneath a shallow cover of soil or superficial material. Elaborate classifications of Karren have been proposed, but four types appear to be particularly common:
(*a*) *Rillenkarren*. These are very shallow grooves or runnels that are often scored in great numbers down sloping limestone faces. They are due to solution by small rills of water and can develop on freshly uncovered limestone within the space of a few years.
(*b*) *Rinnenkarren*. These are much larger and deeper runnels, sometimes approaching half a metre in depth. Individual grooves may be traceable tens of metres down suitable limestone surfaces and after heavy rain each carries a fast-flowing rivulet of water.
(*c*) *Rundkarren*. As the name implies, these are rounded blocky forms that develop initially beneath a soil cover. In many areas of active exhumation the rounded upper surfaces can be observed gradually appearing from beneath the overlying material. The intervening depressions exhibit much less consistency in slope and direction than in the case of Karren produced at the surface; they are due to percolating rather than free-flowing water.
(*d*) *Kluftkarren*. These are the features frequently described in Britain as grikes from the local name used in North Yorkshire where they are very fully developed on the Carboniferous limestone plateaus. Controlled almost entirely by rock structure, they are essentially joints widened by solutional activity until the cleft may finally penetrate to a depth of several metres. Although solution in some instances takes place directly at the surface, a two-stage origin has also been invoked since there are indications that some clefts were initiated beneath a soil cover and are currently being widened by the growth of Rillenkarren along their walls.

Tors Especially fierce controversy has surrounded the origin of tors. Tors are small residual eminences composed of bedrock and rising steeply above the surrounding area. They are particularly characteristic of, but not confined to, areas of granite rock. Three groups of theories may be distinguished, two invoking weathering processes but the third attributing them to surface water flow. The first hypothesis envisages deep chemical rotting followed by exhumation, an idea championed in recent times by Linton. According to Linton structure is a major influence in tor evolution since zones of massive rock prove much more resistant to chemical decay than intervening areas of closely jointed rock (Fig. 7.5). He visualizes a prolonged weathering phase, possibly under a moist tropical climate, during which rotting proceeds to a depth of at least tens of metres in the less resistant zones. With tectonic uplift or some other cause of accelerated erosion the rotted debris is removed to leave the coarsely jointed blocks as upstanding residuals. Support for this hypothesis is found in the irregular weathering front occasionally observed in deep sections, sometimes with large groups of unaltered corestones available as future tor material. A factor which may complicate interpretation of such sections is hydrothermal alteration of the igneous rock during a late stage of its original emplacement; decomposition of this type can be difficult to distinguish from that due to atmospheric weathering. The second hypothesis, supported by many workers familiar with arctic environments, is that similar hilltop features can be produced by intense frost shattering, an idea discussed more fully in Chapter 14. The third viewpoint is that of King who argues that many of the features which have been termed tors and attributed to weathering processes are no more than residuals associated with a late stage in the development of subaerial erosion surfaces. He believes that this is certainly the origin of numerous tor-like features in Africa.

It seems that tors may well illustrate the problems arising from 'convergent processes', that is, very different processes that nevertheless result in similar end products. If the forms are dentical, then by themselves they give no indication of origin.

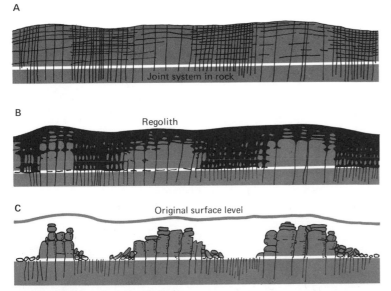

Fig 7.5 Stages in the evolution of tors that originate from deep chemical weathering (after Linton, 1955): (A) Initial form; (B) Deep weathering; (C) Removal of regolith.

A

Joint system in rock

B

Regolith

C

Original surface level

On the other hand, one may suspect that further work will disclose systematic morphological distinctions allowing classification according to genesis. The whole problem would be greatly simplified if constancy of the climatic zones could be assumed. However, the recent geological past has seen such great fluctuations that a single area may have experienced conditions ranging from periglacial to moist warm temperate; it is quite feasible, for instance, that some tors in the British Isles have been initiated by deep rotting during warm inter-glacial periods and exhumed by rapid mass movement during cold glacial episodes.

Etchplains Certain authors have invoked chemical weathering on a much larger scale to explain the evolution of landforms in tropical and subtropical latitudes. They suggest that, during phases of tectonic stability, chemical weathering in hot humid regions may cause deep rotting over very extensive areas. When the equilibrium is disturbed and a new phase of erosion initiated. the first result is stripping of the weak decomposed layer to produce an 'etchplain'. The relief of such a plain would reflect the differing resistance of the rocks to deep chemical decomposition rather than to the processes responsible for exhumation.

Physical properties of the weathered mantle

One of the primary concerns of the geomorphologist must be the physical characteristics of the layer of weathering products that mantles the landscape. Nearly all the material removed from a catchment area must pass through this phase and its development and removal is an integral part of hillslope evolution to be considered in Chapter 8. However, before proceeding to examine the ways in which the weathered mantle or regolith responds to the forces that act upon it on a hillslope, it will be helpful to note some of its physical properties and the methods by which they may be studied.

Thickness and layered structure The thickness of the regolith varies from nil to many metres. Over much of the earth's surface it is between 1 and 2 metres, but occasionally totals in excess of 100 m have been recorded. Many of the greatest values are found in tropical areas in conformity with the general expectation that chemical alteration is likely to be most effective under hot, moist conditions. The character of the regolith also varies greatly with depth. Near the surface the mineral matter is intimately mixed with organic material to form the soil. The ratio of organic to mineral matter decreases with depth, and beneath the soil a second layer consisting very largely of weathered bedrock may be distinguished. At greater depth this in turn gives place to unaltered bedrock. The boundaries between these different layers may be either abrupt or gradational. For instance it is often possible to identify an intermediate zone above the sound rock in which the material, although completely decomposed, retains clear traces of its original structure (Fig. 7.6). Where undisturbed these can be an important guide to the depth to which mass movement, as opposed to weathering, has extended. In very deep regolith it is usually the basal layer of rotted rock that is exceptionally thick, and several workers have attempted further subdivision by recognizing an additional zone near the base characterized by residual corestones. It is obvious that, in any assessment of the overall properties of the regolith, cognizance must be taken of these very significant variations with depth.

Infiltration capacity The maximum rate at which the regolith permits downward percolation of water is referred to as its infiltration capacity. This is an extremely important parameter since it is the major determinant of whether rainwater sinks underground or flows across the surface. When the precipitation intensity surpasses the infiltration capacity, the excess water is unable to percolate downwards and tends instead to generate overland flow. Infiltration rates of the surface layers may be measured by an infiltrometer. This normally consists of an open-ended tube, 0·1 m or more in diameter, partially sunk into the ground so as to isolate a cylindrical core of soil. The rate of percolation by water applied artificially to the surface of the core can then be measured. Usually a container with a mechanism for ensuring a constant head of water is employed, and the rate of water loss from the container is taken as a measure of infiltration. In practice it is found that the rate varies with time. There is often an initial rapid acceptance of water as the pore spaces in the upper soil horizons fill. The magnitude and duration of this early phase depends in part on the original soil moisture content, being greatest where the soil is dry. Thereafter the rate gradually diminishes until, within an hour or so, a relatively steady rate of infiltration is established. It is this constant value that is normally quoted as the infiltration capacity, although when comparison is made with precipitation intensities it is vital to recall the high early values that may be achieved with an initially dry soil. Measured capacities cover a very wide range, but most tend to lie between zero and 100 mm hr^{-1}. Evaluated in this way infiltration is related less to total porosity than to what may be termed 'macro-porosity'. While total porosity refers to the gross volume of voids, macro-porosity includes only the larger pore spaces through which water can move relatively freely. This distinction is well exemplified by sandy soils with a total porosity near the surface of 35 to 50 per cent, and fine-textured soils with a total porosity of 40 to 60 per cent. Yet so many of the pores in the latter case are of small dimensions that the infiltration capacity of a sandy soil is ordinarily very much higher.

Field measurements have usually been confined to a shallow soil layer of 0·2 m or less. One reason is the increasing experimental problems encountered at greater depths, but even more important is the known tendency for water under natural conditions to move laterally as well as vertically; so long as water is confined within a small vertical cylinder, lateral percolation is precluded. Recent research has shown that on normal hillsides an important fraction of the water making its way underground soon moves downslope parallel to the surface. Known as interflow or throughflow this movement can be caused by the presence of a buried horizon having a lower permeability than that of the surface layer. Such a situation is extremely common since the B-horizon of a soil is regularly less permeable than the A-horizon. The significance of this retarded percolation may be assessed by inserting infiltration cylinders to different depths within the soil; those that penetrate an illuviated layer will normally yield lower infiltration capacities than those confined to higher horizons. More direct measurements of interflow are difficult to make, although experiments have been carried out by inserting receptacles at different levels on the wall of a trench dug across a hillslope. On many peat-covered hillsides a system of small circular tunnels has been discovered close to the

Fig 7.6. Deep chemical weathering producing 'spheroids' by selective alteration along intersecting joint planes. The white lines in the above photograph are quartz veins that have proved much more resistant to decomposition than the igneous rocks in which they are set.

junction of the organic and mineral layers. Large volumes of water are conveyed through the tunnel system after heavy rainfall and may be evaluated by diverting the flow along artificial pipes into suitable measuring devices.

Infiltration measurements are essentially a form of point sampling. Experimental data, standardized as to size and depth of the cylinder employed, commonly display large variations according to the nature of the bedrock, slope and vegetation. Even within areas where these three variables appear relatively constant, marked differences may be found indicating a high sensitivity to other irregularly distributed factors. To overcome this problem of 'noise' a large number of samples must be included in any systematic investigation.

Mechanical strength The hillslope regolith is subject to stresses arising in a number of different ways, and its response to these determines both the pattern of movement and ultimately

Fig 7.7. Stress-strain relationships for four idealized materials. In practice many solid materials exhibit properties that are some complex combinations of the forms shown

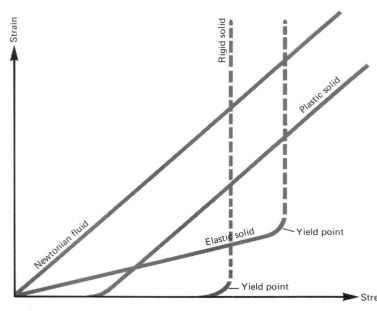

the form of the slope itself. Analysing the mechanical strength of the regolith poses many problems which are still far from complete solution and are beyond the scope of this book. Nevertheless, in view of the importance of the subject and the growing recognition of its significance in geomorphology, at least a brief introduction needs to be given.

The behaviour of materials is normally considered in terms of applied stress and induced strain. Stress is a vector quantity expressed as a force per unit area. Strain is the technical term to denote deformation of a material and is expressed as a fractional change in the original dimensions of the body. Strain may take place in a number of different ways according to the nature of the material; for instance, air normally deforms by turbulent flow, a thin film of water by laminar flow, and a solid by elastic displacement. It is these different behaviours under stress that forms the basis for distinguishing between fluids and solids. A fluid is a substance which cannot sustain shear forces and in which deformation is directly proportional to applied stress. A solid is a substance which possesses both strength to resist a small applied stress and elasticity to restore its original shape after removal of the deforming stress. Three different types of solid may be recognized (Fig. 7.7). In a rigid solid small stresses produce little strain until a critical value, known as the elastic limit, is reached, whereupon abrupt fracturing takes place. In an elastic solid, on the other hand, there is a considerable strain before the point of fracture is reached. In a plastic solid deformation beyond the elastic limit does not produce sudden fracturing but an increase in strain proportional to the increase in stress. These three types of solid are arbitrarily defined and the same material may act in different ways in different circumstances. For instance, granite would normally be considered a rigid solid, fracturing abruptly under the geologist's hammer; under conditions of high temperature and pressure, however, increasing stress produces a strain pattern resembling that of plastic deformation. The duration of stress may also have important consequences, since a force that can be accommodated by elastic strain if applied for a short time may induce 'creep' if applied for a longer period. These examples are given to illustrate the complexity of apparently simple materials. When it is remembered that the regolith consists of an intimate mixture of solids, water and air, in proportions that may vary greatly with time, the difficulty of defining even an approximate mechanical model can be readily appreciated. Many analyses have treated the regolith as being in a solid state of aggregation and possessing properties conforming reasonably closely to those of a plastic solid. Yet a substantial part of true plastic deformation takes place by intracrystalline gliding, whereas in the weathered mantle movement is mainly by intergranular displacements. The physical nature of the movement is therefore very different and the plastic model is by no means entirely satisfactory. A further problem arises from the fact that the mechanical properties vary greatly with depth, particularly near the surface where there are often strongly contrasting soil horizons.

The problems are formidable, not to say daunting. Nevertheless, they must be tackled if a fuller understanding of regolith movement is to be obtained. Much of the initiative has come from engineers who have to consider the practical difficulties of building earthen dams or digging deep motorway cuttings. For such work it is essential to have some knowledge of the way the materials will behave, and investigation of this behaviour is the field of soil mechanics. In this context the word soil is not restricted to the superficial layer in which plants grow but applies to any natural earthen material in which or of which engineering structures are to be built; in addition to superficial sediments and regolith this includes what the geologist would regard as bedrock. The foundations of modern soil mechanics can be traced back to the work of the eighteenth-century French physicist Coulomb. He expounded the general rule that the shear strength of a soil is equal to $c + \sigma \tan \phi$, in which c is cohesion, σ the stress normal to the shear plane and ϕ the angle of internal friction. In bedrock it is the cohesion that is by far the most significant value, but in certain loose, dry aggregates the value of cohesion falls to zero. Each of the values in the Coulomb equation merits further examination.

Although Coulomb originally regarded ϕ as the angle of repose of loose materials, in modern theory it is designated the angle of internal friction and may be measured by the use of direct-shear apparatus. Assuming first that one is dealing with non-cohesive material such as sand or gravel, a sample is placed in a split box and the maximum force that can be transmitted through the sample to a proving ring is measured (Fig. 7.8a). By changing the load on the top of the box, the stress acting normal to the shear plane can be varied. If the results of repeated experiments are plotted as shown in Fig. 7.9a, the angle between the horizontal axis and the line joining the plotted points is the angle of internal friction. For most non-cohesive materials it lies between 30° and 40°. If a suitable cohesive soil is similarly tested in the direct shear apparatus, the plotted data provide a measure of the cohesion of the sample (Fig 7.9b).

As originally formulated by Coulomb, shear strength is directly related to the total stress normal to the plane of shearing. In 1923, however, Terzaghi published a modification of Coulomb's equation which afforded an explanation for certain discrepancies that had arisen between predictions of the equation and measured values. It had been found, for instance, that sand slopes which were stable when dry collapsed when wet. This was first ascribed to a fall in the value of ϕ amounting to as much as 6°. Yet direct shear tests by Terzaghi on wet and dry sand failed to reveal such a difference, and he suggested that Coulomb's equation would remain valid if total normal stress were replaced by effective normal stress. The underlying concept is that the total stress consists of two parts, effective stress and pore stress. The former is the fraction of the total transmitted by the solid-to-solid contacts, while the latter refers to that part transmitted by the pore spaces. Clearly, so long as the pore spaces are filled with air the pore stress is zero and the effective and total stress remain the same. When the pores are filled with water a variable proportion of the total stress is transmitted through the fluid and the effective stress is correspondingly diminished. In consequence the Coulomb equation in modern form is usually written $S = c' + \sigma' \tan \phi'$ where σ' refers to effective rather than total normal stress. This transformation is of critical importance in explaining the diminished shear strength of porous materials as they become saturated; it was the increased pore pressure that started the disastrous flows of the Aberfan tip-heaps in 1966 which resulted in the deaths of 116 schoolchildren. On natural hillslopes regolith that has remained stable for decades or even centuries may suddenly be mobilized by a fall in shear strength after exceptional rainfall.

Fig. 7.8. The essential components of two types of equipment for measuring the shear strength of soil samples.

(a) Shear-box

(b) Tri-axial apparatus

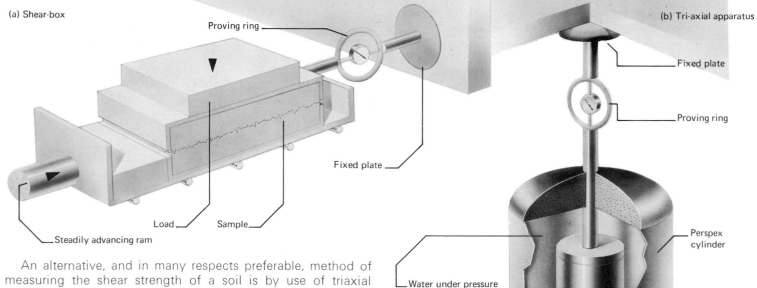

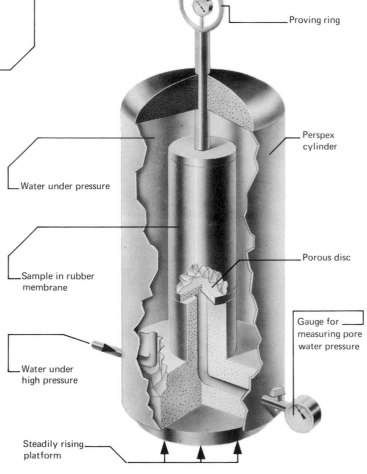

An alternative, and in many respects preferable, method of measuring the shear strength of a soil is by use of triaxial compression apparatus (Fig. 7.8b). The procedure involves enclosing a cylindrical soil sample in an impermeable membrane. The ends of the specimen are capped with porous discs before it is inserted in a pressure chamber filled with liquid. A load is then applied along the axis of the cylinder until failure of the sample occurs. The stress normal to this principal axis is varied by controlling the fluid pressure in the chamber. One great advantage of triaxial over direct-shear apparatus is the ability to control and measure the pore water pressure in the specimen. Two types of test are commonly employed. In the first the water is prevented from draining out of the sample, in the second drainage is permitted as load increases and the amount of expelled water is measured. The cohesion measured in either direct-shear or triaxial apparatus is due to binding forces between individual particles. In rocks it is normally the result of cementing and fusion, but as weathering progresses these bonds are gradually weakened. This is especially clear where the rock breaks down into debris of coarser size than silt. Such material has low cohesion but relatively high values for the angle of internal friction. However, where clay minerals are generated by the weathering process the angle of internal friction is reduced but rather more cohesion is retained. The cohesion value depends on the nature of the clay minerals involved, and because these are so significant in imparting distinctive properties to the regolith they require closer examination.

Clay minerals The term clay is commonly used to designate particles having an upper size limit of 0·002 mm. The grains were at one time simply regarded as very small fragments of other minerals, but many clays were subsequently found to possess properties inconsistent with this view. At the beginning of the present century the Swedish soil scientist, Atterberg, showed that quartz particles ground to the size of 2 microns exhibit no cohesion whereas mica ground to a similar size is highly cohesive. The difference is explained by the presence of electrical charges on the surface of the mica but not on the surface of the quartz. It was inferred that most fine particulate matter in the weathered mantle must be composed of clay minerals having a distinctive crystalline structure and in significant respects resembling the comminuted mica. The use of X-ray diffraction equipment has subsequently shown that there are many different types of clay mineral, each capable of endowing the regolith with unique physical and chemical properties.

The most abundant clay minerals are layered aluminosilicates, often categorized as phyllosilicates owing to their sheet-like structure. There are also hydrous oxide clays composed mainly of aluminium and iron oxides bonded with variable quantities of water. A good example is afforded by gibbsite ($Al_2O_3 . 3H_2O$), one of the principal minerals of bauxite. Of the iron-rich materials the most common is limonite ($Fe_2O_3 . H_2O$) but clays of this composition occur in two distinct crystalline forms known as goethite and lepidocrocite. In addition to these crystalline minerals, compounds may also occur in an amorphous state; one of the most important is allophane, a hydrous aluminosilicate of rather variable composition.

Three major groups of aluminosilicate clay minerals may be distinguished; the kaolinite, montmorillonite and hydrous mica groups. They are all phyllosilicates but differ in crystal structure, surface electrical charges, and resulting mechanical properties.

Crystal structure. Minerals in the kaolinite group consist of alternating layers of octahedral alumina sheets and tetrahedral silica sheets (Fig. 7.10). Adjacent alumina and silica layers, held together by shared oxygen atoms, form what is known as the crystal unit. Since each unit consists of one alumina and one silica sheet, the lattice is often described as being of the 1 : 1 type. The units themselves are held together by linkages between

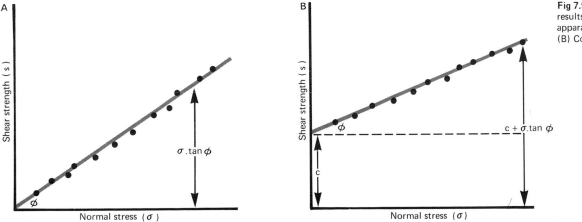

Fig 7.9. Schematic representation of the results of multiple tests with shear-box apparatus: (A) Non-cohesive material; (B) Cohesive material.

Fig 7.10 Diagrammatic sketches of the crystal structure of three common phyllosilicate clay minerals.

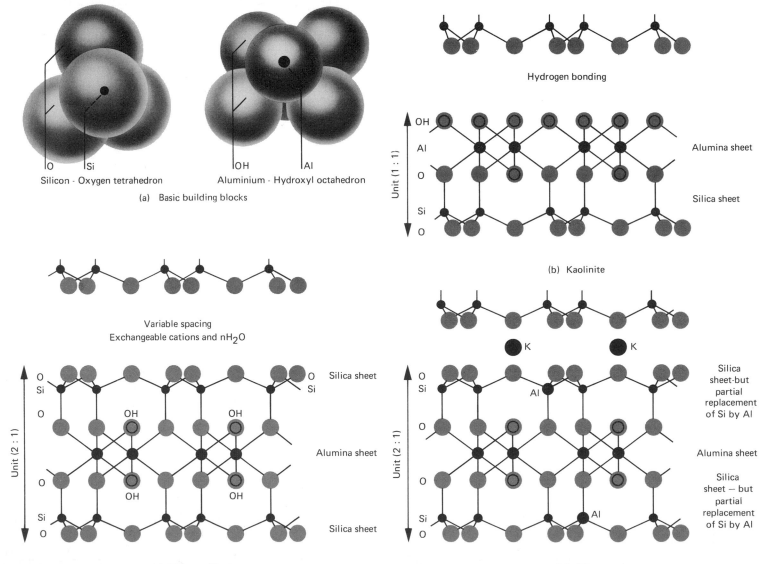

Silicon - Oxygen tetrahedron   Aluminium - Hydroxyl octahedron

(a) Basic building blocks

O   Si

OH   Al

Hydrogen bonding

OH
Al   Alumina sheet
O
Si   Silica sheet
O

Unit (1 : 1)

(b) Kaolinite

Variable spacing
Exchangeable cations and nH$_2$O

O   Silica sheet
Si
O
OH   OH   Alumina sheet
OH   OH
Si
O   Silica sheet

Unit (2 : 1)

(c) Montmorillonite

K   K

O   Silica
Si   sheet-but
partial
replacement
Al   of Si by Al
O
Alumina sheet
O
Silica
sheet — but
partial
replacement
Si   Al   of Si by Al
O

Unit (2 : 1)

(d) Illite

the oxygen and hydroxyl ions, forming a relatively rigid overall structure. The montmorillonite group of clay minerals is composed of crystal units in which two silica sheets enclose a single alumina sheet. On account of this arrangement the crystal lattice is often described as being of the $2:1$ type. The bonding between successive units is extremely weak with the result that the crystals readily expand to admit water and exchangeable cations. The hydrous mica group of clay minerals, of which illite is the commonest example, also possesses a $2:1$ type of lattice. However, in the silica sheet a considerable proportion of the silicon is replaced by aluminium, leaving valences which are satisfied by the incorporation of potassium ions. These lie between the units and make the whole structure more rigid than is normally the case with montmorillonite.

Although it is convenient to divide the phyllosilicate clays into three groups on the basis of their structures, many intermediate forms occur under natural conditions. Individual crystals quite commonly contain several different types of structural unit inter-stratified with each other; these are known as mixed-layer minerals. The very existence of illite as a distinctive clay mineral has recently been questioned since many so-called illites have now been shown to be mixed-layer minerals. Moreover, in many clays ionic substitution is so frequent that the chemical composition varies widely. In montmorillonite, for example, up to 20 per cent of the aluminium may be replaced by magnesium.

The chemically altered mantle of rock waste normally contains a mixed assemblage of clay minerals. It is difficult to generalize regarding the factors which produce one particular assemblage rather than another. Nor is there any set sequence of decomposition products. In the initial phases of weathering parent material may exercise an important control. Granites with a high Al:Si ratio and a low concentration of other metal cations tend to alter to kaolinite. Basic igneous rocks tend to weather to montmorillonite. However, the distribution of clay minerals is clearly influenced by many other factors. Acid soil conditions are conducive to the production of kaolinite, while alkaline conditions favour the evolution of montmorillonite. Regional patterns of dominance by certain clays have been discerned in mid-latitude areas. In the United States kaolinite is particularly abundant in the warm, humid south-east, the hydrous mica group becomes increasingly prominent in the cooler north-east, while montmorillonite reaches its fullest development in the drier west. Another general association is between tropical climates and abundant hydrous oxide clays. This reflects the more rapid leaching of silica in wet low-latitude areas than in most temperate regions. All these relationships between climate and particular clay minerals are indicative of broad trends in the weathering cycle, but at the detailed local scale many other factors are found to operate.

Surface electrical charges. All silicate clay minerals are covered with electronegative charges. These may in part be due to dissociation of the hydroxyl ion into separate oxygen and hydrogen ions. The former are negatively charged and remain part of the basic structure, whereas the latter are positively charged and are loosely held on the external surface. In a strongly acidic environment the hydrogen ions remain adsorbed on the crystal, but under more alkaline conditions may be replaced by other cations such as calcium. Another source of negative charges may be 'broken bonds' at mineral edges and corners. A third factor generating electronegative conditions is ionic substitution, particularly where a high valency cation is replaced by one of lower valency. Examples are afforded by the substitution of Mg^{++} for Al^{+++} in such minerals as montmorillonite, and of Al^{+++} for Si^{++++} in the tetrahedral sheet of some of the hydrous micas.

The number of replaceable ions that can be adsorbed on to a clay mineral reflects its cation exchange capacity and is a readily measured property. Owing to slight differences in structure and composition, samples of clays that are nominally the same yield different values for cation exchange capacity. Nevertheless, the figures for montmorillonite are often ten times those for kaolinite and four times those for a typical hydrous mica. The high value for montmorillonite stems from the widespread ionic substitution, whereas most of the charges in the case of kaolinite arise from dissociation of the hydroxyl. A further factor is the loose structure of montmorillonite which permits water to penetrate between the crystal units. Adsorption can then take place not only on the external surface but also on

Mechanical properties. The contrasts in crystal structure and electrical charge produce important variations in mechanical properties. Some of these result simply from differences in particle size. For example, the rigid kaolinite lattice is associated with particles that are often about 0·002 mm in diameter, whereas montmorillonite crystals rarely exceed 0·001 mm; the hydrous micas tend to be of intermediate size. A complicating factor is the differential expansion which the minerals undergo when wetted. The loosely structured montmorillonite tends to swell the most owing to the penetration of water between the crystal units. A regolith rich in this mineral therefore experiences correspondingly greater volumetric changes. This in turn affects water percolation owing to the tendency for deep surface cracks to develop during dry spells; these may admit large quantities of rainwater before the superficial layers are sufficiently soaked to close the cracks.

The most significant variations, however, result from the different adsorptive powers of the clay minerals. The electrostatic attraction of the typical mineral holds a layer of ions bonded to it. The concentration of ions decreases with distance from the mineral into the surrounding solution. Also bonded to the clay particle is a sheath of water molecules. Those next to the clay are extremely tightly held and give the molecular film properties resembling those of a solid. Further away from the mineral surface the water is still influenced by the electrical charges, but behaves more like a very viscous liquid. Although the thickness of these adsorbed layers may be only 0·005 microns, they exercize a profound influence on such attributes as plasticity and cohesion.

Plasticity was originally investigated by Atterberg. He divided into five stages the range from a clay so saturated with water that it behaves as a liquid to the same clay acting as a solid. The boundaries between his stages have gradually been modified and standardized, but still retain the name Atterberg limits. The three most important are:

1. The liquid limit which is the water content when a pat of the clay placed in a special cup just closes a 2 mm groove after being agitated by a cam which raises the cup and allows it to fall on to a hard rubber base 25 times
2. The plastic limit which is the water content at which the soil begins to crumble when rolled out into long threads 3 mm in diameter
3. The shrinkage limit which is the water content below which further moisture losses cause no reduction in volume.

An important parameter derived from these three measured values is the plasticity index. This is simply the numerical difference between the liquid and plastic limits, both of which are normally expressed as percentages of the dry weight of sediment. It defines the range of water content over which the clay remains plastic. All three Atterberg limits and the plasticity index vary considerably according to the type of clay mineral involved. For instance, montmorillonite remains plastic over a much wider range of water content than does kaolinite, with the hydrous micas as usual occupying an intermediate position. In making such comparisons it is essential to test material of the same grain size since the higher the proportion of fine particles the higher the plasticity index; this is due to the greater surface area on which adsorption can occur per unit weight.

The influence of adsorbed water on the cohesion of clay particles can be demonstrated by mixing them instead in paraffin; the paraffin molecules lack the polarizing properties of water, and a drastic fall in cohesion results. As might be expected montmorillonite clays are normally more cohesive than either kaolinite or illite clays. However, it is not only the mineral type and amount of adsorbed water that affects the shear strength of a clay. Another significant factor is the number and nature of adsorbed cations. The nature of the cations obviously depends upon their availability, which in turn depends upon the weathering processes and the chemical composition of the rock on which they are acting. It is also found that certain cations are adsorbed more strongly than others, and that those which are loosely held may be replaced by others which are more tightly held. A solution of a neutral salt allowed to percolate through a clay will often emerge altered in composition owing to the loss of some of its original cations by adsorption and the addition of others by release from the clay surface. The actual amount of exchange depends on the strength of the solution since an equilibrium tends to be established between it and the adsorbed cations. The order of strength of adsorption when the cations are present in equivalent quantities is normally Al Ca Mg K Na. These reactions are fundamental for the soil scientist since they are a major factor in determining soil fertility. For the geomorphologist their significance lies in the varying physical properties they can confer on what is basically the same clay assemblage. This was demonstrated many years ago by Sullivan who treated nine specimens of an identical clay with different solutions containing only one type of cation. When the specimens were tested it was found that, with the same water content, their strength varied greatly. Of the common cations, sodium and calcium produced relatively weak clays, potassium and hydrogen relatively strong. At certain water values, the potassium-clay was found to have a shearing strength double that of the sodium-clay. On the other hand, sodium-rich clays tend to have a higher liquid limit and only become semifluid with an unusually large water content.

It is evident that the clay minerals constitute a very significant but very variable component of the weathered mantle. Although the foregoing account touches upon only a few of their properties, it demonstrates that many factors influence their behaviour as mobile constituents of a hillslope. In considering the regolith as a whole mention must also be made of the organic component. Concentrated in the soil horizons the complex organic compounds known collectively as humus endow the superficial layers with qualities not found deeper in the weathered mantle. Humus is a colloidal substance having an adsorptive capacity much greater than that of montmorillonite. Its cation exchange capacity is more than twice that of most mineral clays and it has the ability to hold exceptionally large quantities of water. The range of swelling and contraction is correspondingly large, and the presence of humus often confers on the surface horizons a distinctive crumb structure which, in turn, affects their infiltration capacity and general erodibility.

EVANS, I. S., 1969–70. 'Salt crystallization and rock weathering: a review', *Revue Geomorph. dyn.*, 19, 153–77.

GERRARD, A. J. W., 1974. 'The geomorphological importance of jointing in the Dartmoor granite', *Inst. Br. Geogr. Spec. Pub.*, 7, 39–51.

GOUDIE, A., et al., 1970. 'Experimental investigation of rock weathering by salts', *Area 1* (4), 42–8.

GOUDIE, A., 1974. 'Further experimental investigation of rock weathering by salt and other mechanical processes', *Z. Geomorph. Supplementband*, 21, 1–12.

HILLS, R. C., 1970. 'The determination of the infiltration capacity of field soils using the cylinder infiltrometer', *Tech. Bull. No. 3*, Brit. Geomorph. Res. Group.

KING, L. C., 1958. 'The problem of tors', *Geogrl. J.*, 124, 289–92.

KING, L. C., 1975. 'Bornhardt landforms and what they teach', *Z. Geomorph.*, 19, 299–318.

LINTON, D. L., 1955. 'The problem of tors', *Geogrl J.*, 121, 470–87.

PEDRO, G., 1957. 'Nouvelles recherches sur l'influence des sels dans la desagregation des roches', *C. r. hebd. Seanc. Acad. Sci., Paris* 244, 2822–4.

PEEL, R. F., 1974. 'Insolation weathering: some measurements of diurnal temperature changes in exposed rocks in the Tibesti region, central Sahara', *Z. Geomorph. Supplementband*, 21, 19–28.

PELTIER, L. C., 1950. 'The geographic cycle in periglacial regions as it is related to climatic geomorphology', *Ann. Ass. Am. Geogr.*, 40, 214–36.

SKEMPTON, A. W. and R. D. NORTHEY, 1952. 'The sensitivity of clays', *Geotechnique*, 3, 30–53.

STRAKHOV, N. M., 1967. *Principles of Lithogenesis* (trans. J. P. Fitzsimmons), Oliver and Boyd.

SWEETING, M. M., 1972. *Karst Landforms*. Macmillan, 1972.

WIMAN, S., 1963. 'A preliminary study of experimental frost weathering', *Geogr. Annlr.* 45, 113–21.

Selected bibliography

Two relatively modern and complementary texts are C. D. Ollier, *Weathering*, Oliver and Boyd, 1969, and F. C. Loughnan, *Chemical Weathering of the Silicate Minerals*, Elsevier, 1969.

Fluvially-eroded terrains are, in essence, assemblages of hillslopes. An appreciation of the mechanisms responsible for shaping the slopes is therefore critical to an understanding of the landscape as a whole. Yet hillslope studies encounter such formidable problems that for a long time workers were deterred from making the necessary field investigations; the subject was regarded as suitable for theorizing but not for detailed field study. Fortunately the last decade or two has seen a profound change in attitude, and although the problems have not been overcome they are now being more clearly defined and possible solutions actively investigated.

In the general realm of slope studies three particularly significant spheres of investigation may be identified. The first concerns the physical properties of the materials from which the slope is fashioned and has already been discussed in Chapter 7. The second concerns the measurement of slope form. It is clearly vital that some quantitative method of description should be devised, and possible approaches to this topic constitute the subject matter of pp.139–44 below. The third sphere of investigation concerns the nature of hillslope processes. A major difficulty arising in this context is the multiplicity of such processes. Some are chemical, others mechanical; some are confined to the surface layers, others penetrate much more deeply; some go on almost continuously, others are highly intermittent; some are triggered by temperature changes, others by humidity changes. So many different processes operate simultaneously that the slope form cannot be ascribed to any one acting in isolation. The range of processes that assist in the shaping of hillslopes are discussed as the second part of the present chapter on pp. 144-61.

The intricate relationships between form, process and material cannot be stressed too strongly; analysis of individual items under separate headings is purely a matter of convenience. Equally, the dangers of assuming that presentday morphology is due entirely to current processes should be noted. There is abundant observational evidence to show that the form of many hillsides was roughed out by processes no longer operative. It needs to be remembered that many areas of the globe have experienced very significant climatic changes during the last 10 000 years or so.

Measurement of form

Morphological mapping The conventional method of depicting relief is by means of contours. Although these have immense advantages for the cartographer in conveying the maximum information in the minimum space, for the geomorphologist they have severe limitations. This is especially true of contours surveyed as part of a national mapping system. The interval between contours is commonly too large for them to delineate other than the broad pattern of relief. Details of slope form are inevitably lost. Moreover, care needs to be exercized since often no clear distinction is made on topographic maps between accurately surveyed contours and form lines interpolated on the basis of field notes without any actual measurement. Occasionally the crenulations of surveyed contours can be distinguished by eye from the smooth curves of form lines; the contrast illustrates the inadequacy of many topographic sheets for accurate slope studies.

For many decades individual geomorphologists have supplemented the normal published contours by employing special schemes of morphological mapping. Their aim has been to depict the surface form of the ground in much greater detail than is possible by widely spaced contours. Originally there was a tendency for each worker to be highly selective, choosing to depict only those morphological features which he believed to have significance for the geomorphic history of the region under study. More recently it has been argued that the primary aim of morphological mapping should be different. It should endeavour to provide an accurate and full two-dimensional representation of the three-dimensional form of the ground surface. This apparently simple objective has proved remarkably difficult to achieve.

Morphological mapping is founded on the premise that the complex shape of hillslopes and valley sides can be resolved into a finite number of units capable of delimitation in the field. Most schemes that have been proposed involve recognition of two basic types of unit. On the one hand are facets over which the profile slope remains constant, on the other are curved units on which the profile gradient is constantly changing. Curved units are further divided into concave and convex categories. Two types of boundary separating these morphological units are usually distinguished. A break of slope is an angular discontinuity, a change of slope is a more gradual transition from one unit to another. One such scheme employing this general approach is illustrated in Fig. 8.1. Certain terrains lend them-

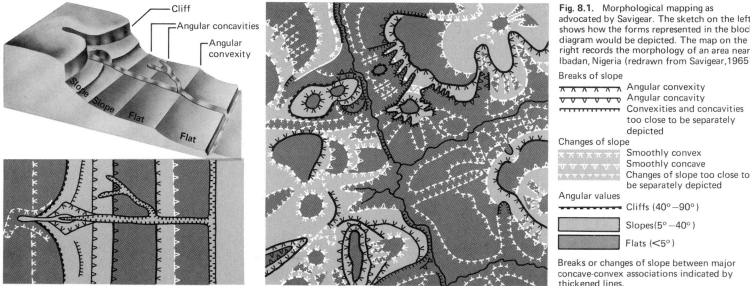

Fig. 8.1. Morphological mapping as advocated by Savigear. The sketch on the left shows how the forms represented in the block diagram would be depicted. The map on the right records the morphology of an area near Ibadan, Nigeria (redrawn from Savigear, 1965).

Breaks of slope

⋏⋏⋏⋏⋏ Angular convexity
ᴠᴠᴠᴠᴠ Angular concavity
⊤⊤⊤⊤⊤ Convexities and concavities too close to be separately depicted

Changes of slope

⋏⋏⋏⋏⋏ Smoothly convex
ᴠᴠᴠᴠᴠ Smoothly concave
Changes of slope too close to be separately depicted

Angular values

⋅⋅⋅⋅⋅⋅ Cliffs (40°–90°)

Slopes (5°–40°)

Flats (<5°)

Breaks or changes of slope between major concave-convex associations indicated by thickened lines.

selves much more readily to depiction in this way than do others. Satisfactory results were obtained in several early studies in the southern Pennines where many of the bold escarpments produced by alternating gritstone and shale outcrops can be resolved into facets separated by breaks of slope. Much less success attended efforts to employ the technique in areas such as clay lowlands dominated by concavo-convex slopes.

In evaluating the success of any mapping scheme a major criterion must be the extent to which the results can be replicated. Experiments have been carried out to test whether rules can be formulated that will ensure different workers producing closely similar maps of the same area. In theory, for instance, contour maps produced by different survey parties should be identical. However, in the case of morphological mapping schemes, no set of field instructions has yet been devised that will guarantee closely comparable maps being produced. In order to achieve uniformity more and more complex rules have been suggested, but these have both failed to attain the desired end and have also added greatly to the range of field measurements that need to be made. The increase in objectivity has not appeared commensurate with the extra effort involved.

Morphological mapping is undoubtedly a very useful technique for the rapid recording of hillslope form in approximate terms. It can greatly supplement the information provided by normal contours, and has important applications, for instance, n the fields of soil survey and terrain analysis. However, its value in detailed quantitative slope studies appears at the present time to be rather limited.

Determination of slope-angle frequencies A totally different approach to hillslope studies has been adopted in a number of methods that endeavour to analyse the frequency with which given slope-angles recur. One procedure is simply a cartographic exercise employing the contours shown on a normal topographic map. The inter-contour distances equivalent to selected critical angles are first calculated and a template scale drawn. Using the scale, shading is then applied to areas where the contours are closer than the selected values; by employing denser shading on steeper slopes a good visual impression of the relief is obtained. The total area falling within each slope-angle class can be measured, and an estimate of the frequency of each class thereby obtained. Because manipulation of the template scale is extremely tedious an alternative photographic technique has sometimes been tried. This fuses contours at a pre-determined critical spacing into a single blurred image. By repeating the photography at different scales a map can eventually be compiled showing the distribution of chosen slope angles. The major limitation of these isoclinal maps, whether constructed manually or by some mechanical means, is that the contours on which they are based contain a strictly limited amount of information about actual surface form. It has to be assumed, firstly, that the ground between contours slopes at a uniform angle and, secondly, that changes in angle occur only along the published contours. Patently, neither of these assumptions is totally valid and the procedure can only yield a first approximation to the true frequency distribution of slope angles.

In theory similar results may be obtained rather more rapidly by using a sampling frame. The gradient can be calculated at randomly selected points by presuming a constant slope between the nearest higher and lower contours. Strahler, who has pioneered many methods of analysing slope forms, employed this technique in studies of water-eroded landscapes in the Appalachians and the mountains of southern California. An important modification that can be readily applied is to make the measurements of slope angle in the field rather than derive them from topographic maps. Admittedly the procedure is more time-consuming but the increased precision often makes it a desirable innovation. To a certain extent the return on the extra effort depends on the nature of the relief. In an accidented area dominated by linear slope profiles the differences between field and map measurements may be quite small. On the other hand in a lowland region with concavo-convex profiles, sharply contrasting results can emerge; this is especially true where the contour interval is large relative to the amplitude of relief.

Profiling A third approach to the analysis of hillslope forms is by measurement of representative profiles. Before the method can be employed two decisions have to be made: how are the lines of the profiles to be selected, and what procedure is to be used in the actual surveying. In the past the choice of profile lines has generally been made subjectively with a particular purpose in mind. This is open to the criticism that the choice may be affected by an unconscious desire to substantiate some preconceived idea. On the other hand, randomly selected survey lines may prove difficult of access and yield profiles which the geomorphologist regards as not strictly comparable. For example, certain of the slopes may be subject to active stream undercutting whereas the majority are free from such effects. Much depends on the purpose for which the study is being made and the nature of the profile analysis that it is hoped to undertake. Profiles might, for example, be chosen by surveying upslope and downslope from randomly selected points. Young, however, has advocated a sampling scheme in which mid-slope lines are first drawn for each catchment halfway between stream and watershed (Fig. 8.2). These mid-slope lines can then be used for sampling purposes, with points selected along them according to a suitable plan. Through the points profiles are surveyed from stream to watershed.

Several different methods have been used for the actual survey. The most popular equipment consists of ranging poles set up at appropriate points, a tape to measure the distances between them and an abney level to determine the corresponding gradients. Normal practice in early studies was to site the ranging poles at obvious changes of slope, ignoring the considerable differences in ground distance that this might involve. Later work has tended to employ a constant ground distance, largely on the score that choosing significant changes of slope can be highly subjective. Any worker who has used uniform measured lengths is aware of occasions when this procedure has smoothed out abrupt breaks of slope; on the other hand, with the measured length kept sufficiently short to ensure accurate representation of all breaks of slope, there appears to be an unnecessary waste of time in surveying lengthy segments of constant gradient. Despite the tediousness, profiles are probably best surveyed by means of short uniform distances, although much must depend on the final type of analysis it is proposed to employ.

Two instruments have been designed specifically for profile surveying. The simplest is the slope pantometer which consists of a four-sided equilateral wooden frame with adjustable corners. Into the frame is set a spirit level which allows the operator to position two of the sides vertically. An angular scale then measures the slope of the remaining two sides, and since the lower is resting on the ground this is the gradient at that particular point. The frame can conveniently be made 1·5 m square so that the measured length is much shorter than is normal with ranging poles and a tape. Speed of operation is high but vegetation and minor surface bumps can cause difficulties. The second instrument is much more complex. Known as a profile recorder its essential components are a set of wheels linked to devices for recording ground distance and changes in slope angle. When the recorder is pushed up or down a slope, the ground distance is measured by a simple mechanism that counts revolutions of the wheels; simultaneously a second mechanism monitors the changing relative levels of the two axles. By internal linkages a continuous profile is automatically drawn on a rotating drum. In the hands of a skilled operator tests have shown that very accurate profiles can be obtained; however, the machine is expensive and only works well on unvegetated or grassy slopes.

Analysis of measured profiles normally involves division of the slope into rectilinear, convex and concave portions. This may be done subjectively, but different workers are liable to vary in their interpretation of both the number of slope units and the positions of the intervening boundaries. It is preferable, therefore, if the division can be carried out in accordance with a set of rules eliminating operator bias. Young has proposed an elaborate scheme, known as the best units system, based upon the variability of angle to be allowed within a single slope unit (Fig. 8.2). He recognizes two basic types of unit, a segment which is a portion of a profile on which the angle remains approximately constant, and an element which is a portion of a profile on which the curvature remains approximately constant. The need is to define in quantitative terms what is meant by 'approximately'. For this purpose Young employs the value known as the coefficient of variation, which effectively expresses the deviation from the mean as a proportion of the mean value itself. This parameter is chosen since it is believed, for instance, that a deviation of 1° is often unimportant on steep slopes but may be highly significant on gentle slopes. A segment is consequently defined as a portion of a slope profile within which the coefficient of variation of angle does not exceed a specified value; an element is defined as a portion of a slope profile within which the coefficient of variation of curvature does not exceed a specified value. At first sight the system, only given in outline here, may seem unduly complicated. However, rigour

demands that all possible circumstances be covered, and the **143** need for such a scheme underlines the complexity of normal hillslope forms. Partitioning of a profile into best units is preferably done by computer since the calculations tend to be extremely tedious.

Provided methods of both surveying and partitioning are the same, there are several different ways in which profiles can be

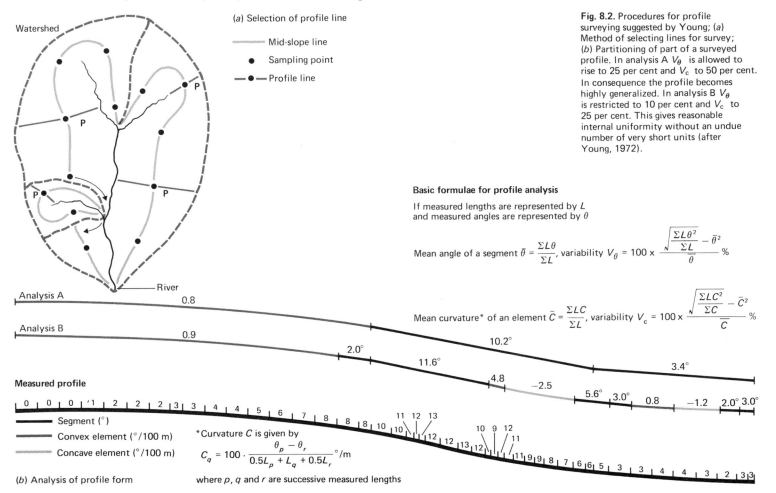

(a) Selection of profile line

— Mid-slope line

● Sampling point

—●— Profile line

Fig. 8.2. Procedures for profile surveying suggested by Young; (a) Method of selecting lines for survey; (b) Partitioning of part of a surveyed profile. In analysis A V_θ is allowed to rise to 25 per cent and V_c to 50 per cent. In consequence the profile becomes highly generalized. In analysis B V_θ is restricted to 10 per cent and V_c to 25 per cent. This gives reasonable internal uniformity without an undue number of very short units (after Young, 1972).

Basic formulae for profile analysis

If measured lengths are represented by L and measured angles are represented by θ

Mean angle of a segment $\bar{\theta} = \dfrac{\Sigma L\theta}{\Sigma L}$, variability $V_\theta = 100 \times \dfrac{\sqrt{\dfrac{\Sigma L\theta^2}{\Sigma L} - \bar{\theta}^2}}{\bar{\theta}}$ %

Mean curvature* of an element $\bar{C} = \dfrac{\Sigma LC}{\Sigma L}$, variability $V_c = 100 \times \dfrac{\sqrt{\dfrac{\Sigma LC^2}{\Sigma C} - \bar{C}^2}}{\bar{C}}$ %

*Curvature C is given by

$$C_q = 100 \cdot \frac{\theta_p - \theta_r}{0.5L_p + L_q + 0.5L_r} °/m$$

where p, q and r are successive measured lengths

— Segment (°)
— Convex element (°/100 m)
— Concave element (°/100 m)

(b) Analysis of profile form

compared. Given that the normal slope is composed of an upper convex element, a lower concave element and an intermediate segment, the relative lengths of each of these units may be analysed. It has been shown that rectilinear segments are much more common than was at one time supposed. They frequently occupy over 10 per cent of the total profile length and occasionally surpass 50 per cent. This is true even in areas such as the chalklands that have traditionally been described as composed of concavo-convex slopes. Concave elements rarely occupy as much as half of a profile and in some situations are completely absent. Convex elements, on the other hand, are virtually always present and, according to Young, frequently occupy over 40 per cent and sometimes over 80 per cent of British profiles. The same author has described slopes in Brazil where more than 95 per cent of the total length is composed of the upper convexity. There seems no doubt that significant contrasts exist in the proportion of profiles occupied by the different basic units, but agreement about the underlying causes has yet to emerge. Another aspect of profiles that is frequently the subject of comparison is the length and angle of the rectilinear segment. One reason for interest in this parameter is the belief that it may represent a limiting gradient under given conditions of process and material. Finally, mention may be made of efforts to analyse the form of concave and convex elements. This has most frequently been done by fitting simple polynomial curves in order to discover whether the elements exhibit close mathematical similarities. The results so far have not proved very illuminating and much controversy has centred round the extent to which apparently curved units are actually composed of several short rectilinear segments.

Particulate movement

Particulate movement refers to the downhill transfer of individual particles moving independently of the rest of the slope materials. It differs from mass movement to be discussed below in that surface debris alone is displaced in this way. Many different processes can give rise to particulate movement, but two important groupings can be recognized. The first encompasses those which induce instability on a free rock face; this leads to the dislodging of fragments that may range in size from individual sand grains to huge boulders tumbling down to accumulate as basal scree. The second grouping includes those processes that operate when heavy precipitation bombards a slope with large raindrops and ultimately generates a surface run-off capable of carrying loose sand and silt particles further downslope. This two-fold division of the processes responsible for particulate movement will be used as the basis for further discussion under the general heads of rockfall and surface wash.

Rockfall Rockfalls are normally limited to free faces that slope at angles of over 40°; a limiting gradient may be specified since granular debris can stand at angles of repose of up to about 40° with the result that dislodged material on lesser slopes will not automatically roll or slide further downhill. The frequency and magnitude of falls may be assessed in a variety of ways. One of the most common is to place some form of trap at the foot of a free face and to measure the amount of debris that is caught in a particular time interval. In a celebrated study of erosional processes at Karkevagge in northern Scandinavia, Rapp employed hessian carpets and wire netting of 10 mm mesh to catch the falling fragments. He emphasized, in addition, the value of simple observations on the volume of fragmentary material resting on snow banks during early summer; under conditions like those of northern Lappland a high proportion of the rockfall appears to be concentrated in late spring and early summer. An extra inventory may be made in late summer, the freshly fallen boulders then being identified by the crushing of the vegetation. Rapp estimated that in an 8-year period of observation the mean annual volume of rockfall from faces with a vertical area of 900 000 m² amounted to some 50 m³. This is equivalent to a rate of retreat of 0·06 mm yr⁻¹. The techniques so far described are suitable for sites where there appear to be numerous small-scale falls occurring each year. More difficult to evaluate is the overall effect of large but infrequent collapses. By collating the observations of residents in the National Parks and Monuments of the south-western United States, Schumm and Chorley have shown that large-scale falls in this arid region are currently a significant mechanism of rockwall retreat. Relatively recent falls can often be located through the contrasting appearance of the newly exposed rock faces; for example, in dry regions they may lack the desert varnish that coats all the older surfaces. However, the actual ages of the falls cannot be defined in this way and care needs to be exercized since, prior to collapse, there is often a gaping joint down which weathering agents can penetrate and alter the rock face.

In the vast majority of rockfalls the initial detachment occurs along a prominent joint system. From an analysis of the stresses in a vertical face Terzaghi has shown that moderately strong unjointed rock should be able to sustain vertical cliffs 1 500 m high. Yet such cliffs do not occur in nature and many slopes of much lower height have collapsed. Terzaghi concludes that limiting heights must be determined by mechanical 'defects' such as joints and faults, rather than by the strength of the rock itself. Failure occurs when the spread of joints through the material so reduces its overall strength that it is no longer able to withstand the shearing stresses. The pattern of joints may be as significant as their frequency. In massive igneous rocks sheeting structures commonly develop parallel to the rock face, while cross-joints divide the sheets into the smaller fragments that constitute the typical rockfall debris. In stratified rocks, on the other hand, the most important single factor is the relationship of bedding and joint planes to the slope (Fig. 8.3). If it is assumed that there is no cohesion across these planes the strata may be regarded as little more than a body of dry masonry. Where the dip is outwards, the component blocks will remain stable so long as their bases are less steeply inclined than the relevant angle of

Fig. 8.3. Schematic representation of the effect of rock structure on the angle of a cliff face. In the case of stratified rocks dipping into the cliff face a critical factor is the relative spacing of the joints (c) and the bedding planes (d). White lines indicate the angle at which slip along joints and bedding planes is assured to start.

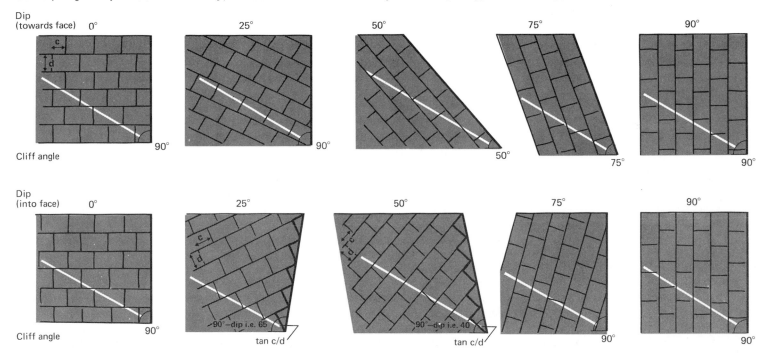

146 friction; the maximum slope angle will be vertical. Where their bases are more steeply inclined than the angle of friction, the blocks will be unstable and will tend to slip downslope; the maximum slope angle will equal that of the dip. Where the bedding dips inwards the maximum possible slope angle will normally be vertical. However, if there are master joints cutting across all the strata, slipping may take place along these with a consequent reduction in gradient. If the joints are slightly offset in the various strata, a crenulate face may develop with the overall slope angle governed by the ratio of the offset distance to the distance between adjacent bedding planes. This theoretical

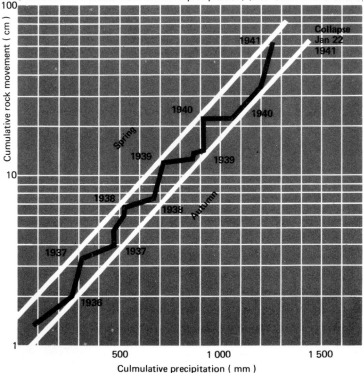

Fig 8.4. The movement of Threatening Rock, New Mexico. The steps in the curve represent the summer periods, while the risers represent the winter periods (i.e. maximum movement but minimum precipitation) (after Schumm and Chorley).

analysis is obviously very idealized. It assumes no cohesion across the bedding planes and a simple regularity of the joint systems. Moreover, in most stratified rocks there are significant lithological changes to be taken into account. However, it does serve to emphasize the critical role played by the jointing system in a typical rockfall.

The incidence of rockfall commonly displays a striking seasonality. Many investigators have found maxima to occur in either autumn or spring, or in some cases in both periods. One of the best documented examples comes from Norway where almost 60 per cent of the rockfalls have been found to occur in the 4 months of April, May, October and November. In high latitudes most workers have associated the peak frequencies with a large number of freeze–thaw cycles. Another potentially important factor, however, is water occupying the joint system. This acts in an analogous fashion to porewater pressure in reducing the mechanical strength of the rock and has been termed cleft-water pressure by Terzaghi. He has argued that the Norwegian data are best explained as due to high cleft-water pressure; in October this might be ascribed to heavy autumnal rain, in April to snowmelt while exits from the joint system are still plugged by ice. Another illustrative example is provided by the measurements made during the 1930s on the sandstone monolith known as Threatening Rock in Chaco Canyon, New Mexico. Resting on a bed of shale, this huge vertical slab, 50 m long, 30 m high and 10 m thick, was slowly moving away from the cliff of which it was formerly part. One steel bar was cemented to the top of the slab and another to the cliff face. For 5 years monthly measurements of the gap were made. When cumulative movement is plotted against date of measurement (Fig. 8.4) five periods of rapid movement and five periods of relative stability can be detected. The intervals of rapid displacement all span the winter season. This is a relatively dry period in Chaco Canyon so that it is difficult to attribute the accelerated movement to wetting of the underlying shale by rain. On the other hand, much of the winter precipitation is in the form of snow and this may be a more effective wetting agent than the brief summer storm. The obvious alternative is that freeze–thaw action during the winter plays a significant role in weakening the structure of the shale. Eventually in January 1941, after the gap had widened by over half a metre, the 30 000-tonne monolith toppled over and was destroyed.

Finally in this discussion of rockfall, reference must be made to the scree slopes that develop as a result of debris accumulation. Most scree profiles are either rectilinear or gently concave, with a maximum angle near the apex of between 30° and 40°. The largest fragments tend to be concentrated around the lower margins of the scree, possibly due to their extra momentum as they fall. It must not be assumed, however, that all scree material is added by simple rockfall. In high latitudes at least an equal volume may be contributed by snow and slush avalanches carrying with them considerable quantities of both coarse and fine debris. Caine has described how, in the South Island of New Zealand, the upper part of the scree is nourished primarily by rockfall whereas the lower part is supplied mainly by spring avalanching. The passage of an avalanche may also assist in redistributing material that had earlier lodged near the top of the scree.

The form of the rock face concealed beneath a scree is generally unknown, although several theoretical models have

been proposed which would provide some guidance as to the likely shape. The simplest case was considered over a century ago by Fisher (Fig. 8.5). He assumed a vertical cliff retreating without loss of angle, together with a constant scree slope. He further simplified the model by assuming no loss of material and the same bulk density for in situ bedrock and accumulated scree. Granted these conditions it can be shown that the buried rock surface should be parabolic in form. Later workers have progressively relaxed some of the constraints imposed by Fisher. For example, cases have been considered where the initial face is not vertical or where there are volumetric changes associated with scree accumulation. These changes may be expressed as the ratio of scree volume to the volume of falling rock, and are conventionally designated by the constant c. In Fisher's original model $c=1$, but several factors can significantly alter this value. The most obvious is the need to take account of the volume of talus voids, thereby raising the value of c. On the other hand, some material may be lost by weathering and removal at the base of the scree, thereby reducing the value of c. The limiting case occurs where virtually all the debris is removed as rapidly as it falls, in which case c will approach zero. It is not too difficult to demonstrate that, so long as c has a positive value, the buried rock face will be convex upwards with its lowest point tangential to the original cliff and its highest point tangential to the ultimate scree surface. If $c=0$, however, the rock face becomes rectilinear and slopes at an angle corresponding to that at which the scree would stand if it were present. This is true whether or not the cliff face maintains its steepness during retreat. The rather surprising prediction is that rockfall with rapid basal removal should tend to produce bedrock slopes inclined at angles corresponding to those of screes. There is observational evidence suggesting that this prediction may be fulfilled in suitable mountain environments, although local factors such as lithological variations often complicate the picture. It also needs to be remembered that rockfall rarely acts alone and there is usually a simultaneous functioning of many different processes.

Surface wash Surface wash is a term denoting the downhill displacement of superficial slope materials by moving water. Two distinct processes may contribute to the displacement,

Fig. 8.5. A model of cliff development where scree accumulates at the foot of a vertical slope. The model assumes no loss of material and no change in bulk density.

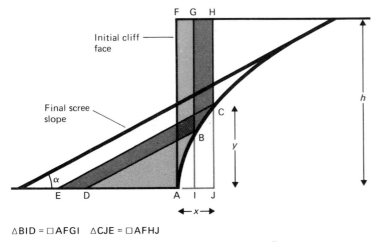

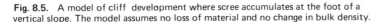

$\triangle BID = \square AFGI$ $\triangle CJE = \square AFHJ$

$\frac{1}{2}y^2 \cdot \cot \alpha = h \cdot x$ $y^2 = 2h \cdot x \cdot \tan \alpha$ $y = (2h \cdot x \cdot \tan \alpha)^{\frac{1}{2}}$

namely raindrop impact and overland flow. Normal raindrops vary in diameter from 0·5 to 7 mm and attain terminal velocities ranging from under 1 m to over 9 m sec^{-1}. From these figures it is clear that the momentum of individual drops differs enormously, and that a heavy thunderstorm is potentially a vastly more effective erosional agent than a period of light orographic rainfall. Drops falling on a flat granular surface commonly produce small impact craters. In addition splashes carry fine particles many millimetres from the original landing point. On a slope there is differential movement in a downhill direction, and both field and laboratory experiments have demonstrated how effective this can be in displacing surface particles. This was apparent in research earlier this century on the badly eroded farmlands of the southern United States. More recent laboratory investigations have shown that fine sand grains can occasionally be splashed as much as a metre in a single leap. It is, of course, true that the effectiveness of raindrop impact depends very much on the density of the vegetation cover. A complete foliage cover can entirely absorb the energy of the raindrops and prevent any dislodgement of the soil particles ; moreover, the cohesion of organically rich soils also acts as a brake on movement. Splash erosion reaches its peak effectiveness on cultivated soils where no protective measures have been taken, and secondly in desert areas where there are considerable areas of bare ground between individual plants. It may also be significant on patches of bare ground beneath a forest cover owing to water dripping from leaves and branches ; large drops formed in this way can approach their maximum theoretical velocity after falling distances of some 7 or 8 m. Nearly all workers agree that splash erosion, if it acted alone, would produce convex hill crests.

In general overland flow is more difficult to study than stream flow. Fortunately scientists in many diverse fields have endeavoured to investigate its frequency and physical characteristics, so that a very large body of data has now been collected. The theoretical basis of surface run-off was analysed in a geomorphological context by Horton. He pointed out that surface discharge is primarily a function of rainfall intensity, infiltration capacity and position on the slope. So long as the soil is able to absorb all the precipitation no flow across the ground is to be expected ; on the other hand, if the precipitation rate exceeds the maximum infiltration rate, the excess volume will be discharged across the surface. As pointed out in Chapter 7, methods for the field measurement of infiltration capacity have been devised. Yet the wide variety of factors influencing local infiltration rates makes direct comparison with recorded precipitation intensities rather hazardous. As a general indication of the magnitudes involved, infiltration capacity normally lies between zero and about 100 mm hr^{-1}, while heavy storms may produce precipitation intensities in the range 20 to 50 mm hr^{-1} ; the implication appears to be that on many slopes direct surface run-off is only a rare occurrence. The third factor emphasized by Horton is position on the slope in relation to the nearest divide. He argues that, given the same excess volume to be evacuated from all points on a slope, the actual discharge will increase as a function of distance from the watershed. On a rectilinear slope this means that the depth of flowing water will increase progressively in a downhill direction. There will also be an increase in velocity so that the ability of the water to dislodge and transport particles will be much greater near the foot of a slope than at its summit. If the surface particles are accorded a measurable resistance to displacement, it follows that near the crest they may remain undisturbed whilst lower down the slope they are picked up and transported. Horton contends that it should be possible to identify what he terms a 'belt of no erosion' adjacent to the watershed ; while the concept seems sound, a better descriptive name would be a belt of no sheetwash erosion since many other processes are capable of operating in the summit zone.

The term sheetwash refers to one particular form of overland flow in which the water moves as a thin and relatively uniform film. As discharge increases flow tends to be concentrated into a series of slightly deeper channels and is known as rillwash. This in turn grades into gully flow and ultimately into channelled stream flow. The nature of fluid movement in overland flow has occasioned much discussion, some authors contending that sheetwash is primarily laminar, others that it is entirely turbulent. Emmett has examined the whole question in a series of carefully designed experiments using both laboratory flumes and specially prepared hillslope plots. In both cases artificial rain was applied at known rates and detailed measurements made of depth and velocity of flow. By varying the rate of water-application changes in depth and velocity were induced and from the body of data collected in this way an attempt was made to assess whether flow was laminar, turbulent or a mixture of the two. Theoretically, as discharge rises, depth increases more slowly in laminar flow than in turbulent flow. The relevant relationships are $D \propto Q^{0.33}$ for laminar flow and $D \propto Q^{0.6}$ for turbulent flow, where D is depth and Q is discharge. In the laboratory flumes the transition from laminar to turbulent flow was readily identified as the exponent in the above formulae jumped quite abruptly from the lower to the higher value. On the field plots the change was much more difficult to detect, probably owing to surface roughness, but Emmett concluded that flow is primarily laminar near the crestline of a ridge and progressively more turbulent as discharge increases further down the slope.

The transporting action of overland flow has been studied by sinking a long shallow trough across a hillslope. The lip of the trough is so positioned that it traps surface water and debris, but excludes water moving through the soil; a cover is fitted to prevent extraneous sediment being introduced by rainsplash. One difficulty with such an installation is that it gives no indication of the area of slope from which the trapped material is derived. To overcome this problem and provide additional information a series of troughs may be inserted at intervals up a hillside. An alternative technique for assessing the effect of surface wash is to measure directly any changes in slope elevation. This is most often done by driving rods through the soil into the underlying parent material. Provided the rods are firmly fixed, the length of the protruding end may be monitored to determine the rate of surface change. In one such study in New Mexico a loose washer was slipped over the top of each stake; the washer provided a firm base on which to make the measurements and also allowed ready identification of any site where deposition had taken place. Another possible procedure is to use a specially constructed horizontal frame, supported at each corner by a deeply buried leg; the changing elevation of the ground may then be monitored by means of light-weight rods slotted through the frame.

There seems little doubt that surface wash is an important slope-forming process in arid and semi-arid areas (Fig. 8.6).

Contributory factors include the sparse vegetation cover, permitting a relatively uniform and unimpeded flow of water, and the very intense precipitation of local convectional storms. Several studies have shown that the effectiveness of overland flow depends as well on the surface being rendered suitably friable. By comparing the form of clay slopes in Colorado during different seasons Schumm has identified an annual cycle. Winter frosts weaken the surface structure so that individual particles are readily dislodged by raindrop impact in spring and early summer. At the same time the precipitation compacts the surface, thereby enhancing run-off which, during late summer, cuts shallow rills that persist until the return of winter frosts. Measurements with stakes have shown that certain clay and shale slopes in the arid south-western United States are retreating remarkably rapidly under the influence of such an annual cycle. Values for surface lowering of over 2 mm yr^{-1}, and occasionally over 5 mm yr^{-1}, have been recorded in a number of separate investigations.

The magnitude of the erosion measured on bare arid slopes dwarfs that recorded on vegetated temperate slopes. Stakes have almost invariably proved too crude a means of determining the minute amount of surface lowering in the latter areas, and the only information on the subject comes from the use of sediment traps. Rarely have values as great as 0·1 mm yr^{-1} been recorded and a more representative figure would appear to be 0·01 mm yr^{-1}. It is evident that there must be huge contrasts in the effectiveness of sheetwash erosion. The dominant factor seems to be the protection afforded by the plant cover. This acts in a variety of ways: it intercepts precipitation, protects from raindrop impact, increases surface roughness, and induces an absorbent crumb structure in the soil. It might be anticipated that the importance of surface wash would also diminish under a tropical forest cover, but current evidence on this point seems rather equivocal. The major proponent of overland flow in tropical humid conditions has been Rougerie. By means of stakes he has recorded rates of surface lowering in the Ivory Coast comparable to those in arid regions. Potentially favourable factors include high intensity rainfall, thin cover of leaf litter, and low permeability among many latosols. Yet it needs to be remembered that most of the work by Rougerie was concentrated in areas of semi-evergreen seasonal forest and his results

Fig 8.6. Under the influence of surface runoff in semiarid regions, badland slopes such as those pictured above recede with exceptional rapidity. The contrast between the rounded crest and gulleyed margin of each individual ridge underlines the way in which the dominant process may change with passage down the slope.

may not be applicable to all tropical environments. At one time surface wash under equatorial evergreen forest tended to be dismissed as unimportant, but since about 1960 there seems to have been a gradual shift of opinion favouring it as a possibly significant process in low latitude regions.

Theoretical models of slope development through the action of surface wash have been formulated by several workers. Emmett, for example, has argued that laminar flow will produce a downhill increase in gradient and turbulent flow a downhill decrease. If he is correct in supposing there is a progressive downslope change from laminar to turbulent flow, the resulting profile should be concavo-convex in form. It is widely agreed that turbulent sheet-floods by themselves tend to produce a

concave profile. Carson and Kirkby have argued that different circumstances will produce different degrees of concavity. They contend, for instance, that slopes formed from coarse-grained bedrock will tend to be more strongly concave than those formed from fine-grained material. An effective vegetation cover will generally reduce the concavity so that grass-covered clay slopes, if moulded mainly by sheetwash, will be relatively rectilinear. If a factor for rainsplash erosion is introduced, their model predicts that surface wash acting over a long period of time should ultimately produce concavo-convex hillslopes.

Mass movement

Mass movement, sometimes referred to as mass wasting, may be defined as the downhill transference of slope materials moving as a coherent body. The layer moving in this way may range in thickness from a few centimetres to many metres; its velocity can be as little as a millimetre a year or several kilometres an hour. Given these and other contrasts, it is small wonder that many different classificatory schemes have been suggested. Here the classification proposed by Carson and Kirkby will be followed. This distinguishes three main types of movement: slide, flow and heave (Fig. 8.7). In slide all differential movement is concentrated at the base of the mobile layer; a clearly defined shear plane separates the undeformed moving mass from the stable material beneath. In flow the velocity attains its maximum value at the surface and diminishes

Fig 8.7 Schematic representation of the forces operating on any slope material, and of the three types of mass movement most commonly induced.

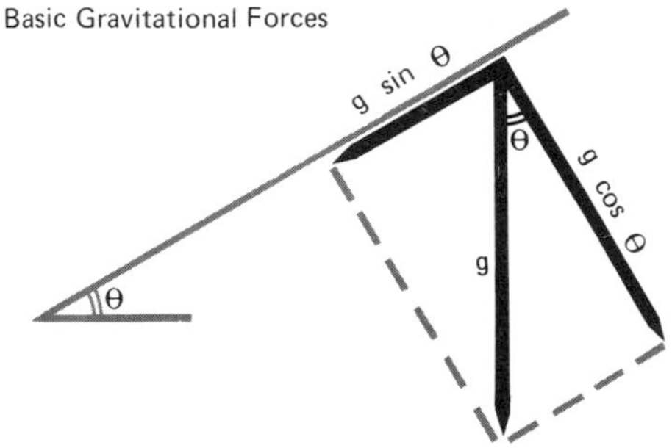

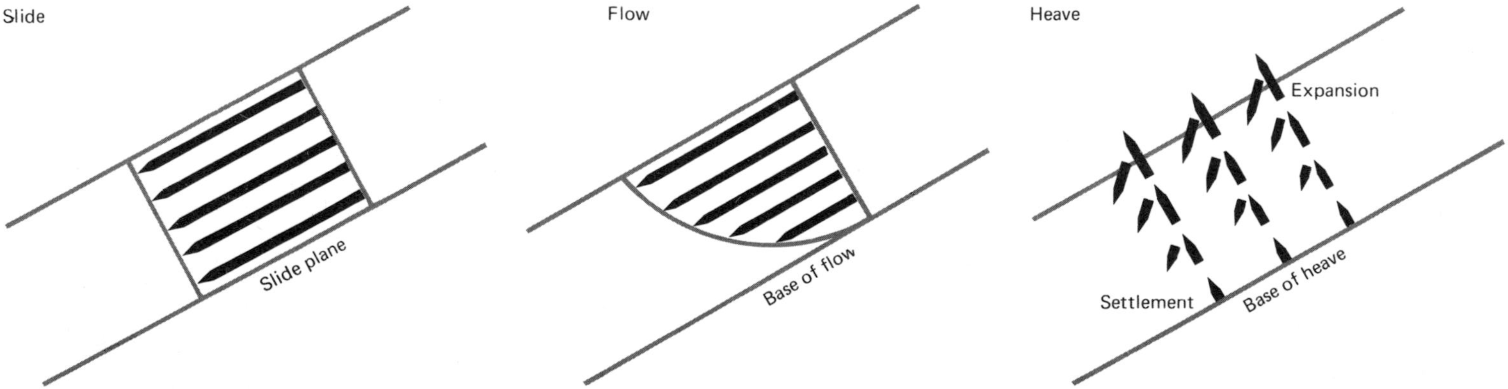

Basic Gravitational Forces

Slide

Flow

Heave

152 with depth; this implies internal shear which is found to be greatest near the base of the moving layer. In heave the surface layers expand and contract in a direction normal to the slope; by itself this does not involve downhill movement, but under the influence of gravity repeated heaving leads to downslope transport. Since mass movement almost invariably involves more than one of these three types acting concurrently, the classification is conveniently depicted in the form of a triangular diagram (Fig. 8.8). Superimposed on the figure are lines indicating the relative displacement rates and moisture values associated with each process.

Soil creep Soil creep is the slow downhill movement of the regolith that results from constant minor rearrangements of the constituent particles. It may be regarded as the product of three separate mechanisms, heaving, continuous creep and biological

Fig 8.8 Classification of mass movement processes (after Carson and Kirkby).

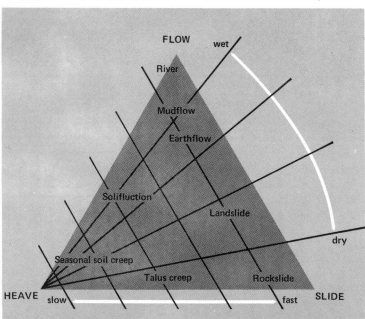

activity. Of these three, heaving by alternate expansion and contraction is probably the most important; it is occasioned in two principal ways, either wetting and drying or freezing and thawing. Continuous creep refers to the behaviour of moist clay materials under gravitational shear stresses; although very small, these may cause such soils to deform slowly and move downhill without actual heaving. Biological agents promote creep by repeated disturbance of the regolith; among the contributory processes are the growth and decay of plant roots, the burrowing activities of soil fauna, and the trampling of animals.

The heaving mechanism, so critical in determining the overall rate of creep on most slopes, is conveniently discussed under the three heads of frequency, amplitude and nature of materials. Frequency is primarily a function of climate. Freeze–thaw cycles will be most numerous where the ambient temperature regularly crosses the freezing point; cycles of wetting and drying will occur in greatest number where heavy precipitation alternates with periods of powerful desiccation. In both cases the upper layers of the soil will experience effective oscillations more frequently than the lower layers; the theoretical result should be faster movement at the surface than at depth. The amplitude of heaving may be expected to have similar consequences. This is because expansion or contraction at any point in the regolith will be transmitted upwards so that movement at the surface becomes an integration of all the individual movements at depth. Other things being equal, the greater the magnitude of the heave the faster the rate of downhill creep. Finally, the nature of the material composing the regolith will have a profound effect on any heaving that occurs. Here it is necessary to distinguish between moisture and temperature controls. Wetting and drying will produce much greater volumetric changes where the regolith contains a high proportion of clay; moreover, as explained in Chapter 7, certain clay minerals are much more prone to swelling than others so that the mineralogical composition is a further important factor. Freezing and thawing is similarly affected by grain size, but in this case a relatively high proportion of silt promotes the greatest degree of expansion and contraction (see p. 297).

Predictions based on the foregoing arguments have been partly confirmed by field measurements. Determining the rate of soil creep presents many problems and no entirely satisfactory technique has yet been devised. Methods that have been tried can be divided into three major groups. The first involves use of rigid surface pegs. In some cases these have simply been inserted to a shallow depth and the positions of the protruding ends resurveyed at intervals by reference to some fixed point. One difficulty encountered with this method is that the pegs tend to tilt and the measured displacements are open to several different interpretations. A variant of the simple peg, known as a T-bar, has been employed by a number of workers. This consists of a vertical shaft surmounted by a horizontal arm so designed that its angle can be very accurately determined by means of a spirit level. The basic idea is that both position and tilt should be measurable. Apart from the practical problem of disturbance by animals and birds, a more fundamental objection is the indeterminacy of the differential movement that must occur between deforming soil and rigid stake. The second group of methods aims to overcome this basic problem by drilling a small hole and inserting some form of flexible piping that will readily deform during downhill creep. Plastic tubing has been the most favoured material, normally employed in combination with some device for measuring its changing angle at depth. The best instrument for the latter purpose is a specially designed inclinometer that records electrically the deviation of the tube from the vertical. For a full assessment of creep, the changing position of the top of the insert must also be fixed; this may be done by measurement either from a stable point on the slope or from a long datum stake set centrally down the tube itself. The latter is only feasible where a large-diameter tube has been used and where the rate of creep is slow. In all these experiments the assumption is made that the piping is sufficiently flexible to move and deform with the soil and not be by-passed. In order to test whether this is likely some workers have advocated filling small vertical holes with dyed sand which can be excavated with a trowel after a suitable lapse of time. Movement of the column of sand can then be compared with that of nearby flexible piping. The third group of methods involves tracing various types of buried marker. Young initiated a technique that has been adopted in many subsequent studies. A shallow pit is dug and into a vertical face aligned directly downslope a grid of horizontal pins,

a few centimetres long, is inserted. The positions of the ends of the pins are surveyed with respect to some fixed point, the face is covered with paper, and the pit carefully filled in. After an appropriate interval the hole is excavated again and the positions of the pins resurveyed. With care it may be possible to repeat the procedure several times using the same set of pins, but each time the pit is dug out there is an obvious risk of accidental disturbance. A consequent limitation of Young-pits, as they have been termed, is that measurements can only be made at infrequent intervals, and it is correspondingly difficult to relate creep to such transient soil conditions as moisture content and temperature. Alternative forms of buried marker include small cones, sometimes connected by means of fine wires to a reference shaft above ground level. Although the lengthening of the wires permits regular monitoring, individual markers seem to behave so unpredictably that movements other than true creep often seem to be involved.

All the measuring methods so far devised suffer from the shortcoming that the very means employed is liable to affect the natural processes it is desired to observe. Insertion of stakes and augering of small holes will disrupt the original soil structure; thermal relations in the soil will be affected by metal bars and hollow plastic tubes; most important of all, the passage of water through the regolith is liable to be altered by almost any of the techniques described. Despite these limitations the results obtained in many parts of the world are beginning to yield a tolerably consistent pattern. On moderately steep slopes the uppermost soil layers commonly creep downhill at a mean rate of 1–2 mm yr⁻¹ in humid temperate regions, 3–6 mm yr⁻¹ in humid tropical regions, and 5–10 mm yr⁻¹ in semi-arid regions with cold winters. An alternative way of expressing creep measurements is by the volume of material passing through a unit length of the imaginary plane trending along the contour and plunging normal to the ground surface. This takes account of the velocity variations with depth. Less consistency is apparent in these figures (Table 8.1), and in a few instances workers have actually recorded movement which does not diminish with depth but which shows two distinct peaks, one normally at the surface and the other below the densest part of the root zone. The assumption in such cases is that the root system is acting as a restraining influence and there seems little doubt that further research will emphasize the close relationship between vegetation and creep movement.

Almost all investigators have concluded that creep acting alone tends to produce convex slope forms. Various arguments have been adduced to substantiate this view, although it is often expressed in the simple statement that the amount of soil to be transported increases downslope and therefore a progressively steeper gradient is necessary to ensure its disposal. If it is postulated on theoretical grounds that the rate of transport by soil creep should be proportional to the sine of the slope angle, it can be shown by a simple mathematical model that a rectilinear profile will be progressively altered to a convex form. This will be achieved by a reduction of gradient appearing first at the crest of the slope and then extending towards its base. Subsequent evolution will include a diminishing maximum angle and slope curvature, but with no tendency for a concave element to develop. The relative importance of soil creep in fashioning hillslope form has still to be firmly established, but a number of recent writers have questioned the primary role sometimes accorded to it in the past.

Solifluction As a process characteristic of periglacial regions solifluction will be discussed in greater detail in Chapter 14. In this section it will suffice to note that accelerated movement of the arctic regolith is ascribable both to increased freeze–thaw activity and to exceptionally high seasonal values for soil moisture content. The frost action contributes to faster creep, the

Table 8.1 Observed volumetric rates of soil creep.

Area		Slope angle (degrees)	Downhill creep (cm^3 cm^{-1} yr^{-1})
British Isles	=N. England	25	0·5
	*Scotland	17	2·1
	+Wales	Various	0·49—6·52
North America	=Washington D.C.	17–28	0·2
	*Maryland	17	1·3
	*Ohio	20	6·0
	=New Mexico	45	5·6
	*California	19	650·0
	++Puerto Rico	17·5–19·5	7·5–9·1
	**Puerto Rico	7	1·52
Other regions	=Northern Territory (Australia)	16	4·4–7·3
	=New South Wales (Australia)	15	1·9–3·2
	=Tatar (USSR)	22	5·7–8·4
	=Malaya	10	12·4

*Compilation by Carson and Kirkby 1972
=Compilation by Young 1974
+Slaymaker, Trans. Inst. Brit. Geogr. 1972
++Lewis 1974   **Lewis 1975

soil water to periods of flowage. In addition, water during winter may be preferentially concentrated into ground-ice lenses that substantially lift the overlying soil. In theory this elevation is likely to be in a direction normal to the surface, while fall during melting will be vertical. The process may be regarded as a seasonal heave of exceptional amplitude, and is supplementary to any displacement that may be induced by freeze–thaw cycles of much shorter duration. The relatively high rate of movement means that solifluction is more readily measured than simple soil creep. Surface stones being transported on a mobile solifluction sheet have been marked with paint and their changing distances from a fixed point regularly surveyed. Short stakes have been driven into the ground and their downslope movements monitored. To determine the velocity distribution at depth, plastic tubing has been employed in the same fashion as that described for soil creep. These procedures have all served to confirm the general rapidity of mass movement by solifluction. Surface movement is highly seasonal but mean annual rates commonly lie between 10 and 50 mm, while downslope volumetric transport often amounts to as much as 50 cm^3 cm^{-1} yr^{-1}. One characteristic of solifluction that has emerged in many studies is the wide range of values encountered even on the same section of slope. This is probably associated with marked variations in the composition, and therefore in water content, of the mobile layer.

In theory solifluction might be expected to produce convex slopes similar to those predicted for creep. However, several workers have found little correlation between slope angles and rates of movement in arctic areas, so that there must be some hesitation in applying the same arguments as were earlier adduced for creep. It may well be that variations in lithology, particularly in terms of susceptibility to frost action, assume such importance in periglacial regions that simple models predicated on a homogeneous bedrock have little practical application.

Talus creep The fragments comprising open-work scree are subject to small intermittent displacements which together constitute the movement known as talus creep. Dislodgement of individual particles may be initiated by ice wedging, temperature oscillations, wetting and drying, avalanching, the impact of additional falling debris and a variety of other factors. The relative importance of each differs from locality to locality. The rate of talus creep has been assessed by laying lines of painted stones across the surface of a scree and repeatedly surveying their positions with respect to some reference point on nearby bedrock. In practice this procedure usually reveals wide variations in downslope movement. Some stones remain completely stable for several years while others nearby may travel a metre or more. This erratic behaviour arises from constant minor readjustments, sometimes triggered by virtually random factors; however, in arctic and alpine environments the single most important trigger is almost certainly ice wedging.

Rockslide The term rockslide denotes a large-scale slope failure in hard, jointed rock where there is some disintegration of the moving body as it passes downslope. On one hand the rockslide grades into rockfall, with some authors wishing to distinguish an intermediate category of rock avalanche; on the other it grades into landslide where the whole mass moves as a more coherent unit. Terzaghi has analysed the conditions which seem to conduce to rocksliding. As in the case of rockfall, he emphasizes the critical importance of the joint system, contending that it is the spread of continuous joints through a massive rock that ultimately weakens the structure and leads to sliding. Contributory factors include variations in the cleft-water pressure and gradual chemical weathering of the sound rock between joint planes. Once the stresses exceed the shear strength along any plane in the rock, movement will ensue. The failure will tend to be progressive, since weakening in any one part will throw additional stresses on others. If water in the joint system is a major factor, failure will often spread upwards from an initial fracture at the base of the slope since it is here that hydrostatic pressure reaches its greatest value. Terzaghi has estimated that, with a random joint pattern, hard massive rock should be able to sustain slopes at angles of up to about 70°. This figure is derived primarily from measurements on the shear strength of crushed rock aggregate, but receives strong support from observations made when the cuttings for the Panama Canal were being designed. It was then reported that in central America rocks such as massive granite and quartzite sustain

Fig 8.9 A rockslide in Montana triggered by the same earthquake as that responsible for the floodplain scarp illustrated in Fig 4.8. Note how part of the 28 million cubic metres of debris, possibly supported on a cushion of air, rode up the opposite valley side to a height of 130m. In the foreground the lake dammed by the slide reached a depth of over 55m. (Photo J. R. Stacey, U.S.G.S.)

high natural faces at angles of at least 70° without loss of stability.

Landslide A landslide is a form of rapid mass movement in which nearly all the differential displacement is concentrated along a basal slip plane. Landslides assume many different forms depending on the nature of the bedrock and regolith involved. One possible classificatory scheme suggested by Skempton is based upon the shape of the sliding mass. Using the ratio of maximum thickness to hillslope length, he distinguishes two major classes: surface slides in which the ratio, expressed as a percentage, lies between 2 and 5 per cent, and deep slides in which it lies between 10 and 30 per cent. In the former case the plane of failure is so shallow that most of the slide material normally consists of regolith, whereas in the latter it plunges so deeply that a high proportion of the moving body is usually composed of bedrock.

Analysis of the stresses associated with shallow slides is reasonably simple (Fig. 8.10). It is rendered much easier if the mobile layer is assumed to be uniform in thickness, and to rest on a slope of constant angle and infinite extent. This dispenses with the need to consider side and end effects, and may be justified on the grounds that a shallow slide has very great length in comparison with its depth. The shear stress acting along the plane of the slide is readily computed for a regolith column of unit width, and can be compared with the shear strength of the material as measured in a triaxial test. At the time of failure the two values should be approximately equal. The shear strength is reduced to a minimum when the regolith is completely saturated and porewater pressure attains its greatest value (see p. 131). This is one reason why the great majority of shallow slides takes place after periods of heavy rain.

In practice shallow sliding often occurs on slopes of lower angle than the foregoing analysis would suggest. Two reasons for this seeming anomaly may be adduced, both related to the special properties of materials composed of clay minerals. The first is an important distinction between what are termed normally-consolidated and over-consolidated clays. During sedimentation the basal layers of a clay stratum are subject to progressive compaction as extra material is added above; water is expelled and the clay particles more densely packed. Eventually the stratum may be so deeply buried that it is subject to immense compressive forces from the overlying beds. If these are later removed by erosion, the clay is described as over-consolidated. Its physical properties differ from those during the accumulation phase since the particles remain more tightly packed and the water content lower. Removal of the super-incumbent weight leads to fissuring and jointing, with local shearing sometimes indicated by slickensides. Just as with hard rock penetrated by joints, the overall strength of the clay will be less than that of any small unfissured sample.

The second reason for shallow clay slides on relatively gentle slopes is an important distinction between the properties known as peak and residual strength. During laboratory testing, the original strength of a sample is often high but once shearing has started the strength diminishes before finally settling to a more constant residual value. The change is associated with a reorientation of the flaky clay particles parallel to the direction of shear and a simultaneous fall in cohesion. This means that a clay, if forced for any reason to shear past its peak strength, will thereafter be a much weaker material. It also follows that, if one section of a clay slope is forced past its peak, additional stress will then be thrown on some other section, possibly causing the peak to be surpassed there also. In this way progressive failure may reduce the shear strength to its residual value along the entire length of the slip surface. In such circumstances the most significant mechanical property for a long-term stability is the residual rather than the peak strength.

As an illustration of the importance of residual strength, reference may be made to studies of slopes on the London clay in south-eastern England. It has been found that these slopes are only stable at angles of less than about 10°. This figure is the same as that predicted from measurements of the residual

158 strength of the clay and leads to the conclusion that it is the limiting value for long-term stability. Of course, this does not mean that all slopes will lie at 10° since other processes may operate to lower them; however, if stream undercutting steepens a slope beyond its critical gradient, sliding will then function to reduce its angle to the stable value.

Analysis of the stresses in deep slides is more complex, and no attempt will be made here to do more than introduce the basic principles involved. Observation shows that the slip plane beneath deep slides tends to approximate an arcuate form. The moving mass in consequence often exhibits a degree of rotation, and the term rotational slip has appropriately been applied to such cases. Deep slides along arcuate failure planes

Fig 8.10 Diagramatic representation of the stresses that operate in a shallow slide when side- and end-effects are ignored.

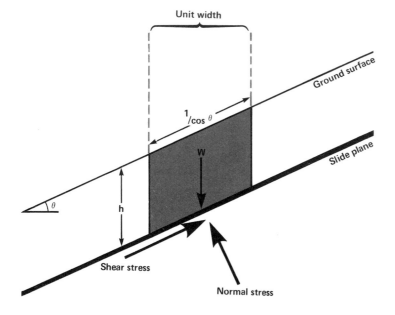

If w is unit weight of soil, total weight (**W**) of column = **h.w.** Acting over an area $\frac{1}{\cos}\theta$ the vertical stress = **h.w.**cos θ. This may be resolved into a shear stress equal to **h.w.**cos θ . sin θ and a normal stress equal to **h.w.**cos$^2\theta$.

are especially characteristic of thick clay strata. One of the earliest scientific investigations was undertaken in Sweden where the authorities were concerned at the number of slides affecting the national railway system. The report of the Royal Swedish Geotechnical Commission in 1922 provided what was then the fullest evaluation of the stresses involved, and its basic concepts are still employed in the so-called Swedish method of stability analysis. The initial premise is that the slide surface conforms to a circular arc, with the centre of the imaginary circle defining the axis of rotation (Fig. 8.11.) Where the axis lies outside the points at which the slip plane intersects the ground, all the weight of the overlying mass is contributing to the potential failure; where the axis lies inside the points, the tendency towards rotation is resisted by the weight of the toe of the slide. For purposes of computation the whole mass is divided into vertical slices and the stress contribution of each is separately calculated. The aggregate stress may then be compared with the shearing resistance along the arc due to the strength of the materials involved. Allowance must again be made for porewater pressure and the particular properties of over-consolidated clays. The greatest practical difficulty lies in determining along which of the many possible slip planes movement is most likely. This was originally done by tedious trial-and-error procedures, but is now normally carried out by computer.

Under natural conditions rotational slips are commonly associated with permeable caprocks overlying thick clay or shale strata. They are therefore especially characteristic of scarp faces where undercutting at the foot of the slope steepens the gradient until the cuesta summit becomes unstable. The precise form of the slip plane is much influenced by structural details, and is by no means always a simple circular arc. Nevertheless, backward tilting of the resistant caprock is commonplace, sometimes resulting in an ill-drained hollow in which organic matter begins to accumulate. Radiocarbon dating may then be employed to give an estimate of the minimum age of the slide. Detailed mapping in the British Isles has shown that many of the scarps of central and southern England have been severely affected by landsliding, much of it probably induced by late-Weichselian periglacial conditions when the strength of the materials was severely reduced by their high water content.

Earthflow Although many landslides involve a minor flow component, particularly along the saturated toe of the moving mass, other forms of displacement take place mainly by flowage and it is these that are normally categorized as earthflows. They vary greatly in nature, from minor disturbances in which the original turf cover remains intact apart from small marginal ruptures, to spectacular events in which massive bodies of water-saturated material progress rapidly downhill. The more mobile forms are sometimes designated mudflows, but here that term will be reserved for the sediment-laden streams that thicken in consistency as they proceed across the surface of desert fans.

Rates of earthflow movement range from a fraction of a metre per month to several kilometres per hour. Primary controls include the nature of the materials involved and their water content. An example of relatively slow flowage is provided by long clay tongues descending the edge of the Antrim plateau in north-eastern Ireland. The material comes from weathering of Liassic shales and includes varying proportions of montmorillonite, illite and kaolinite. The rate of movement has been assessed by surveying the positions of specially installed pegs at monthly intervals. To provide more complete records an instrument was devised to monitor continuously the displacement of an individual peg. By these techniques a close association between rainfall and velocity of flow was demonstrated (Fig. 8.12). A second association was found between patterns of instability within the tongues and local concentrations of montmorillonite.

The most startling illustrations of rapid earthflow are found in certain distinctive clay deposits in eastern Canada and Scandinavia. Laid down during the post-glacial marine transgression, these sediments have the remarkable property known as high sensitivity. Sensitivity refers to the shear strength of a remoulded sample of a clay compared with its strength in the undisturbed state. The ratio of these two values may exceed 8 in a highly sensitive clay, and in extreme cases surpasses 25. In effect, the clay can be transformed from a weak solid into a viscous fluid without any change in water content. Once this alteration has taken place, rapid flow can occur down quite gentle slopes. Disastrous earthflows have affected a number of settlements along the valley of the St Lawrence and its major tributaries; in 1955, for instance, part of the small town of Nicolet was carried into the local river. The fundamental problem presented by such occurrences is the cause of the change of state in the clay. Liquefaction appears to be progressive, spreading quickly through the body of the clay, and is probably triggered by either localized internal shearing or basal sliding; the whole process is a further example of variations in shear strength occasioned by internal structural rearrangements.

Movement of material in solution

Many studies of the dissolved solids in streams have shown that a significant proportion of the lowering of the land surface must be taking place by chemical solution. In eastern Australia Douglas has concluded that, with annual run-off values between 100 and 1 500 mm, as much as half the material discharged from a catchment is carried in solution. A broadly similar conclusion

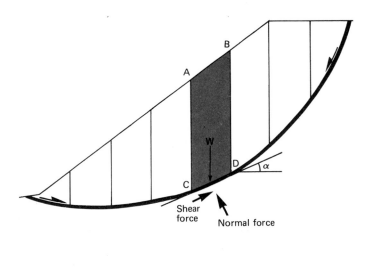

For column ABCD　　shear force = W cos α
　　　　　　　　　　normal force = W sin α
Total shear force along arc = Σ W sin α
(but note the effect of toe where shear force operates in reverse direction)

may be inferred from data collected on many rivers in the United States. In a detailed investigation of a small catchment in Maryland a group of American workers has shown that, with a run-off averaging about 150 mm yr^{-1}, the material carried seawards in solution is almost five times that carried in suspension. In this particular instance the writers comment that, although the landforms of the catchment suggest major mechanical erosion, it is actually chemical denudation that is dominant at the present time.

Despite this evidence geomorphologists have generally neglected the influence of solution on the development of hillslope forms. One reason is the difficulty of ascertaining the precise source of the solutes carried by streams. It has often been assumed that a high proportion comes from groundwater stored deep within the bedrock and therefore having little direct influence on slope development. Even when the falsity of this argument has been exposed, assessment of the volume of solutes transported by interflow has proved a most formidable undertaking. The difficulties of directly evaluating interflow have already been indicated, and without frequent measurements of both volume and solute concentration it is impossible to compute the total quantity of material being carried downslope in this way.

Among the more significant controls are the availability of the various elements in the parent rock, their solubility, and the volume of water passing through the regolith. Of critical importance is the rapidity with which percolating water equilibrates with the local soil environment. Experimental evidence suggests that approximate equilibrium may be established in a matter of a few hours or at most a day or two, and this tends to be confirmed by field measurements which often reveal interflow water as almost fully saturated with respect to the abundantly available elements. In certain limestone areas, for instance, virtual saturation with $CaCO_3$ has been demonstrated within little more than a metre of the surface and Williams has estimated that as much as 25 per cent of the total solution may take place actually at the surface. Analysis of interflow on non-calcareous rocks shows the percolating water to be rapidly enriched with other elements. Particular interest attaches to silica which has been found to remain more constant in concentration than almost any other

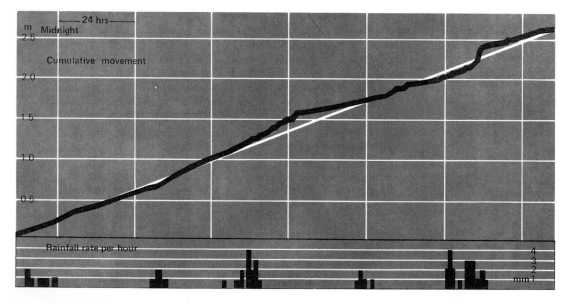

dissolved material. Even during flood discharge the concentration of silica in streams falls only slightly, the implication being that the water has picked up this solute in a relatively short space of time. In the United States the median value of silica in groundwater samples is about 17 ppm and in stream-water samples about 14 ppm. In effect, the low variability in concentration means that the total amount of silica removed from any catchment tends to be a direct function of run-off.

More data are required concerning the variation in chemical activity at different points on a slope. Formulation of the catena concept by soil scientists is itself indicative of the variability that is often encountered. Only with such additional information will it be possible to predict with any confidence the way in which solution contributes to the fashioning of slope form.

CAINE, N., 1969. 'A model for alpine talus slope development by slush avalanching', *J. Geol.*, 77, 92–100.

CAMPBELL, I. A., 1974. 'Measurement of erosion on badland surfaces', *Z. Geomorph. Supplementband.* 21, 122–37.

CLEAVES, E. T., et al., 1970. 'Geochemical balance of a small watershed and its geomorphic implications', *Bull. geol. Soc. Am.*, 81, 3015832.

DOUGLAS, I., 1973. 'Rates of denudation in selected small catchments in eastern Australia', Univ. Hull Occ. Pap. in Geog. No. 21.

EMMETT, W. W., 1970. 'The hydraulics of overland flow on hillslopes', *Prof. Pap. U.S. geol. Surv.* 622-A.

LEWIS, L. A., 1974, 'Slow movement of earth under tropical rain forest conditions', *Geology*, 2, 9–10.

LEWIS, L. A., 1975. 'Slow slope movement in the dry tropics : La Paguera, Puerto Rico', *Z. Geomorph.*, 19, 334–9.

MOSLEY, M. P., 1973. 'Rainsplash and the convexity of badland divides', *Z. Geomorph. Supplementband*, 18, 10–25.

PRIOR, D. B. and N. STEPHENS, 1972. 'Some movement patterns of temperate mudflows : examples from north-eastern Ireland', *Bull. geol. Soc. Am.*, 83, 2533–44.

RAPP, A., 1960. 'Recent development of mountain slopes in Karkevagge and surroundings, northern Scandinavia', *Geogr. Annlr.*, 42, 71–200.

ROUGERIE, G., 1956. 'Etudes des modes d'erosion et du faconnement des versants en Cote d'Ivoire equatoriale', *Slopes Comm. Rep.*, 1, 136–41.

SAVIGEAR, R. A. G., 1965. 'A technique of morphological mapping', *Ann. Ass. Am. Geogr.*, 55, 514–38.

SCHUMM, S. A., 1964. 'Seasonal variations of erosion rates and processes on hillslopes in western Colorado', *Z. Geomorph. Supplementband.* 5, 215–38.

SCHUMM, S. A. and R. J. CHORLEY, 1964. 'The fall of Threatening Rock' *Am. J. Sci.*, 262, 1041–54.

SCHUMM, S. A. and R. J. CHORLEY, 1966. 'Talus weathering and scarp recession in the Colorado plateaus', *Z. Geomorph.*, 10, 11–36.

SKEMPTON, A. W., 1953. 'Soil mechanics in relation to geology', *Proc. Yorks. geol. Soc.*, 29, 33–62.

SKEMPTON, A. W., 1964. 'Long-term stability of clay slopes', *Geotechnique*, 14, 77–102.

STRAHLER, A. N., 1950. 'Equilibrium theory of slopes approached by frequency distribution analysis', *Am. J. Sci.*, 248, 800–14.

TERZAGHI, K., 1962. 'Stablity of steep slopes on hard unweathered rock', *Geotechnique* 12, 251–71.

WILLIAMS, P. W., 1968. 'An evaluation of the rate and distribution of limestone solution and deposition in the River Fergus basin, western Ireland' in 'Contributions to the study of Karst' (Ed. P. W. Williams and J. N. Jennings) Aust. Nat. Univ. Canberra.

YOUNG, A., 1974. 'The rate of slope retreat' in 'Progress in geomorphology', *Inst. Br. Geogr. Spec. Pap.*, 7.

Selected bibliography

After years of neglect hillslopes have, since 1970, been the subject of two excellent texts, both with very full bibliographies : A. Young, *Slopes*, Oliver and Boyd, 1972, and M. A. Carson and M. J. Kirkby, *Hillslope form and process*, CUP, 1972.

In the previous chapter the surface movement of water over hillslopes was briefly discussed. Augmented by discharge from underground sources, the surface flow is gathered into stream channels which are shaped and moulded by the downstream passage of the water. This fashioning may be achieved either by erosion of bedrock or more commonly by picking up and redepositing fragmental materials that form the walls and floors of the channels. All these processes require energy which the streams possess by virtue of their elevation above sea-level. This potential energy is converted into kinetic energy as the water flows down the channel, the amount of kinetic energy available for modifying the stream bed varying greatly at different river stages. As the water-level rises during a flood, increasing stresses are placed upon the channel material which, if loose, may be set in motion when a critical stress is reached. The detritus entrained in this way is transported downstream, being either removed from the river basin entirely or re-deposited when the stream lacks the power to carry it further.

In the following pages attention will be directed in turn to the erosional and transporting actions of a stream, to the forms and patterns assumed by stream channels and finally to the landforms that result from fluvial deposition. However, as an essential preliminary, the basic principles of stream flow will first be discussed.

Stream hydraulics

Basic concepts The flow of fluids presents a subject of immense range and complexity. Much of it lies beyond the scope of this book but is very adequately covered in numerous volumes on fluid mechanics. For present purposes it is only necessary to outline certain basic concepts required for a proper understanding of the geomorphological activity of streams. Figure 9.1 illustrates the terminology most frequently employed in the description of streams and the channels they occupy.

Since water is a Newtonian fluid it is unable to resist shearing stresses when in static equilibrium and any deformation it suffers is directly proportional to the applied stress (Fig. 7.7). The rate of deformation or flow is a function of the internal friction or viscosity of the material. Viscosity may therefore be defined as the ratio of shear stress to the rate of shear strain. Its value for water at 0°C is 0·018, at 10°C 0·013 and at 20°C 0·010 poise. Strictly speaking the value defined in this way should be designated the dynamic viscosity to distinguish it from another fre-

frequently employed measure known as the kinematic viscosity. The latter is the dynamic viscosity divided by the density, and for stream water normally has a value of about 0·01 cm² sec⁻¹. Both the dynamic and kinematic viscosity are calculated on the assumption that flow is laminar. Where flow is turbulent a third and rather different form of viscosity becomes important. Known as the eddy viscosity, this is normally much greater than the dynamic viscosity. It is of great importance in increasing the shear stress that a flowing stream can exert upon the materials composing its channel. It should be noted, however, that it is not an intrinsic property of the fluid but varies with the turbulent flow of the water.

Fluids may exert pressure in two ways, known as static and dynamic pressure. Static pressure, well exemplified by the sensation felt when diving, is proportional to the height of the overlying column of fluid. Dynamic pressure results from movement of the water. It is due to the kinetic energy possessed by the moving water and is equal to $\frac{1}{2}mV^2$, where m is mass and V is velocity. The total energy possessed by a small mass of moving water may be computed by adding the energy equivalents of the static and dynamic pressure to the potential energy. Expressed in symbolic terms for a single point this may be written as

$$p+\frac{V^2}{2g}+h=\text{total energy}=\text{constant},$$ where p is pressure, V is

velocity, g is gravity and h is height above an arbitrary datum. Known as the Bernoulli equation, this is strictly applicable to an ideal fluid with zero viscosity. In a stream the initial potential energy is converted into kinetic energy, which in turn is dissipated in overcoming the frictional resistance represented by the viscosity of the water. The conversion of the energy to heat is exactly analogous to the more familiar case of frictional heat generated between two solids. In this way the available energy diminishes in a downstream direction. Figure 9.2 illustrates the application of the Bernoulli equation to stream flow. The line depicting total energy less the frictional loss is commonly called the energy grade line, and the distance between it and the channel bed indicates the quantity known as specific energy.

Nature of stream flow Virtually all stream flow is turbulent, that is, characterized by irregular eddying movements superimposed on the overall downstream motion. Only at exceedingly low velocities in a shallow straight channel could the flow be laminar. The best-known expression for defining the transition velocity from laminar to turbulent flow is the Reynolds number (Fig. 9.3). This dimensionless figure is computed by dividing the product of the velocity and depth by the kinematic viscosity. With values of kinematic viscosity commonly around 0·01 cm² sec⁻¹ insertion of typical values for velocity and depth (or hydraulic radius) yields Reynolds numbers far in excess of the 2 000 usually regarded as about the upper limit for laminar flow.

Two types of turbulent flow may be distinguished. By far the most common is tranquil or streaming flow but in certain circumstances this may be transformed into rapid or shooting flow. As the name implies, shooting flow involves marked acceleration of the water and occurs where there is a pronounced constriction of the channel. This is normally confined to rapids where the stream is flowing either on bedrock or around large boulders since a channel moulded in finer detrital material tends to have a relatively uniform cross-section. The visual characteris-

Fig. 9.1 Conventional terminology in describing stream channels.

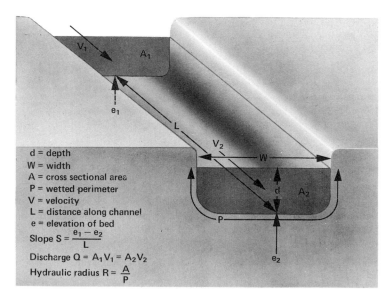

d = depth
W = width
A = cross sectional area
P = wetted perimeter
V = velocity
L = distance along channel
e = elevation of bed
Slope S = $\frac{e_1-e_2}{L}$
Discharge Q = $A_1V_1 = A_2V_2$
Hydraulic radius R = $\frac{A}{P}$

tics of shooting flow are a fast, streamlined appearance with little disturbance on the upstream side of any obstacle and stationary oblique rolls on the downstream side. In tranquil flow, on the other hand, obstacles tend to produce disturbed water on the upstream side and standing cross-waves on the

Fig. 9.2. Application of the Bernoulli equation to stream flow and derivation of the subcritical and supercritical flow states.

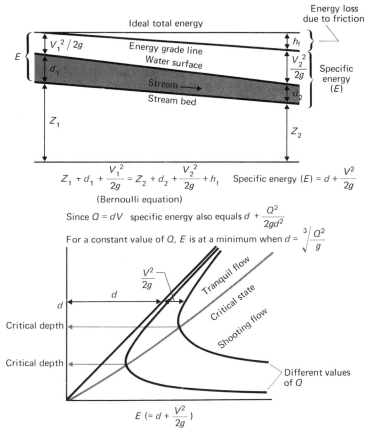

$$Z_1 + d_1 + \frac{V_1^2}{2g} = Z_2 + d_2 + \frac{V_2^2}{2g} + h_f \quad \text{Specific energy } (E) = d + \frac{V^2}{2g}$$

(Bernoulli equation)

Since $Q = dV$ specific energy also equals $d + \dfrac{Q^2}{2gd^2}$

For a constant value of Q, E is at a minimum when $d = \sqrt[3]{\dfrac{Q^2}{g}}$

For a given discharge, there are both supercritical and subcritical depths and velocities

At the critical depth $d^3 = \dfrac{Q^2}{g}$. Since $Q = dV$, $d^3 = \dfrac{d^2 V^2}{g}$, $d = \dfrac{V^2}{g}$, $V = \sqrt{gd}$

i.e. Froude number 1.

downstream side. The velocity at which transition from tranquil **165** to shooting flow occurs is defined by a second dimensionless figure known as the Froude number. This is calculated by dividing the mean velocity by $\sqrt{gD}$, where g is gravity and D is depth. When the Froude number is less than unity the flow is tranquil, but when above unity the flow is shooting. The value $\sqrt{gD}$ also defines the speed at which ripples move across shallow water, and this explains why surface disturbances are not transmitted upstream when the Froude number exceeds one. The existence of different types of turbulent flow results from the fact that, for given values of specific energy and discharge, there are two alternatives, either a shallow high-velocity form or a deeper low-velocity form (Fig. 9.2). The flow may change quite rapidly from one form to the other The transition from tranquil to shooting flow is accompanied by a sharp drop in water level, whereas the opposite change generates the feature known as a hydraulic jump. This is the prominent reversed gradient, often foaming at the surface, where the water reverts to slow deep flow.

In both types of turbulent flow random eddies cause the water velocity at a point to fluctuate. Velocity as measured with a normal flow meter is time-averaged and the actual maximum and minimum values can differ from the mean by 50 per cent or more. The degree of turbulence at any point may be assessed with sensitive instruments that record the changes in velocity along three separate coordinates, one parallel to the general flow direction and the other two at right angles. The variability, commonly expressed in the form of the standard deviation, is then employed as a measure of the intensity of turbulence (Fig. 9.4). In a straight segment of channel, the intensity is least near the surface and increases to a maximum near the bed; in a sinuous segment, the zone of greatest intensity tends to be displaced towards the outer and deeper side of each curve.

Two types of shear layer are found within a flowing stream. The most important occurs along the channel boundary. Although turbulence is greatest close to the stream bed, the water actually in contact with the channel is stationary. Above this may come a very thin film in which movement is almost entirely laminar, followed in turn by a buffer layer in which there is mixed laminar and turbulent motion. Details of the laminar sublayer are much influenced by the roughness of the channel materials. In general the sublayer can only develop fully where it is thicker than the diameter of the boundary grains; if the grain sizes are too great rough turbulent flow predominates right down to the channel bed. The second type of shear layer is associated with the phenomenon known as separation. This occurs where an abrupt change in channel alignment causes the water to break away from the solid boundary. Immediately downstream from large obstacles or sharp bends, separation produces a linear zone of intense turbulence known as the free shear layer; systematic eddying patterns are a common accompaniment.

Relationship of flow velocity to channel characteristics

Engineers concerned with such practical tasks as the regulation of river courses and the construction of canals have enunciated a series of empirical formulae that relate flow to such variables as channel slope, shape and roughness. The earliest important formula was introduced by the French engineer Chezy in 1768 when he designed a canal for the Paris water supply. He argued that flow velocity would be proportional to the square root of the product of the slope and the hydraulic radius. The actual velocity would depend upon a coefficient, often referred to as the Chezy coefficient, incorporating a value for channel roughness. In the nineteenth century elaboration of the relationships in the Chezy equation led to formulation of the now widely employed Manning equation $V = R^{\frac{2}{3}} S^{\frac{1}{2}}/n$. In the Manning equation n is a coefficient indicative of channel roughness. The term roughness in this context refers not only to the size of bed material, but also to the sinuosity of the channel and the presence of such obstructions as reeds and fallen trees. The value of n can be estimated with considerable precision by experienced engineers, and since it may be as low as 0·025 for a clean straight channel but over 0·075 for a winding overgrown channel, the velocity of water flowing in the same size and slope of stream course can vary by a factor of more than 3.

Erosion

Bedrock channels Although erosion of channel bedrock must presumably be a fundamental mechanism in subaerial denudation, it has been the subject of relatively little detailed investigation. Three separate but interrelated processes are commonly invoked: corrosion, corrasion and hydraulic action. Corrosion refers to the chemical activity of the flowing water, and is therefore closely related to chemical weathering. Changing river stage not only subjects part of the rock surface to periodic wetting and drying but also promotes many further weathering processes. These can be particularly effective since loosened grains are constantly swept away to expose fresh rock. Of course, if the water is already saturated with respect to certain compounds it may be chemically less aggressive than the precipitation from which it is derived. Yet there seems little doubt that chemical decomposition is a vital precursor to much corrasional and hydraulic activity.

Corrasion refers to the wearing away of bedrock by the scraping and gouging effect of rock debris in transit along the

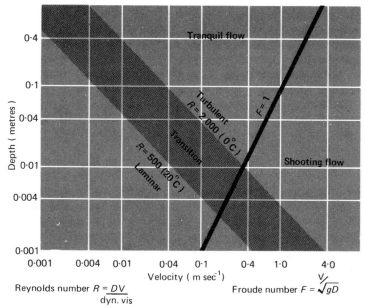

Fig 9.3. Criteria for the classification of different types of stream flow.

Reynolds number $R = \dfrac{DV}{\text{dyn. vis}}$

Froude number $F = \dfrac{V}{\sqrt{gD}}$

stream channel. The size, shape, composition, number and velocity of the fragments all influence the corrasive power of the debris. In general the bed load sliding and rolling along the channel floor will be most effective, and as long ago as 1753 Brahms expounded a so-called sixth-power law stating that the weight of the largest moveable fragment varies as the sixth power of the velocity. As the calibre of the load increases in this way, so also does the velocity of individual particles with the result that corrasion is enormously enhanced during periods of spate. The clearest testimony to this abrasion is the presence

of smoothed and rounded rock surfaces, often moulded into deep potholes. Unlike mobile channel sediment, bedrock forms impose persistent flow patterns within the stream. An initial hollow can induce a swirling motion by which rock fragments are consistently made to impinge against the walls and floor; ultimately the depression will be transformed into a prominent pothole, sometimes with internal spiral flutings. High velocity flows are required for such action to be fully effective, and the best examples often lie so far above normal stream level that they can only have been formed during periods of spate. Most of

Fig. 9.4. Schematic diagram illustrating variations in mean velocity, sediment concentration and intensity of turbulence (i.e. deviation of velocity from local mean value).

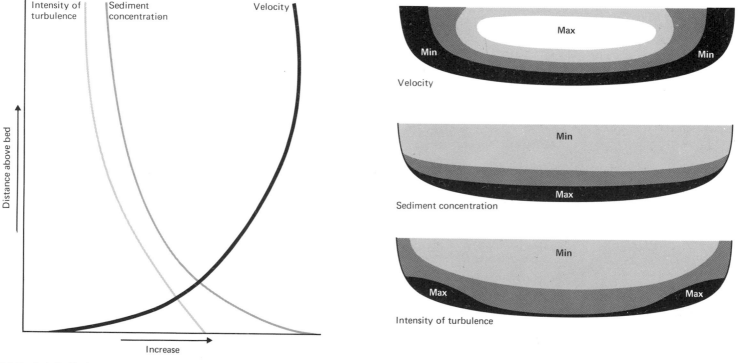

(a) Vertical distribution

(b) Cross-sectional distribution

the time the hollows remain dry, their floors occasionally littered with the well-rounded pebbles sporadically used as abrasive tools.

Hydraulic action refers to the mechanical work of the water alone. Although the momentum of the water may suffice to displace small joint-bounded fragments of the bedrock, especially if already loosened by weathering, it is doubtful whether many virgin rocks can be effectively attacked in this way. However, one process which may locally be important is the so-called cavitation that arises from abrupt changes in pressure as water accelerates and decelerates. According to the Bernoulli equation (p. 164) any increase in velocity must lead to a reduction in pressure. If the acceleration is sufficient, the pressure can drop to the point where numerous small bubbles form and create the impression that the water is foaming. As the pressure rises again the bubbles implode and emit powerful shock waves which hammer adjacent solid surfaces. The effect of cavitation has long been known to engineers concerned with wear on ships propellers and turbine blades where extremely high velocities are involved. It has also been invoked to explain deep pits gouged from solid concrete spill-ways on certain dams, one such example attaining a depth of almost 0·5 m in a single day. It is not easy to prove a particular erosional form as due to cavitation, and the smooth surface of a typical pothole contrasts with the fretted appearance of a badly eroded propeller; the difference may be due to cavitation acting alone on the propellers but in conjunction with a sediment load on the stream channels. Although the importance of cavitation is difficult to assess directly, the conditions under which it is a potential means of erosion can be derived from the Bernoulli equation. Assuming localized acceleration of the mean stream velocity by a factor of 2, water 1 m deep is capable of producing cavitation when flowing at an average rate of about 9 m sec^{-1}. In order to attain this rate in an open channel a sustained gradient of over 5°, or a vertical fall of about 4 m, is necessary. Although these calculations indicate no more than the general order of magnitude, they suggest that cavitation may be effective along certain steep rapids and at the foot of substantial water-falls.

Alluvial channels The concept of erosion in alluvial channels is rather different from that in bedrock. Erosion in the latter case is an irreversible process since any later aggradation will involve replacement of the bedrock by detrital sediments. Erosion in alluvial channels, on the other hand, normally occurs during periods of spate with any mobilized debris replaced during the falling river stage by similar material from higher upstream. In the vertical plane one may distinguish for any reach of channel three possible long-term conditions: aggradation; degradation; and equilibrium. All three represent the net effect of a large number of much shorter cycles, each involving both erosion and deposition; it is the relative magnitude of these last two that determines the long-term effect. In the horizontal plane the amount of channel erosion determines the potential rate of lateral stream migration.

Research has concentrated on the factors that conduce to short-term erosion of the channel floor and sides. The general field of investigation is known as loose boundary hydraulics and because conditions in streams are so complicated much of the experimental work has been done in laboratory flumes where far greater control of the many variables is possible. Attention has been directed in particular to determining two values known as the critical erosion velocity and the critical tractive force. The term critical in these expressions refers to the minimum values of velocity and tractive force required to set particles of a specified size in motion. In an attempt to assess critical erosion velocities Hjulstrom in 1935 collected the results of over 30 experiments and plotted the data in the form of a graph. In this way he showed that the most readily moved material has a diameter of 0·25 mm and is set in motion by water flowing at 0·2 m sec^{-1}. Both above and below this size the critical erosion velocity rises. In the case of the larger particles this is due to the greater force required for lifting and rolling, in the case of the smaller particles to the greater force required to overcome electro-chemical cohesion. A year after Hjulstrom's work was published, a German engineer named Shields gave an account of his experimental work on the flows required to set cohesionless particulate material in motion. Concentrating on the concept of critical tractive force, he demonstrated that the ratio of the tractive force to the resistance of sand grains is a function of the ratio of the grain size to the thickness of the laminar sublayer. His work has since become a standard reference for engineers and has the advantage of being based on fundamental physical principles; on the other hand it does not yield reliable figures for material less than about 0·2 mm in diameter.

Figure 9.5 is a composite diagram based on the work of both Hjulstrom and Shields. Although conceptually very important, it must be emphasized that its application to ordinary stream channels is fraught with difficulty. The diagram refers to smooth but not to irregular beds and if, for instance, there are sand ripples on the stream floor the grains tend to move before the predicted critical velocity is reached. Fine cohesive material tends to break away in lumps rather than be entrained grain-by-grain. Perhaps most important is the relatively unsorted character of alluvium when compared with the uniform grains employed in most flume experiments. In 1914 Gilbert recorded that the addition of fine material to previously uniform coarse debris not only increases the total transported load but also the amount of coarse debris that is moved. Later workers have noted that, in

mixed sediments, some of the coarse grade may actually begin to move before the fine. Sundborg has argued that a major factor explaining these apparent anomalies may be the effect of the diverse particle sizes on flow patterns near the stream bed. From an analysis of published data he distinguishes three situations: in coarse mixed material greater than 6–8 mm in diameter it is the finer fraction that is first entrained; for mixed sediments ranging between 6–8 mm and 0·8–1·0 mm it is the medium grain size; and for sediments less than 0·8–1·0 mm it is usually the coarsest material. He concludes that, although data for well-sorted sediments indicate that the most readily entrained grains lie in the size range 0·2–0·5 mm, for more poorly sorted sediments it is particles in the 1–7 mm range that are most easily set in motion.

Field investigations to test critical erosion velocities have been made by placing selected cobbles on small floats so designed that they rise to the surface once the stone starts to move. In one such experiment in California, cobbles ranging in weight from 9 to 156 kg were employed and the stream velocity was

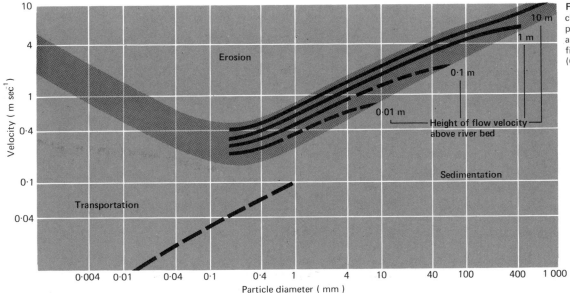

Fig 9.5. Stream velocities required to set channel material in motion. For larger particles the data are from Sundborg (1956) and are based on the work of Shields; for finer particles the data are from Hjulstrom (Geogr. Annlr., 1935).

measured at the moment each float was released during a flood. The critical velocities, varying from 1·3 to 3·0 m sec⁻¹, proved to be in reasonable accord with the values predicted by flume experiments.

The amount of erosion that will take place is not directly predictable from the critical erosion velocity. Inasmuch as channel enlargement reduces the water velocity a self-regulating mechanism may be envisaged. Several workers have claimed that severe vertical scour occurs during spate discharge; for instance, the channel of the Colorado at Yuma has reputedly been deepened by as much as 8 m during peak flow. However, in a survey of the evidence available in 1954 Lane and Borland emphasized the patchy distribution of deep scour features and suggested that marked overdeepening is of relatively rare occurrence. More recent studies have tended to confirm this view and to indicate abrupt channel constrictions as the only regular locations for deep scouring. Lateral erosion of stream banks, particularly on the outside of meanders, has been evaluated by the insertion of long stakes. The length of the protruding ends can be measured at intervals and the rate of channel migration calculated. Many of the experiments carried out in this way have underlined the importance of weathering in preparing fine bank materials for rapid removal. A bank may be eroded much more rapidly in winter after freeze–thaw cycles have rendered the surface friable than at the same discharge in summer.

Sediment transport

Basic principles Once set in motion sediment may be moved downstream either as bedload or as suspension load. The distinction between these two parts of the total sediment discharge is blurred by the exchange that takes place between them; an individual particle may become temporarily suspended, but then fall back to the stream floor to form part of the bedload. Nevertheless, the distinction is important since different principles underlie movement of the two types of load. If one imagines a bed of fine but non-cohesive material, a stream flowing at below the critical erosion velocity will leave all the particles undisturbed. As the velocity increases a few particles will slide or roll a short distance and then come to rest again; meanwhile others will be disturbed and make similar downstream movements. Further increases in velocity will lead to more and more grains being set in motion until a complete layer a few particle diameters thick will become mobile. The grains roll, slide or skip down the channel with their immersed weight borne mainly by contact with the bed. Meanwhile some of the dislodged particles, particularly the finer ones, will be lifted into the fluid, their immersed weight supported by turbulent upward currents within the water. These particles constitute the suspension load and may be transported long distances without any contact with the stream bed. Considerable differences exist between the bedload and suspension load in terms of particle size, total weight and downstream velocity. The finer particles in the suspension load may move at a velocity approximating that of the water, whereas the bedload as a whole travels much more slowly.

In principle, movement of stream sediment is due to the drag or shear stress exerted by the flowing water. The precise mechanism by which the available power represented by the moving water is converted into the work of transportation still defies full understanding. One cause of the difficulty is the complex nature of the variables involved. For instance, the two most fundamental variables appear intuitively to be stream velocity and bed material. Yet in considering the former attention must be directed not only to mean velocity but also to the velocity profile and degree of turbulence; similarly the bed material must be described in terms of size, shape, density and cohesion. The interdependence of the variables further complicates the subject to the point where it is difficult to derive laws of any great generality from experimental work. On the other hand, a deductive approach from fundamental physics needs to make so many simplifying assumptions that the results can rarely be applied directly to natural streams. An added problem is the uncertainty attaching to many field measurements, particularly those relating to bedload, so that theory cannot always be tested against observation with any real confidence.

Theoretical models Several workers have endeavoured to compute sediment transport rates from basic physical principles.

Fig. 9.6. A brief synopsis of the main arguments deployed by Bagnold in predicting likely sediment discharge from commonly measured streamflow parameters. No single diagram can do justice to Bagnold's elegantly argued hypothesis and, for further details, reference should be made to the original work.

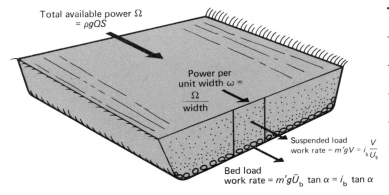

Total available power Ω = $\rho g Q S$

Power per unit width $\omega = \dfrac{\Omega}{\text{width}}$

Suspended load work rate = $m'gV = i_s \dfrac{V}{U_s}$

Bed load work rate = $m'g\bar{U}_b \tan \alpha = i_b \tan \alpha$

Symbols used

ρ	= fluid density	$m'g$	= immersed weight of load
g	= gravity acceleration	V	= fall velocity of suspended solids
Q	= discharge	i	= transport rate
S	= slope	$\bar{U}$	= mean transport velocity ($\bar{U}_s$ assumed to equal water velocity)
ω	= stream power per unit width	e	= efficiency

$\tan \alpha$ = friction coefficient ($\alpha \simeq 33°$)

Subscripts s and b refer to suspended load and bedload respectively

For both types of load, work rate = available power × efficiency,

$$\therefore i_b \tan \alpha = e_b \omega \quad \text{or} \quad i_b = \frac{e_b \omega}{\tan \alpha}$$

and, assuming available power for suspended load is now $\omega(1 - e_b)$

$$i_s \frac{V}{\bar{U}_s} = e_s \omega (1 - e_b) \quad \text{or} \quad i_s = \omega \frac{e_s \bar{U}_s}{V}(1 - e_b)$$

By summation total load = $\omega \left(\dfrac{e_b}{\tan \alpha} + \dfrac{e_s \bar{U}_s}{V}(1 - e_b) \right)$

Assuming $e_s = 0.015$ and $(1 - e_b) = {}^2/_3$, the latter equation simplifies to

$$\omega \left(\frac{e_b}{\tan \alpha} + 0.01 \frac{\bar{U}_s}{V} \right)$$

For purposes of comparison with gauging-station records (See bar graph) only the suspended material was considered

$$i_s = \omega \left(0.01 \frac{\bar{U}_s}{V} \right)$$

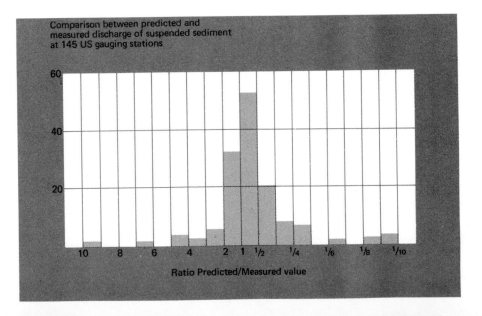

Comparison between predicted and measured discharge of suspended sediment at 145 US gauging stations

Ratio Predicted/Measured value

172 Bagnold, for example, has argued that the stream may be regarded as a transporting machine and that the concept of efficiency, long familiar to engineers dealing with machines, can appropriately be applied to sediment transportation. The basic equation may be expressed as Rate of doing work=available power×efficiency. The available power in the case of a stream is clearly the rate of liberation of kinetic energy as the stream flows down its channel. This energy is available for the two separate mechanisms of bedload and suspension load transport. Space does not permit a full recapitulation of Bagnold's arguments and it must suffice to note that, from a few fundamental principles, he derives for the total transport rate the relatively

simple expression $w\left(\dfrac{e_b}{\tan\alpha}+0.01\dfrac{\overline{U_s}}{V}\right)$ (Fig. 9.6).

This formula can then be tested against data from either flume experiments or natural streams. Bagnold found the predicted values correspond closely to observed flume values over a wide range of flow conditions and grain sizes; the only pronounced discrepancy occurs with a very fine suspension load. The lack of reliable bedload measurements for natural streams means that the data from sediment gauging stations have to be compared with predicted suspension loads. Again a reasonable correspondence is found although some of the discrepancies are much larger than those associated with the flumes. Observed values lower than predicted values may be explained in several ways, such as a general shortage of suspendable material or a flow temporarily below the critical level needed to lift available particles into suspension. More puzzling are instances where the actual load greatly exceeds the predicted capacity of the river. There is no entirely satisfactory explanation of these cases, but Bagnold suggests that the discrepancy may occasionally result from sampling procedures in which part of the bedload has been counted as suspension load.

Bagnold's analysis is an outstanding example of the deductive approach in fluvial geomorphology. Based upon the application of a few simple physical principles, its formulation was entirely independent of observational data. Its success in forecasting transport rates under a wide range of artificial and natural conditions is eloquent testimony to its general soundness. Nevertheless it also pinpoints some of the limitations in our present knowledge of sediment transport by rivers. Doubt persists regarding the precision of certain gauging techniques; the relationships that apply for sand and silt are not necessarily true for finer cohesive particles; in natural streams unlimited availability of transportable solids cannot be assumed; the constant discharge employed in many laboratory studies is uncharacteristic of natural streams where variations due to changing stage can be of prime importance.

Field investigations For field investigations proceeding beyond the routine gauging techniques described in Chapter 6 a prime requirement is some means of labelling sediment so that its actual movement either on the stream bed or suspended in the fluid can be monitored. Two methods have been commonly employed, one involving fluorescent tracers, the other radioactive tracers. The former have the advantage of being cheap and safe but normally demand the recovery of samples so that the counting of fluorescent grains can take place under ultra-violet light. The passage of radioactive tracers can be monitored by drawing detectors across the stream bed without the need for removing samples, but on the other hand the equipment is relatively expensive and stringent health safeguards need to be observed. In a study of the slow-flowing River Idle in Nottinghamshire sand-sized tracers incorporating the radionuclide Sc^{46} were released into the water and their downstream movement plotted by lowering detectors to the stream bed. In the course of a 3-week experiment the mean velocity of the water varied between 0·3 and 0·6 m sec^{-1}, while the tracers moved at an average rate of 0·064 m hr^{-1}. A flood led to a pronounced acceleration of the tracers with their velocity rising to over 2 m hr^{-1} for a period of $4\frac{1}{2}$ days. In a similar 13-day study of the North Loup River in Nebraska when the water velocity averaged about 0·7 m sec^{-1}, sand-sized tracers were found to move at a mean rate of about 1·15 m hr^{-1}.

Sands coated with fluorescent dyes have been employed in a study of suspended sediment movement along a gravel-bed river in Colorado. Three distinctive colours were used for identifying three grain sizes, 0·15–0·3, 0·3–0·52, 0·52–1·29 mm. The suspension load was analysed by means of a depth-integrated sampler operating some 800 m below the point at which the labelled sands were introduced into the stream; in addition a pump was used to abstract material from just above the bed layer. The rate of travel for the fastest particles was found to be inversely proportional to the square of their diameters. Although the mean water velocity was 1·4 m sec^{-1}, particles of 0·86 mm diameter travelled at 0·015 m sec^{-1}, and particles of 0·3 mm at 0·135 m sec^{-1}. Only the smallest grains of 0·15 mm approached the velocity of the transporting fluid and it seems clear that the coarser particles must have spent a considerable part of their time on the stream bed.

Other equipment used in the experimental study of bedload movement includes movie cameras, acoustic apparatus and highly sensitive echo-sounders. Photography is often ruled out by turbidity, while acoustic equipment allowing the user to listen to the impacts between moving particles is only appropriate to channels with coarse bedload and cannot be adequately calibrated to provide volumetric measurements. Ultrasonic echo-sounders developed since about 1960 can detect changes in bed elevation of less than 10 mm, and an array of such instruments mounted above a sand-bed channel can monitor the downstream passage of both ripples and dunes. How good an estimate of transportation rates this provides remains uncertain, although it probably indicates the minimum possible value. Any assumption that bedform movement can be directly translated into volumetric terms ignores two potential sources of error. If numerous particles leap forward several ripples at a time, the true transportation rate may greatly exceed that computed from bedform migration. The same would be true if the mobile layer were significantly thicker than the height of the bedforms. The retrieval of cores from streams seeded with tracers offers the potential of testing this latter point. In the case of the River Idle mentioned above tracers were found to extend 0·12 m below the channel bed and in the case of the North Loup 0·44 m below the bed.

Theoretical computations, sediment-gauging techniques, experiments with labelled particles and measurements of bedform migration, all provide estimates of transportation rates. However, even replicated experiments using the same measuring method occasionally provide data varying by 20 per cent or more, and discrepancies of that order are commonplace when comparisons are made between different methods. Much further work is evidently needed before even the basic elements of sediment transportation by natural streams are fully understood.

Attrition The attrition suffered by the transported load has proved remarkably difficult to measure, largely because it can very rarely be differentiated from a simple sorting effect. Downstream rounding of pebbles from a single identifiable source leaves little doubt regarding the effectiveness of attrition on coarse particles, but much more doubt attaches to fine material. Most information has come from laboratory experiments. Kuenen, for instance, has employed a large circular dish coated with concrete. An electrically-driven rotating paddle drives the water and sediment around the dish and the distance travelled by individual particles is estimated by coating grains with fluorescent paint and measuring their velocities under ultraviolet light. Kuenen concluded that the attrition of fine quartz material is extremely slight. For instance, after grains 1.5 mm in diameter had travelled 120 km at 0·45 m sec^{-1}, it was found that the weight loss averaged 0·000067 per cent per kilometre. The same worker has estimated that to reduce a quartz cube of 0·4 mm sides to a sphere would require river transport of several million kilometres. Attrition appears to be much less effective in the case of fine-grained than coarse-grained debris, while another significant variable is undoubtedly the hardness of the transported material.

Channel forms

Stream channels and the flows that occupy them are immensely varied. The channel may be single-thread or anastomosing, deep or shallow, straight or meandering, steep or gentle, while the flow may be steady or flashy, intermittent or perennial. Two distinct approaches may be discerned in attempts to analyse the relationship between channel form and discharge. The first is devoted to examining temporal variations at a single point of such parameters as stream width, depth and velocity. These constitute the 'at-a-station relationships' and reflect the basic form of the channel at the point of measurement. The second approach involves analysing the downstream changes in such

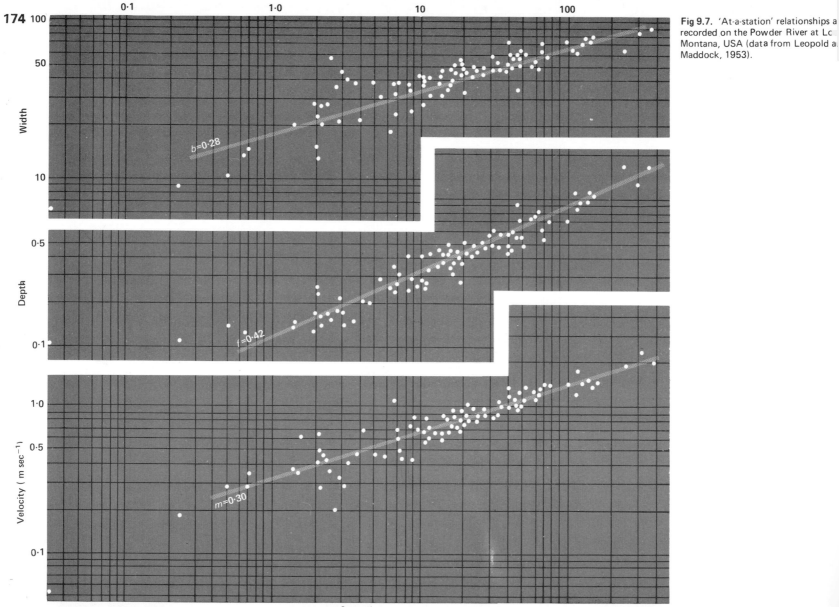

Fig 9.7. 'At-a-station' relationships a recorded on the Powder River at Lc Montana, USA (data from Leopold a Maddock, 1953).

Width

$b=0.28$

Depth

$f=0.42$

Velocity (m sec^{-1})

$m=0.30$

Discharge (m^3 sec^{-1})

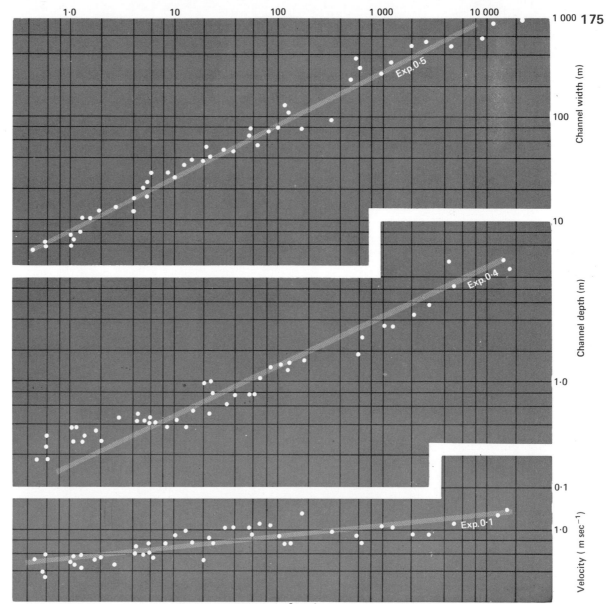

Fig 9.8. The relationship of variations in width, depth and velocity to downstream increases in discharge within the Missouri and Mississippi basins, USA (data from Leopold and Maddock, 1953).

properties as channel width, depth and slope.

Channel form may be regarded as reflecting the interaction between fluid stress and boundary resistance. Fluid stress varies with stream stage, whereas boundary resistance remains relatively constant. Intuitively, it might be expected that the fashioning of the channel would occur primarily during bankfull discharge when fluid stress attains its maximum value. For this reason the dominant discharge responsible for moulding channel form is often equated with bankfull discharge, although the possibility that the increased frequency of a slightly lower stage might be equally effective should not be overlooked. The interval at which bankfull discharge recurs appears to vary slightly from area to area, but most studies have suggested a periodicity of between 6 and 18 months.

At-a-station relationships Leopold and Maddock used the term hydraulic geometry in 1953 to denote the relationships between those stream parameters that are dependent on discharge. For instance, as discharge rises at a station there is a corresponding increase in stream width, depth and velocity. If the data for these three variables are plotted on logarithmic graph paper (Fig. 9.7) the points are found to lie close to a straight line indicating that each tends to be a power function of discharge. One may, therefore, write three equations: $w = aQ^b$; $d = cQ^f$; $v = kQ^m$; in which Q is discharge, w is width, d is mean depth, and v is mean velocity, while a, b, c, f, k and m are empirical constants. Since Q is itself the product of w, d and v, it follows that both $a \times c \times k$ and $b + f + m$ must equal unity. As might be expected the actual values of the constants differ from one station to another. For example, in a steep-walled channel rising discharge will be accompanied by an increase in depth but little change in width; conversely, in a gentle-sided channel it will be accompanied by a rapid increase in width and a slower increase in depth. The nature of the bed material seems in practice to be a major determinant of whether width or depth increase more rapidly. Studies in the United States have shown exponent b to be lower where the bed materials are cohesive than where they are non-cohesive. On a regional scale this often means that b has a lower value in humid than in arid and semi-arid areas.

The three parameters considered above are so clearly inter-related as to form a natural grouping for at-a-station studies. However, it is possible to inquire how certain other parameters change as discharge fluctuates. For instance, Leopold, Wolman and Miller have plotted suspension load as a function of discharge and have found an approximately linear relationship when logarithmic scales are used. This, of course, is no more than a sediment-rating curve with the shortcomings already discussed (p. 110). Nevertheless the authors suggest that, at least as a first approximation, one may write $G = pQ^j$ where G is sediment load and p and j are empirical constants; values of j are said to lie commonly in the range 2·0 to 3·0.

Downstream variations in channel properties In 1953 Leopold and Maddock prepared a series of graphs designed to examine how the three parameters of width, depth and velocity normally change along the length of an individual stream channel. Using logarithmic scales they plotted values for each against mean annual discharge; the latter was employed in place of the theoretically preferable bankfull discharge because its value was more readily available at most gauging stations. Although there was some scatter among the points, the authors generally found a simple straight-line relationship to characterize each drainage basin (Fig. 9.8). This suggests that a series of equations may be formulated analogous to those for at-a-station records. The exponents, however, are found to differ from those of a single cross-section, the calculated values commonly approximating to $b = 0.5$, $f = 0.4$ and $m = 0.1$. These figures indicate that a channel accommodates the downstream increase in discharge mainly by an increase in width, with a rather smaller change in depth. Surprise is sometimes expressed that velocity also increases in a downstream direction since rapid flow is regarded as typical of the upland stream cascading down a series of rapids. However, careful collation of mean flow rates at bankfull discharge usually indicates a down-valley acceleration. The increase is generally small and the scatter of points rather wide; nevertheless, it remains true that the mean velocity of the water, although a relatively conservative value, tends to exhibit a slight downstream growth. ·

A further aspect of the changes in the three parameters so far

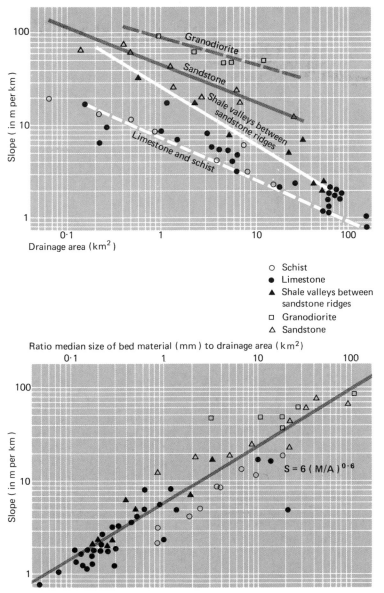

Fig 9.9 Channel gradients in parts of Virginia and Maryland: (A) Plot of channel slope against drainage area, showing a wide overall scatter of points that appears to be due, at least in part, to contrasts in geology: (B) Plot of channel slope against the ratio of the size of bed material to drainage area (M/A). This appears to lead to a considerable reduction in the scatter of points (data from Hack, 1957).

discussed is their cyclical regularity along certain limited reaches. In its most fully developed form this gives rise to the feature known as a pool-and-riffle sequence. Shallow gravel bars or riffles alternate with long deep pools in which the bed material is noticeably finer. At low flow the stream traverses the riffles with broken surface and then passes more slowly through the pools; at high flow these contrasts may be totally obliterated by the greater depth of water. Recognizing the favourable environment for the incubation of fish eggs created by aeration of water flowing over the riffle, Stuart became concerned at the absence of such bedforms in artificial channels below many Scottish hydro-electric dams. He found that piles of gravel placed at intervals of 5—7 times the channel width were soon shaped by the flowing water into stable forms closely resembling those in a natural pool-and-riffle sequence. The spacing employed by Stuart is typical of natural channels, but its basic cause remains obscure. Pools and riffles are characteristic of channels with poorly sorted bed material and are best developed where much of the sediment is coarser than sand. Sections show the larger debris on the riffles to be a superficial cover that is often only one or two particle-diameters thick. In part this may be due to a winnowing effect as the water flows more rapidly over the riffle, but the regular spacing remains a puzzle and the whole phenomenon requires much further investigation.

A fourth parameter of very considerable interest is the downstream change in channel slope. Although the seaward decline in gradient has elicited comment for several centuries, mathematical analysis is a more recent development. Two separate but related approaches may be discerned. The first has been concerned with the relationship between channel slope and other stream parameters, the second with the overall form of river long profiles. Into the first category comes the research in which measured slope has been plotted against discharge and size of bed material. As long ago as 1877 Gilbert maintained that his flume studies showed slope to be inversely proportional to dis-

178 charge, and in elaboration of this statement many workers on natural streams have plotted these two parameters in graphical form. As a result a relationship of the form $S=tQ^z$ has been claimed, although in practice catchment area has often had to be substituted for discharge owing to the lack of suitable data. The scatter of points has generally proved to be rather wide and even if the mathematically best-fit line is calculated the exponent z is found to vary considerably from one catchment to another. This clearly implies the intervention of factors other than discharge and many workers have argued that the most important is likely to be the calibre of the bed material. In 1957 Hack made· a detailed study of 15 streams in Virginia and Maryland, measuring in each catchment stream length, drainage area, channel slope, channel cross-section and size of bed material. Employing drainage area as a measure of discharge, he plotted this

against channel slope and found a wide scatter of points (Fig. 9.9). Classification of the points according to geology considerably reduced the scatter, leading to the inference that rock type is an important factor in controlling stream gradient. Further analysis suggested that rock type acts through the calibre of load supplied to the stream, and Hack concluded that the relations were best expressed by the general formula

$$S=k\left(\frac{M}{A}\right)^{0\cdot6}$$ in which S is channel slope, M the median size of the bed material, A the drainage area and k a constant. Employing similar arguments to those of Hack, Brush in 1961 analysed data from 16 streams in central Pennsylvania. Graphing channel slope against stream length, which he used as a substitute for drainage area, he managed to distinguish three groups of plotted points (Fig. 9.10). Each group was identified

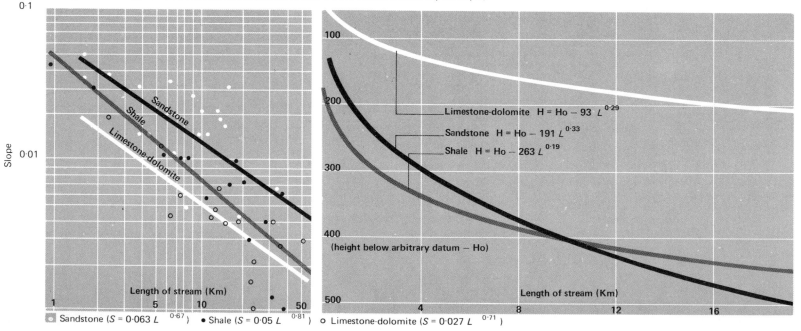

Fig 9.10. Analysis by Brush of channel gradients in central Pennsylvania: (A) Plot of channel slope against stream length for sandstone, shale and limestone-dolomite streams; (B) Hypothetical long profiles, calculated from regression lines in (A) and related to arbitrary datum, H_o

Limestone-dolomite $H = H_o - 93\ L^{0\cdot29}$

Sandstone $H = H_o - 191\ L^{0\cdot33}$

Shale $H = H_o - 263\ L^{0\cdot19}$

(height below arbitrary datum — H_o)

Length of stream (Km)

Slope

Length of stream (Km)

■ Sandstone $(S = 0.063\ L^{0\cdot67})$ ● Shale $(S = 0.05\ L^{0\cdot81})$ ○ Limestone-dolomite $(S = 0.027\ L^{0\cdot71})$

with a specific bedrock lithology, and was found to conform to a relationship defined by the formula $S = aL^{\frac{k}{4}}$ in which S is channel slope and L is stream length. Statistical analysis confirmed the variation of slope with length to be significantly different for the three rock types. Nevertheless the exponents in the equations do not differ greatly, so that the slope changes are rather alike regardless of lithology; it is the absolute value of slope for any given stream length that depends on lithology. Brush examined the possibility of the geological control being exercized through the size of the bed material, but found that the introduction of a factor for mean particle size only slightly reduced the scatter of the plotted points.

The works of Hack and Brush exemplify two possible approaches to the examination of downstream changes in channel slope. Their studies have now been repeated with slight

variations in many different areas and environments. A general **179** relationship of the form $S = tQ^z$ has been established, but many factors other than discharge have been shown to play a significant role. At the same time as channel slope was being analysed in this way other workers were concentrating on the overall form of river profiles. Kidson, for instance, showed that the profile of the Exe in south-western England can be described by a series of intersecting logarithmic curves (Fig. 9.11). He employed an equation which may be simplified to the general form $y = a - k\,logx$, where y is elevation, x is the distance from a theoretical source, and a and k are constants. In practice each concave segment of the profile was separately treated and the constants yielding the closest fit to the measured points independently computed. This procedure was followed in the belief that each segment represents the product of a stable

Fig. 9.11. The long profile of the middle and upper Exe as surveyed and interpreted by Kidson. It has been partitioned into nine concave segments, each treated as part of a former profile that may be reconstructed by extrapolating mathematically the short reach that survives. Also shown for one such reconstructed stage (the Westermill) is a succession of benches believed to have been eroded at or close to the contemporaneous sea-level.

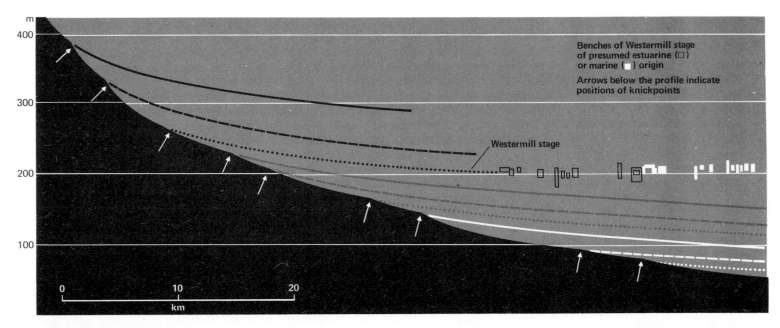

erosional phase ultimately terminated by uplift and initiation of a new knickpoint; by extrapolation of the fitted curves it was hoped to reconstruct the earlier profiles. Although this technique has often been used with little regard to the underlying cause of the concavity, it is of interest to compare the results with those of workers concerned with the relationship between slope and other stream parameters. In order to do this it is necessary to reduce both sets of data to the common measurements of elevation and distance from source. This can be done, for instance, in the case of the investigations by Brush where he expressed the relationship in the form $S = aL^k$. By integration it is possible to convert this to the form $H = H_o - \dfrac{a}{k+1}\left(L\right)^{k+1}$ (Fig. 9.10). The basic similarity of this curve to that fitted by Kidson is evident, both yielding results that indicate some form of logarithmic relationship between elevation and downstream distance. At the same time there is a fundamental disagreement between those workers who regard each steepening of a stream profile as a knickpoint initiated by a fall in base level and those who argue that in many cases abrupt changes in gradient are simply attributable to passage from one bedrock to another.

Channel Patterns

A great variety of channel patterns exists on the earth's surface. In some ways this simple observation is rather surprising. As Scheidegger has remarked, were water to flow in a straight channel from high to low point we would probably regard it as quite natural and self-explanatory. The fact that streams are often highly sinuous and may constantly divide into subsidiary channels to form anastomosing systems obviously requires some explanation. In order to analyse the patterns they must be measured and classified, but this in itself presents problems. Although there is often a clear-cut distinction between a single-thread and a braided stream, or between a straight and a meandering channel, these are end members of spectra that include numerous intermediate forms. These latter should not be forgotten in the course of the following discussion where emphasis will naturally tend to focus on the more striking examples of the phenomena. The regional distribution of channel patterns also begets a tendency to regard one type as the norm and any other

as exceptional. For instance, the person living in north-western Europe might well regard the single-thread channel as typical and the braided channel as unusual, but this is scarcely the view that would be adopted by an inhabitant of, say, Iceland or the Great Plains of the United States where anastomosing channels are commonplace (Fig. 9.12). This regional differentiation is itself significant since it points to common causal factors of widespread extent.

Braided channels Among factors that have been invoked to explain channel braiding are steep gradient, large sediment load, coarse non-cohesive bed material, widely fluctuating discharge and rapid aggradation. Whilst each may be present in a particular case, none appears to offer an adequate explanation by itself and usually a combination of factors is responsible. Coarse, poorly sorted bed materials combined with a sharply fluctuating discharge seem to be particularly conducive to braiding. This is probably because such conditions lead to the frequent deposition of coarse debris on the channel floor during falling river stages. Field observation suggests that this material then constitutes the nucleus of a gravel bar which grows by later accretion at its downstream end. Once vegetation is established such bars may be transformed into relatively permanent features, but the majority are subject to alternating erosion and deposition which completely modifies their appearance between successive floods. It has often been claimed that braiding results from overloading which in turn leads to aggradation, but it now seems that braiding may persist for long periods on some streams with little or no permanent deposition.

The frequent oscillation between erosion and deposition may lead to remarkably rapid shifting of either individual braids or the entire river bed. The former case is particularly well exemplified by glacial outwash streams, and in one detailed investigation in the valley below the Emmons Glacier on the flanks of Mount Rainier in Washington the channel alterations were so rapid that Fahnestock was able to use time-lapse photography as part of his study. Shifts of whole stream courses are especially common where rivers emerge from mountains on to gentler plainlands; for instance, the River Kosi, renowned for its devastating flooding at the foot of the Himalayas in northern India, has moved

Fig 9.12. An aerial view of the anastomosing pattern formed by meltwater streams issuing from Vatnajökull in south-eastern Iceland.

laterally more than 100 km during the last two centuries.

Another characteristic of braided channels deserving brief mention is their downvalley gradient. This has been investigated by several workers who have endeavoured to compare braided and single-thread channels by plotting bankfull discharge of the two types against channel slope. As discussed earlier, such a procedure discloses an inverse relationship of the general form

wide. For their discharge,
atively steep gradients. Yet
ather than increase when a
te braids. The reason is the
nels. The total width of the
depth decreases, with the
. 166) the enhanced value
R.

nnels Few single-thread
served sinuosities ranging
ely meandering (Fig. 9.13).
d for measuring channel
loyed is the ratio between
ratio of straight or winding
1·5; the latter figure has
lower limit for meandering,
der system the ratio may
n occasionally arise with
e no obvious valley length
ler belt axis is appropriately

eams the line of maximum
m one side of the channel
y seen in pool-and-riffle
nately on each side of the

stream bed. In such c... urrent does not flow straight down the channel but impinges against each bank in turn. These transverse movements have been invoked by some workers as

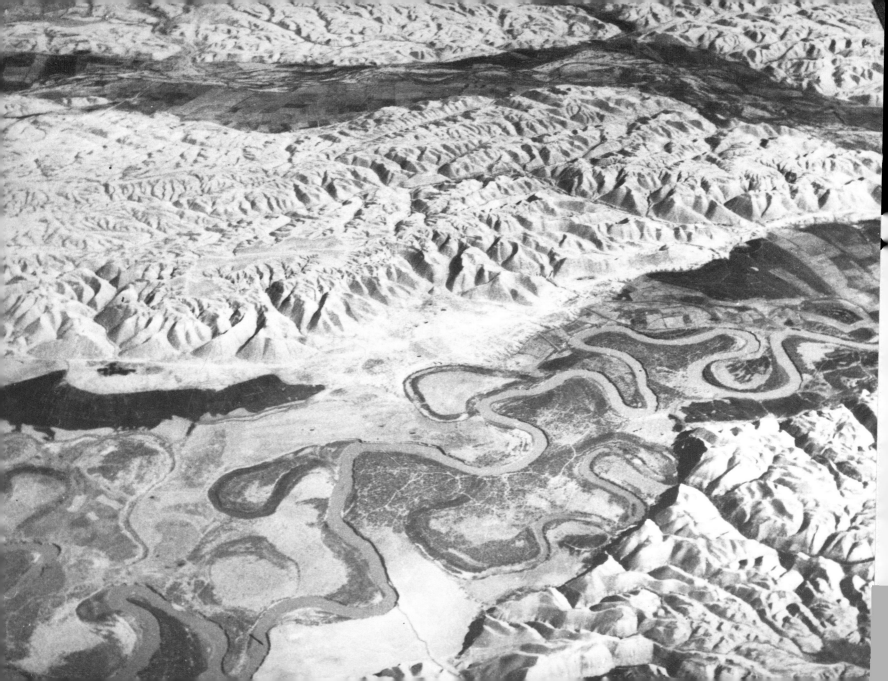

Fig 9.13. An intricate series of meanders traced by the river Jordan across its floodplain north of the Dead Sea. The mobility of the river channel is indicated by the large number of abandoned loops, several of which are quite fresh although others are now marked merely by the pattern of vegetation.

the prime cause of meandering; there certainly appear to be grounds for viewing the pool-and-riffle sequence as analogous to the regular deeps and shallows of the meandering stream. This, of course, does not explain meander development since the underlying cause of riffles is not yet understood. As the most distinctive of all stream patterns meanders have attracted much attention and two approaches in particular have yielded significant results.

The first approach has been concerned with the geometry of the pattern and has evolved a widely used terminology (Fig. 9.14). Analysis of active meander belts on alluvium has demonstrated a number of simple mathematical relationships,

particularly between channel width and meander wavelength, amplitude and radius of curvature. It is generally agreed that the relationships are best expressed in the form $L=aW^b$, $A=cW^d$ and $R=eW^f$, but in each case the exponent is so close to unity that simple linear equations are almost equally appropriate. As pointed out earlier, channel width is related to stream discharge and it follows that many meander belt dimensions are in turn a function of discharge. This fact has been used by several workers in computing the volume of water responsible for shaping abandoned meander traces. Dury has made the most elaborate calculations in a study of streams in both western Europe and North America that have apparently shrunk to a fraction of their former size. The basis for recognizing these streams was that the meanders they now describe are much smaller than those described by the valleys in which they flow. Two categories of meander were identified, stream meanders adapted to present conditions of discharge, and valley meanders

Fig 9.14 (A) Terminology employed in describing meanders; (B) Examples of the simple geometrical relationships of meanders in alluvium.

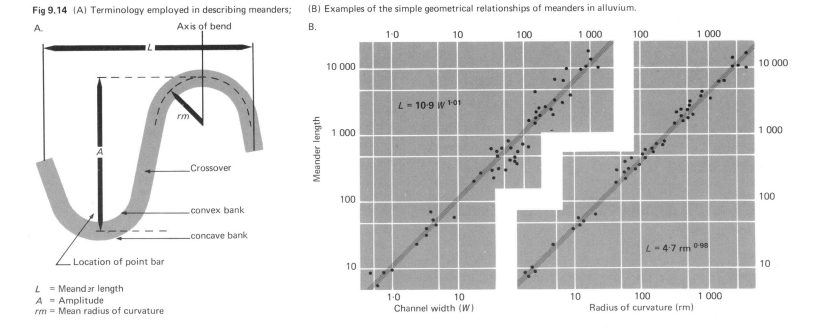

A.

Axis of bend

L

rm

A

Crossover

convex bank

concave bank

Location of point bar

L = Meander length
A = Amplitude
rm = Mean radius of curvature

B.

$L = 10.9\ W^{1.01}$

$L = 4.7\ rm^{0.98}$

Meander length

Channel width (W)

Radius of curvature (rm)

184 shaped during former periods of much greater discharge (Fig. 9.15). Dury argued that if channel width is a power function of discharge with an exponent of 0·5, and wavelength a linear function of width, then the relationship between wavelength and discharge may be expressed by the formula $l \propto q^{0.5}$. If the same arguments apply to valley meanders so that $L \propto Q^{0.5}$, it follows that $\dfrac{Q}{q} = \left(\dfrac{L}{l}\right)^2$.

In many examples the ratio between the wavelength of stream and valley meanders proved to be about 5, so that the inferred reduction in bankfull discharge would be as much as 25 times. Possible causes of reduction of this magnitude were discussed by Dury, with preference being accorded to decreased precipi-

tation, increased temperature and diminished storminess. Although the influence of this last factor on rainfall intensity means that it is unnecessary to envisage a former mean annual precipitation 25 times its present value, the change required by Dury's analysis remains extremely formidable and often lacks supporting evidence from the fields of pollen analysis and sedimentology. In consequence many workers have queried his basic assumptions with attention focusing in particular on whether stream and valley meanders are strictly analogous. Whereas the former are normally developed on alluvium, the latter are sculpted from bedrock. Inasmuch as the pattern reflects the interaction between flowing water and channel materials, the contrasting properties of bedrock and alluvial

Fig 9.15. A mass plot of the wavelength of valley and river meanders against drainage area (after Dury, 1965).

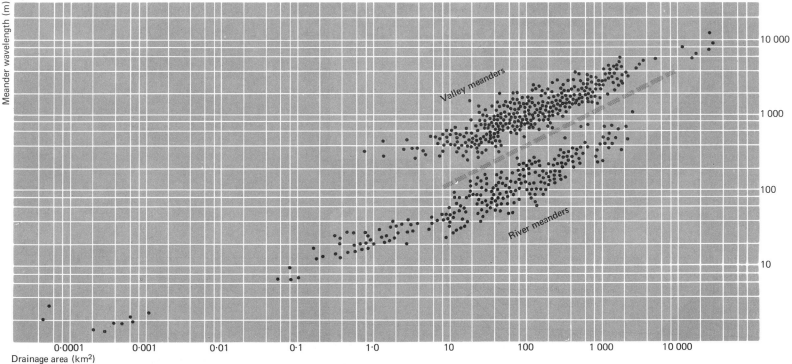

deposits might be expected to produce dissimilar patterns. The idea that bedrock and alluvial meanders have different dimensions is a very old one. Winslow early contested the view of Davis that entrenched meanders are simply floodplain meanders that have been vertically incised, claiming that in many instances the amplitude and wavelength are inconsistent with this interpretation. More recently Hack has shown that the wavelength of a meander train in Michigan changes as the stream passes from bedrock on to alluvium, while Tinkler has maintained that entrenchment on certain Texan streams is accompanied by an increase in meander wavelength. One may perhaps conclude that hypotheses founded purely upon dimensional similarities need to be treated with great caution.

The second approach to meander studies has been concerned with the nature of the flow round the curves. Many investigators have shown that, on the outer side of each curve, there is a super-elevation of the water surface that increases directly with velocity and tightness of the bend. This generates return currents at depth so that the flow assumes a helical form. Eventually a flow separation may develop, with a free-shear layer characterized by spiral vortices extending outwards from the inner concave bank. This reduces the effective width of the main current and concentrates flow against the downstream section of the curve. It is a common observation that the thalweg or line of maximum depth hugs the outer side of each meander and passes from one side of the channel to the other near the crossover point. There is, however, an important lack of congruence between the channel and flow patterns. The thalweg approaches closest to the bank at the downstream end of each meander, and crosses the centre-line of the channel downstream from the point of inflection. Erosion is thus concentrated on the outer bank below the axis of the meander while deposition occurs in the form of a point bar at a corresponding position on the inner bank. Such observations help to explain the downvalley migration of the channel bends, but not the beautifully regular form of the fully developed meander train. Bagnold has indicated one factor which might contribute to the dimensional relationships between channel width and meander curvature. He points out that total flow resistance round a 90° bend in a pipe is known to be minimized when the ratio of the radius of

curvature to pipe diameter lies between 2 and 3. Analogous relationships in open channel flow might account for meander wavelengths commonly tending towards 8–12 times the channel width, but it is not immediately apparent why the river should minimize the flow resistance, nor why meanders of different degrees of sinuosity should develop. In other words, the analogy with pipes may explain certain details of form but not the overall pattern. Langbein and Leopold, on the other hand, have focused attention on the overall pattern. They maintain that the length of channel between any two given points on a valley floor is independently controlled so that the problem resolves itself into the property possessed by meanders which would not be possessed by any other pattern composed of the same length of channel. They argue on general grounds that meanders tend to a form which may be described as a sine-generated curve. In such a curve the direction at any point is a sine function of distance along the meander (Fig. 9.16). It has the property of minimizing the sum of the squares of the deviations in direction along each successive unit length. Comparing this family of curves with actual meanders the authors demonstrate a close fit to examples ranging from laboratory flumes to the Mississippi river and from channels in alluvium to those in resistant bedrock. The arguments deployed constitute one part of a more general thesis that there is a tendency towards a minimization of the variance of certain stream parameters such as energy expenditure per unit area of channel. The planimetric data seem to confirm applicability of the general thesis to meander shape, but it is also claimed to apply to both bed shear, defined as stream depth times slope, and a friction factor defined as bed shear divided by the velocity squared. In order to test this wider hypothesis Langbein and Leopold selected five streams in Wyoming where meandering and straight reaches of channel occur in close proximity. Slope, depth and velocity were measured at closely-spaced stations during periods of nearbankfull discharge. The variance of the three values was calculated for each straight and meandering reach and, as might be expected, proved to be much greater on the curved than on the straight sections. However, when the data were combined to yield values of bed shear and the friction factor DS/v^2 the variance along the curved reaches proved to

186 be appreciably lower than along the straight. Since reduction in the variance of these parameters had previously been postulated as a general rule of stream behaviour, this demonstration may be held to support the inference that a meandering pattern is possibly more stable than a straight or sinuous one.

Landforms resulting from fluvial deposition

Of the many landforms produced by stream deposition, three that are illustrative of the range of processes involved will first be discussed, namely floodplains, alluvial fans and deltas. Attention will then be turned to river terraces in which active accumulation has ceased and the previously deposited sediments are suffering dissection.

Floodplains Defined as areas of alluvial land periodically inundated by the streams which they fringe, floodplains have too often been dismissed as flat monotonous features to have attracted the study they deserve. In practice detailed survey will often reveal subdued relief forms providing important clues to the processes by which the floodplain has been constructed.

Two major groups of processes may be distinguished. The first includes deposition on the floodplain surface by waters escaping from the confines of the channel during periods of spate. As they spread out across the floodplain, such waters lose velocity and deposit as a thin sheet of fine sediment material previously carried in suspension. The second group of processes derives from the shifting of stream channels. As a channel loop moves laterally it leaves behind it pointbar deposits which are approximately equal in volume to the material eroded from the concave bank. Since these deposits consist primarily of the traction load of the stream they are usually much coarser than the overbank sediments. The importance of channel migration is clearly attested on many floodplains by the numerous meander scrolls visible on aerial photographs. On the other hand, where lateral stream shifting is less active, fine overbank deposits may provide a relatively uniform and featureless cover. The balance between overbank sedimentation and migration of meanders is thus a highly significant control of floodplain form.

Distinctive features to be found on floodplains include levees, sandsplays, floodbasins and oxbows. Levees are ridges

Fig. 9.16. The geometry of meanders: (A) The relationship between a hypothetical meander and the sine curve from which it is derived; (B) − (D) Comparisons of actual river meanders with matching sine-generated curves. Above, for each segment of river the sine-generated curve is shown by a dashed line. Below, the corresponding sine curve is shown by a continuous line while the actual river meander is depicted by plotted points representing the direction of flow as a function of distance along the channel (in part after Langbein and Leopold, 1966).

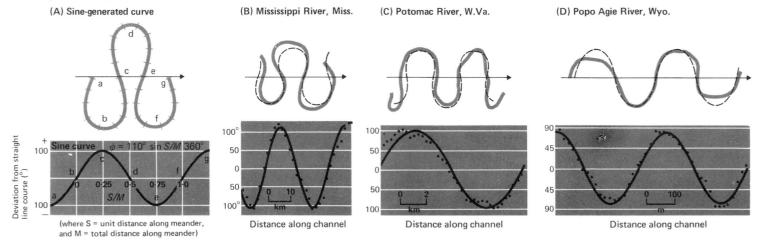

(A) Sine-generated curve   (B) Mississippi River, Miss.   (C) Potomac River, W.Va.   (D) Popo Agie River, Wyo.

Sine curve $\phi = 110° \sin S/M \ 360°$

(where S = unit distance along meander, and M = total distance along meander)

Deviation from straight line course (°)

Distance along channel   Distance along channel   Distance along channel

trending parallel to the stream channel and in the case of a large river may rise more than 5 m above the general floodplain level. Composed of the coarser fraction of the suspension load deposited as soon as the water escapes from its channel, they normally consist of fine sand and silt. Even where relatively continuous their line is interrupted at intervals by transverse depressions, called crevasses, through which the water first spills as the river level rises during floods. The fast-flowing water passing through a crevasse carries relatively coarse sediment which is spread out across the adjacent floodbasin as a lobate sandsplay. Having spilled over the levees, water may be diverted many kilometres downstream before it is able to regain the main channel. Re-entry will often be deferred until the confluence of a tributary which may itself have been diverted by levee development along the trunk river; tributaries occupying such a lateral position have been termed yazoos after the stream of that name on the Mississippi floodplain. It is implicit in such cases that the meander belt of the major river does not occupy the full width of the floodplain. With aggradation concentrated close to the meander belt the position of the trunk river may ultimately become unstable. A lateral shift of the whole meander train then occurs and the river comes to occupy one of its former floodbasins. Known as avulsion, this process has characterized the growth of many of the great floodplains of the world. It is also possible that a form of avulsion in which a rapidly aggrading river spills from its original valley into a neighbouring lower one, offers a better explanation for certain river diversions than the traditional concept of capture.

Within the confines of the meander belt there is normally a downvalley progression of individual channel loops. In some cases the rate of movement is too uniform to permit the development of many meander cut-offs, but in others overrunning of one loop by another is so common that the floodplain is littered with oxbow lakes in various stages of infilling. When a cut-off occurs, the abandoned segment of channel is first isolated by deposition of bed material across its ends and then infilled with fine overbank sediments mixed with the organic residues of plants growing in the lake. Former oxbows may so diversify the composition of an alluvial sheet that the orderly downstream progression of meander loops is interrupted and many new

instances of cut-off induced. It has been pointed out that two types of cut-off can be distinguished. The first is the neck cut-off where the new channel develops across the narrowest point of the meander core, and the second the chute cut-off where the stream assumes a fresh course along the depression between successive scroll bars. In general the former type is characteristic of clay and silt floodplains, the latter of sand and gravel floodplains where overbank deposits are relatively insignificant.

In an instructive study of the Missouri floodplain downstream from Kansas City, Schmudde has emphasized the need for a close watch on the sequence of events during a rising flood stage. Using the local term of bottom for floodplain fragments isolated by the river, he recognizes loop bottoms surrounded by meanders and long bottoms flanked by relatively straight channel reaches. Natural levees are virtually absent and the most obvious characteristics of the bottoms are their relatively steep downvalley gradients and slightly corrugated cross-valley forms. The steepness arises from the fact that, where the river is flowing obliquely across the valley, the bank on the downstream side is consistently 1 m or more above that on the upstream side. Consequently during a flood it is the lower end of each bottom that is first inundated by sluggish backwater. Only when the river rises to substantially higher levels does water spill directly down the valley. Zones of rapid flow may then attain the velocity necessary to scour elongated furrows, while across intervening areas of slower flow there is simultaneous deposition. In this way the surface of the floodplain becomes corrugated with a relief amplitude between 1 and 2 m. The forms and processes described by Schmudde probably have their counterparts in many smaller floodplains, and it is certainly true that individual floodplain segments often display gradients steeper than those of the valley as a whole. A simple model of regular and evenly distributed overbank deposition is clearly inappropriate to such cases.

Sections through floodplains in temperate latitudes often display a metre or more of fine sediments resting on coarse sediments. This succession is not necessarily indicative of a long-term change in fluvial regime; the contrasting materials may simply represent deposition by the separate mechanisms of overbank sedimentation and channel migration. It is even possible to visualize a steady-state condition under which a thick layer of sand and gravel might be progressively built up beneath a thin capping of clay. Nevertheless, the upward change from coarse to fine sediments beneath many modern floodplains seems to result from major climatic fluctuations. The basal sands and gravels often contain structural and organic evidence of aggradation under periglacial conditions. They also display minor abandoned channels indicative of deposition by braided rather than single-thread streams. It is believed, for instance, that the Mississippi assumed its meandering habit less than 10 000 years ago, and that the great thickness of coarse sediments beneath the modern floodplain is attributable to a formerly braided river (Fig. 9.17). The familiar floodplain of cool temperate latitudes with its sluggish meandering stream is often a recent addition to the landscape. In Australia Schumm has detected a rather different sequence of conditions leading to accumulation of the Murrumbidgee riverine plain in New South Wales. The modern river has a deep, narrow and meandering channel, but traces of at least two earlier channel forms can be discerned. The oldest comprises a series of shallow, wide and straight channels attributable to rivers carrying a coarse sandy bedload during a period of relative aridity. Of intermediate age are channel forms resembling those of today but with very much larger overall dimensions; the rivers are believed to have been transporting fine clayey sediment but with discharges computed from meander sizes and application of the Manning equation (p. 166) up to five times that of the modern drainage. Schumm ascribes these large meandering channels to a period in which the total rainfall greatly exceeded that of today.

These illustrations have been included to emphasize the importance of considering not only the current processes but also the evolutionary history of a floodplain. In some cases the actual surface is reworked with such frequency by the migrating river that no features of great antiquity can survive. Schmudde records that one-third of the entire alluvial surface of the lower Missouri floodplain has been worked over in 75 years, while Leopold, Wolman and Miller estimate that a much smaller floodplain in Maryland will be completely reworked in about 1 000 years. By contrast the lower Ohio channel seems to have shifted very little during the last millenium so that there are presumably great variations from one example to another.

Alluvial fans Composed of detrital sediment and shaped like a segment of a low-angle cone, alluvial fans are found in almost all climatic environments. Although best known from arid and semi-arid areas, the single most important factor in their development is the abrupt juxtaposition of mountain and lowland. Where a stream emerging from the upland loses its transporting capacity all or part of its load is deposited. The size of the resulting fan is often directly related to the area of the upland catchment. Along a continuous mountain front the fans may coalesce to form a compound feature termed a 'bajada' in arid regions. Alluvial sediments sometimes accumulate to total thicknesses of over 300 m, the greatest values being attained a short distance below the apex of the fan. The steepest surface gradients also occur near the apex where they may locally surpass 10°, but over most large fans the gradients are rather lower than this and angles of under 5° are more normal.

Recent studies have shown that many individual fans are constructed not only by normal streamflow but also by viscous debris flows or mudflows in which the sediment concentration is so high that the overall density sometimes exceeds $2 \cdot 0$ g cm^{-3}. Debris flows, unlike streams, do not generally give rise to selective deposition but result in ill-sorted accumulations that frequently terminate on the fan surface as slightly elevated lobes. Cobbles and boulders can be transported and often appear to become concentrated around the margins of a lobe. The area over which normal stream deposition occurs is determined in part by percolation of water into the fan and in part by diminishing surface gradients. The location of accumulation frequently shifts as old courses are abandoned and new ones adopted. The surface of a fan may consequently be divided into active and disused washes. The former have a micro-relief up to 1 m in amplitude, but after abandonment this gradually diminishes and in an arid environment the undisturbed gravel acquires a coating of desert varnish. Thereafter weathering and soil-forming processes are as important as mode of deposition in explaining the character of the surface sediment. From this character, or from the nature of the soil and vegetation cover in humid regions, relative ages may be assigned to different parts of

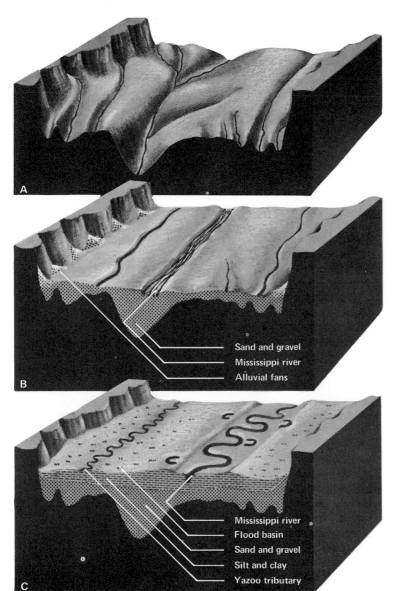

Fig 9.17. Stages in the recent evolution of the lower Mississippi floodplain. (A) Entrenchment during lowered sea-level *c.* 20,000 years B.P. (B) Rapid aggradation by heavily laden, braided river; tributaries building alluvial fans *c.* 10,000 B.P. (C) Slow aggradation by modern meandering river: levees, backswamps, yazoos (after Fisk).

Sand and gravel
Mississippi river
Alluvial fans

Mississippi river
Flood basin
Sand and gravel
Silt and clay
Yazoo tributary

a fan. The actual speed at which the surface of a typical fan is worked over by migrating washes is more difficult to assess. There is no doubt that many fans are very active features, and in California one third of a new 3-km length of road across a fan was affected within 50 years.

The cyclical evolution of fans has provoked much discussion. Whereas many modern examples have an apex which grades smoothly into the valley dissecting the mountain front, others are deeply trenched by the main watercourse. These fan-head trenches appear to indicate that aggradation has ceased and a phase of erosion has supervened. When a mountain front is elevated by faulting, the initial response may be fan accumulation but, with further erosion of the upland, declining stream gradients may result in dissection of the previously deposited materials. Although such a long-term cycle of growth and decay seems plausible, identification of the stage reached by an individual fan poses many problems. It is tempting to associate a smooth surface lacking a fan-head trench with vigorous aggradation but in some instances it seems that as much material as is added at the apex by the main feeder is removed from the toe by numerous smaller watercourses. In other words an equilibrium stage is reached with neither net growth nor net loss but with a form resembling that of a young aggrading fan. Some workers have similarly stressed the need for care in interpreting fan-head trenches. Debris-flows tend to build steeper surfaces than stream deposition so that if both processes are operative the debris-flow material may be subject to almost immediate stream dissection; trenching on this interpretation has little to do with the age of the fan. However, the depth of dissection arising in this way is probably limited, and deep trenches seem more likely to result from changed environmental controls. Since many fans owe their initiation to faulting, they are particularly prone to continued tectonic movement and many aspects of their morphology, including fan-head trenches, have been ascribed to this cause. A further significant factor may be climatic change. This is especially true in arid regions where even modest changes can result in substantial fluctuations in stream activity. Alluvial fans are evidently complex features subject to a wide variety of controls and it is not always easy to relate details of form to one particular control. Indeed, laboratory simulation has suggested there may even be threshold gradients capable of triggering successive phases of aggradation and trenching without any radical environmental alterations.

Deltas Deltas, like fans, are sedimentary accumulations sited where streams lose part of their transporting capability. However, because deltas are built out into water the factors influencing their growth are rather different. The initial requirement is that the volume of material brought down by the river should exceed that removed by marine or lacustrine agencies. Given that pre-requisite, at least four major factors influence the form of the resulting delta: the density and salinity of the water body, the character of the river load, the nature of the coastal processes, and the growth habits of colonizing vegetation. Although delta forms are traditionally subdivided into arcuate and bird's-foot types, in practice the diversity of factors controlling sedimentation results in a multitude of different shapes.

The density and salinity of the receiving water affects the rate at which the incoming load is deposited. If a muddy river flows into highly saline water, the clay fraction of the suspension load flocculates into larger particles which rapidly settle to the bed. The outstanding illustration is the Terek delta which at one time prograded into the Caspian Sea at about 300 m yr^{-1}. Dense, heavily laden rivers may, by contrast, plunge beneath the surface of fresh-water lakes and carry virtually all their sediment to the deeper parts of the basin. The proportion of in-flowing sediment that is incorporated in a delta thus varies widely from one river to another, irrespective of such factors as wave action and tidal range. This is not to deny the importance of coastal processes which are often the single most important influence on form. In a low-energy environment where the water body is incapable of redistributing the sediment, long embankments or levees may grow outwards from the land on both sides of the major river channels. The process resembles an inflowing jet of water decelerated along its margin with attendant deposition of the load. Bars may also develop on the channel bed and this appears to be a primary cause of bifurcation into separate distributaries. Weaknesses in the levees may be revealed during periods of flood, the overflow and enlargement of any breaches being known as crevassing. In this way a composite bird's-foot delta evolves with relatively slow infilling of the inter-channel regions.

In an environment where coastal processes are capable of redistributing all the sediment supplied by a river, long isolated levees are precluded and delta progradation takes place along a much broader front. The very varied depositional processes include not only overbank fluvial sedimentation but also wave and current action along the coastal margin. Much of the actual prograding takes place by the construction of delicate beach ridges or the growth of off-shore bars with enclosed lagoons. The basins surrounded by levees and barrier beaches then become the sites of accumulation for flood deposits. Fluvial, lacustrine and marine sediments frequently interdigitate. In addition to normal river floods, there may also be occasional large-scale inundations by the sea during storm surges; these can be highly destructive of all earlier depositional forms.

It has sometimes been claimed that deltas are confined to coasts of small tidal range. This is patently untrue with the Ganges delta prograding into the Bay of Bengal where the tidal range exceeds 3 m and the Irrawaddy delta advancing seawards where the tidal range is over 5 m. In fact, tidal scouring can have an important effect on surface detail, since the induced currents may be able to mobilize the fluvial sediments accumulating along the distributaries. As on the Ganges or Mekong delta, deep channels flaring seawards then become characteristic, together with complex waterway networks shaped to accommodate the regularly reversing ebb and flow currents. An important secondary influence is the nature of the vegetation cover. In low-latitude areas mangroves tolerant of periodic inundation by salt or brackish water trap sediments with their long stilt roots. The rich organic mud becomes the habitat for a varied burrowing fauna that promotes rapid assimilation of newly deposited sediment. Mangrove growth may encroach so much on minor waterways that eventually flow is concentrated in a few relatively large channels. The surface level of the islands between the channels is concurrently built up until they can be colonized by fresh-water swamp vegetation. On deltas of higher latitudes the earliest colonizers are usually grasses such as *Spartina* or rushes of the *Juncus* family; these also trap sediment after the occasional marine or fluvial inundation. By contrast, drier habitats on the distributary levees and sandy beach ridges or cheniers may soon be occupied by trees; the derivation of the word 'chenier' from the French for oak itself testifies to the characteristic vegetation on the Mississippi delta. Ultimately the woodland may spread across the former basins as the land level is raised and the various environmental zones migrate seawards.

It is not unusual to find one segment of a delta stationary or even retreating while another is prograding rapidly. This arises in part from periodic changes in distributary patterns, many examples of which are now well documented. The best-known case is that of the Mississippi which has constructed five separate deltas in the last 5 000 years (Fig. 9.18); moreover, if strenuous artificial efforts were not being made to regulate the river, an increasing proportion of the discharge would now be using the Atchafalaya channel and starting to build a sixth delta. Prior to development of the Sale-Cypremort delta, sea-level had been rising so rapidly that deltaic landforms were submerged almost as quickly as they were constructed; it follows that most of the surface features of the other great marine deltas of the world must be similarly youthful.

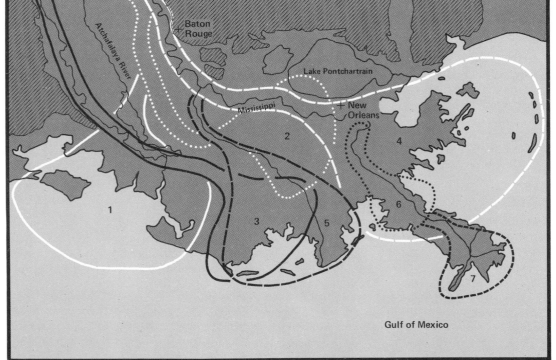

Fig. 9.18. The component deltas of the Mississippi delta system, together with an indication of their individual ages.

	Delta	Years B.P.
1	Salé-Cypremort	5 400 – 4 400
2	Cocodrie	4 600 – 3 600
3	Teche	3 900 – 2 750
4	St. Bernard	2 850 – 1 650
5	Lafourche	1 900 – 700
6	Plaquemines	1 150 – 500
7	Balize	500 –

192 River terraces Fluvial sediments deposited along a valley floor may later be dissected so as to form flat or gently shelving terraces raised above the level of contemporary flooding. Terraces so formed vary in morphology, structure and origin. One consequence is that many different criteria have been used in attempts at classification. A frequently-employed division is into paired and unpaired terraces. The former result from rapid vertical incision of a river into an old valley floor; the latter are produced by concomitant vertical and lateral movements of a river, terraces being left first on one side and then on the other at slightly different levels. A second frequent division is into rock and aggradational terraces. A rock terrace consists of a bedrock platform overlain by fluvial deposits no thicker than could be moved during a single migration of the river. An aggradational terrace, on the other hand, is underlain by river materials so thick that they imply an important period of accumulation prior to the start of downcutting. A third classificatory system is based upon the presumed initiating factor. This normally distinguishes two primary causes of terrace formation, changes in base level and changes in climate. A variation in either of these controls is capable of inducing river incision and the abandonment of an old floodplain. By combining the three methods of subdivision, which appear to be largely independent of each other, eight different types of terrace might be distinguished.

In the past there has undoubtedly been a tendency to emphasize base-level changes at the expense of other initiating factors. Both eustatic falls in sea-level and localized tectonic uplift have regularly been invoked to explain the dissection of former floodplains. The assumption in such models is that a terrace diverges downstream from the modern river, attaining its maximum relative altitude close to the old coastline (e.g. Fig. 9.11). There is no reason to doubt that such terraces exist, but it is becoming increasingly apparent that they represent only a fraction of the total number. Dating by means of organic materials often shows that an aggradation took place when sea-level was below that of the present day, and that the subsequent downcutting occurred when sea-level was rising. In such cases it seems clear that the fundamental control is less likely to have been a change in base level than a change in climate.

The rapid climatic fluctuations of the Pleistocene epoch were particularly favourable for terrace development. The growth and decay of ice-sheets induced sea-level oscillations affecting alike the lower river courses in all latitudes. Simultaneous variation in the supply of sediment and in the transporting capacity of the streams modified fluvial activity nearer the headwaters of many catchments. In detail the effects in the upper reaches are likely to have varied according to the particular climates experienced. In regions that suffered periglacial conditions accelerated hill-slope movements and highly seasonal stream regimens appear to have led to temporary aggradation; in modern desert areas periodic cooler and wetter phases may well have induced spasmodic downcutting by the better nourished streams; in equatorial areas less extreme environmental fluctuations probably had comparatively little influence on fluvial activity. It is evident that no single model of terrace development can cover all such circumstances, and instead attention will be focused on two illustratative examples.

The first example concerns the evolution, since the Eemian interglacial, of the Avon valley in the English Midlands (Fig. 9.19). Four terraces of this age are normally identified, No. 4 being the highest and No. 1 the lowest. The gravels of all but No. 3 have yielded organic materials indicative of intensely cold conditions. No. 3, on the other hand, has yielded remains of hippopotamus and other mammals that seem to imply aggradation during an interglacial phase. The relationship between terraces No. 3 and No. 4 has been a matter of dispute, and underlines the need to draw a clear distinction between terrace form and the age of the underlying gravels. Most workers have held that, following an initial phase of valley deepening, a major aggradation began below the level of the No. 3 terrace and culminated at the height of the No. 4 terrace; the total thickness of the deposits exceeded 15 m. The ensuing episode of down-cutting was interrupted by a brief period of lateral erosion during which No. 3 terrace was fashioned from the earlier fluvial sediments. On this interpretation the aggradation began during the Eemian period and terminated in early Weichselian times. After the fashioning of No. 3 terrace, renewed downcutting ended with accumulation of the gravels of No. 2 terrace. Radio-carbon assays indicate this terrace deposit to be of mid-Weichselian age, with dates ranging from approximately

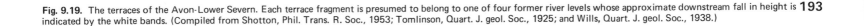

Fig. 9.19. The terraces of the Avon-Lower Severn. Each terrace fragment is presumed to belong to one of four former river levels whose approximate downstream fall in height is indicated by the white bands. (Compiled from Shotton, Phil. Trans. R. Soc., 1953; Tomlinson, Quart. J. geol. Soc., 1925; and Wills, Quart. J. geol. Soc., 1938.)

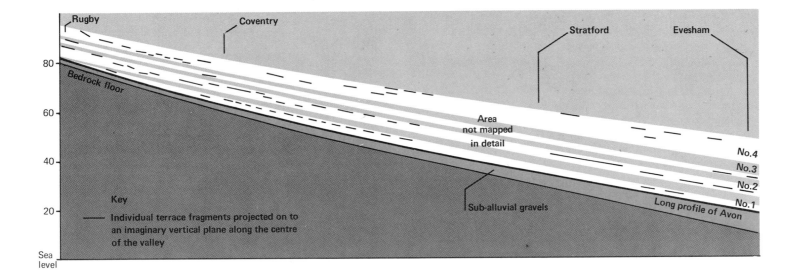

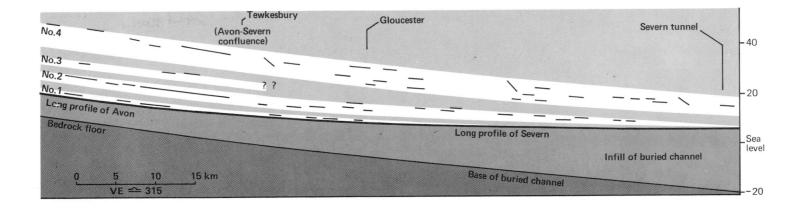

38 000 to 28 000 years B.P. On two further occasions during late-Weichselian times the Avon deepened its valley as a prelude to brief aggradational episodes; the first ended with the deposition of No. 1 terrace, the second with accumulation of the modern floodplain.

The Avon terraces bear no consistent relationship to the modern floodplain. The No. 4 terrace, for instance, lies some 12 m above present river level at Rugby but 24 m above it near Tewkesbury; on the other hand, the No. 2 terrace between the same two points maintains a more or less constant altitude of about 6 m. Presumed correlatives of the Avon succession have been traced by Wills to the head of the Severn estuary 60 km below Tewkesbury. The equivalent of No. 4 terrace converges on the modern alluvial flats until no more than 12 m above them at the entrance to the estuary. The equivalent of No. 2 terrace similarly declines and is virtually at the level of the flats where the river debouches into the estuary. The correlative of No. 1 terrace plunges below the level of modern alluvium in the vicinity of Gloucester. A deep sediment-filled channel has been traced from near Tewkesbury downstream to the estuary where it descends to a depth of at least -22 m. Further west the exact line of the channel is only imperfectly known, but it clearly falls to at least -30 m within a further 20 or 30 km.

The passage of many of the terraces beneath alluvium, and ultimately below sea-level, conceals much of the evidence necessary to relate them to former marine levels. However, certain inferences can be drawn from the apparent ages of the terraces and known glacio-eustatic oscillations (see pp. 336–44). During the last interglacial, when the gravels of the No. 3 terrace began to accumulate, sea-level around the Severn estuary probably differed little from that of today. By the time the aggradation responsible for the No. 4 terrace was complete, a marked refrigeration of the climate had taken place and presumably sea-level had fallen. It may even be speculated that the initial result of the climatic change was gravel accumulation, but that this was superseded by incision as the effect of the eustatic change worked upstream. Later aggradational phases undoubtedly occurred when sea-level was lower than today, but the evidence as yet is inadequate to relate them to either climatic or eustatic fluctuations.

The second example to be discussed is the remarkable succession of very recent cut-and-fill phases in the arid and semi-arid areas of the south-western United States. Considerations of both geography and dating mean that in this case the influence of sea-level change can be safely discounted. Within the last few thousand years there have been numerous oscillations between aggradation and the cutting of steep-walled ravines or arroyos. Most workers have invoked climatic change as the primary cause, but have failed to agree about the precise mechanism involved. Some have argued for arroyo-cutting during wetter interludes, others during drier interludes. Neither hypothesis receives unequivocal support from palynological data, and several workers have maintained that channel incision begins during periods of more intense summer storminess. The most recent episode of arroyo-cutting dates from about 1870 and is responsible for converting many shallow washes, formerly crossed by pioneer settlers without difficulty, into steep-walled channels incised over 10 m below the original valley floor. Climatic statistics do not entirely support the ascription of this phase of arroyo-cutting to increased storminess, and overgrazing by stock has been claimed as an important contributory factor. However, that could not apply to earlier episodes of incision and it must be admitted that the overall controlling factors are still far from clear. It has even been suggested that cut-and-fill sequences do not reflect climatic oscillations but are an expectable consequence of random meteorological events. Arroyo-cutting would then be caused by 'super-floods' with a very long recurrence interval; aggradation, on the other hand, would take place during intervening periods of less extreme conditions. One result of this cataclysmic hypothesis might be different cut-and-fill sequences in different valley systems, since 'super-floods' could be due to highly localized storms. Two final points of very general application are worthy of mention. Deposition in one area obviously implies erosion in another, so that there will often be sharply contrasting conditions within a single catchment. Secondly, once certain threshold values have been exceeded, a long-standing equilibrium may be destroyed and only restored after many consequential changes. For example, if a protective soil cover is gullied during an exceptional storm, the underlying material may be highly susceptible to removal by rainfall that had previously been quite impotent. The soil remnants on the inter-gully areas may be continually undercut so that their protective role is effectively ended.

ALEXANDER, C. S. and N. R. NUNNALLY, 1972. 'Channel stability on the Lower Ohio River', *Ann. Ass. Am. Geogr.* 62, 411–17.

ALLEN, J. R. L., 1965. 'A review of the origin and characteristics of recent alluvial sediments', *Sedimentology*, 5, 89–191.

BAGNOLD, R. A., 1960. 'Some aspects of the shape of river meanders', *Prof. Pap. U.S. geol. Surv.*, 282-E.

BAGNOLD, R. A., 1966. 'An approach to the sediment transport problem from general physics', *Prof. Pap. U.S. geol. Surv.*, 422-I.

BEATY, C. B., 1974. 'Debris flows, alluvial fans and a revitalized catastrophism', *Z. Geomorph. Supplementband* 21, 39–51.

BRUSH, L. M., 1961. 'Drainage basins, channels and flow characteristics of selected streams in central Pennsylvania', *Prof. Pap. U.S. geol. Surv.* 282-F.

CRICKMORE, M. J., 1967. 'Measurement of sand transport in rivers with special reference to tracer methods', *Sedimentology*, 8, 175–228.

DENNY, C. S., 1967. 'Fans and sediments', *Am. J. Sci.*, 265, 81–105.

DURY, G. H., 1965. 'Theoretical implications of underfit streams', *Prof. Pap. U.S. geol. Surv.*, 452-C.

FAHNESTOCK, R. K., 1963. 'Morphology and hydrology of a glacial stream – White River, Mount Rainier, Washington', *Prof. Pap. U.S. geol. Surv.*, 422-A.

FISK, H. N., 1947. 'Fine-grained alluvial deposits and their effects on Mississippi river activity', *Mississippi River Comm. Waterways Experiment Station*.

HACK, J. T., 1957. 'Studies of longitudinal stream profiles in Virginia and Maryland', *Prof. Pap. U.S. geol. Surv.* 294-B.

HACK, J. T., 1965. 'Post-glacial drainage evolution and stream geometry in the Ontonagon area, Michigan', *Prof. Pap. U.S. geol. Surv.*, 504-B.

HOOKE, R. Le B., 1968. 'Steady state relationships on arid region alluvial fans in enclosed basins', *Am. J. Sci.*, 266, 609–29.

HELLEY, E. J., 1969. 'Field measurement of the initiation of large bed particle motion in Blue Creek near Klamath, California', *Prof. Pap. U.S. geol. Surv.*, 562-G.

KENNEDY, V. C. and D. L. KOUBA, 1970. 'Fluorescent sand as a tracer of fluvial sediments', *Prof. Pap. U.S. geol. Surv.* 562-E.

KIDSON, C., 1962. 'The denudation chronology of the River Exe', *Trans. Inst. Br. Geogr.*, 31, 43–66.

KUENEN, Ph. K., 1959. 'Experimental abrasion 3. Fluviatile action on sand', *Am. J. Sci.*, 257, 172–90.

LANE, E. W. and W. M. BORLAND, 1954. 'River-bed scour during floods', *Trans. Am. Soc. civ. Engrs.*, 119, 1069–80.

LANGBEIN, W. B. and L. B. LEOPOLD, 1966. 'River meanders – theory of minimum variance', *Prof. Pap. U.S. geol. Surv.*, 422-H.

LEOPOLD, L. B. and T. MADDOCK, 1953. 'The hydraulic geometry of stream channels and some physiographic implications', *Prof. Pap. U.S. geol. Surv.*, 252.

SAYRE, W. W., 1965. 'Transport and dispersion of labeled bed material, North Loup River, Nebraska', *Prof. Pap. U.S. geol. Surv.*, 433-C.

SCHICK, A. P., 1974. 'Formation and obliteration of desert stream terraces – a conceptual analysis', *Z. Geomorph. Supplementband*, 21, 88–105.

SCHMUDDE, T. H., 1963. 'Some aspects of land forms of the Lower Missouri river floodplain', *Ann. Ass. Am. Geogr.*, 53, 60–73.

SCHUMM, S. A., 1968. 'River adjustment to altered hydrologic regimen – Murrumbidgee River and paleochannels, Australia', *Prof. Pap. U.S. geol. Surv.*, 598.

STUART, T. A., 1953. *Spawning Migration, Reproduction, and Young Stages of Loch Trout*, HMSO, Edinburgh.

SUNDBORG, A., 1956. 'The river Klaralven. A study of fluvial processes', *Geogr. Annlr.*, 38, 125–316.

TINKLER, K. J., 1971. 'Active valley meanders in south-central Texas and their wider implications', *Bull. geol. Soc. Am.*, 82, 1783–800.

WORSSAM, B. C., 1973. 'A new look at river capture and at the denudation history of the Weald', *Rep. Inst. geol. Sci.*, 73/17.

Although first published in the mid-1960s the basic reference on streams remains : L. B. Leopold, M. G. Wolman and J. P. Miller, *Fluvial Processes in Geomorphology*, Freeman, 1964. A valuable introductory text is provided by M. Morisawa, *Streams: Their Dynamics and Morphology*, McGraw-Hill, 1968. A more advanced book (written from an engineering viewpoint but with much of interest for the geomorphologist) is H. W. Shen, *River Mechanics*, Shen, 1971.

Rates and patterns of landscape evolution

In Chapters 8 and 9 hillslopes and streams have been discussed with, primary emphasis on the relationship between form and process. The time-scale employed in the discussion has intentionally been kept short, even though many of the landforms might also be viewed as the culmination of a long and involved geomorphic history. At the same time, little has been said about the interaction between hillside and valley floor despite the fact that both are obviously part of a more comprehensive denudation system. The main aim of the present chapter is to link these two themes by considering the long-term evolution of the landscape as a whole. The method adopted is to review a number of models of landscape evolution, but before these can be satisfactorily evaluated it is desirable to have at least a rough idea of the rates at which denudation systems commonly operate.

Rates of denudation

Measurements of current sediment discharge The second half of the twentieth century has seen an immense increase in the systematic monitoring of streams. The most frequently measured parameter is water discharge, but an increasing number of stations take regular readings of sediment and solution load. The main reason behind this upsurge is practical rather than scientific; it stems from the need of densely populated and highly industrialized nations to assess both the quantity and quality of their water resources. With the growth in stream monitoring by civil authorities has come an awareness by geomorphologists of the significance of the collected data. The techniques for measuring instantaneous sediment discharge described in Chapter 6 need no further elaboration here, but the uncertainties in many of the procedures need to be borne in mind when using the results. Moreover, care needs to be taken that figures from different sources are strictly comparable. Often no attempt is made to measure the bedload, although sometimes a rather arbitrary weighting is applied to the suspension load as an allowance for the unmeasured fraction. Similarly some estimates of gross sediment yield include the solution load whilst others do not. These factors can introduce significant differences between apparently compatible sets of data and render regional comparisons very hazardous.

On a global scale Holeman has collated records from 12 major rivers discharging very large quantities of suspended sediment (Table 10.1). Exceptionally high rates are associated with some of the rivers of south-eastern Asia, parts of the Yellow River catchment currently being lowered at 2·7 mm yr⁻¹. The figures in the table should not be regarded as typical of the globe as a whole since the rivers were selected on the basis of their size and high sediment discharge. Nevertheless, in aggregate their catchments cover over 15 per cent of the world's land surface so that, if the figures are reliable, the mean rate of surface lowering they provide, namely 0·15 mm yr⁻¹, is almost certainly accurate to at least one order of magnitude. No allowance is made for bedload or solution load which could significantly increase the calculated values.

On a continental scale, the network of stations maintained by government agencies in the United States provides one of the most comprehensive coverages yet available. It is estimated that, from the conterminous United States, 446 million tonnes of suspension load are currently being discharged into the surrounding oceans each year. This is equivalent to 64·7 tonnes from each square kilometre of the land surface, and assuming a density of 2·7 g cm⁻³ represents a mean lowering of 0·024 mm yr⁻¹. However, this average value conceals marked regional variations. For example, the streams draining to the Atlantic seaboard carry suspended sediment equivalent to an annual lowering of their catchments of 0·006 mm. The corresponding figure for the whole Mississippi basin is 0·034 mm, while for a group of catchments on the north Californian coast it is over 0·4 mm. These are conservative values since they include no allowance for either bedload or solution load. The true rate of surface lowering could well be 50 per cent higher than the quoted figure.

Comparisons may also be made of regional denudation rates

Table 10.1 Suspended sediment discharge of twelve of the world's major rivers (modified and adapted from Holeman 1968). The figures in brackets are values given by Meybeck and Carbonnel (Nature 1975) for the corresponding discharge of dissolved solids.

River	A Catchment area (10^3 km²)	B Mean water discharge (10^3 m sec⁻¹)	C Suspended sediment discharge (10^6 tonnes yr⁻¹)	D Sediment yield (tonnes km⁻² yr⁻¹) (col C/col A)	E Rate of surface lowering (mm yr⁻¹)
Yellow	672	3·3	1 820	2 708	1·00
Ganges	956	11·7	1 450 (76)	1 517	0·56
Brahmaputra	666	19·2	725 (80)	1 088	0·40
Yangtze	1 942	22·0	500 (100)	257	0·09
Amazon	5 775	175·0	500 (290)	87	0·03
Indus	969	5·6	435	449	0·17
Irrawaddy	430	13·6	300	698	0·26
Mississippi	3 269	18·4	300 (142)	91	0·03
Magdalena	240	7·5	250	1 041	0·39
Mekong	795	11·0	170 (60)	214	0·08
Colorado (at Grand Canyon)	357	0·5	129	361	0·13
Red	119	2·2	120	1 008	0·37
Total	16 190	290·0	6 699	414	0·15

on a single rock type. Several workers have attempted to assess solution rates in limestone areas where this is presumably the most effective means of surface lowering (Table 10.2). With one exception all the values lie within the relatively narrow band $0.015-0.083$ mm yr^{-1}. If, as is commonly the case, limestone is regarded as a relatively resistant material many other rocks can presumably be lowered at a faster rate.

Sediment traps All particulate sediment in transport along stream channels must eventually be deposited. Whilst it may temporarily come to rest within the channel itself, more permanent deposition normally takes place in either a lake or the sea. These sites of long-term accumulation may be described as sediment traps. They vary enormously both in dimensions and in the length of time for which they have been the recipients of detrital material. At one end of the scale are small, artificial reservoirs that have been acting as sediment traps for no more than a decade or so. At the other end are the ocean basins which have been receiving the products of continental erosion for millions of years. The basic principles of study are the same for all. The amount of sediment must be determined, the potential source area defined, and the period of accumulation delimited. Thereafter it is a relatively easy matter to compute the mean rate of surface lowering represented by the accumulated sediment. However, each of these preliminary calculations may introduce some degree of uncertainty into the final result. The subaqueous position and high water content of the sediments makes their total volume difficult to assess accurately; the period for which a sediment trap has been operating may be subject to doubt; while a final possible source of error is leakage from one sediment trap to another. The last is particularly important in the oceanic environment where marine processes, besides supplying a potential extra source of sediment, can redistribute the material introduced by rivers. Despite these limitations, sediment

Table 10.2 Estimates of current rates of solution in limestone areas.

Area		Rate of solutional lowering (mm yr^{-1})
British Isles	*Yorkshire	0.049–0.050
	*Derbyshire	0.075–0.083
	*East Anglia	0.025
	*South Wales	0.018
	*Co. Clare	0.055
Continental Europe	+Poland	0.032
	*Swiss Alps	0.015
	=Austrian Alps	0.028
	*Yugoslavia	0.077–0.080
North America	*Jamaica	0.072
	++California	0.017–0.021
	**Arctic Canada	0.002

(Data from *compilation by Sweeting 'Karst Landforms' (Macmillan) 1972; **Smith I. B. G. Spec. Pub. 4 (1972); +Pulina Geogr. Polinica (1972); =Bauer Erdkunde (1964); ++Marchand Am. J. Sci. (1971).

traps still afford one of the best means of assessing denudation rates.

In a few areas small reservoirs have been constructed for the specific purpose of trapping and measuring sediment. These are particularly appropriate in regions of intermittent stream flow where the total volume of debris brought down during the wet season, or even during an individual storm, can be readily assessed once the water in the reservoir has evaporated. Installations of this type have been employed in Israel and the western United States, but normally reliance has to be placed on information coming from reservoirs designed for other purposes. A good example is afforded by Lake Mead on the Colorado River in Arizona. Impounded in 1935 behind the 180 m Hoover Dam this was the subject of a detailed investigation by the United States Geological Survey in 1948. The volume of sediment was computed by comparing bathymetric charts accurate to ± 0.6 m with earlier contoured maps. A further check on sediment thickness was provided by low-frequency echo-sounders capable of penetrating the superficial cover but subject to reflection from the original bedrock surface. It was found that in 14 years debris had accumulated near the lake head to a thickness of 83 m, while even at the foot of the dam it had reached about 30 m owing to the action of turbidity currents. To determine the density and composition of the highly fluid sedimentary mantle, 46 cores and over 300 surface samples were recovered for analysis. From the results it was computed that 1 775 million tonnes of particulate material had been fed into the lake by the Colorado in a period of 14 years, an average annual rate of 127 million tonnes. This figure agrees remarkably well with the 129 million tonnes of suspension load recorded at an old-established gauging station a short distance higher upstream in the Grand Canyon. The similarity of the two values suggests that bedload transportation by the Colorado is small in comparison with the suspension load. A simultaneous survey of the water chemistry of Lake Mead tended to confirm the solution load measurements made at the Grand Canyon gauging station, although chemical reactions over the bed and around the margins of the lake made precise calculations difficult. It was estimated that, within the 14-year period investigated, the overall denudation rate of the Colorado catchment upstream from Lake Mead approximated 0.16 mm yr^{-1}.

The techniques adopted on Lake Mead are only appropriate where massive volumes of sediment are involved. Under normal circumstances the cover on a reservoir floor is too thin for indirect measurement, and advantage has to be taken of periods when the water is drained for engineering purposes. Even then measuring the volume of new sediment poses substantial problems. Sampling procedures must take account of the many variations in the character of the material. Among factors needing careful evaluation are the water content, the percentage of organic matter, and the volume of desiccation cracks. In each of these respects fine muds differ from coarse silts and sands. Given the complexity of the floor deposits, an estimate of the sediment yield with an accuracy of ± 10 per cent demands a detailed and painstaking survey. In the British Isles a study of Strines reservoir in the southern Pennines has revealed a mean surface lowering at a rate of 0.013 mm yr^{-1} for the last 87 years, while a study of Cropston reservoir in the flatter Midlands has provided a corresponding figure of 0.0012 mm yr^{-1} for the last 95 years.

Assessment of the volume of sediment contained in natural lakes usually depends either upon coring techniques or upon echo-sounding at frequencies capable of penetrating to the underlying rock. The length of time during which accumulation has been taking place is often very problematical. In glaciated areas it is frequently assumed that lacustrine sedimentation began immediately the basin was divested of its ice cover. In such circumstances outwash from the retreating ice would probably lead to rapid accumulation at first, with a gradually decreasing rate as the climate ameliorated. Extensive coring would then be necessary to divide the basin fill into separate sedimentary units in order to eliminate the effects of glaciation. Few reliable surveys of such a nature have yet been concluded, and most studies have relied instead upon dating horizons within a relatively small number of cores and from this deducing relative rates of sedimentation at different periods. Dating has usually been by means of pollen analysis and radiocarbon assays, but a method of increasing importance is based upon palaeomagnetism. This uses not only long-term polarity reversals but also the virtually continuous shifts of the magnetic poles which are clearly imprinted on many undisturbed lacustrine sediments.

The most important form of sediment trap within the ocean basins is the deep-sea fan. In many instances these are clearly built of detritus emanating in the main from a single continental catchment. The primary mechanism of transfer from the continental shelf to the deep ocean floor is the turbidity current, capable of carrying debris 100 km or more in a single movement. Admittedly there are many possible causes of loss or accession of material between the river mouth and the fan surface, but in general the volume of a fan appears to offer a reasonable guide to the rate of mechanical denudation provided that the date of initiation can be firmly established. The latter often proves a difficult task, although circumstances occasionally lead to a fan being fed with debris for only a short period. It has been argued that this happened in the case of Navy fan off the coast of southern California. Owing to details of the submarine topography, the supply of debris was restricted to a brief Weichselian interval of very low sea-level. The total volume of sediment is about 56 km^3 derived from a catchment of 6 500 km^2; if accumulation took place for the estimated 15 000 years, the rate of mechanical denudation must have been about 0·6 mm yr^{-1}. As more information is gained by deep-sea drilling, the age of constructional submarine forms should be established with greater precision. Current programmes of oceanographic research are already beginning to shed further light on continental erosion rates. In a study of the huge Ganges-Brahmaputra fan which covers an area of 3 000 km by 1 000 km in the Bay of Bengal, geophysical surveys have shown that the sediments often exceed 3 km in thickness. The age of rocks recovered from the base of the fan suggest it was initiated no earlier than Miocene times. Making the necessary allowances for density variations, it has been calculated that some 8 million km^3 of rock have been eroded from the Ganges-Brahmaputra basin in a period of less than 20 million years. The total area of the catchment is now just over 2 million km^2, but of this amount a quarter is composed of alluvial floodplain and other depositional forms. From the remainder a volume of rock averaging 5·3 km thick has been removed, yielding a long-term rate of

Fig 10.1. A rim-to-rim view of the Grand Canyon showing, in the distance, the remarkably smooth plateau surface formed by the Permian Kaibab limestone (compare with Fig 10.2).

mechanical denudation of just over 0·26 mm yr^{-1}. However, within the fan a major unconformity points to intermittent deposition that may well be related to separate phases of Himalayan orogenesis. On the basis of known geology periods of rapid uplift have been assigned to middle Miocene and early Pleistocene times; if the part of the fan above the unconformity is entirely Pleistocene in age it implies a recent denudation rate of about 0·7 mm yr^{-1}.

Calculations based upon geological relationships Many assessments of long-term denudation rates have been based upon geological arguments. These normally centre around the destruction of some dateable topographic feature or the dissection of some parent material of known age. In each case the time constraints are provided by geological evidence. For example, where horizontal marine strata overlie rocks severely deformed in an orogenesis, the time interval represented by the unconformity indicates the maximum period taken to destroy any mountains that were formed; using this approach it has been argued that on a continental margin a major mountain chain can be completely erased in a period of 20 to 40 million years. Likewise, where plateaus are formed by dissection of uplifted strata, incision of the drainage cannot have begun prior to the deposition of the youngest continuous bed. In both these examples basic principles define the maximum period available for fluvial erosion; in many instances supplementary evidence permits a closer definition of the varying rate at which denudation has operated.

In the foregoing pages methods of assessing current erosion rates have been described from the Colorado valley in the United States and the southern Pennines in Great Britain. It is instructive to consider the light shed by geological considerations on the long-term erosion rates in these two areas. On the simple principle enunciated in the preceding paragraph, excavation of the Grand Canyon (Fig. 10.1) cannot

have begun before Mesozoic times since its lip is formed by the Permian Kaibab limestone (Fig. 10.2). Beyond the lip are dissected remnants of formerly continuous marine strata of which the youngest are no older than Tertiary in age; nearly all workers have assented to the proposition that this constitutes the earliest possible date for initiation of the canyon. Even so, much uncertainty still surrounds the exact period at which the Colorado first assumed its present course and began incising its valley. Where the river now emerges from the Grand Canyon near Lake Mead extensive beds of mid-Cenozoic age, some of them little more than 10 Ma old, bear no sign of accumulation close to a major sediment-carrying river; on this evidence it seems improbable that the major incision began more than 10 million years ago. Further time constraints are provided by lava flows within the canyon, one of which has been dated as 3·3 Ma old and lies only 100 m above present river level. In other words, of a total incision of 1 500 m, over 90 per cent may have been achieved in an interval of less than 7 million years at an average rate of about 0·2 mm yr^{-1}. Although it is difficult to compare this figure with a denudation rate for the whole catchment, all assessments point to the Colorado basin as an area of rapid geomorphic change.

In the southern Pennines the rocks at present undergoing erosion are of Carboniferous age, but it has been suggested that a chalk cover formerly extended across the whole region (Fig. 10.3). The minimum feasible elevation for the Cretaceous rocks is provided by the 600 m plateau of Kinderscout, and if the general argument is sound it implies that during the last 70 Ma denudation has removed the Mesozoic cover and penetrated at least 450 m into the underlying strata. Although such a sequence of events may be correct, it still provides a very inadequate basis for attempts to assess local denudation rates. For that purpose a more closely defined chronological framework is required. This is not easily obtained in the southern Pennines, even though many workers have identified one or more erosion surfaces truncating the dome of Carboniferous rocks (Fig. 10.4). The most prominent is the undulating Upland Surface which bevels the limestone in the core of the dome at an elevation of between 300 and 360 m. Within solution hollows on this plateau lie detrital sediments containing distinctive floral remains of middle or upper Cenozoic age. On the assumption that the Upland Surface was originally fashioned by subaerial processes some 5 Ma ago, the subsequent incision by the streams has been achieved at a mean rate of no more than 0·03 mm yr^{-1}.

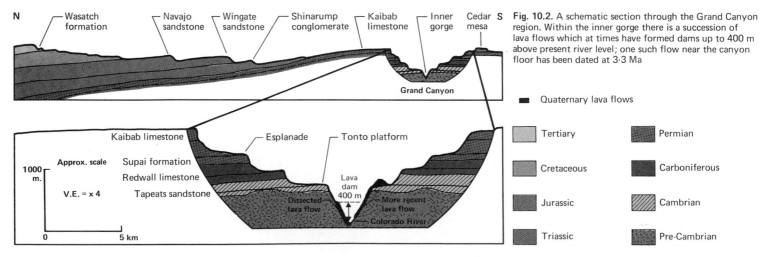

Fig. 10.2. A schematic section through the Grand Canyon region. Within the inner gorge there is a succession of lava flows which at times have formed dams up to 400 m above present river level; one such flow near the canyon floor has been dated at 3·3 Ma

Spatial and temporal contrasts in denudation rates

Regional variations With the increasing number of sediment-gauging stations widely distributed throughout the world, it is possible to start evaluating some of the environmental factors that control sediment yields. In recent years several workers have claimed to be able to discern global patterns related to either climate or topography or some combination of the two (Fig. 10.5). Corbel, for instance, has published an analysis of 80 catchments. He views annual temperature and precipitation as the dominant climatic factors, relief amplitude as the dominant topographic factor. In each precipitation class the denudation rate is believed to vary inversely with temperature; at the same time in each temperature class it varies directly with precipitation. This means that for the same topographic form, the highest rates are reckoned to be in cold wet regions and the lowest in hot arid regions, with as much as a hundredfold difference between the extremes. Mountainous regions undergo

erosion at a rate between two and five times that of plainlands in the same climatic class, but the very highest values are reserved for areas currently experiencing glaciation. Multiplying the area of the continental surfaces falling into each environmental class by the appropriate erosion rate, Corbel calculates the global sediment yield from nonglacial lands as $3\,824\times 10^6$ m³ yr⁶, equivalent to a mean annual lowering of about 0·028 mm. Of the gross total he estimates 48 per cent is transported seawards in solution, 52 per cent as particulate matter.

Another French worker, Fournier, has plotted the sediment yield from 78 individual catchments against selected precipitation parameters such as annual total, seasonality and intensity. He finds particular significance attaching to a weighted measure of seasonality defined as the square of the precipitation in the rainiest month divided by the mean annual total. When this measure is plotted against sediment yield, the catchments appear to fall into four distinct categories: (a) Gently sloping temperate regions; (b) Gently sloping tropical,

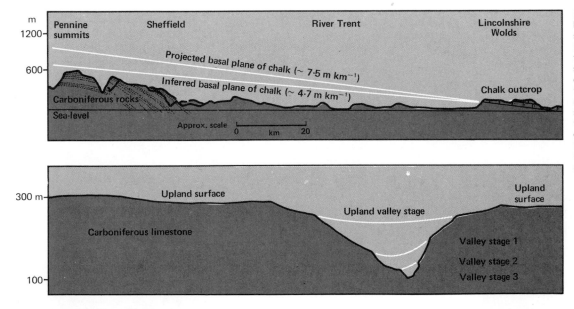

Fig 10.3. The geomorphic evolution of the Pennines as visualized by Linton who suggested that the present drainage was initiated on a former cover of chalk. On the top diagram, the upper line represents a simple extrapolation of the present dip of the chalk in Lincolnshire, the lower line a position for the chalk base which Linton regarded as more likely. On the bottom diagram, detailed stages in the denudational history subsequent to the unroofing of the Derbyshire dome are depicted. From the valley cross-profiles it is inferred that the region has been subject to intermittent uplift and rejuvenation of the drainage.

Fig 10.4. The undulating plateau of the southern Pennines truncating the limestone outcrop in the core of the Derbyshire dome (cf. Fig 10.3). Steeply incised valleys, like Dovedale here, have been held to indicate the polycyclic evolution of the landscape, although Pitty has offered an alternative interpretation for the origin of the upland plateau (p. 223).

sub-tropical and semi-arid regions; (c) Steeply sloping humid regions; (d) Steeply sloping semi-arid regions. By assigning all major drainage basins to one of these categories and calculating for each the seasonality measure referred to above, Fournier estimates the mean rate of lowering for all continental areas as 0·4 mm yr⁻¹, a figure more than a full order of magnitude greater than that given by Corbel. Equally striking is a conflict over regional variations. The highest sediment yields according to Fournier occur in tropical monsoon and savanna regions, although the latter according to Corbel are among the areas of least erosion. Conversely, Fournier ranks cold regions as zones of minor sediment yield whereas Corbel argues that they suffer above-average denudation.

Russian workers have also been active in the analysis of gauging records. The best known is Strakhov who emphasizes the wide range of values for surface lowering implied by data from 60 of the world's largest rivers. At one end of the scale are the Yenesei and St Lawrence basins being lowered at 0·0015 mm yr⁻¹; at the other the Ganges, Brahmaputra and Mekong basins being lowered at 0·38 mm yr⁻¹. He maintains that two fundamentally different zones can be identified. The first comprises the temperate moist belt of the northern hemisphere, generally receiving 150–600 mm annual rainfall and defined on its southern margin by the 10°C mean annual isotherm; within this zone denudation rates rarely exceed 0·006 mm yr⁻¹. The second zone embraces the moist tropical and sub-tropical regions with annual rainfall in excess of 1 200 mm and mean annual temperature above 10°C; here the denudation rate normally lies between 0·02 and 0·04 mm yr⁻¹, but in large parts of south-east Asia surpasses 0·35 mm yr⁻¹. Strakhov regards relief as a further major factor, holding that the rate of erosion varies directly with the amount of tectonic uplift. The lowest rates are therefore found in high-latitude low-relief areas, the highest in tropical mountainous regions. He also emphasizes the importance of seasonality in the rainfall regime, arguing

that a prolonged dry spell thoroughly loosens the surface soil which, in consequence, is more easily removed during later downpours. From the available data Strakhov formulates the additional rule that chemical denudation, despite regional variations, tends to increase with rising sediment yield. Lowland rivers normally carry more material in solution than in suspension, while the reverse is true of mountain streams. Yet most upland rivers carry more solutes per unit area of catchment than do plainland rivers. The total sediment yield from all non-glaciated continental areas is put at $6\,752 \times 10^6$ m³ per annum, implying a mean denudation rate of approximately 0·05 mm yr⁻¹. About 28 per cent of this aggregate value is attributable to chemical solution, although the figure varies regionally from 20 per cent in Asia to 42 per cent in Europe. Strakhov agrees reasonably closely with Corbel on the general magnitude of global denudation but almost exactly reverses his climatic weighting; on the other hand, Strakhov's regional distribution pattern resembles that of Fournier but with the absolute values much reduced. It is worth noting that Strakhov bases his estimates on data compiled as long ago as 1950, and doubt must now attach to the reliability of some of the values he employs.

From regional statistics in the United States the American geomorphologist Schumm has identified relief and run-off as the major determinants of sediment yield. He contends that the yield is an exponential function of relief when the latter is defined as basin relief divided by basin length. He also argues that the highest denudation rates occur in semi-arid areas with adequate rainfall to promote active run-off but too little to sustain a dense cover of protective vegetation. The actual precipitation values necessary to achieve maximum erosion will depend upon temperature (Fig. 10.6); with a mean annual temperature of 10°C the required rainfall is 300 mm, but with the temperature raised to 20°C precipitation must exceed 600 mm. Schumm concludes that the maximum feasible denudation rate approximates 1·0 mm yr⁻¹ and that the mean rate probably lies between 0·03 and 0·09 mm yr⁻¹, both figures relating to combined suspension and solution loads.

Some of the more reliable measurements of recent years have been collated by Young who stresses the apparent overriding importance of topography. Grouping the data into two classes,

Fig 10.5 An interpretation of the views of Corbel, Fournier and Strakhov on current rates of erosion (for purposes of comparability a density of 2·7gm $^{-3}$ has been assumed for the material removed).

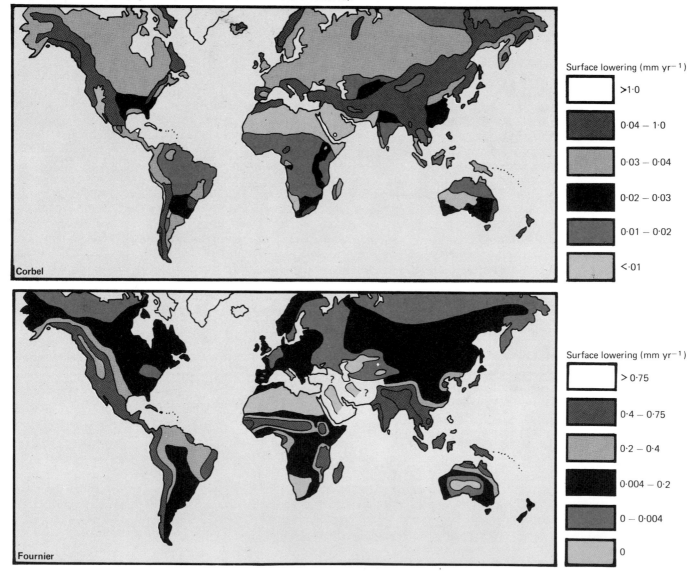

Surface lowering (mm yr^{-1})

>1·0

0·04 — 1·0

0·03 — 0·04

0·02 — 0·03

0·01 — 0·02

<·01

Corbel

Surface lowering (mm yr^{-1})

> 0·75

0·4 — 0·75

0·2 — 0·4

0·004 — 0·2

0 — 0·004

0

Fournier

normal and steep relief, he shows that they are separated by an order of magnitude in terms of denudation rate (Fig. 10.7). For areas of normal relief he quotes a median rate of 0·046 mm yr⁻¹, and for areas of steep relief a median rate of 0·5 mm yr⁻¹.

This review of work by a wide variety of authors reveals some glaring discrepancies in both postulated controls and overall rates of denudation. Nevertheless certain common strands can be discerned. Virtually all workers agree that there are large regional disparities and that one major factor is relief. There also appears to be growing consensus that a representative figure for denudation in regions of moderate relief is 0·05 mm yr⁻¹ and in mountainous regions 0·5 mm yr⁻¹. What is still far from clear is the nature and degree of climatic control.

Temporal variations As more accurate assessments of current denudation rates become available, it is extremely tempting to extrapolate the values backwards in time. Yet such a procedure is fraught with difficulty. In the geologically recent past the impact of Man has been of profound importance. There is scarcely a river basin in the world where he has not had a marked influence on presentday geomorphic processes. **209** Through all stages of the hydrological cycle his influence, although often difficult to evaluate, is undoubtedly present. Apart from attempts to alleviate aridity by such techniques as cloud-seeding, other activities like reservoir construction and urbanization have had unplanned but measurable effects upon local climates. Even the chemical composition of the rainfall has been altered so that downwind from many industrial areas the water is demonstrably more acidic. Yet the greatest changes wrought by Man arise from his modifications of land-use. These affect not only the nature of the sediment supply to the streams but also the stream regimes themselves.

Man first had a significant impact on the natural erosional processes at the time of the Neolithic revolution. Land under crop production is laid bare for at least a short period each year during which it is subject to increased rain splash and surface wash. This is a particularly severe problem on steep slopes, well exemplified by the experience of farmers in the Tennessee valley during the first half of this century. Even under mature crops, areas of bare soil may still lie open to attack. A further

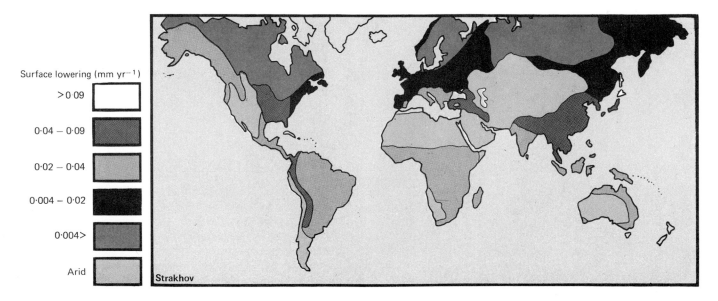

Surface lowering (mm yr⁻¹)

>0·09

0·04 – 0·09

0·02 – 0·04

0·004 – 0·02

0·004>

Arid

Strakhov

210 consideration is the disturbance by ploughing which tends to accelerate the rate of creep on cultivated hillslopes. Although the conversion of land from forest to permanent pasture for stock-raising may have less obvious and immediate consequences, in theory it will tend to modify hillslope processes by its influence on such factors as rainfall interception, soil organic activity and leaching. It has even been claimed that in upland Britain deforestation has promoted extension of the blanket peat; this would clearly alter the hillslope processes quite drastically. In less humid regions overgrazing can produce dramatic changes when the protective grass sod is broken and areas of soil are laid bare to erosion by both wind and rain.

The magnitude of the changes arising from these agricultural practices is still not firmly established. Ideally the impact of human activity might be regarded as a third variable to be considered alongside relief and climate as determinants of current erosion rates; ultimately some form of multivariate analysis may be possible but at present the data are inadequate for such a refined approach. Strakhov comments on what he regards as the anomalously high denudation rates on the European plainlands and in the Mississippi valley. He contends

that these show agricultural activities to multiply natural erosion rates two, three or four times. This approach presupposes that the writer can accurately forecast the natural rate, but the continuing conflict over the significance of various climatic controls indicates the need for caution. In the long term a more reliable approach may be that of comparing sediment yield from two much smaller catchments similar in all respects save land-use; for example, one drainage basin may be subject to arable farming while the other remains under woodland. In pursuit of this idea suspension load data have been compiled for a series of small catchments along the middle Atlantic seaboard of the United States. These suggest that conversion from forest to arable cultivation in this area increases the sediment yield ten-fold. Yet the severity of the enhanced erosion will obviously depend on local farming practices. Sediment yield from cultivated land in parts of Mississippi has been estimated at up to 100 times that from forested land, whereas from mixed farmland in Oregon the corresponding figure is nearer three times. The impact of grazing has been similarly assessed by controlled experiments in an arid region of Colorado. Of eight minor catchments, four were left open to grazing animals and four were fenced off. By means of small reservoirs specially constructed as sediment traps, those protected from grazing were found to produce much less material than those which remained unfenced.

Urbanization and industrialization also transform the natural processes. When the ground is first disturbed during construction work the rate of sediment yield is often enormously increased. Mean lowering of some small catchments in the Washington metropolitan area of the United States has temporarily attained a rate as high as 5 mm yr^{-1}. Once building is complete the extensive cover of concrete and other resistant materials causes a rapid fall in sediment production. However, an equally important consequence of the impermeable cover is the change in river regime. The proportion of rainfall entering the stream channels as surface run-off is greatly increased and the result is a much more flashy flow. This effect has long been recognized by engineers concerned with the design of artificial conduits, but its geomorphic implications have yet to be fully investigated. More frequent flooding has been reported down-

Fig 10.6. The relationship between denudation rates and recipitation totals as computed by Schumm.

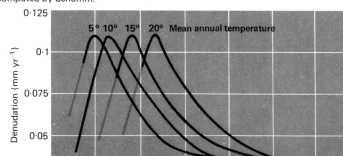

stream from many urban areas and in theory the character and velocity of the sediment load should be altered. It must also be remembered that, in many agricultural regions, tile drains and other artificial means of removing water contribute their share to the change in river regime.

The use of natural stream channels for the disposal of urban

and industrial waste leads to a radical alteration in the composition of river water. One manifestation is the dramatic change in the quality of lake waters in many parts of the world. The deterioration of Lakes Erie and Ontario, for instance, has necessitated international legislative action. In cases of bad pollution evaluation of chemical denudation may be virtually impossible. In agricultural regions lime added to the fields to combat leaching is much more readily mobilized than bedrock calcium carbonate, a potentially significant factor when attempting to assess solution rates in limestone areas. It must also be remembered that chemical and biochemical activity can modify processes within the stream channels themselves; for example, waters enriched with plant nutrients may promote algal and other aquatic growth that would not otherwise be present.

The effects of human interference demand great care in the interpretation of current sediment yields. A second factor rendering simple extrapolation very hazardous is climatic change. As already pointed out, erosion rates appear to vary significantly under different climatic regimes. Since most regions have been subject to major Pleistocene fluctuations, it must be presumed that they have experienced varying rates of denudation; in the future it may prove increasingly feasible to identify dateable horizons within sediment traps and thereby relate sediment production to known climatic oscillations. A third consideration in the long-term extrapolation of current erosion rates must be the effect of declining relief. The control exercised by relief has been amply demonstrated, and if there is neither tectonic uplift nor climatic change it must be assumed that over a long period the rate of sediment yield will gradually decline. A curve depicting the change is likely to have a negative exponential form with a reduction to very low values once the major relief features have been destroyed.

Models of landform evolution

The cycle concept of W. M. Davis It was the American worker Davis who had the most profound effect upon the early development of geomorphology in the English-speaking world. In a series of extremely influential publications between

Fig 10.7 Rates of surface lowering as measured by various workers and analysed by Young. On the right are shown the median values and inter-quartile ranges for steep and normal relief respectively (after Young, 1972).

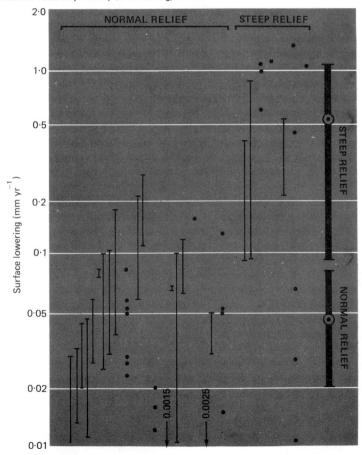

212 1889 and 1934 he argued that, although changes within a lifetime may be too slight for measurement, landscapes can be arranged in evolutionary sequences illustrative of cyclical changes that take place over a much longer period (Fig. 10.8*a*). The ordering of landscapes in this manner rests upon certain fundamental premises which are difficult to substantiate by direct observation but which Davis sought to justify by a simple deductive approach. Perhaps the single most crucial concept is that of base-level or a level below which streams cannot erode their valley floors. The majority of writers have construed this as a simple extension of sea-level beneath the continents but, thus defined, it has little practical value and several early workers sought an alternative formulation in the idea that there is a critical minimum gradient below which a stream course cannot be reduced. Conceived in this way base-level would slope gently upwards towards the interior of the continents. No satisfactory definition of this limiting gradient has ever been proposed, and Davis was compelled to supplement the idea of base-level with a further fundamental concept, that of 'grade'. He argued that, once uplift of a land mass is complete, the sub-

sequent erosional history may be divided into two distinct phases. At first rapid stream incision produces narrow steep-sided valleys. However, a point is reached at which the speed of downcutting abruptly diminishes and vertical erosion is in large measure replaced by lateral erosion with widening valley floors and ever more extensive floodplains. Treated as the transition from a youthful to a mature landscape stage, the change was equated with the attainment by the streams of a notional state termed grade.

For purposes of defining grade Davis generally had recourse to the idea of a balance between erosion and deposition, or between ability to do work and the work that needs to be done. Implicit in his writing was the idea that, in the early stages of downcutting, more energy is available than is being employed in simple transportation and the surplus is expended in corrasion of the valley floor. With incision of the valley the load supplied to the stream increases while the available energy diminishes and eventually a balance is struck between these two quantities. Thereafter any changes are relatively slow and depend on a gradual reduction in load as the valley-side slopes become

Fig 10.8 The cycle concept of W M Davis. He normally described the sequence as depicted on the left although the available relief, that is, the altitudinal difference between the initial surface and the level at which floodplains begin to form, will clearly have an important effect on the nature of landform evolution.

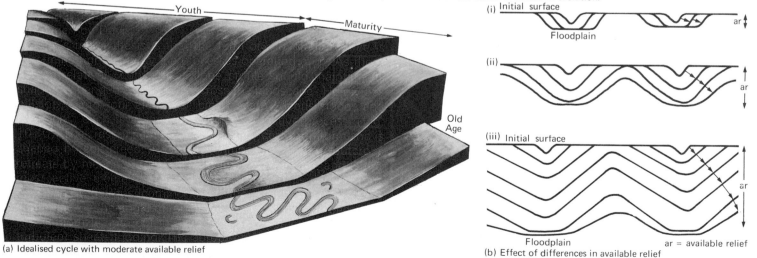

(a) Idealised cycle with moderate available relief

(b) Effect of differences in available relief

ar = available relief

more and more gentle. Unfortunately the idea of a balance between available energy and work to be done is both vague and confused. Work to be done is commonly assumed to refer to sediment transportation, while energy expenditure is regarded as divisible between that function and corrasion of the stream bed. Such a division is obviously unreal because, on the one hand, corrasion itself is dependent on sediment movement and, on the other, it is impossible to envisage conditions in which all energy is devoted to transportation and there is no possibility of erosion. Owing to these and other weaknesses, several workers have contended that the concept of grade is best totally abandoned. Whilst this may be so, it is worth noting that there is an indisputable decrease in potential energy as the streams incise their valleys, accompanied presumably by a reduction in the rate of downcutting. Whether the change from vertical incision to lateral erosion is abrupt or progressive is much more open to question. In either case, if the independent speed of valley-side retreat remains constant, the reduced pace of downcutting will lead to much broader transverse profiles. In this limited respect one of the basic postulates of the Davisian model may be accepted, even though much of the original explanation is rejected.

The change from a mature to a senile landscape was regarded by Davis as being characterized by a floodplain sufficiently broad to accommodate a fully-developed meander belt. In his analysis of landform evolution Davis made frequent reference to stream patterns, but later work has shown many of his inferences to be totally unjustified. For instance, free-swinging meanders are not necessarily a sign of old age as brief consideration of the meanders on the lower Mississippi immediately demonstrates (see p. 188). On the other hand, it is difficult to contest his general thesis that, with prolonged tectonic stability, the amplitude of relief will continue to decline until eventually an extensive surface of low relief or peneplain is formed. Once again the fundamental idea seems sound, but many of the embellishments are highly questionable.

For purposes of exposition Davis made many simplifying assumptions which were later to become the target of critical comment. One of the most important was the separation of uplift and erosion into distinct episodes. He thereby envisaged initial uplift taking place without appreciable erosion, and later erosion proceeding under conditions of prolonged tectonic stability. It must be emphasized that he recognized the unreality of this assumption and in some of his writings in German specifically discussed the likely effects of simultaneous uplift and erosion. From modern work on the relationship between relief and denudation rates, it might be anticipated that the greater the total amount of uplift, the greater the effect of contemporaneous erosion. However, current rates of uplift so exceed those for denudation that it is clear erosion does not normally constitute a limit for continued surface elevation. Indeed, it seems in this respect the model proposed by Davis may be more realistic than some of his sternest critics have been prepared to admit. Greater weight seems to attach to the argument that stability is unlikely to endure long enough to permit reduction of an upland area to a peneplain. Accumulating data on crustal mobility strengthen this belief, and underline the need to consider persistent tectonic activity as a factor in landform evolution. It has been pointed out that, despite numerous supposed uplifted examples, the only stage in the Davisian cycle not represented on the earth's surface at the present day is the peneplain. This has led many geomorphologists to argue that the sequence of stages envisaged by Davis rarely reaches culmination in a typical peneplain, and that alternative explanations should be sought for the presumed elevated examples.

One factor to be borne in mind when considering the complete sequence from uplifted mountain mass to peneplain is isostatic adjustment. The contrast of large Bouguer anomalies over modern mountain chains but small anomalies over ancient denuded remnants implies deepseated changes within the crust and upper mantle. On the assumption that a crustal 'root' of density $2 \cdot 7$ g cm^{-3} is replaced by mantle material of density $3 \cdot 4$ g cm^{-3}, the nominal amount of rock to be removed in erasure of a mountain chain is increased almost four-fold. This is obviously an important consideration in estimating the speed with which erosion can destroy such a feature. Of equal importance is the nature of the movement associated with isostatic adjustment. Two possibilities present themselves. The first is that uplift takes place continuously pari passu with denudation. The second is that it occurs spasmodically and represents periodic adaptation to stresses which slowly accumulate as the surface layers are removed by erosion. The evidence is far from conclusive but many geomorphological investigations have revealed marginal benches suggestive of episodic uplift long after the main mountain-building phase has ceased; this could well be due to isostatic adjustment triggered by denudation. A further guide is the incomplete isostatic compensation of certain cordilleran remnants, suggesting that erosion has as yet failed to activate the final uplift required for equilibrium to be restored.

In expounding his ideas of cyclical landform development,

Davis devoted considerable attention to the evolution of valley-side slopes. He contended that during youthful stream incision the valley sides will be steep, the controlling factor being the relative rates of fluvial downcutting and backwearing by slope processes. However, once rapid valley deepening has ceased slope processes will become almost the sole infuence on form leading to a gradual decline in angle (Fig. 10.9). On steep youthful slopes crags often mark the outcrop of more resistant rocks and debris moving down to the streams is coarse. With the passage of time moderately angled and more regular hillsides become mantled with finer detritus until, in old age, the very gentle peneplain slopes are covered with deep

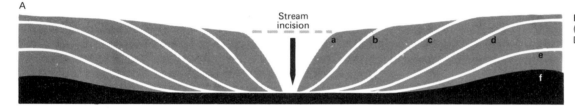

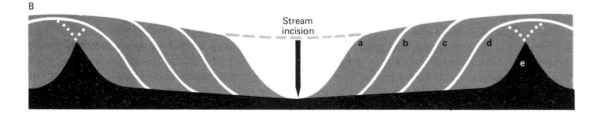

Fig 10.9 Three models of valley-side evolution: (A) W M Davis; (B) W Penck as interpreted by Davis; (C) W Penck from an original diagram.

and almost immobile waste. Unfortunately Davis tended to restrict his discussion of hillslope processes to generalities and provided little guidance to his thinking on the complex relationship between process and form. Moreover he attempted to apply the notion of grade to valley sides and, as already seen, this is a concept that defies satisfactory definition. In consequence most attention has been directed at his attempts to integrate slope development with other aspects of landscape evolution. He suggested that, in many instances, the rectilinear valley sides of the youthful stage will just be consuming the last remnants of the original surface when the streams come to the end of their active downcutting. He acknowledged, however, that these two critical stages might not be reached simultaneously, since their timing is dependent on such factors as amount of uplift and drainage density. It has been estimated that, given normal stream spacing, simultaneity is likely when uplift is between 60 and 90 m. Particular interest attaches to the sequence of landform development that follows uplift in excess of this. Regular valley spacing and constant slope angles will theoretically induce an accordance of interfluve heights that may be maintained long after the original surface has been consumed (Fig. 10.8b). This implies that evenness of divides is not by itself an adequate criterion for the reconstruction of an uplifted surface; identifiable fragments of the old surface must be preserved.

Davis examined the effect of repeated minor uplift in considerable detail, arguing that the majority of landscapes are polycyclic in origin. However, his discussion concentrated on changes along the valley floor and paid relatively little attention to the effects on valley-side slopes. He visualized a relative fall in base-level as initiating a knickpoint which slowly migrates upstream. In the course of this migration the valley floor is dissected and the former floodplain deposits are left as elevated terraces. This model of rejuvenation, affording one possible interpretation of long profiles and associated terraces, has been widely employed in the study of valley-floor evolution (see, for instance, Fig. 9.11). However, extension of these ideas to encompass evolution of the valley as a whole presents formidable problems. Recent minor uplift, suitably attested by river terraces, has presumably had little direct effect upon valley-side evol-

ution; the very presence of the terraces indicates isolation of the higher hillslope from events on the valley floor. On the other hand, it is difficult to believe that over a longer period of time major changes in the rate of river downcutting will not be reflected in the form of valley cross-profiles. If a moderately long phase of stability is assumed to have resulted in gentle gradients close to the river, rejuvenation will produce a new generation of steep slopes whose initial angle will be determined by the relative rates of stream incision and slope backwearing. As the speed of downcutting slackens, the charactistic angle will decline and the segment of the valley cross-profile directly related to the new phase of stability will progressively broaden. The resultant 'valley-in-valley' form is often regarded as diagnostic of long-term intermittent uplift, but the question that Davis failed to answer concerns the subsequent evolution of such profiles. Implicit in much later work has been the idea that the older slopes above each hillslope convexity are virtually fossilized. One common practice has been to interpolate smooth curves between paired breaks of slope (see, for example, Fig. 10.3). It is easy to fall into the trap of assuming these 'reconstructions' accurately depict the valley as it existed at some time in the past. Such an assumption can only be founded in the belief that slopes do not continue to evolve once the valley floor has been rejuvenated. There appears to be no logical reason for a cessation in slope development and any satisfactory model must incorporate the possibility of continued evolution.

Finally, it is worth recalling that the periodicity of major climatic oscillations during the Pleistocene epoch seems to have been of the order of 10^5 years. Davis gave no precise figures regarding the length of time required for an erosion cycle to run its full course, but on the basis of known denudation rates it can hardly be less than 10^6 years. This means that any area will almost certainly have experienced numerous climatic fluctuations in the course of a single cycle, with consequent changes in the erosional processes. It follows that landscapes, as well as being polycyclic, must also be viewed as 'polygenetic' in origin.

This discussion has endeavoured to highlight certain strengths and weaknesses in the Davisian model of landscape evolution. The historical approach to landform description and explanation is a fundamental one to which Davis made an outstanding contribution. Nevertheless, many workers have felt unable to accept his model in its entirety and have proposed a wide range of modifications to remedy what they saw as short-comings. Others have argued that there are basic inadequacies which demand formulation of completely new models. It is to the more important of these alternative schemes that attention must now be directed.

The geomorphological concepts of W. Penck The objectives of Penck, a German worker who published his most important work, *Die Morphologische Analyse*, in 1924, differed fundamentally from those of Davis. He regarded surface morphology as the result of competition between crustal movement and denudational processes, and as a geologist viewed landform analysis as a potential tool for deciphering recent tectonic history. His ideas have given rise to much controversy, partly because he used such obscure language that it is often difficult to tell exactly what he intended and partly because the book was still incomplete when he died so that certain critical chapters are missing.

Penck argued that the primary control of landform evolution is the relationship between the rate of river downcutting and the rate of tectonic uplift. This relationship in turn influences the form of the valley-side slopes. He contended that three major crustal states may be distinguished. The first is that of stability in which no active displacement is occurring; this he regarded as characteristic of extensive continental areas at the present time. The second is domed uplift which starts by affecting a relatively small area but later expands to encompass much more extensive regions. The third is intense crustal upheaval characteristic of mountain areas where tangential compression is combined with regional arching. An important aspect of Penck's thesis was that the rate of crustal movement varies greatly from time to time; if it happens to remain constant for a prolonged period, the rate of river downcutting will adjust until an equilibrium is achieved. For example, if uplift proceeds more rapidly than stream incision, river gradients will steepen to the point where downcutting keeps pace with the crustal movement; con-versely, if the rivers can cut down more rapidly than the land is being elevated, the gradients will be reduced to the point where the two rates are equalized. Such adjustments will take some time to achieve and if the speed of tectonic movement undergoes frequent changes disequilibrium will be common.

Penck maintained that conditions of equilibrium and dis-equilibrium will produce contrasting slope forms. So long as the rates of uplift and downcutting remain equal – a condition termed gleichformige Entwicklung – slope profiles will tend to be rectilinear; if downcutting accelerates – a condition termed aufsteigende Entwicklung – increasing steepness of the newly-formed slopes will produce a convex profile; if down-cutting decelerates – a condition termed absteigende Entwick-lung – declining steepness of the newly-formed slopes will produce a concave profile. The final possibility is that of tectonic stability. It is here that much misrepresentation of Penck's ideas has occurred (e.g. Fig. 10.9B). He has often been quoted as believing in the parallel retreat of slopes, but his actual views appear to have been very different. He argued that the steep segments of a slope profile retreat more rapidly than the gentle segments. He considered the simplest case of a steep rectilinear face rising directly from the valley floor. Weathering attacks all parts of the face equally so that it tends to retreat without change of angle. At its foot there emerges a sloping surface, designated a Haldenhang, across which debris is transported to the stream. This less steep unit is itself subject to gradual replacement from below by an even more gentle slope. The eventual effect is a concave profile in which the constituent elements appear to migrate upwards and away from the valley floor. As Fig. 10.9c shows, Penck even visualized an originally convex slope evolving towards a concave form. This is achieved by the steep basal portion overrunning the more gentle upper section, and by the simultaneous development of a Haldenhang. There seems no doubt that, irrespective of the original form, Penck envisaged stable conditions leading to concave profiles. Although the intersection of such profiles might be expected to produce sharpcrested interfluves, he argued that in practice they would be rounded by weathering.

The most fundamental difference from the model proposed by Davis lies in its emphasis on concurrent uplift and erosion.

A surface of low relief may evolve in two contrasting circumstances. That which develops during prolonged tectonic stability is characterized by widespread gentle slopes of concave form; to such a feature Penck gave the name Endrumpf. That which forms during slow but accelerating uplift has broad convex interfluves and is termed a Primarrumpf. Implicit in this concept is the notion that accordant crestlines may be produced by a mechanism other than dissection of an uplifted erosion surface. Penck elaborated on the idea of a Primarrumpf in considering landform development on an unheaved dome. He assumed that movement would gradually accelerate and spreads outwards from an initial centre. Any point on the periphery of the dome would first experience slow uplift and then progressively faster movement. During the slow phase an annular Primarrumpf would tend to form, only to be dissected as uplift accelerated and the dome expanded. Penck maintained that ultimately a succession of concentric benches might be formed, constituting the assemblage of features he labelled a Piedmonttreppen. Most workers have found Penck's explanation of Piedmonttreppen inadequate, although the elevated massif with a series of marginal benches has certainly been recognized as a common relief form.

For many years the ideas of Penck received scant support from English-speaking geomorphologists. The obscure language in which they were expressed contributed to this neglect, but a secondary factor was reliance on an interpretation of Penck's views offered by Davis. There can be no doubt that the American writer misconstrued some of these views, and this finally led to a mistaken account of Penck's ideas on slope development being castigated as 'one of the most fantastic errors ever introduced into geomorphology'. More recently workers have returned to the original German exposition and have found it thought-provoking even though at times muddled and contradictory.

The cycle of erosion according to L. C. King The ideas of the South African geomorphologist, King, are more closely related to those of Davis than those of Penck. In his book *South African Scenery* King acknowledges his debt to Davis and expounds a model of landform development which differs from

that of the American mainly in its ideas about slope evolution. He believes in the paramountcy of cyclic erosion in the development of continental landscapes, and argues that the chief defect of the Davisian concept is the absence of any idea of parallel slope retreat.

Like Davis, King believes that once an area has been uplifted, knickpoints generated on the rivers near the coast begin to migrate far inland. During this youthful stage rapid incision produces gorges but as the speed of downcutting declines the valley-side slopes are reduced to a stable angle determined by lithology and the physical agencies acting on them. During maturity downcutting virtually ceases and is replaced by lateral corrasion. The chief means of landscape change is the migration of the valley-side slopes away from the rivers without significant change of angle. At their base develop relatively smooth flat areas known as pediments. These have gentle concave profiles which may locally attain an angle of over 10° but generally do not surpass 5°. Gradual extension of the pediments eventually produces a pediplain. This surface of low relief is composed of a series of subdued intersecting concavities above which rise occasional steep-sided residuals. In the final stage of an erosion cycle protracted weathering slowly transforms the divides into broad convexities so that the senile landform ultimately comes to resemble the peneplain envisaged by Davis.

King's major thesis is that the form of a migrating slope is determined primarily by the processes operating on it. As these remain relatively constant, a slope profile should maintain its shape unchanged as it moves across the landscape. He divides hillslopes into four elements: crest; scarp; debris slope; and pediment; each is associated with a distinctive process (Fig. 10.10). The crest or summit area is usually convex in form and is shaped by soil creep. The scarp composed of rocky outcrops retreats by means of rockfall, landslips and gullying, and is the most active element of the whole profile. The debris slope consists of detritus coming from above and its gradient is controlled by the angle of repose of the material. The pediment is a smooth feature cut in solid rock and fashioned mainly by turbulent sheet-flooding. The abrupt change of gradient at the base of a debris slope denotes a sudden transformation in the processes shaping the landscape.

As a South African, King formulated his ideas in a predominantly semi-arid area that has suffered little recent geological deformation. His attempts to extend the ideas to other regions of the world have aroused much controversy. He maintains that the 'four-element hillslope is the basic landform that develops in all regions of sufficient relief and under all climates wherein water-flow is a prominent agent of denudation'. Sufficient relief is important since its absence will lead to elision of the scarp and debris slope. In such cases the crest will grade directly into the pediment to yield a concavo-convex profile which is subject to very slow denudation. Weak bedrock may yield a similar profile. King contends that climate is not a fundamental control but influences only the degree to which the various slope elements are developed. There is a 'basic homology of landforms' and it is misleading to magnify the differences due to climate. The most efficient erosional system operates under conditions of heavy intermittent rainfall. With increased humidity rapid weathering tends to blanket the landscape with a thick immobile regolith; with increased aridity the water is unable to move all the debris to the stream channels and large fans along the edge of the uplands severely restrict pedimentation. Of course, whether to stress the similarities or dissimilarities of slopes in different environments is largely a matter of choice. Moreover, attribution of a common slope form to a single process can lead to semantic problems; the term pediment, for example, has been extended by some writers to basal concavities in humid temperate regions despite the fact that fashioning by sheet-flood erosion has never been demonstrated for such areas.

A vital consequence of the King model of slope evolution is that scarps can continue to migrate unaltered in form even when isolated from the modern valley floor by renewed uplift. In essence the landscape consists of a massive mobile staircase eating back into the upland regions. By contrast with the Davisian model, a prolonged phase of stability does not necessarily lead to the destruction of earlier surfaces but simply adds another tread to the staircase. Only when a particular scarp has migrated across the whole of a land surface is the evidence for earlier evolutionary phases erased. Because the stairway of scarps is regarded as mobile rather than static, each pediment surface may be viewed as having both a local age, that is the time when it was first formed at the particular point under consideration, and a regional age referring to its date of initiation either at the coast or on the flanks of a crustal fold. In practice King normally employs regional ages inferred either from superficial deposits or from coastal sedimentary sequences. By this means he identifies three major surfaces on the African continent: the Gondwana surface of Jurassic age, the African surface of Cenozoic age, and the post-African surface of late Cenozoic age. His contention that landform evolution is fundamentally similar in different environments has also encouraged him to seek analogous histories for other continents. The great age assigned to the Gondwana surface in Africa is particularly intriguing because it implies that, at the time of formation, the southern continents may still have been joined

Fig 10.10 Hillslope development according to L.C. King: the morphological elements (A) of a hillslope formed under the action of running water (B) and mass movement under gravity (C) (after King, 1962).

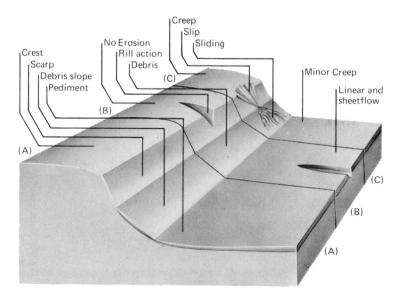

together as Gondwanaland. King has therefore paid special attention to the lands of the southern hemisphere where he finds striking similarities with the evolutionary sequence in Africa. Later claims of analogous sequences in both Europe and North America have led him to draw the obvious inference that rhythmic global tectonics have controlled topographic development on all the major continents.

The ages assigned by King to the earlier of his cyclic surfaces are greater than many geomorphologists believe likely. Measurements suggest erosional processes operate too fast to permit preservation of such ancient features. However, if King's model of landscape evolution is correct, computed rates of denudation could be misleading. With downwearing as the dominant process, current rates of sediment yield would render preservation of elevated landforms for 100 Ma or so most unlikely. On the other hand, scarp recession as envisaged by King would permit removal of immense quantities of material without effacing remnants of the earliest surfaces. Yet it is difficult to deny that a serious conflict still exists. This may be resolved in the long term by better dating methods than those currently available. For the present, however, it is informative to enquire how far King's ideas receive support from more detailed local studies. Regions that might be expected on theoretical grounds to preserve ancient landscapes are continental interiors subject to a little tectonic uplift, and with climates of small erosional potential. One such area is the arid heart of Australia where mapping and dating is greatly assisted by the presence of duricrusts. The most distinctive is a silcrete caprock up to 9 m thick which required for its formation a more humid climate than today and a stable plainland over which the crust could develop. Suitable conditions seem to have occurred only once in the continental heartland and can be dated with some confidence since the crust is formed in part from Cretaceous rocks and is locally overlain by Miocene lacustrine deposits. Rising above the silcrete caprock are bevelled summits believed by certain workers to have been fashioned in a period of erosion extending back continuously as far as Palaeozoic times. The general time-scale invoked by King appears to receive support from these Australian studies. Another intensively studied continental interior is that of North America. Here the whole landscape appears vastly more mobile with numerous Cenozoic interludes of both accumulation and erosion. Increasing doubt has been cast upon the preservation of features earlier than mid-Cenozoic in age, and some workers have contended that no part of the landscape can be older than Pliocene. The contrast with the Australian interior might be explained by greater crustal instability and more humid climatic conditions. Yet difficulties arise when the same arguments are applied to Africa. For example, remnants of the Gondwana surface have been identified by King within 200 km of the coast in Lesotho, Angola and Cameroon. Elevations range from 1 500 to 3 000 m and local precipitation totals from 500 to 2 500 mm yr^{-1}. In Lesotho the mean temperature of the coldest month falls below 5°C, while in Angola there are pronounced seasonal variations in rainfall. These are all factors regarded by one author or another as conducive to rapid erosion, and most estimates of current denudation rates seem inconsistent with the preservation of really ancient erosion surfaces.

By emphasizing the homology of all water-eroded landscapes King tends to minimize the significance of climatic change. On the other hand, the great age he assigns to cyclic landforms increases the likelihood that many have experienced profound and long-lasting climatic variations. These would include not only the recent Pleistocene fluctuations, but also the more enduring changes associated with the slow drift of the continents across the surface of the globe. If validated King's thesis would clearly have immense and exciting implications, but it must be admitted that there are formidable objections to its general acceptance. At best the evidence seems to point to huge contrasts in the length of time for which uplifted erosion surfaces may be preserved.

Climatic variants of the cyclic models Both Davis and King argued that their models of landscape evolution could be adapted to a wide variety of climatic environments with only minor modifications. Nevertheless, as early as 1909 Davis was convinced of the need to formulate a separate cycle for arid regions, and since that date many writers have proposed variants of the original Davisian cycle which they believe appropriate to specific climatic zones. Peltier has argued that at least nine 'morphogenetic regions' should be distinguishable, each in theory characterized by a unique combination of frost action, chemical weathering, mass movement and pluvial erosion. Yet it is worth noting that Peltier's scheme starts from an assumption, that different processes will produce different forms, which has so far received little rigorous testing. Peltier himself has shown by morphometric studies that the relationship between mean hillside slope and channel spacing varies from one environment to another, while Chorley and Morgan have demonstrated that a number of morphological distinctions between England and the south-eastern United States may be ascribed to climatic rather than lithological contrasts. However, much more research is required to establish the full relationship between climate and morphology.

Davis was persuaded to recognize a distinctive desert cycle of erosion by three considerations: the absence of the normal base-level control in areas without perennial streams draining to the coast; the increasing importance of wind action in intensely arid regions; and finally the belief that a unique combination of processes might lead to slope retreat without angular decline. In formulating details of the cycle Davis was much influenced by his familiarity with the block-faulted landforms of the south-western United States. He maintained that the lack of drainage to the coast will result in much internal redistribution of sediment. The floor of any structural basin will act as a local base-level, but instead of being stable its elevation will slowly rise as more and more detritus accumulates. Integration of adjacent basins may take place either by capture or by overflow of the sedimentary fill. The margins of uplifted areas will retreat as steep scarps fringed by progressively widening pediments. The amplitude of relief will steadily decline, and as the power of running water diminishes the relative importance of wind will increase, primarily as a transporting medium but also as a local abrasive agent at the foot of bedrock slopes. This represents a further noteworthy weakening of the normal base-level control. Nevertheless, the original accidented topography will ultimately be reduced to a surface of low relief diversified by occasional steep-sided residuals.

The value of recognizing a distinctive cycle of desert erosion is certainly open to question. Aridity is a climatic state that can befall any region irrespective of structure or relief; in consequence deserts are extremely diverse, topographically, geologically, biologically, and even climatically. Whereas extreme dryness may retard the denudational processes and thereby fossilize any inherited landscape, semi-aridity often leads to rapid transformation of such a landscape. Climatic fluctuations thus become a major factor in considering the evolution of desert regions. Most arid areas at the present day bear the imprint of earlier more humid conditions. Among the clearest signs are well-integrated drainage systems in regions now almost totally dry, and indications of formerly more extensive lakes. Examples are afforded by Lake Eyre in Australia, Lake Chad in Africa, and Great Salt Lake in North America, all of which have temporarily overflowed and established connections with the sea. Many workers have questioned whether a desert cycle can ever run its full course without significant interruptions during more humid interludes, and some have contended that it would be better to regard full aridity as a special condition associated with an unusual combination of erosional and depositional processes. Normal base-level control is at least temporarily suspended and pedimentation appears to become particularly prevalent. Yet the most distinctive characteristic is the increasing importance of wind as a transporting agent. In the recent past there has probably been a tendency to underrate the significance of wind action. In part this stems from the selection by Davis of the south-western United States, with relatively few dune areas, as a model for desert landscape evolution, and secondly in his relegation of wind action to the later stages of the cycle of erosion. A valuable corrective is to examine satellite imagery of the world's deserts where it can be seen that individual dune areas not infrequently cover 10 000 km^2 or more.

A second climatic zone in which the cycle of erosion has been deemed to vary significantly from those outlined by Davis and King is the humid tropics. Although King's ideas have received wide support in the drier tropics, many workers have maintained that they pay too little attention to rapid chemical weathering to be applied unaltered to hot moist conditions. Most models of landscape evolution assume an approximate equality between the rates of weathering and surface lowering. However, in many humid tropical areas the mantle of rotted bedrock locally surpasses 50 m in thickness, and in such cases it appears that the downward advance of the weathering front must have outpaced removal of the regolith. A number of geomorphologists have suggested that two relatively independent surfaces, the ground surface and the weathering front, may interact to produce a distinctive set of landforms. For instance, at the end of a long phase of stability surface processes will alter the ground elevation very little, but chemical action may continue to depress the weathering front. Tectonic uplift, on the other hand, will lead to stripping of the regolith at a much faster rate than it is being generated by further rotting. If the weathering front is approximately level, the stripping will eventually produce an 'etchplain' of low relief. However, the junction between sound and rotted bedrock is often very irregular so that partial removal of the regolith tends to expose cores of unaltered rock between more extensive areas of decomposed material. The appropriate model of landform evolution depends to a certain extent upon the agencies of surface denudation. Pugh has argued that in West Africa pedimentation is the normal process and he envisages a fall in base-level starting a new cyclic surface which rapidly dissects the old regolith (Fig. 10.11). Where the new surface encounters buried masses of sound rock it may first sweep past to leave prominent residuals. Eventually, however, the margins of the residuals will themselves be subject to pedimentation, finally resulting in a pediplain that transects both sound and decomposed rock and leaves upstanding only the largest and most resistant of the original unaltered masses. Pugh distinguishes two types of residual. Bornhardts are rounded hills characterized by curved sheeting structures, whereas kopjes are irregular in outline and possess a dominantly rectangular jointing system. The difference between the two is usually attributable to geological contrasts, bornhardts forming in massive rock and kopjes in more closely jointed rock. Detailed studies along the edges of residuals occasionally reveal what appears to be a second phase of chemical weathering. Narrow zones of decomposed rock are being removed by intermittent surface flow to produce annular hollows. The sides of the rotted zone slope very steeply and in extreme cases excavation results in an overhang around the foot of the residual.

A further way in which chemical processes affect tropical landform development is by the generation of lateritic crusts. In areas of deep weathering the upper layers of the regolith might be expected to succumb particularly rapidly to stripping since they would presumably be the most thoroughly decomposed. However, in many regions it has been noted that the surface horizons of a thick regolith are converted into a hard lateritic crust pari passu with undercutting by a retreating scarp. This appears to be due to improved drainage and aeration as the local amplitude of relief increases, and it has the effect of making decomposed rock in tropical areas much more resistant than might at first be supposed. Receding scarps steepened by their cap of hardened crust can even produce landforms reminiscent of mesas fashioned from horizontally stratified bedrock.

Fig. 10.11. Evolution of tropical areas by deep chemical weathering followed by pedimentation.

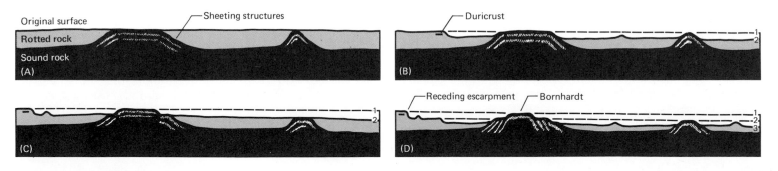

A third climatic zone for which a distinctive cycle of erosion has been proposed is the periglacial, but discussion of this topic is deferred until Chapter 14.

Non-cyclic concepts of landscape evolution Reaction against the limitations of the Davisian erosion cycle has led to reassessments more radical than those discussed so far. Particularly significant has been the development of the 'dynamic equilibrium' concept with its emphasis on landforms adapted to presentday conditions rather than inherited from the past. Increasingly geomorphologists have tended to concern themselves with the relationship between form and process and to adopt an agnostic attitude towards landform history. This is not to deny the existence of a prolonged period of landform evolution, but to contend that its details are now lost beyond reconstruction. Two instances of the contrast with Davisian thought may be given. Davis and his disciples maintained that certain parts of the landscape are so fossilized after rejuvenation that, by climbing to a broad upland interfluve, one might literally stand upon a preserved segment of a landscape dating back a million years or more. The geomorphologist who views landforms as constantly evolving argues that such complete inactivity is impossible. Precipitation over such a long period of time must inevitably have had some effect, at the very least in the form of deep chemical rotting. If such rotted rock is not in evidence it means that material has been removed and therefore the landscape is not fossilized. In the field of channel patterns a similar sharp conflict may be discerned. Davis tended to explain modern forms by reference to presumed conditions in the past, interpreting meanders as developments arising from chance perturbations in a youthful stream course. Yet such explanations are demonstrably inadequate for the beautifully regular pattern of meanders in which geometry is closely related to current discharge. In the foregoing examples, the weaknesses of the Davisian argument are readily apparent. Yet the question remains whether they are sufficient to undermine the whole cyclical concept.

The most extreme advocate of a non-cyclic approach to landform studies has been the American geomorphologist, Hack. Whereas Penck may be said to have viewed landscape as the result of competition between uplift and erosion, Hack views it as the product of competition between the resistance of crustal materials and the forces of denudation. The evenness of Appalachian ridge crests which Davis explained as due to rejuvenation of an old peneplain, Hack sees as the manifestation of equal resistance to the forces of erosion. He argues that the orderliness of stream organization first discerned by Horton will naturally lead to a regularity in the overall pattern of relief. Within a single climatic region where stream and slope profiles are both controlled by the nature of the bedrock, similar geological conditions should produce similar topography. By extension of this argument, Hack explains the even-crested ridges of the Appalachians as a simple reflection of uniform bedrock resistance along the strike of major sandstone beds. A corollary of this view is that very little can be inferred about their evolution. The point has been well illustrated by Small who considers the simple case of streams cutting down through a succession of thick clays and sandstones. While being excavated in the sandstones the valleys will be narrow and steep-sided, but in the clays they will be broad and gently sloping. Examining these valleys at any stage in their history, a geomorphologist could infer from their present form nothing about their previous evolution. If one imagines the same succession of rocks uni-clinally tilted, the effect of downcutting will be to produce asymmetric valleys which preserve their shape while migrating down-dip, and so again reveal nothing about their history. These two illustrations assume that downcutting can proceed without hindrance, but in practice base-level must clearly constitute a limit. This will inevitably lead to progressive modification of valley cross-profiles, but does not alter the basic thesis that they still represent an equilibrium between erosional forces and bedrock resistance. As the land surface is lowered the available energy diminishes and the new forms are merely a response to the changing conditions; once again no inferences can be drawn about landform history. Admittedly, in an area uplifted after protracted stability the former slopes may be briefly preserved before succumbing to reinvigorated erosion. Yet Hack maintains that such conditions are too rare and too fleeting to constitute a suitable framework for normal landform studies.

As mentioned earlier, investigations into current rates of denudation have generally cast doubt on the antiquity of landforms. Characteristic sediment yields equivalent to surface lowering of 0·05–0·5 mm yr^{-1} represent long-term changes at a rate of 50–500 m per million years. Of course these figures are means for whole catchment areas and do not necessarily apply, for instance, to interfluve crests; yet it is difficult to conceive of rapid erosion continuing to affect valley floors and slopes but leaving watershed areas totally intact. This specific question has been considered by Pitty in a discussion of the limestone plateau of the southern Pennines (see p.204). He endeavours to show that there is no satisfactory source for the calcium carbonate currently carried in the streams other than the interfluve regions and concludes that the plateau surface must be subject to lowering at a rate of approximately 0·08 mm yr^{-1}.

Fig. 10.12. A sketch of Pitty's concept of solutional lowering producing a bevelled surface across a domed limestone outcrop.

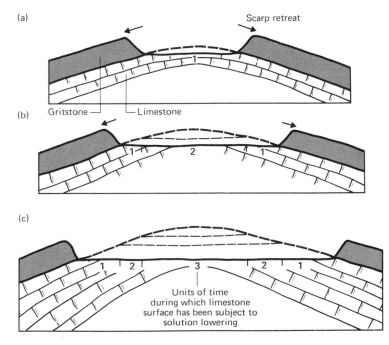

(a)

Scarp retreat

Gritstone — Limestone

(b)

(c)

Units of time during which limestone surface has been subject to solution lowering

If his arguments are valid, it is difficult to see how the Upland Surface can be 5 Ma old and still preserve details of its original form. Pitty, in fact, has proposed a model whereby steady solution acting on a progressively unroofed limestone dome might produce a central plateau surface (Fig. 10.12). If his model is applicable to the southern Pennines the dated organic materials preserved in solution hollows on the Upland Surface are presumably not in their original position; that they have been subject to some unknown amount of solutional lowering is indicated by their generally disturbed condition. It is worth recording that the concept of dynamic equilibrium has recently been applied to landforms in two other British areas mentioned in this book, Wales and the Weald, in each case casting legitimate doubt on the traditional denudation chronology.

It is intrinsic to Hack's argument that landforms adapt rapidly to changing environmental controls. Whilst that may be a tenable position, if the thesis is pushed to extremes it verges on the absurd since it would preclude, for instance, the identification of formerly glaciated areas. It seems that it may be necessary to seek some compromise between the over-simple cycle concept as outlined by Davis and the agnostic attitude adopted by Hack. One possible compromise lies in the recognition of water-eroded landscapes as complex assemblages of features that adjust to change in the controlling factors at different rates. The variety of so-called relaxation rates requires that any evolutionary approach should adopt a time-scale appropriate to the particular feature under investigation. For instance, many rivers of North America and Europe changed from a braided to a meandering channel style less than 10 000 years ago; it is totally inappropriate to seek to explain their present style by reference to a cycle of erosion initiated a million or more years ago. On the other hand, flat-topped interfluves may evolve so slowly that, although modified in detail, they still preserve in essentials the form attained as part of an ancient landscape prior to rejuvenation. It has become fashionable to distinguish between 'time-bound' and 'timeless' studies as if there were some rigid natural separation. For the practising geomorphologist, however, it is essential to recognize that environments are constantly changing and that very few landforms show an instantaneous adaptation to new conditions;

224 to a certain extent all landforms are prisoners of their own evolutionary history. Some of the basic precepts of dynamic equilibrium are very difficult to test and there is the risk that in individual circumstances they may be assumed rather than demonstrated; the mere fact that a particular process can be shown to be operating is no guarantee that it is responsible for fashioning the landforms with which it is associated.

If it is accepted that aspects of the landscape may evolve independently of each other, the assumption that all water-eroded landscapes are tending towards a unique end-product, such as a peneplain, must obviously be open to question. It is incontestable that, without external changes, there must be a progressive diminution in available energy. How this energy is distributed within the geomorphic system may provide an important guide to the long-term trends in landform evolution where no rigid cyclical pattern is envisaged. In discussing meanders in Chapter 9 it was pointed out that there may be a tendency towards the minimization of variance in energy expenditure per unit area of a stream channel. This may be viewed as part of a more general thesis that there is a progressive change towards maximum entropy. Entropy in this context is a measure of energy distribution and reaches its maximum when there is an equal probability of given energy levels being encountered anywhere in the system. By implication there is also a trend towards minimum free energy capable of doing geomorphic work. Although the Davisian peneplain might be regarded as a manifestation of a condition of maximum entropy, it is only one possible form among many.

AHNERT, F., 1970. 'Functional relationships between denudation, relief and uplift in large mid-latitude drainage basins', *Am. J. Sci.*, 268, 243–63.

CHORLEY, R. J. and M. A. MORGAN, 1962. 'Comparison of the morphometric features, Unaka Mountains Tennessee and N. Carolina, and Dartmoor', *Bull. geol. Soc. Am.*, 73, 17–34.

CORBEL, J., 1964. 'L'erosion terrestre, etude quantitative', *Ann. Geogr.*, 68, 97–120.

CUMMINS, W. A. and H. R. POTTER, 1967. 'Rate of sedimentation in Cropston reservoir, Charnwood Forest, Leicestershire', *Mercian Geol.*, 2, 31–9.

CURRAY, J. R. and D. G. MOORE, 1971. 'Growth of the Bengal deep-sea fan and denudation in the Himalayas', *Bull. geol. Soc. Am.*, 82, 563–72.

FOURNIER, F., 1960. *Climat et erosion: la relation entre l'erosion du sol par l'eau et les precipitations atmospheriques*, Presses Univ. de France.

HACK, J. T., 1960. 'Interpretation of erosional topography in humid temperate regions', *Am. J. Sci.* 258A, 80–97.

HOLEMAN, J. N., 1968. 'The sediment yield of the major rivers of the world', *Wat. Resour. Res.*, 4, 737–47.

KING, L. C., 1962. *Morphology Of The Earth*, Oliver and Boyd.

LEOPOLD, L. B., 1973. 'River channel change with time: an example', *Bull. geol. Soc. Am.*, 84, 1845–60.

LEOPOLD, L. B. and W. B. LANGBEIN, 1962. 'The concept of entropy in landscape evolution', *Prof. Pap. U.S. geol. Surv.*, 500-A.

LINTON, D. L., 1956. 'Geomorphology' in *Sheffield and Its Region*, Brit. Ass. Adv. Sci.

LUSBY, G. C., et al., 1963. 'Hydrologic and biotic characteristics of grazed and ungrazed watersheds of the Badger Wash basin in western Colorado', *Wat. Supply Pap. U.S. geol. Surv.*, 1532-B.

MEADE, R. H., 1969. 'Errors in using modern stream-load data to estimate natural rates of denudation', *Bull. geol. Soc. Am.*, 80, 1265–74.

MOORE, P. D., 1973. 'The influence of prehistoric cultures upon the initiation and spread of blanket bog in upland Wales', *Nature Lond.*, 241, 350–3.

NORMARK, W. R. and D. J. W. PIPER, 1972. 'Sediments and growth pattern of Navy deep-sea fan, San Clemente basin, California borderland', *J. Geol.*, 80, 198–223.

PITTY, A. F., 1968. 'The scale and significance of solutional loss from the limestone tract of the southern Pennines', *Proc. Geol. Ass.*, 79, 153–77.

PUGH, J. C., 1966. 'The landforms of low latitudes' in *Essays in Geomorphology* (Ed. G. H. Dury), Heinemann.

SCHUMM, S. A., 1965. 'Quaternary paleohydrology' in *The Quaternary of the United States*, Princeton Univ. Press.

SMALL, R. J., 1970. *The Study of Landforms*, CUP.

SMITH, W. O., et al., 1960. 'Comprehensive survey of sedimentation in Lake Mead, 1948–49', *Prof. Pap. U.S. geol. Surv.*, 295.

STRAKHOV, N. M., 1967. *Principles of Lithogenesis* (trans. J. P. Fitzsimmons), Oliver and Boyd.

TWIDALE, C. R. and J. A. BOURNE, 1975. 'Episodic exposure of inselbergs', *Bull. geol. Soc. Am.*, 86, 1473–81.

WALSH, P. T., et al., 1972. 'The preservation of the Neogene Brassington formation of the southern Pennines and its bearing on the evolution of upland Britain', *J. geol. Soc. Lond.*, 128, 519–59.

WORSSAM, B. C., 1973. 'A new look at river capture and at the denudation history of the Weald', *Rep. Inst. geol. Sci.*, 73/17.

YOUNG, A., 1958. 'A record of the rate of erosion on Millstone Grit', *Proc. Yorks. geol. Soc.*, 31, 149–56.

YOUNG, A., 1969. 'Present rate of land erosion', *Nature Lond.*, 224, 851–2.

YOUNG, A., 1974. 'The rate of slope retreat' in *Progress in Geomorphology*, Inst. Brit. Geogr. Spec. Pub. No. 7.

Selected bibliography

The basic models of landform evolution outlined in the foregoing pages are elaborated in : W. M. Davis, *Geographical Essays,* Dover Publ., 1954 (reprint of 1909 edition) ; W. Penck, *Morphological Analysis of Landforms: A Contribution to Physical Geology* (trans. H. Czech and K. C. Boswell), Macmillan, 1953, and L. C. King, *Morphology Of The Earth*, Oliver and Boyd, 1962.

The distinctive geomorphic processes of arid and tropical regions are discussed in R. U. Cooke and A. Warren, *Geomorphology in deserts*, Batsford, 1973 and M. F. Thomas, *Tropical geomorphology: a study of weathering and landform development in warm climates,* Macmillan, 1974.

Like other geomorphological features, glacial landforms may be studied either as they are developing at the present day or in the 'fossil' state after the processes responsible for their formation have ceased to operate. Direct observation of landform development at the sole of a glacier presents obvious problems and many features are known only in fossil form. However, from a study of modern glaciers important inferences may be drawn about the fashioning of the underlying land surface, and advances in glacial geomorphology have often been stimulated by glaciological research.

The area of the continents currently mantled by ice approaches a total of 15 million km². Of this figure by far the greatest part is concentrated in two regions, Antarctica with a cover of 12·5 million km² and Greenland with a cover of 1·7 million km². In addition to these huge ice bodies there are countless smaller masses ranging in size from those nestling in minor niches on steep mountain slopes to others blanketing thousands of square kilometres of lowland. The variety is immense and offers for study an extremely wide range of glacial environments. The purpose of the present chapter is to examine those characteristics of modern ice bodies believed to have particular significance for the moulding of the subglacial land surface.

Classification of glaciers and ice-sheets

Morphological classification Although all ice-masses possess certain common properties, they assume a wide variety of forms depending on their volume and relationship to the underlying relief. It is therefore useful to essay some form of morphological classification even though the boundaries between different categories often become rather blurred. The traditional two-fold division is between valley glaciers in which the ice is laterally confined by rock walls, and ice-sheets in which the ice is so much thicker that its surface form shows considerable independence of topographic control. Four common variants of the first class are: (*a*) cirque glaciers; (*b*) Alpine valley glaciers; (*c*) outlet glaciers; and (*d*) piedmont glaciers. Cirque glaciers are relatively small and occupy uniquely shaped rock-walled basins. Many examples are little more than 1 km in length but all are characterized by distinctive patterns of movement. Alpine valley glaciers are fed by ice issuing from one or more cirques and may attain lengths in excess of 100 km. Their surfaces exhibit many contrasting forms ranging from smooth to deeply crevassed slopes, from gentle longitudinal gradients to precipitous ice falls composed of séracs. Outlet glaciers differ from the Alpine type in being nourished by an ice-sheet, and although this may have important consequences for the glacier regime, in terms of pure morphology the distinction is relatively slight. Piedmont glaciers, on the other hand, have greatly distended snouts where they protrude beyond the limits of constraining valley walls and the ice is able to spread freely over an adjacent lowland. The change in thickness and direction of ice movement often produces a very intricate crevasse system.

Ice-sheets are divisible into two major classes based upon size. The smaller are known as ice-caps and the larger as true ice-sheets. Ice-caps are relatively thin. They tend to be confined to a single topographic region which permits division into upland and lowland types. In an upland location the ice generally rests upon an undulating plateau surface and feeds a group of peripheral outlet glaciers. A lowland ice-cap occupies an area of subdued relief little above sea-level, and owing to its topographic situation cannot usually support major outlet glaciers. Both types share a general domed form comparable to that displayed on a much larger scale by true ice-sheets. These latter are of sufficient thickness and extent to bury completely bedrock topography ranging from plainlands to true mountain ranges. Their upper surfaces resemble very broad domes in which gradients steepen towards the edges. Analysis of marginal slopes along both the Antarctic and Greenland ice-sheets has disclosed a considerable degree of conformity. In the absence of gross disturbance by partially buried mountain ranges, it is found that the ice surface rises to 1 000 m in the first 50 km from the edge, to 2 000 m in the next 150 km, and to 3 000 m in a further 350 km. These figures provide one possible guide to likely gradients around the edges of active Pleistocene ice-sheets; on the other hand it has been claimed that, for unexplained reasons, the marginal slopes of the Laurentide ice-sheet in North America were slightly lower.

Thermal classification Although ice temperatures vary widely, they are conventionally used as a basis for dividing valley glaciers and ice-sheets into two simple classes. The first are termed 'temperate' and consist of ice at its pressure melting point. The second are designated 'cold' and are composed of ice below its pressure melting point. There can be no doubt regarding the significance of temperature as a criterion for classification. Among the temperature-dependent properties of an ice-mass are the strain rate and the depth of meltwater penetration. Under the same stress ice will deform much more rapidly near its pressure melting point than at lower temperatures. The increased strain rate must affect both the velocity of an ice-mass and its surface form. On a temperate glacier water may flow deep below the surface, gradually enlarging any tunnel system by thawing the ice-walls; in a cold glacier, on the other hand, heat lost to the ice will ensure that meltwater refreezes within a short distance of the surface.

It must be stressed that the temperature relationships within ice-sheets are much more complex than the foregoing statements might at first sight seem to imply. Indeed, the whole concept of temperate ice-masses has been assailed by some workers who have argued that it is almost possible to envisage a large dynamic body of ice in which temperature is everywhere nicely adjusted to the pressure melting point. Interpreted literally the idea would require ice near the surface to be at 0°C, for its temperature to fall slowly during burial and to rise again during exhumation. It seems that some latitude must be allowed from this idealized picture and Russian glaciologists have suggested a maximum permissible deviation, below 15 m from the surface, of 1°C from the true melting point. The need to qualify the value for the upper 15 m reflects the depth to which seasonal temperature changes are known to penetrate.

However, it is the classification of apparently cold ice-masses that poses the greatest problems. Near the surface the dominant control is the temperature of the freshly fallen snow; at the South Pole this may be as low as −50°C and over wide areas of both Antarctica and northern Greenland is −28°C or less. At the base of these great ice-sheets there is an input of heat from three sources, the geothermal flux, friction with the underlying bedrock and shearing within the ice itself. On this basis temperature might be expected to vary directly with depth, and several workers computed theoretical temperature profiles for thick ice-sheets before deep drilling became a feasible proposition. Their forecasts have now been vindicated by drilling projects which have succeeded in penetrating to bedrock in both Greenland and Antarctica (Fig. 11.1). In 1966 a borehole at Camp Century in northern Greenland was sunk through 1 387 m of ice permitting direct measurement of a complete temperature profile. It was found that from −24°C at a depth of 10 m the temperature fell slightly to −24·6°C at 154 m but then rose steadily until at the base of the hole it was −13°C. This is still much lower than the pressure melting point, but the results did seem to indicate that in a thicker ice-sheet higher basal temperatures might be encountered. Ample

confirmation was supplied in 1968 when a borehole at Camp Byrd in Antarctica penetrated 2 164 m of ice. From surface values of −28°C the temperature fell to −28·8°C at 800 m, but then rose to −13°C at 1 800 m. Below this depth it was found impossible to take direct readings but extrapolation of the measured curve suggested that the basal ice would be very close to its pressure melting point. This interpretation was validated by the presence of a thin film of water at the bottom of the hole and thereafter it became impossible to treat the Antarctic ice-sheet as a simple 'cold' ice-mass. Even before this observational evidence became available, several workers had estimated that the basal ice in Antarctica was at pressure melting point over about half its total area.

These findings do not nullify the concept of a thermal classification of ice-masses. The idea is undoubtedly an important one, but it must be recognized that temperature distributions are often more complex than unduly simple models might suggest.

Subglacial topography

Although the form of the bedrock relief beneath a small glacier may be inferred from simple surface observations with little risk of gross error, a major ice-sheet conceals the subglacial topography so effectively that elaborate techniques are required to determine the position of the interface between rock and ice. Seismic survey was pioneered on a large scale by the Wegener expedition to Greenland during the years 1929–31. By detonating a charge in the upper layers of the ice and recording the compressional waves reflected from the bedrock surface, it is possible to compute from the travel time the depth to the interface. By using an array of instruments the elevation of a number of points can be determined simultaneously and a map of the buried topography produced. After allowance for the effect on wave velocity of density and temperature variations, the results obtained by seismic techniques should be accurate to ±5 m. Although suitable for most ice-masses, the method has proved less satisfactory in Antarctica where cold dry snow in the surface layers attenuates the compressional waves and makes the record difficult to interpret. The problem has been partially overcome by detonating charges at depths of 50 m or

Fig 11.1 Temperature profiles of the first boreholes to penetrate the total thickness of the Greenland and Antarctic ice-sheets.

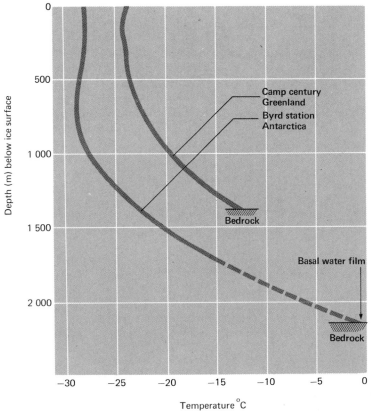

so, but in a hostile environment the inevitable delays and extra equipment are a heavy price to pay.

An alternative method to seismic survey is gravimetry. As explained in Chapter 1, it is possible to eliminate one by one the sources of variation in the local gravity field. On the gently sloping surface of an ice-sheet, any residual variation will mainly be due to unevenness of the bedrock relief. There is no isostatic compensation for purely local topographic features and given two points of equal elevation but with different thicknesses of ice cover, higher gravity values will be recorded where the ice is thin than where it is thick. The rapidity with which a survey can be carried out makes the method particularly useful for filling in detail between seismic stations. On a smooth ice-sheet away from the complications introduced by large nunatak mountain ranges, gravimetry should be capable of yielding measurements of ice thickness accurate to ± 20 m.

In the 1960s a new survey method was devised. It had previously been noted that planes equipped with radar altimeters were obtaining reflections not only from the ice surface but also from deep within the ice. It was soon realized that this second reflection was coming from the ice-rock interface and afforded a potential means of surveying bedrock relief. Special echo-sounding equipment was developed to operate at frequencies affording maximum penetration of the radio-waves through ice; in Antarctica a frequency of 35 MHz has often been employed. The instruments have been adapted for use from either aircraft or surface vehicles. Since absorption of radio-waves is much lower in cold than in temperate ice, the method is particularly suited to work in Antarctica where other techniques have proved less satisfactory. It also has the great advantage of giving a continuous profile. The velocity of radio-waves in ice is dependent on density so that some allowance has to be made for assumed ice density; nevertheless, the majority of recent results are believed to be accurate to ± 10 m. Although valley glaciers pose problems of reflection from valley walls, strong directional antennae can be fitted to surface vehicles and consistent measurements obtained within 50 m of the glacier edge.

International cooperation in the study of ice-sheets, combined with the technical advances outlined above, has provided a much greater basic knowledge of subglacial topography than was available a few years ago. Investigations in Antarctica have disclosed a bedrock relief consisting of two contrasting parts, an undulating plain in the eastern section of the continent and a mountain range in the smaller western section. Much of the plain is blanketed beneath 3 000 m of ice (with the thickness possibly exceeding 4 000 m in places), but over the mountains more variable values are found. Were the ice-sheet to melt and isostatic recovery to take place, western Antarctica would become a mountainous archipelago, while the eastern part of the continent would be converted into an elevated plateau. The hypsographic curve for the new land surface would not differ fundamentally from those of the currently unglaciated continents. Greenland exhibits subglacial relief consisting of an extensive central plain near sea-level and fringing mountains to both east and west. The ice over the central plain commonly exceeds 2 500 m in thickness. It is characteristic of both Antarctica and Greenland that the main ice-mass rests in a broad, shallow basin fringed by more or less continuous hills and mountain ranges; a similar form presumably characterized the areas depressed beneath the Pleistocene ice-sheets of North America and northern Europe.

Mass balance studies

For any ice-mass in equilibrium with its environment it is a necessary condition that the gain by means of accumulation should be equal to the loss by means of ablation. Accumulation must include all processes by which material is added to the ice-mass; the most important is that of direct precipitation but others that may locally be effective include avalanching, wind blowing and refreezing of run-off from surrounding land surfaces. Ablation must encompass all processes by which material is removed; the most obvious is simple melting but others include sublimation and iceberg calving. The time-period to which measurements of accumulation and ablation refer must also be specified. In practice this is usually the balance year which begins in late summer after ablation has reached its greatest extent and the ice surface is therefore at its lowest point, and continues to the time of the corresponding state in the next calendar year. Owing to slight annual variations each balance year will not necessarily be of 365 days duration, although this should be the average over a long period.

Field measurements Measurement of the gross accumulation rate presents such formidable problems that most investigations concentrate instead on the more easily calculated net value. The net annual accumulation refers to that part of the material added during the balance year that is still present at the end of the ablation season. It can be assessed by measuring the annual incremental volume over that part of the ice-mass known as the accumulation area. On any steady-state ice-mass there must by definition be an accumulation area balanced by an ablation area; the boundary between the two is known as the equilibrium line (Fig. 11.2). Within the accumulation area various further subdivisions are often present. In the coldest regions the summer

temperatures may remain below freezing point so that no meltwater is produced; this region is referred to as the dry snow zone. It is commonly flanked by the percolation zone in which slightly less extreme conditions generate summer meltwater that percolates downwards and refreezes. Close to the margins of the dry snow zone the quantity of water will be small and the depth of percolation shallow, but with increased summer warmth the whole thickness of the annual accumulation layer may be permeated with meltwater. Saturation often leads to slush avalanching in a region designated the slush zone; this is the highest area where loss of material takes place by surface run-off. Such varied conditions clearly demand careful selection of measuring sites if an accurate assessment of net annual accumulation is to be made. The usual procedure is to dig pits or take cores at sample sites and then to identify the annual

Fig. 11.2. Schematic representation of factors in the nourishment of an ice-mass. The same basic principles apply to all ice-masses from a small cirque glacier to a large ice-sheet.

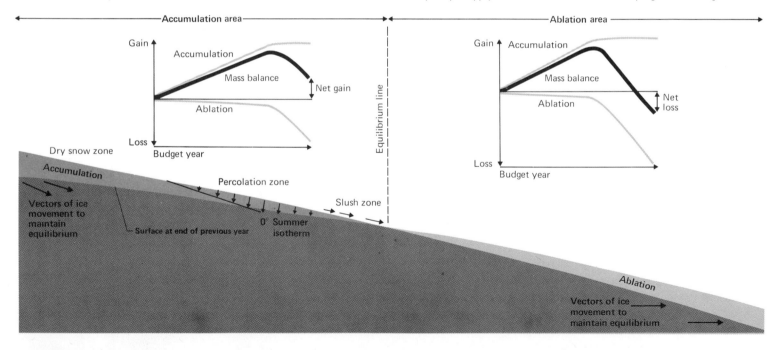

incremental layers. This is normally facilitated by thin dirt bands which mark the snow surface at the end of each balance year. These are particularly conspicuous close to rock outcrops which can supply the necessary fine debris. Another useful marker horizon is the compact dense ice that results from partial melting and refreezing during the summer season. On the other hand, in a dry snow zone identification of annual layers may prove much more difficult. In central Antarctica, for instance, the most valuable marker is often a layer of coarsely crystalline snow known as 'depth hoar' which appears to form each autumn just below the surface. In all net accumulation studies measurement of the annual layers must be followed by determination of snow density so that the thicknesses can be converted into water equivalent values.

To strike a balance with net accumulation, ablation must be measured over the entire ablation area. The aim is to determine the amount of material lost from the surface of dense ice while ignoring the melting of any superficial snow cover. Stakes are normally drilled through the snow cover during spring deep into the underlying ice. The distances between the tops of the stakes and the surface of the ice are then measured and the observations repeated at intervals until the close of the summer season when fresh snow begins to accumulate once more. Many stakes are needed for an accurate assessment of ablation, but the figure obtained, once converted into a water equivalent value, can be directly compared with that for net accumulation.

Analysis of mass budgets If an ice-mass is in equilibrium the net accumulation and ablation totals should theoretically be equal. In practice exact equality is rarely found. Several possible reasons may be advanced. The first relates to uncertainty over the calculated figures which are often based upon inadequate sampling. A second arises from minor year-to-year variations in both accumulation and ablation; whilst the former can sometimes be evaluated by study of the incremental layers, the latter can only be established by protracted field observations. A third possible reason is intrinsic to the ice-masses themselves. An ice-mass adjusts only slowly to changed climatic conditions so that complete equilibrium may be a relatively rare occurrence. The actual rate of adjustment differs significantly from one glacier to another, generally being much longer in the case of a large ice-sheet than in a small valley glacier.

Given methods of sufficient accuracy it should be possible to categorize ice-masses as having either a positive net balance when accumulation exceeds ablation, or a negative net balance when the reverse relationship holds. The two most important climatic parameters are precipitation and temperature and a change in either may disturb an ice-mass from its equilibrium state. For example, refrigeration will lead to a positive net balance occasioned primarily by a reduction in summer ablation. Immediate effects will be a raising of the surface level in the accumulation area, a lowering of the height of the equilibrium line and a slight elevation in the ablation area. Equilibrium can only be re-established when the additional net accumulation is compensated by increased loss and this will normally be achieved by advance of the ice margin to produce a larger ablation area. A similar pattern of changes will attend an increase in snowfall without any modification in temperature. Nye has considered the theoretical effect of a positive net balance and has shown that a 'kinematic wave' of increased discharge will be propagated across the ablation area. This wave moves approximately four times as fast as the ice itself and on reaching the terminus induces first thickening and then advance. A negative net balance normally causes the whole ice-mass to thin, with the largest change occurring near the terminus. Theory predicts, and observation confirms, that minor changes in the net balance can result in large shifts in the position of the ice front.

The intricate nature of the climatic controls, and the sensitivity of glaciers to these controls, can result in very complex patterns of ice advance and withdrawal. During the first half of this century when the majority of Alaskan valley glaciers were shrinking a few were growing slightly and one, Taku Glacier, actually advanced about 5·5 km. Variations of this type may arise from such factors as different temperature and precipitation relationships on opposite sides of a watershed, or different relative sizes of accumulation and ablation areas on adjacent glaciers. Although deviations from strict synchroneity can be very important in short-term studies, it remains true that the major advances and retreats of the Pleistocene ice-sheets seem to have been broadly contemporaneous, even on a global scale.

Before leaving the subject of mass balance studies it is worth examining in a little more detail the case of Antarctica. Particular interest attaches to this ice-sheet in view of its sheer size. Ignoring the effect of isostatic compensation, it is currently estimated that release of all the water locked up in Antarctica would raise sea-level by well over 50 m; conversely, an increase in mean ice thickness of just over 100 m would cause sea-level to drop 4 m, enough to disrupt shipping and radically alter many coastlines of the world. It has even been suggested that cyclical growth and melting of the ice-mass could exercise a major control on world weather patterns. Considerable practical importance therefore attaches to the present mass budget. Unfortunately measurement presents exceptional difficulties. Much of the continent is a dry snow accumulation zone with poorly defined incremental layers. It is generally agreed, however, that the continental interior is one of the driest areas on earth, its precipitation being equivalent to a water depth of no more than 50 mm. Near the margins of the ice-sheet the corresponding value is about 500 mm yielding a mean for the whole continent of approximately 150 mm. The most formidable problems of assessment attach to ablation which takes place mainly by loss of material from floating shelves where the ice-sheet projects into the Antarctic Ocean. Iceberg calving is the most prolific source of ablation accounting for as much as 90 per cent of the total loss. Individual bergs usually measure a few hundred metres in length, but occasionally gigantic fragments over 100 km long break away from the front of a shelf. Photographic survey can be employed to monitor calving, but less readily assessed is any melting that takes place beneath the shelves. Another potential source of error is the loss due to blowing of unconsolidated snow into the sea. Aeolian redistribution of snow is a particular feature of Antarctica, and it has been estimated that each year some 10^{11} tonnes are swept into the surrounding ocean by powerful katabatic winds blowing down the flanks of the ice-sheet. In the circumstances conclusions about the mass balance must be tentative but most workers believe that the ice volume is either stable or growing very slightly. It should be stressed that an ice-mass like that of Antarctica responds extremely slowly to changed en-

vironmental controls. Following a major climatic alteration it would probably be several thousand years before equilibrium was fully restored. It may also be noted that the unusual forms of ablation around Antarctica detract from its value as a model of the ice-sheets that developed in the northern hemisphere during the Pleistocene epoch; in this respect Greenland, with a conventional ablation zone locally 150 km broad, may provide a better analogue than does Antarctica.

Ice movement

The problem of ice movement can be approached from two independent but related viewpoints. Laboratory experiments may first be performed on ice to determine some of its physical properties, particularly the rate at which it deforms under a given stress. This approach has the advantage that such variables as temperature and pressure can be strictly controlled and information may be obtained regarding the true physical nature of ice deformation. Nevertheless, any predictions derived from laboratory experiments require testing on actual glaciers and ice-sheets, and here conditions are usually much more complex than those envisaged in the rather idealized laboratory models. The greatest difficulty in the field lies in obtaining a full three-dimensional picture since it is only the surface layer that can be readily observed.

Before examining the laboratory and field methods in more detail, it is worth making some general observations which may seem obvious on reflection but which can easily be overlooked. The movement of any ice-mass consists of a transfer of material from the accumulation to the ablation area. The maximum discharge must occur at the equilibrium line and in the case of a steady-state ice-mass must equal the net accumulation. By the same token discharge through any cross-section in the accumulation area must equal the net addition above that section; in the ablation area it must equal the net loss below that section. In a steady-state ice-mass the constant form of the accumulation area can only be maintained by a downward inclination of the velocity vectors; similarly in the ablation area the vectors must normally be inclined upwards relative to the ice surface.

234 Laboratory studies Ice is a crystalline solid which, in a temperate glacier at least, is close to its melting point. It is now recognized that it behaves neither as a very viscous Newtonian fluid nor as a brittle solid. Instead a given stress induces deformation which first increases in rate and then settles down to a relatively steady value. In a single crystal this plastic flow or creep can be observed to take place by the slipping or gliding of layers one over the other parallel to the basal plane. It is possible to induce gliding on other planes but the required stress is very much greater. It follows that the orientation of crystals within an ice-mass is an important determinant of its overall physical properties. Observation shows that crystals are not randomly aligned in glacier ice but exhibit varying degrees of preferred orientation with the c-axes perpendicular to any planes of shearing. Since crystals in the firn and shallow ice layers do not display this preferential alignment it is presumably induced by movement taking place at depth. A similar effect can be observed in the laboratory when a block of polycrystalline ice is subjected to simple shear stress. From an initial random distribution a preferred alignment gradually develops by intergranular movement and recrystallization. As it does so the strain rate increases and accords more closely with that measured in a single ice crystal; this is presumably because a higher proportion of the movement is taking place by gliding parallel to the basal planes.

Plotting the values for shear stress and strain rate obtained in numerous laboratory experiments it is found that a simple power relationship exists between them (Fig. 11.3). This is conventionally expressed in the form known as Glen's power law : $\dot{\varepsilon} = k\sigma^n$, where $\dot{\varepsilon}$ is the strain rate, σ the stress, and k and n are constants. The value of k varies with temperature, but n is relatively independent of temperature. In most studies n is assumed to be either 2·5 or 3, although a higher figure may be appropriate under large stresses where a high proportion of the crystals are preferentially aligned ; similarly, a lower figure better fits the data where only small stresses are involved. In order to convert Glen's flow law into a form in which it can be applied to a glacier or ice-sheet it is necessary to evaluate the stresses that will then operate. It is easiest to take the example of a simple parallel-sided slab of ice resting on a uniform slope (Fig. 11.4). Shear stress is then clearly a function of two readily measured parameters, slope and thickness. However, it needs to be remembered that its value will reach a maximum at the base of the ice and decline to zero at the surface. Yet the upper layers of ice are carried forward on the lower so that the actual movement will be greatest at the surface and decline with depth. It can be shown by integration that in the simple example considered, and when n in Glen's flow law has a value of 3, the surface velocity is proportional to the third power of the sine of the slope angle and to the fourth power of the thickness. Where the slab is not of uniform thickness so that the slopes at the base and surface of the ice are unequal, the situation is more complex. In such cases the basal shear stress can be shown to be propor-

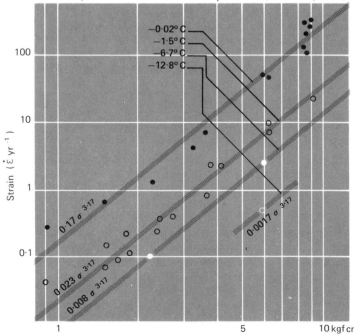

Fig 11.3 The results of laboratory experiments to measure the strain rate of ice when subject to varying shear stresses. Ice was tested at four different temperatures. The plotted points conform to the general relationship $\dot{\varepsilon} = K\sigma^n$ known as Glen's power law (after Glen, 1955).

tional to the sine of the surface slope. A corollary is that movement should be in the direction of the surface slope and not the basal slope; this offers the theoretical explanation for the observed ability of ice to scour deep rock basins and apparently move uphill. A further complication, and a closer approach to reality, is introduced by considering the effect of an irregular valley long profile. The gradient of the valley floor will normally consist of steeper and more gentle reaches. Nye has analysed the effect of slope variations and has shown that they should produce two types of flow which he terms extending and compressive (Fig. 11.5). Compressive flow with thickening of the ice will occur where the bedrock slope is reduced; shear planes will tend to develop curving upwards from the base towards the surface. Extending flow will be found where the ice accelerates down a steeper slope; shear planes will here curve downwards from the surface towards the bed.

Field investigations In field investigations (Fig. 11.6) it is essential to realize that the observed motion is normally composed of two parts. The first derives from internal deformation of the ice and is the part that has been considered previously from an experimental standpoint. The second is the movement of the ice-mass as a whole relative to its bedrock floor; this component of the total displacement is termed sliding or slip. There is a very long tradition of measuring glacier movement, particularly in the Swiss Alps, and one of the most famous early observers was Agassiz who inserted a series of stakes in the surface of the Unteraar glacier and was able to demonstrate that the central zone moves more rapidly than the margins. The techniques employed have changed little in principle although improved surveying equipment has increased the potential accuracy. The standard procedure is to bore stakes into the surface layer and to monitor both horizontal and vertical displacements by observation from a theodolite set up at convenient points on the valley walls. The majority of measurements have been made in the ablation area rather than the accumulation area, and few glaciers have been surveyed along their full length.

The measured rates vary greatly from one ice-mass to another. The interior of Antarctica with its lack of fixed reference points presents particularly severe problems of measurement, but surface velocities almost certainly average less than 20 m yr^{-1}. In the peripheral area the ice accelerates to over 100 m yr^{-1}, and in a series of ice streams attains rates of over 500 m yr^{-1}. These ice streams are radiating zones, sometimes delimited by promi-

Fig. 11.4. Basic analysis of the shear stresses operating on a parallel-sided slab of ice resting on a slope of angle α. By application of Glen's power law it is relatively easy to compute the velocity profile which, in such a case, is essentially a function of slope angle and ice thickness.

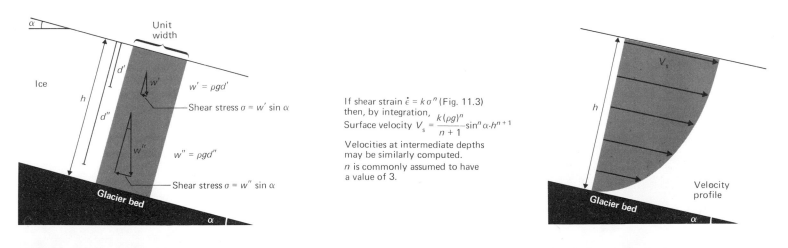

If shear strain $\dot{\epsilon} = k\sigma^n$ (Fig. 11.3) then, by integration,
Surface velocity $V_s = \dfrac{k(\rho g)^n}{n+1}\sin^n\alpha \cdot h^{n+1}$

Velocities at intermediate depths may be similarly computed.

n is commonly assumed to have a value of 3.

$w' = \rho g d'$
Shear stress $\sigma = w' \sin \alpha$
$w'' = \rho g d''$
Shear stress $\sigma = w'' \sin \alpha$

Ice

Unit width

Glacier bed

V_s

Velocity profile

Glacier bed

nent lines of crevasses, in which the ice moves much faster than the adjacent parts of the sheet. The reasons behind their siting remain uncertain. In many cases they appear to be related to depressions in the subglacial topography, and it is possible that the greater ice thickness permits the basal layers to reach pressure melting point. Alpine valley glaciers have generally been recorded as moving at velocities between 20 and 200 m yr^{-1}, but accelerating to rates in excess of 1 km yr^{-1} down the steeper slopes. The fastest rates are usually found on outlet glaciers descending from the Antarctic and Greenland ice-sheets; in the former area the cold nature of the ice precludes exceptionally high values, but in Greenland velocities in excess of 7 km yr^{-1} have been recorded for such outlet glaciers as the Jakobshavn Isbrae.

The foregoing figures relate to surface layers near the centre line of the glacier. Although velocities close to the rock margins are much reduced, they are of particular interest as a means of evaluating local slip (Fig. 11.7). For accurate assessment of side-slip, stakes need to be inserted within 1 m or so of the ice edge; in some instances they have been installed horizontally with their ends virtually touching the valley wall. The slip velocity is often expressed as a percentage of the maximum velocity in the same cross-section, and values have been found to range from under 10 to over 65 per cent. More difficult to measure is the basal slip at the sole of the ice. Velocities at depth

within an ice-mass can be ascertained but expensive and difficult procedures are involved. They normally entail drilling a hole to the base of the ice and inserting some form of flexible casing. After surveying the position of the hole an inclinometer must be lowered down the casing to measure the slope and alignment of the pipe at various depths. If a cap is placed over the hole and the measurement procedures repeated after an interval of a year or more, a new profile of the casing can be drawn and the velocity variations with depth computed. This technique was first adopted in a study of the Aletsch Glacier in Switzerland and has subsequently been repeated for a number of North American glaciers. They have all conformed to the theoretical expectation that the surface layers move more rapidly than the lower, and that maximum differential movement is concentrated near the base. Another method of observing basal slip has been to use tunnels excavated for commercial purposes beneath several glaciers. These have the advantage of permitting a relatively continuous monitoring of movement at the sole of the ice. Expressed as a percentage of the maximum movement in the profile, the amount of slip in temperate glaciers has been found to vary from 10 to 75 per cent with a mean value of about 50 per cent. Even on the same ice-mass the figure can vary appreciably, indicating the complex relations that exist at the sole of a glacier.

Slip generally appears to increase significantly with any

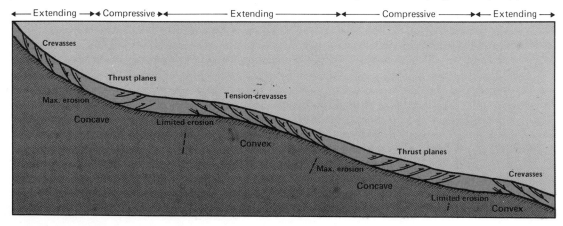

Extending → ◄ Compressive ► ◄ Extending → ◄ Compressive → ◄ Extending →

Crevasses

Thrust planes

Max. erosion

Concave

Limited erosion

Tension-crevasses

Convex

Max. erosion

Concave

Thrust planes

Crevasses

Limited erosion

Convex

Fig. 11.5. Simplified representation of Nye's idea of extending and compressive flow relate to convexities and concavities of the bedrock profile. One factor not depicted is the accumulation/ablation ratio; for the same long profile extending flow will theoretically be more common in the accumulation area compressive flow in the ablation area.

thickening of a basal water film. At least part of the movement probably takes place by regelation around small irregularities in the bedrock floor. The increased pressure on the onset side of any protuberance leads to local melting. The water then flows to the lee side where it refreezes under conditions of diminished pressure. The latent heat which is released may be conducted through the protuberance to assist further melting on the upstream side, but the poor thermal conductivity of rock means that small projections will be more effective than large ones. Both theoretical arguments and observations conduce to the view that basal slip is likely to be much faster in a temperate than in a cold glacier; and many workers have contended that a cold glacier is characterized by an absence of basal slip, the lowest

ice layers being frozen directly on to bedrock.

A final aspect of ice movement requiring examination is the temporal variation which many glaciers exhibit. By means of lasers it has been shown that the movements conventionally measured over periods of weeks, months or even years are actually composed of an immense number of discrete but extremely small displacements. This jerky motion is more intense in summer than in winter, and tends to reach a maximum during warm weather and after heavy rainfall. As seasonal temperature fluctuations are confined to the upper layers of the ice it seems likely that the acceleration is due primarily to thickening of the basal water film during ablation. Support for this view has come from observations in subglacial tunnels where direct

Fig. 11.6. Three examples of field measurements on glaciers in western North America: (A) Variations in surface velocity, movement being expressed as a proportion of the maximum velocity along the centre line of the glacier (after Meier, 1960); (B) Velocity changes with depth (after Paterson, 1969); (C) Seasonal fluctuations in surface velocity and meltwater discharge. Both values increase in summer and decline in winter, but are significantly out of phase (after Hodge, 1974).

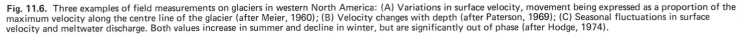

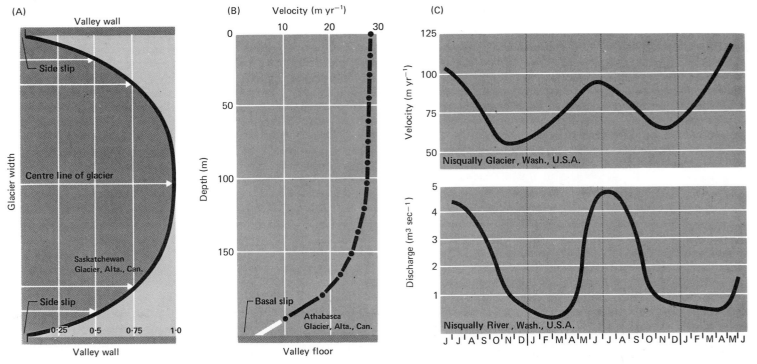

measurement of water pressure can be made. During summer the pressure rises substantially and the effective weight of the ice acting on the underlying rock is reduced. This buoyancy reduces the friction and the glacier slides much more rapidly over its bed.

Longer term velocity variations known as surges have now been recorded from a number of glaciers. These temporary accelerations may last from a few months to several years. During that time ice which has lain almost stagnant for decades may suddenly thicken and move as a wave several kilometres down the glacier. In a well-developed surge the ice can travel 20 m in a day, with the aggregate movement in a year or two exceeding 10 km. Few examples have been studied in detail, although during 1956–7 a mass transfer of ice on the Muldrow Glacier in Alaska resulted in surface elevations being lowered over 50 m in the upper reaches and raised by a similar amount in the downstream reaches. The underlying cause of surging remains in dispute. It seems, however, that certain glaciers have a tendency to become increasingly unstable until some unknown factor triggers a surge.

Glacial meltwater

Meltwater streams In the interior of Antarctica where the temperatures even in summer do not reach freezing point there is obviously no production of surface meltwater. However, this is a relatively unusual circumstance and on the majority of ice-masses meltwater is produced in great quantities. On un-compacted snow in the accumulation area meltwater frequently percolates downwards and refreezes; in the ablation area, on the other hand, it commonly nourishes surface flow that persists throughout the summer season. Marked diurnal fluctuations are characteristic, with peak discharge in the early evening and minimum discharge in the early morning. As on a normal water-eroded landscape both sheet and channelled flow may be found, but surface lowering takes place by melting rather than abrasion. In this way gullies are occasionally eroded to depths of 10 m or more. On the steep ice-slopes near the margin and terminus of a glacier high gradients and smooth channel walls combine to produce exceptionally fast flow. A further character-istic is the tendency for meltwater streams to plunge below the surface so that individual supraglacial courses are often of very restricted length. Once the water has descended down a crevasse or moulin its englacial and subglacial course is difficult to follow. Attempts to use tracer dyes have not been entirely successful and great uncertainty still surrounds the precise routes followed between surface sinks and resurgences near the glacier terminus. Being confined within tunnels and shafts the meltwater may be under very considerable hydrostatic pressures and these are sometimes manifested in the violence with which water is ejected at the snout. By means of a specially constructed laboratory in a tunnel beneath the Glacier d'Argentiere in the French Alps, records of water pressure have been made over a period of several years. During winter pressure is relatively steady at 9–10 kgf cm^{-2}, but in summer it fluctuates daily between about 6 and 12 kgf cm^{-2}, presumably due to melting on the glacier surface. The measurements imply the existence of a water table within the glacier and help to explain the ability of meltwater to flow uphill and score channels on the up-valley side of buried bedrock ridges.

Constrictions in the passages through the ice locally ac-celerate the water to extremely high velocities. Conversely large englacial cavities act as storage chambers in which the meltwater is virtually immobile. Such cavities have been encountered during a number of tunnelling and drilling oper-ations and their control on meltwater discharge has been demonstrated for the Athabasca Glacier where large inter-mittent floods occur with no apparent relationship to ambient temperatures; it appears that local blind passages and cavities must fill with water and give rise to a sudden spate at the snout when they are temporarily integrated into the drainage system.

The typical convex cross-profile of a valley glacier tends to shed meltstreams to a lateral position where they are forced to turn and flow parallel to the valley walls. Often the glacier

Fig 11.7. Two photographs illustrating conditions at the margin of a Swiss valley glacier. (Left) The irregular gap along the valley wall which rapidly channels any surface meltwater beneath the ice. (Right) Detail of the rock-wall showing both the smooth face produced by abrasion, and the very rough surface resulting from freeze–thaw activity where there is no direct contact with the glacier ice. The scratch marks indicate the importance of side-slip as rock fragments such as that shown in the photograph are dragged along the valley wall.

margins are deeply crevassed and a proportion of the meltwater disappears without even reaching the ice edge. Nevertheless, surface melt augmented by run-off from adjacent valley sides commonly sustains truly marginal drainage. An individual stream rarely flows any great distance before disappearing down a passage beneath the ice margin, but its place is soon taken by a new stream. The water passing beneath the ice does not immediately proceed to the valley floor but is often channelled roughly parallel to the glacier edge in what is termed a sub-marginal position. The channels of the marginal drainage may be excavated in either ice or bedrock, or may have one wall composed of rock and the other of ice. It is not uncommon to find streams passing from one material to the other and back again within the space of a 100 m or less. A typically irregular long profile produces sharp contrasts in velocity. In places the meltwater may be ponded back into shallow lakes, while in others it plunges down precipitous slopes. An added characteristic of marginal stream courses is their general instability occasioned by down-valley movement of the glacier, by melting of the surface ice, and by changes in the pattern of crevasses.

Ice-dammed lakes Impounded against the margins of modern glaciers are hundreds of small ice-dammed lakes. They are most commonly situated at the lower ends of tributary valleys where the main valley is occupied by ice. Less frequently they occur where a major drainage artery is obstructed by a powerful glacier issuing from a tributary valley. A third situation in which they are sometimes found is around the frontal edge of a glacier in a hollow between the ice margin and an abandoned morainic ridge. Examples of ice-dammed lakes occur in virtually all regions of temperate glaciers. The most famous instance in the Alps is the Marjelensee held up against the Aletsch Glacier. This lake was 1 600 m long in 1878 but since that date it has gradually declined and owing to shrinkage of the impounding glacier is now virtually extinct. The great majority of modern ice-dammed lakes are small, often less than 1 km² in extent. Among the larger examples are Lake George in Alaska which in 1965 covered some 120 km², and Graenalon in Iceland which is usually credited with an area of 18 km².

One difficulty in assigning sizes to particular lakes is the frequent fluctuations they exhibit. Although a few lakes spill across bedrock cols away from the ice edge, a situation tending to stabilize the water-level, most discharge through subglacial or englacial passages. Water-level may then fluctuate wildly, with icebergs formerly floating near the shoreline often left stranded on the hillside by a sudden fall. At the same time the discharge of water from the glacier snout will be extremely erratic. A mere trickle may suddenly give way to an immense spate, often known by the Icelandic term jokulhlaup. The fast-flowing meltwater has great transporting powers and besides moving exceptionally large boulders may also carry huge icebergs torn from the glacier front. In some jokulhlaups momentary discharges have been estimated to exceed 100 000 m³ sec⁻¹.

Records are now available of the changes in level of many ice-damned lakes. Considerable diversity exists. Some lakes discharge regularly each summer whilst others continue to fill for several years and are evacuated only at infrequent intervals. Variations in lake height are often recorded in minor strandline features closely analogous to those seen in flooded gravel pits after a fall in water level. Much doubt still surrounds the precise mechanism by which discharge takes place. Thorarinsson has argued that when a lake reaches a critical depth the water will float the ice barrier and then escape beneath. One difficulty with this hypothesis is that the lifted ice margin might be expected to sink back and reseal the lake basin. In such circumstances the water-level would oscillate only narrowly about the depth required for flotation. To overcome this problem it has been suggested that the barrier is unlikely to sink back in exactly its original position so that water will continue to discharge until the level of the limiting rock sill is reached. Liestol has emphasized that once water finds an escape route, the passage it uses will be speedily enlarged by melting; even if the ice falls back to its earlier position water will still be able to find its way out. Attempts to test the flotation hypothesis by surveying the elevation of a point on the ice dam to see whether it is lifted during the period of discharge have so far proved inconclusive. A totally different mechanism has been proposed by Glen. He contends that, owing to the differing densities of ice and water, stresses at the base of an ice cliff might be adequate to force a breach in the dam. For this mechanism to be really effective the lake would need to be at least 150 m deep, at which point deformation could certainly be expected; conceivably the water might then gain access to other passages through the ice leading ultimately to an escape at the glacier snout. However, the need for very deep water seems to limit the applicability of the hypothesis. A third possibility is that changes within the impounding ice periodically lower the level of the barrier. Movement of the ice may introduce deep crevasses, or meltwater already within the ice may enlarge its tunnel system to tap the lake. Once the lake water has started to escape, the passage it is using will tend to be enlarged by melting.

Rates of erosion

Many different techniques have been employed in efforts to determine the rates of erosion associated with modern glaciers. Some have been directed to assessing the overall speed with which the subglacial land surface is being lowered, while others have concentrated on the effectiveness of certain specific processes. In the first category comes the measurement of sediment load being carried by meltwater issuing from a glacier snout. Many early workers commented on the milky appearance of such water and ascribed it to the high content of fine 'rock flour'. Tarr described how, in Alaska, from a sample collected in a pail, several centimetres of mud and sand settled out within a few minutes. In the same area Reid estimated the amount of silt being removed each year from the Muir Glacier to be equivalent to a subglacial rock layer 19 mm thick. Such a figure could only be regarded as a very rough approximation since it was not based upon systematic sampling. A more determined attempt to obtain representative values was made in 1939 by Thorarinsson who found that the meltwater from the Hoffellsjokul in Iceland was discharging sediment equivalent to a bedrock lowering of $2 \cdot 8$ mm yr^{-1}. Since 1950 many further investigations in various parts of the world have nearly all confirmed that the rate of surface lowering is much faster in glaciated than in non-glaciated areas. It should be emphasized, however, that the measurement of total sediment load on meltwater streams presents such severe problems that the figures obtained so far indicate only the general order of magnitude. Among the difficulties encountered are the extremely rapid variations in water discharge, the

common tendency towards channel braiding, and the need for an assessment of bedload transport which may be relatively large on certain meltwater streams. Further complications are introduced by morainic accumulation around the snout of the glacier (which should be added to the sediment transported by the meltwater) and by material supplied from the valley walls above the glacier (which should be subtracted if the effect of glacial erosion is to be isolated). One of the most detailed investigations so far undertaken is that by Corbel in the French Alps. After a study covering a period of several years, he concluded that the floor of the St Sorlin Glacier is being lowered at a rate of $2 \cdot 2$ mm yr^{-1}, just over half this figure being recorded as discharge in the meltwater stream and just under half as morainic deposition. By comparing this value with that from adjacent unglaciated valleys, he showed that the St Sorlin valley is being lowered many times faster. It seems that the mean rate for erosion by active glaciers probably lies in the range 1–5 mm yr^{-1}, which is substantially higher than almost all the figures quoted in Chapter 10 for stream denudation.

The overall values discussed above are the result of glacial and meltwater erosion operating both beneath and around the margins of an ice-mass. Direct observation of the relevant processes is normally restricted to marginal locations. Glacial erosion has been studied by inscribing symbols on rock faces in front of an advancing glacier and observing their state of preservation when the ice later withdraws. The classic experiment of this type was the cutting of a cross 3 mm deep in front of the Dachstein glacier in Austria. This was completely obliterated and replaced by freshly polished rock after brief coverage beneath the glacier snout. Similar observations have been repeated many times since, although the tendency for glaciers to retreat during the present century has limited the opportunities for such experiments. Wherever the technique has been employed the results have implied that rapid erosion is taking place.. However, important limitations are that the rock face is never deeply buried, is abundantly supplied with meltwater and experiences relatively wide temperature fluctuations. In other words, the results are not necessarily typical of the subglacial environment as a whole. Attempts have recently been made to assess erosion by bolting specially prepared rock slabs

242 to the valley floor near the head of long subglacial tunnels. Boulton and Vivian have recorded that basalt and marble slabs attached in this position and overrun by the ice for a period of 3 months were abraded to mean depths of 1 and 3 mm respectively; during the same interval basal slip amounted to 9·5 m.

The rapidity with which marginal meltwater streams may incise their courses into solid rock is attested by numerous observations. The most famous are those of Tarr and von Engeln in their explorations of the Alaskan glaciers around the turn of the present century. They record how meltwater may carve gorges many metres deep within a period of a decade or less. Flint quotes an example of a small stream first superimposed on to resistant bedrock in 1935 which, by 1941, had excavated a gorge 15 m wide and more than 8 m deep. In favourable circumstances downcutting at a rate of 1 m yr^{-1} does not seem exceptional, and is adequate testimony to the power of fast-flowing meltstreams. There seems no reason why equally rapid meltwater erosion should not occur beneath the ice, although direct observational evidence is relatively limited. Artificial tunnels have occasionally offered opportunities for studying subglacial streams. Great seasonal variations in discharge have been recorded, with particularly violent flow characteristic of summer. Near Zermatt potholes 2·3 m wide and 2 m deep have been seen to form in ophiolites in a period of 3 years, and experimental quartzite blocks have been scored to a depth of 100 mm along the course of another stream in under 5 years.

BOULTON, G. S. and VIVIAN, R., 1973. 'Underneath the glaciers', *Geogr. Mag.*, 45, 311–19.

CORBEL, J., 1962. *Neiges et glaciers*, Colin Armand.

FLINT, R. F., 1971. *Glacial and Quaternary geology*, Wiley.

GILBERT, R., 1971. 'Observations on ice-dammed Summit Lake, British Columbia, Canada', *J. Glaciol.*, 10, 351–6.

GLEN, J. W., 1954. 'The stability of ice-dammed lakes and other water-filled holes in glaciers', *J. Glaciol.*, 2, 316–18.

GLEN, J. W., 1955. 'The creep of polycrystalline ice', *Proc. R. Soc.*, A 228, 519–38.

GOW, A. J., et al., 1970. 'Antarctic ice sheet: preliminary results of first core hole to bedrock', *Science N.Y.*, 161, 1011–13.

HANSEN, B. L. and C. C. LANGWAY, 1966. 'Deep core drilling in ice and core analysis at Camp Century, Greenland, 1961–1966', *Antarct. J. U.S.*, 1, 207–8.

HARRISON, W. D., 1975. 'Temperature measurements in a temperate glacier', *J. Glaciol.*, 14, 23–30.

HODGE, S., 1974. 'Variations in the sliding of a temperate glacier', *J. Glaciol.*, 13, 349–69.

LIESTOL, O., 1955. 'Glacier-dammed lakes in Norway', *Norsk. geogr. Tidsskr.*, 15, 122–49.

MATHEWS, W. H., 1974. 'Surface profiles of the Laurentide ice sheet in its marginal areas', *J. Glaciol.*, 13, 37–43.

MEIER, M. F., 1960. 'Mode of flow of Saskatchewan glacier, Alberta, Canada', *Prof. Pap. U.S. geol Surv.*, 351.

NYE, J. F., 1952. 'The mechanics of glacier flow', *J. Glaciol.*, 2, 82–93.

POST, A., 1960. 'The exceptional advances, Black Rapids and Susitna Glaciers', *J. geophys. Res.*, 65, 3703–12.

ROTHLISBERGER, H., 1972. 'Water pressure in intra- and subglacial channels', *J. Glaciol.*, 11, 177–203.

SAVAGE, J. C. and W. S. B. PATERSON, 1963. 'Borehole measurements in the Athabasca glacier', *J. geophys. Res.*, 68, 4521–36.

STENBORG, T., 1969. 'Studies of the internal drainage of glaciers', *Geogr. Annlr.*, 51A, 13–41.

THORARINSSON, S., 1953. 'Some new aspects of the Grimsvötn problem', *J. Glaciol.*, 2, 267–75.

VIVIAN, R., 1970. 'La nappe phreatique du glacier d'Argentiere', *C. r. hebd. Seanc. Acad. Sci. Paris*, 270, 604–6.

VIVIAN, R. and G. BOCQUET, 1973. 'Subglacial cavitation phenomena under the Glacier d'Argentiere, Mont Blanc, France', *J. Glaciol.*, 12, 439–51.

VIVIAN, R. and J. ZUMSTEIN. 'Hydrologie sous-glaciaire au glacier d'Argentiere (Mont Blanc, France)', Union Geodesique Symp. on Hydrology of Glaciers, 1974.

Selected bibliography

Almost all issues of the *Journal of Glaciology* contain papers adding significantly to our knowledge of modern ice masses. A valuable review is provided by W. S. B. Paterson, *The Physics of Glaciers*, Pergamon, 1969.

3|12

Glacial erosion and deposition

The preceding chapter has emphasized the contribution to an understanding of glacial landforms made by glaciological studies. Yet it needs to be remembered that the processes of erosion and deposition operating deep beneath an ice-sheet have never been directly observed. Admittedly in recent years much information has come from artificial tunnels excavated beneath cirque and valley glaciers, but how representative these facilities are of conditions beneath thicker ice-masses it is difficult to ascertain. Laboratory experiments such as those to test the effect of sliding ice on prepared rock samples, and theoretical models of likely circumstances at the base of an ice-mass have also contributed significantly to our present knowledge. Yet it is the study of landforms and deposits in formerly glaciated areas that has often provided the clearest guide to the way an ice-mass with its attendant meltwater can modify the landscape across which it passes. It is to the interpretation of these formerly glaciated regions that the present chapter is primarily directed.

Glacial erosion

Glacial erosion is generally attributed to two separate processes, plucking and abrasion. The former refers to the quarrying effect when moving ice freezes on to bedrock and pulls out a block which it then carries away. Abrasion is due to the grinding effect of the debris being transported in the sole of the ice-mass and has been likened to the action of sandpaper. The existence of two distinct processes is manifest in the erosional form known as the roche moutonnée. The onset side of a roche moutonnée is rounded and scored by abrasion while the lee side is angular and blocky due to the removal of joint-bounded fragments by plucking. There has been considerable discussion regarding the relative effectiveness of these two processes.

Many minor landforms bear testimony to the abrasive action of ice armed with rock debris. The most common is the striated pavement where fine lines have been scored parallel to the direction of ice movement. With diminishing size, striations grade into polished surfaces on which the individual striae can only be detected with a hand lens. With increasing size they appear to grade into various forms of fluted surface. Grooves 1 to 2 m deep and up to 100 m long have been described from Ohio, while in the Mackenzie valley in Canada giant examples attain depths of 30 m and can be traced for distances in excess of 10 km. Features of this dimension stand out conspicuously on aerial photographs and give certain areas of the Canadian Shield a prominent lineated appearance. The mechanism involved in fashioning such large-scale grooves obviously differs in detail from that

Fig 12.1 The distinctive form of Half Dome rising 1 480m above Yosemite valley floor, Sierra Nevada, California. The position of the 'missing' half coincides with a zone of well jointed rock which succumbs much more readily to weathering and erosion than the rest. The massive structure of the rock in the vertical face contrasts with the superficial sheeting on the dome itself.

responsible for fine striations, but both bear witness to the importance of glacial abrasion.

Many workers have contended that plucking is quantitatively the more effective means of erosion. Supporting evidence comes from the large blocks of unweathered rock incorporated in both glacial and glacifluvial deposits. Moreover, rocks of similar mineralogical composition but with contrasting joint systems have been found to respond very differently to glacial erosion. It seems unlikely that abrasion alone would give rise to such differences, and they are much more readily explicable in terms of quarrying action. A striking example is afforded by the Yosemite valley in California (Fig. 12.1) where massive granite has resisted glacial erosion but granite with closely spaced joints has been deeply quarried. Cliffs over 1 000 m high mark the junction between the two rock types but the most spectacular manifestation is Half Dome (Fig. 12.2). Here almost 50 per cent of an originally hemispherical mountain has been removed while the rest remains virtually untouched. It is small wonder that Matthes who made the outstanding scientific study of Yosemite valley was a great proponent of the efficacy of glacial plucking.

It is worth noting, however, that the concept of plucking involves certain difficulties. In the previous chapter it was suggested that a temperate glacier is characterized by a thin film of water separating the ice from the underlying rock, while a cold glacier is characterized by an absence of basal sliding because the ice is frozen directly to bedrock. Neither appears suited to plucking. From a consideration of subglacial temperature regimes, Boulton has suggested that plucking is only effective where there is refreezing of water previously released by basal melting. He envisages the existence of distinct thermal zones at the sole of many ice-masses (Fig. 12.3). In certain zones the heat generated at the base of the ice will exceed that which can be conducted upwards so that melting must occur. The surplus water produced in this fashion then migrates to other zones where it refreezes on to colder basal ice, raises the tem-

Fig 12.2. Yosemite valley in California, a classic example of a glacial trough. The sheer face of El Capitan on the left is some 925m high; the post-glacial fill on the valley floor at the foot of El Capitan is over 300m thick (see Fig 12.8). In the centre background is the summit of Half Dome.

perature of that ice and so ensures continued basal slip. It is in these latter zones that plucking will be particularly effective, being facilitated not only by adhesion but also by the wedging effect of any water that penetrates into the joint system. Boulton's hypothesis is a variant of an older view that highly localized pressure variations can induce continuous melting and refreezing. Several workers who have examined the rock-ice interface at the end of glacier tunnels have commented on the presence of clear ice, a phenomenon which they ascribe to refreezing of water released from localized areas of excess pressure. Kamb and La Chappelle even identified a continuous regelation layer up to 29 mm thick lying between the normal glacier ice and the bedrock surface. By this extension of Boulton's argument, it is possible to visualize plucking as an effective process over a wide area of a glacier base. A further factor that may assist plucking is the physical rupture of rock owing to the stresses that arise from differences in ice pressure on the onset and lee side of obstructions. It has even been suggested that this may be more significant in the development of roches moutonnées than simple freeze–thaw activity.

Cirques Being amongst the most distinctive of all landforms, these steep-sided semicircular basins of glaciated highlands have long been accorded special descriptive names in a wide variety of languages. For instance the local name in Scotland is corrie, in Wales cwm, in Norway botn and in Germany Kar, but there is an increasing tendency to adopt the French term cirque in recognition of its use in a scientific context by de Charpentier as long ago as 1823. Many techniques may be employed in the study of cirques, but attention here will be confined to two separate approaches. The first is investigation of modern cirque glaciers, the second is morphometric analysis based upon comparison of shape, aspect and elevation of the landform itself.

An early investigation that was to have a profound effect on views about cirque development was made by Johnson who, at the beginning of the present century, was lowered down the bergschrund at the back of a cirque glacier on the northern face of Mount Lyell in California. He found that where the bergschrund intersected the headwall at a depth of some 40 m the rock surface displayed much evidence for active freeze–thaw processes. Huge angular boulders that had tipped forward to rest against the glacier were sheathed in clear ice. Similar material could be seen in the enlarged joint system of the headwall, amidst many partially dislodged blocks. Long icicles attested to the presence of migrating water which had later frozen. Johnson inferred that there must be many freeze–thaw cycles operating at the base of the bergschrund, possibly at a regular daily interval during summer. Later workers have confirmed the substance of his observations but cast doubt upon his interpretation. Thermographs installed for lengthy periods have shown that air temperature variations in a bergschrund are generally of low amplitude and take place only slowly (Fig. 12.4). Even in summer there is no regular diurnal fluctuation, and intervals when the air is sufficiently warm to induce melting are infrequent. As a modification of the original Johnson hypothesis, Lewis suggested that most of the ice observed at the base of a bergschrund is water which has trickled down the backwall from the surface and frozen on entering regions of increasingly low temperature. However, by itself such a process would provide little basis for continued freeze–thaw activity since the headwall would presumably become coated with ice which would then function as a protective shield. Recently, therefore, less emphasis has been placed on the frost-sapping mechanism at the base of a bergschrund, and many workers have come to regard such a process as restricted to a shallow zone near the uppermost edge of the glacier. In effect this means that frost-shattering, which is undoubtedly rife on the rockwall above a cirque glacier, extends beneath the ice surface to a critical depth where the temperatures are too stable for the process to contine; the bergschrund often extends down below this critical depth.

Although headwall sapping around the upper margins of a cirque glacier is clearly an important process, it does not explain many of the distinctive features of the cirque. In particular it does

Fig. 12.3. Schematic representation of the different thermal regimes that may be encountered at the base of an ice mass.

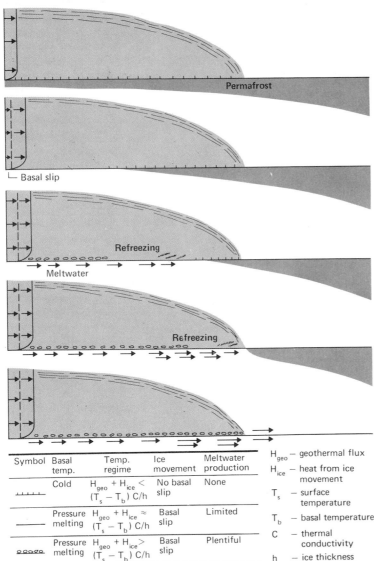

Symbol	Basal temp.	Temp. regime	Ice movement	Meltwater production
┴┴┴	Cold	$H_{geo} + H_{ice} <$ $(T_s - T_b)\ C/h$	No basal slip	None
———	Pressure melting	$H_{geo} + H_{ice} \approx$ $(T_s - T_b)\ C/h$	Basal slip	Limited
ᴼᴼᴼᴼᴼ	Pressure melting	$H_{geo} + H_{ice} >$ $(T_s - T_b)\ C/h$	Basal slip	Plentiful

H_{geo} — geothermal flux
H_{ice} — heat from ice movement
T_s — surface temperature
T_b — basal temperature
C — thermal conductivity
h — ice thickness

not account for the scouring of the rock basin which appears to be due primarily to the nature of the ice movement. This has been investigated on several occasions by the digging of tunnels, one of the most informative being that excavated under the leadership of Lewis in Vesl-Skautbreen in Norway during the summer of 1951. It had previously been postulated that the distinctive surface banding to be seen on many cirque glaciers is due to annual accumulation layers tilted in the ablation zone so as to dip back into the glacier. This seemed to imply some form of rotational movement, a hypothesis which received strong support when the walls of the tunnel were examined (Fig. 12.5). Stratification in the ice due to annual accumulation layers was found to dip towards the backwall at an angle of about 28° near the tunnel entrance. Traced deeper into the glacier the dip gradually declined until finally its direction was reversed and the stratification near the headwall was almost parallel to the rock face. At the rock face itself a narrow zone of bubble-free ice, almost certainly attributable to regelation, was rendered dark by its high content of transported debris. Elsewhere the ice was remarkably clean with no sign of detritus migrating from the headwall into the main body of the glacier. The disposition of the accumulation layers, deformation of the tunnel wall and displacement of surface stakes, all served to confirm that there is a degree of rotational movement. The motive force appears to be the excess of accumulation near the headwall, combined with maximum ablation near the toe. The intrinsic steepening of the glacier which this engenders requires slow rotation to maintain equilibrium. The internal deformation is relatively small, being restricted in the main to frictional retardation close to the headwall. It is now widely held that cirques are initiated by thick snow accumulating either in a suitable valley head or in a depression enlarged by nivation beneath a snow patch. When the snow can accumulate to a thickness of about 30 m the basal layers are transformed into true glacier ice. Thereafter abrasion and plucking at the base of the rotating ice, augmented by freeze–thaw processes at the top of the newly formed headwall, ensure development of the typical cirque form.

In considering cirque morphology, it must first be emphasized that great variations in size can be found, ranging from the enormous Walcott cirque in Antarctica which is 16 km wide and has a backwall 3 000 m high to more typical examples which are 1–2 km wide and have backwalls some 300 m high. The question immediately arises as to how far the shape varies with size and to what extent it is influenced by such factors as geological structure. Various parameters have been devised to measure cirque form. Manley examined the ratio between cirque length and height and found that it normally lies between 2·8 and 3·2. He concluded that the nature of the local rock has relatively little effect on morphology, and this was confirmed during a more detailed analysis of Scottish cirques undertaken by Haynes. She surveyed the long profiles of 67 cirques and showed that they generally conform to a relatively simple logarithmic curve of the family $y = (1-x)e^{-x}$. The consistency of form appears to justify the belief that process is the dominant control with rock structure playing only a subsidiary role. On the other hand, true cirques are confined to lithologies that can sustain very steep headwall slopes, which in practice often

means areas of igneous or metamorphic rock. The shales and mudstones of the Southern Uplands in Scotland are noticeably lacking in cirques, whereas the form is well exemplified in the nearby English Lake District. Many observers have commented on the way rock structure influences morphological detail even if it does not determine overall form. Planes of weakness control the size and shape of blocks that are removed, and the slope at the foot of the headwall often consists of a staircase of treads and risers defined by bedding and joint systems. Equally conspicuous is the preferential widening of joints aligned parallel to the headwall; this suggests that rock dilatation may be a significant accessory process in cirque enlargement.

It has long been recognized that cirques display distinctive patterns of elevation and orientation. The former has been invoked as an indication of the altitude of Pleistocene snowlines since the initiation and development of cirques requires conditions normally found only near the snowline. However, a

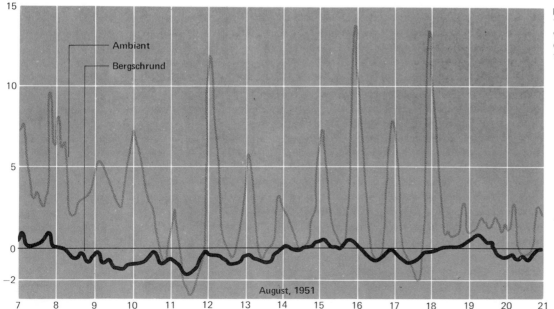

Fig. 12.4. Temperatures recorded in Tverrabreen bergschrund, Norway, showing the very pronounced damping effect when compared with the outside atmospheric temperatures (after Battle, 1960).

major difficulty is knowing whether any particular group of cirques developed contemporaneously, and if so at which period. The method is most satisfactorily employed where former glaciers never developed beyond the cirque stage. Flint, for instance, has used the technique in reconstructing the late Pleistocene snowline in the Rocky Mountains and in the Cascade Range in the western United States. Yet even where a mountainous area has later been covered by an ice-sheet a regular pattern of cirque-floor elevations can often be discerned. In the Scottish Highlands, for example, cirque floors display a systematic rise from west to east (Fig. 12.6). In this and other cases the parallelism with modern isohyets has been held to support the view that the cirques lie close to a critical Pleistocene snowline, probably dating from the onset of the last glaciation. Further support comes from the simultaneous reoccupation of certain groups of cirques during climatic fluctuations at the close of the last glaciation.

In the northern hemisphere cirques are preferentially orientated towards the north and east, and in the southern hemisphere towards the south and east. The obvious association is with contrasts in insolation and this has usually been regarded as the primary cause of the preferred alignment. During summer suitable valley heads would retain thick snow banks on shaded slopes while corresponding sites in the full sun would be completely bare of snow. Nivation processes beneath the snow banks would slowly erode the rock floors until true cirque glaciers were able to develop. It might at first seem that similar conditions should exist at higher elevations on the sunny slopes, but this is not necessarily true; by localizing the preservation of snow banks the shadow effect can concentrate the attack of nivation processes in a way that is not normally possible on sunny slopes. A second influence stressed by certain writers is the effect of wind carrying snow on to the lee face of a mountain mass so as to leave the windward face relatively bare. Snow blown in this way will tend to collect in sheltered hollows and nourish the snow banks which are an essential preliminary to full cirque development.

Glacial troughs The amount of topographic modification achieved by valley glaciers and ice-sheets varies enormously from one locality to another. Some valleys that must have carried considerable quantities of ice emerge relatively unscathed at the end of a glaciation, while others such as the Lauterbrunnen in Switzerland (Fig. 12.7) and the Yosemite in California are almost completely transformed. Yet the greatest changes are probably wrought where an ice stream manages to carve a completely new depression across a mountain mass, sometimes removing 1000 m or more of solid rock in the process. In the following account attention will first be directed to the

Fig. 12.5. Sections through Vesl-skautbreen showing flow lines and computed positions of ablation surfaces at 10-year intervals (based upon velocity measurements at glacier surface and along length of tunnel). Ablation surfaces as observed on the tunnel wall are shown below (after McCall, I960).

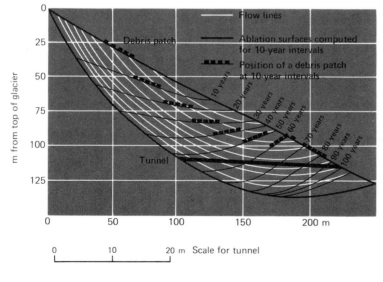

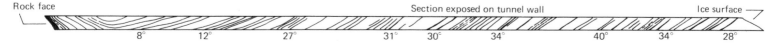

distinctive features of a valley reshaped by an alpine glacier, and then to the forms produced by concentrated linear erosion beneath an ice sheet.

Glacial erosion affects both the transverse and longitudinal profiles of a valley. The transverse profile is conventionally described as U-shaped although in relatively few cases do the valley walls approach the vertical. In Sweden Svensson found a transverse section through the Lapporten valley to conform quite closely to a parabola, but surprisingly few attempts have yet been made to apply similar morphometric analyses. In part this probably reflects continuing uncertainty as to the degree of glacial erosion which many valleys have suffered. It is by no means unusual to find reaches of typical U-shaped profile

separated by reaches in which a V-shaped profile dominates. In some instances the V-shaped segment is merely an incision into a rock bar on the floor of a trough and can be explained by subglacial meltwater activity. However, in other cases the variation in cross-profile affects the full depth of the valley and it is difficult to avoid the conclusion that ice has selectively widened certain reaches but has left others almost untouched. Another difficult problem concerns the extent to which conversion into a U-shaped cross-profile has been accompanied by valley deepening. Rock basins testify to the ability of a glacier to deepen its bed, and hanging tributary valleys sometimes bear witness to the amount of vertical downcutting that has taken place (Fig. 12.8). However, the testimony of hanging valleys is

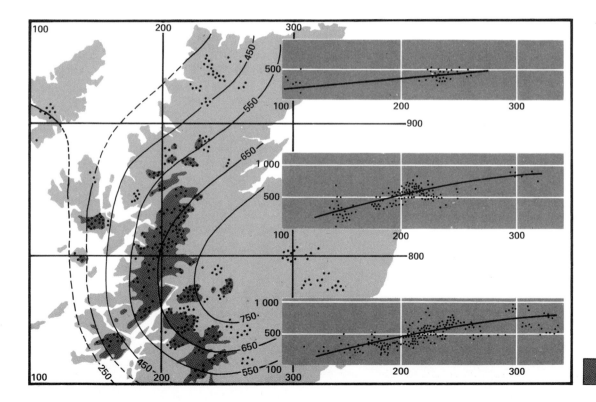

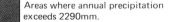

Areas where annual precipitation exceeds 2290mm.

Fig 12.7. The glacial trough of Lauterbrunnen in Switzerland. Although excavated mainly in sedimentary rather than igneous rocks, it nevertheless displays a form similar in many respects to the Yosemite valley pictured in Fig 12.2.

by no means always clear. On the one hand widening of the main valley without any deepening can produce a limited amount of hang, while on the other the tributary valleys may themselves have been subject to an unknown amount of glacial erosion. At one time it was argued that the gentle benches surmounting many Alpine troughs are remnants of the pre-glacial valley profile which might be reconstructed by extra-polating the remaining fragments towards the valley centre. Most Alpine workers strongly contest this view and are reluctant to admit that there has been wholesale deepening of the valley floors. Among the more plausible alternatives is that the Alps have long been subject to intermittent uplift and that the troughs already had a valley-within-valley form before being modified by glacial action.

The longitudinal profile of a glacially eroded valley is charac-teristically irregular. Not only does the downstream gradient change dramatically but there are often reaches of reversed slope enclosing deep rock basins. The basins may still be occupied by lakes but many have been partially or totally infilled with sediment; for example seismic investigations on the floor of the Yosemite trough have revealed a detrital accumulation exceeding 600 m in thickness. The factors determining where pronounced deepening of the valley floor takes place have been the subject of many theories. It has been argued that an increase in erosive capacity leads to the development of a step immediately downvalley from a confluence of two glaciers. Many examples might be quoted to uphold this thesis, but the most consistent support comes from the excavation of trough headwalls just below the coalescence of two or more cirque glaciers. It needs to be remembered that the valley is to the glacier as the channel is to the stream, and that just as the channel enlarges below a tributary confluence so the cross-sectional area of the valley may be expected to increase in the analogous position. It seems that a part of this increase is often achieved by valley deepening. However, not all examples of

steps in the long profile can be ascribed to greater ice discharge and alternative explanations must be sought. Many workers have endeavoured to relate them to changes in lithology, although several different arguments have been deployed. The simplest envisages selective bedrock erosion by the ice, often ascribing particular importance to the spacing of the joint systems as an aid to plucking. Where such an argument appears inappropriate, recourse may be had to the idea of pre-liminary deep weathering preparing the bedrock for removal; this might involve either decomposition during an interglacial phase or frost-riving during the cold period immediately pre-ceding ice advance. Another explanation sometimes offered is that where a valley is laterally constricted the glacier enlarges its cross-sectional area by eroding its floor more deeply. Despite this wide variety of possible explanations many specific rock steps still defy a full understanding, and a rather different approach to the problem has been adopted by Nye. He emphasizes the potential effect on ice movement of any initial profile irregularities. Employing the idea of extending and com-pressive motion outlined in Chapter II, he suggests that each concave segment will tend to be subject to rapid erosion by compressive flow while any convexity will escape relatively lightly under extending flow. On this basis it can be argued that irregularities tend to be self-enhancing and that any factor capable of initiating them, including those discussed earlier in this paragraph, may ultimately be responsible for large rock bars, steps and basins.

Troughs fashioned by ice accumulating beyond the limits of the original watershed can be divided into several different types. The simplest comprises those shaped by outlet glaciers descending from the margins of an ice-cap or ice-sheet. These show few significant differences from the alpine type, al-though many outlet glaciers are unusually steep and fast-flow-ing. A second type is associated with glacial diffluence in which an individual valley or outlet glacier divides into separate branches. This arises where a plentiful supply of ice fills the available valley to a level at which part of the glacier can escape across the surrounding watershed into an adjacent catchment. Numerous examples can be seen in presentday glaciated areas such as Alaska or Greenland. At first the diffluent lobe will carry

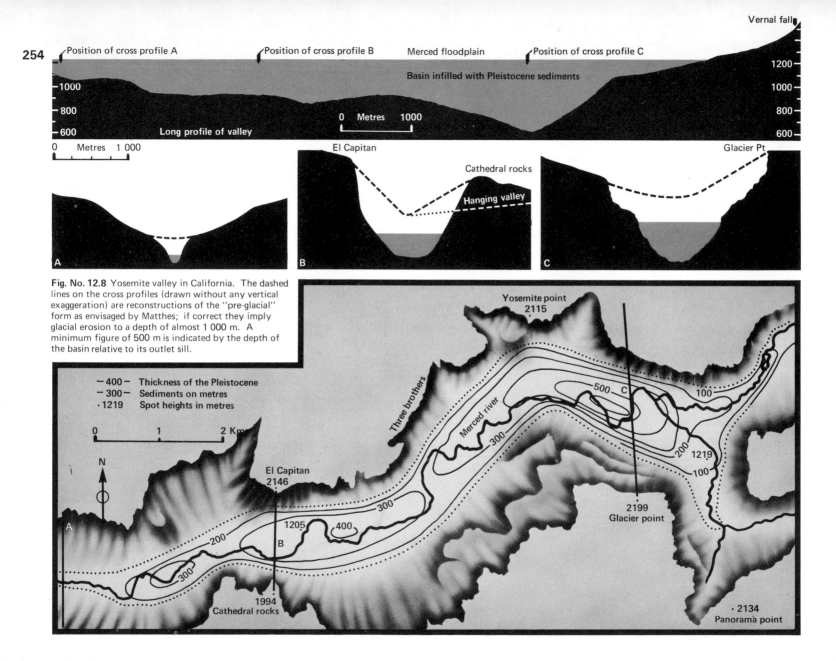

254

Position of cross profile A   Position of cross profile B   Merced floodplain   Position of cross profile C   Vernal fall

Basin infilled with Pleistocene sediments

1200

1000

1000

800

800

600

Long profile of valley   0   Metres   1000   600

0   Metres   1 000

El Capitan   Cathedral rocks   Glacier Pt

Hanging valley

A   B   C

Fig. No. 12.8 Yosemite valley in California. The dashed lines on the cross profiles (drawn without any vertical exaggeration) are reconstructions of the "pre-glacial" form as envisaged by Matthes; if correct they imply glacial erosion to a depth of almost 1 000 m. A minimum figure of 500 m is indicated by the depth of the basin relative to its outlet sill.

—400—   Thickness of the Pleistocene
—300—   Sediments on metres
·1219   Spot heights in metres

0   1   2 Km

N

Yosemite point
2115

Three brothers

Merced river

500   C

100

300

200

1219

100

200

El Capitan
2146

1205

B

400

300

2199
Glacier point

A

300

1994
Cathedral rocks

·2134
Panorama point

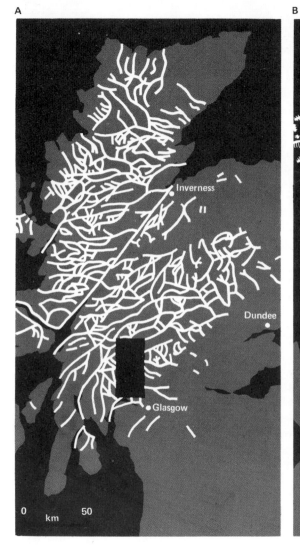

A

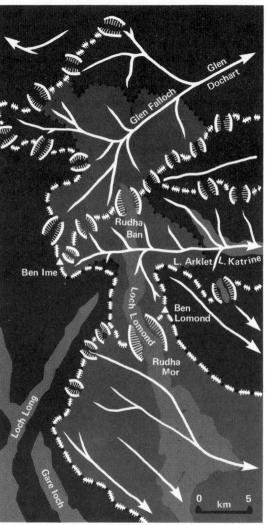

B

Fig 12.9 Aspects of glacial erosion in the **255** Scottish Highlands: (A) Glacial troughs (after Linton, quoted by Clayton, 1974). Note in particular the anastomosing pattern in the most intensely glaciated areas; (B) Scouring of the Loch Lomond trough by southward-moving ice from original eastward-trending river catchments. Glacial erosion at both Rudha Ban and Rudha Mor is estimated to have removed 500 m of rock (after Linton and Moisley, 1960).

 Present Lomond catchment

Reconstructed preglacial watershed

Glacially scoured breaches in preglacial watershed

 Preglacial drainage lines

a relatively small proportion of the discharge, but by vigorous erosion the route may be so deepened that eventually it is used by a much higher percentage of the ice.

A third type of trough is fashioned beneath an ice-cap or ice-sheet. If accumulation increases beyond a certain point valley glaciers ultimately merge to form a single continuous ice-mass. Although the domed form of the new ice-mass may demand a broadly radial pattern of outflow, this does not mean that the bedrock topography ceases to exert any influence. There is undoubtedly differential streaming within many ice-masses, with the maximum velocities concentrated along pre-existing valleys. However, instead of conforming completely to the original stream pattern the ice may also gouge new routes across interfluves and impose on the underlying relief a system of 'iceways' adapted to the transference of ice from the accumulation to the ablation area. Numerous breaches may be torn in existing watersheds until the preglacial relief pattern is modified almost beyond recognition (Figs. 12.9 and 11). Master troughs carved by the more actively eroding ice streams can attain dimensions comparable to those of the original water-eroded valleys. It has been suggested that Loch Lomond occupies such a trough, bulldozed through an earlier watershed when vigorously moving ice was diverted southwards from its original outlet; in the process the rock surface was locally lowered some 500 m. Being almost entirely the result of glacial erosion, large troughs of this origin tend to be straight or gently curving with the classic U-shaped form uninterrupted by numerous tributary valleys. In recent years Sugden has argued that the concept of ice streaming offers a better explanation for many glacial troughs than does the idea of diffluence to which they had previously been attributed. He contends, for instance, that many of the Cairngorm troughs in Scotland could only have been fashioned when the area was almost entirely buried by ice; it is much more likely that the troughs continued to be shaped beneath an ice-sheet than that their development was confined to the relatively short periods of growth and decay. The idea that a cover of sluggish ice preserves broad interfluves relatively intact, while streaming ice simultaneously scours deep troughs, may help to explain the frequently observed contrasts between upland and valley morphology. Sugden has hypothesized that the basal ice along the main valleys may be at pressure melting point while over the adjacent uplands it is sufficiently cold to be frozen to the bedrock.

In the case of a large ice-sheet the ice-shed may eventually be displaced many kilometres from the original watershed. This situation is found today beneath the ice-sheets of Greenland and Antarctica, and its former existence in areas of Pleistocene glaciation is attested by erratics carried from source regions on one side of a mountain chain to be incorporated in glacial deposits on the far side. In Scandinavia erratics on the Norwegian coast can be identified as coming from Sweden, while in northern Scotland a distinctive augen-gneiss in glacial deposits on the Atlantic coast is derived from outcrops on the opposite side of the Northwestern Highlands. Passes on the divide through both the Scandinavian and Scottish mountains appear to have been fashioned into typical glacial troughs by ice streaming through them towards the Atlantic coastlands.

A final reference must be made to fjords which are essentially troughs partially submerged by the sea. They provide some exceptionally fine examples of the features already described as diagnostic of troughs. For instance, many possess unusually steep sides and prominent rugged headwalls. However, their most distinctive attribute is undoubtedly the deep rock basin lying submerged near their seaward end. Separated from the open ocean by a shallow bar or threshold, the basins often descend many hundreds of metres below present sea-level. As the Weichselian fall in sea-level did not exceed about 100 m, and in any case most fjord areas must have been isostatically depressed, it is difficult to avoid the conclusion that much of the glacial erosion took place below sea-level. Assuming a density for the ice of 0·9 g cm^{-3} a glacier 1 000 m thick would still be in contact with its bed when in water almost 900 m deep, and there is no reason for believing it to be incapable of eroding at such a depth. Fjords may therefore be regarded as the product of thick, active ice-streams descending from high coastal plateaus and mountains. The threshold appears to signify a rapid decline in the erosional power of the ice, possibly resulting from accelerated basal melting at the point of contact with sea water.

Erosional forms in lowland areas Compared with the erosional forms of mountain areas those in lowland regions have often been rather neglected. In part this may be due to evidence that ice can move across an area without disrupting the underlying surface at all. This is attested by till resting directly on such delicate materials as stratified sand and silt, laminated clay and, perhaps most diagnostic of all, fossil soil profiles. Although preliminary freezing may have rendered each of these more resistant than it appears today, it remains true that an ice-sheet operates neither uniformly nor universally as a scouring agent. On the other hand, there is abundant evidence that many lowland areas have been significantly degraded by the passage of an ice-sheet. The resultant forms vary according to the nature of the bedrock.

On the igneous and metamorphic basement rocks of Canada and Scandinavia there is occasionally large-scale grooving. More characteristic, however, is seemingly haphazard relief with numerous lakes and marshy areas set amidst low, craggy hills of irregular outline. A rather similar 'knock and lochan' topography is found on the coastal lowlands of Lewisian gneiss in north-western Scotland. Faults, shatter belts and other lines of weakness have been selectively eroded so that, except in such small-scale features as striated pavements and roches moutonnées, direct glacial lineation is not very obvious. Estimates of the amount of glacial erosion involved in the production of the present relief vary widely. It has been claimed that, over part of the Canadian shield, features exhumed in pre-glacial times from beneath a cover of Ordovician rocks are still largely unaltered despite the passage of the continental ice-sheet. On the other hand, the relationship in Massachusetts between the present surface and older sheeting structures has been held to indicate a lowering of 3–4 m by abrasion on the onset side of the average rock knob, and of up to 30 m by plucking on the lee side. Intensive local scouring is implicit in the depth of certain lakes on the Canadian Shield; the floor of Great Slave Lake, for instance, lies 600 m or more below water-level. In Scotland it is difficult to envisage parts of the knock and lochan relief being produced without at least 50 m of glacial erosion.

In lowland areas with isolated resistant outcrops set amidst much weaker strata, upstanding knolls are frequently fashioned into streamlined forms known as rock drumlins. Their onset sides are usually steep and craggy, their lee sides gentler and more smoothly rounded. Long curving depressions are often scored along both margins. Rock drumlins occur in considerable numbers throughout the Scottish lowlands where intrusive igneous rocks provide the resistant cores to the hills. Where less resistant rock is preserved on the lee side, the term 'crag and tail' may appropriately be used. It is worth emphasizing that the true crag and tail, such as that on which Edinburgh is built, can only be produced where there is a significant degree of glacial erosion.

In scarp-and-vale areas the nature of glacial modification seems to depend in part on the direction of ice movement relative to the strike of the rocks. The chalk cuesta of eastern England is believed to illustrate the effect of overrunning by obliquely moving ice (Fig. 12.10). South-west of Hitchin, beyond the generally accepted limit of the glaciation, the chalk forms the prominent escarpment of the Chiltern Hills with its crestline frequently exceeding 200 m in height. To the north-east, by contrast, the escarpment is much reduced in elevation and is progressively set back many kilometres from the western edge of the outcrop. The typical cuesta form is lost, being replaced by a low and relatively even-crested plateau locally plastered with drift. The view that the morphological contrasts between glaciated and unglaciated segments of the cuesta are due to ice scouring receives support from the very chalky nature of the till sheet in southern East Anglia. Even stronger confirmation is to be found in quarries east of Hitchin. Here outcrops that at first sight appear to be composed of in situ bedrock are in fact composed of gigantic slabs of chalk tens of metres in length and many metres thick piled one on top of the other. The frequency of such enormous erratics is difficult to assess, but other examples are known from cuestas that have been overrun by ice. Further north, for instance, Lincoln Edge has obviously suffered significant modification to judge from the widespread distribution of huge Lincolnshire limestone erratics. It has even been suggested that the straightness and regularity of the Edge may be due to trimming of former limestone promontories by ice moving parallel to the strike. Where the ice advances at right angles to the trend of a cuesta, different erosional forms can be

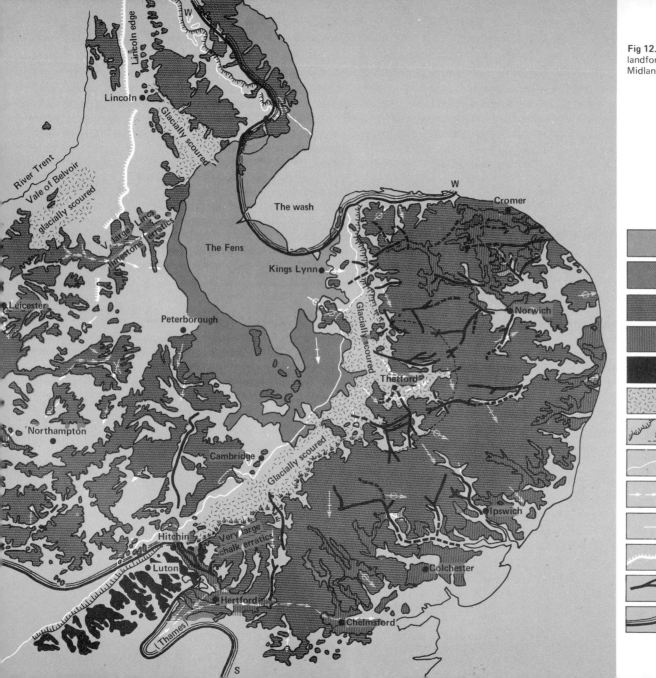

Fig 12.10 Selected aspects of the glacial landforms of East Anglia and the East Midlands.

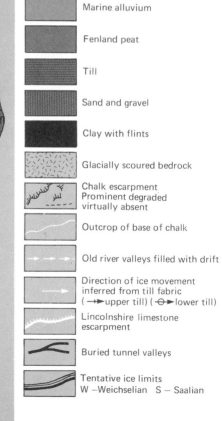

	Marine alluvium
	Fenland peat
	Till
	Sand and gravel
	Clay with flints
	Glacially scoured bedrock
	Chalk escarpment Prominent degraded virtually absent
	Outcrop of base of chalk
	Old river valleys filled with drift
	Direction of ice movement inferred from till fabric (→▶ upper till) (⊖▶ lower till)
	Lincolnshire limestone escarpment
	Buried tunnel valleys
	Tentative ice limits W –Weichselian S – Saalian

Map labels:

Lincoln edge
Lincoln
River Trent
Vale of Belvoir
glacially scoured
Glacially scoured
The wash
W
Cromer
The Fens
Kings Lynn
Norwich
large Lincs limestone erratics
Leicester
Peterborough
Thetford
'Northampton
Cambridge
Glacially scoured
Ipswich
Very large chalk erratics
Hitchin
Luton
Colchester
Hertford
Chelmsford
(Thames)
S

fashioned. These are well illustrated in Southern Ontario where the Laurentide ice-sheet overwhelmed the Niagara limestone escarpment (Fig. 12.11). Instead of passing uniformly over the crest of the scarp, the ice developed a series of more active streams, each of which scored a deep re-entrant in the face of the hills. The effects on the Niagara scarp may be compared with those on an even greater scale along the Allegheny scarp further east. Here the ice moving out of the Lake Ontario basin encountered a massive, partly dissected cuesta. Ice-streams developing along the axes of pre-glacial valleys scored the exceptionally deep rock basins in which the Finger Lakes now lie. Seneca Lake occupies a basin more than 300 m deep and is overlooked by hanging valleys which suggest an even greater amplitude of glacial erosion.

On relatively featureless lowlands erosion often seems to have been concentrated beneath lobes at the edge of the continental ice-sheet. In the Netherlands three or four deep rock basins have been identified. Each is elongated parallel to the trend of ice movement and seems to have developed beneath a lobe of the Saalian ice-sheet (Fig. 12.12). Although the Netherlands is known today for its low relief, by the time the Scandinavian ice reached this area a fall in sea-level might have induced sufficient rejuvenation to promote ice streaming. However, an alternative interpretation is to view the basins as variants of the depressions commonly found near the snouts of certain Alpine glaciers and known by the German term Zungenbecken. The very deep lakes along the northern and southern edges of the Alps mark the positions of such basins and their consistent siting seems to reflect very active erosion where piedmont glaciers formerly spread out on to lower ground. The Malaspina Glacier at the present day is underlain by a similar basin and clearly there are processes inducing intense scouring where ice initially fans out in a divergent pattern. Strongly compressive flow associated with radial shear planes passing upwards from the base of the ice appears the most likely explanation, and this might also be invoked where divergent movement occurs within the terminal lobe of a continental ice-sheet.

Glacial deposition

According to its location, detritus being transported by an ice-mass may be divided into three broad categories: englacial; supraglacial; and subglacial. The relative quantity of material in each category varies widely. Englacial detritus is normally very sparse and in an ice-sheet or large valley glacier deep cores usually encounter great thicknesses of almost pure ice. The concentration of englacial debris only attains significant values if compressive flow is tending to carry material from the sole towards the surface. This is sometimes very apparent near the snout of a glacier where debris can be seen moving up shear planes towards the ablating surface above. Supraglacial material, other than fine wind-blown particles, may be virtually absent on a large ice-sheet with few nunataks. On a valley glacier, by contrast, the ice surface is sometimes completely masked by avalanching from nearby rock faces. Often the supraglacial debris is concentrated into lateral moraines along the sides of a glacier, or into a medial moraine below the point where two glaciers merge. The tendency is for the material to become more widely distributed in the ablation zone. This is due primarily to its insulating effect which results in the debris gradually coming to occupy an elevated position in relation to the surrounding clear ice. When this happens the material is prone to slip sideways and so to be redistributed over the ice surface. Subglacial detritus is more difficult to trace and measure. However, when holes are drilled to the sole of a glacier the basal ice is nearly always dirty in appearance and often contains relatively large rock fragments. There seems little doubt that the subglacial load is extremely widespread, although rarely aggregating to any great thickness.

Deposition of the transported detritus is inevitably a complex process involving many different mechanisms. That fraction deposited directly from the ice is normally termed till. Its most characteristic attributes are poor sorting and general lack of stratification. In addition the larger constituent fragments, although rarely well rounded, often have their edges and corners blunted by abrasion; occasionally a prominently striated face has been imposed by contact with the underlying bedrock. Nevertheless it is difficult to specify a typical mechanical composition since this varies widely according to the rock types over which the ice has moved. There are, for instance, great contrasts between the glacial deposits on the clay lowlands of

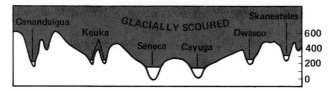

Fig. 12.11. Glacial landforms around the southern and western margins of Lake Ontario. The inset shows a schematic section through the Finger Lakes, emphasizing the efficacy of ice-streaming and, in this limited area, the concurrent scouring of the interfluves. Further south and west interconnected troughs were gouged by the ice but the plateau surface itself was relatively little affected.

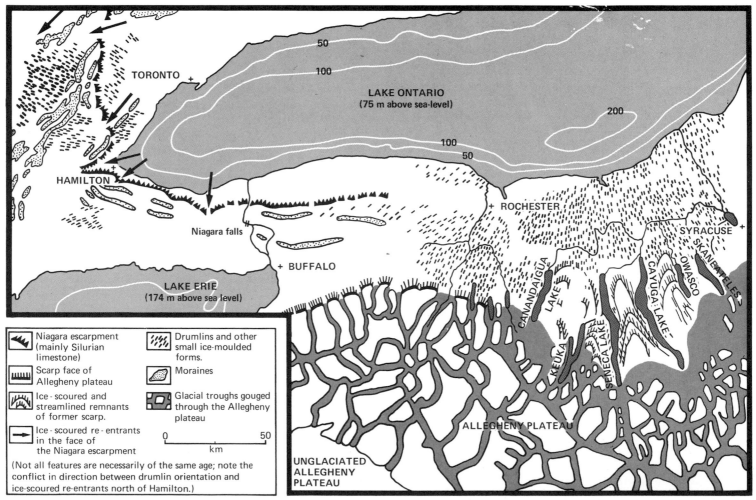

TORONTO +

LAKE ONTARIO
(75 m above sea-level)

50
100
200
100
50

HAMILTON +

+ ROCHESTER

SYRACUSE +

Niagara falls

+ BUFFALO

LAKE ERIE
(174 m above sea level)

CANANDAIGUA LAKE
KEUKA
SENECA LAKE
CAYUGA LAKE
OWASCO
SKANEATELES

ALLEGHENY PLATEAU

UNGLACIATED
ALLEGHENY
PLATEAU

Niagara escarpment
(mainly Silurian
limestone)

Scarp face of
Allegheny plateau

Ice - scoured and
streamlined remnants
of former scarp.

Ice - scoured re - entrants
in the face of
the Niagara escarpment

Drumlins and other
small ice-moulded
forms.

Moraines

Glacial troughs gouged
through the Allegheny
plateau

0   50
km

(Not all features are necessarily of the same age; note the
conflict in direction between drumlin orientation and
ice-scoured re-entrants north of Hamilton.)

Inset (top):
Canandaigua   GLACIALLY SCOURED   Skaneateles
Keuka   Owasco
Seneca   Cayuga
600
400
200
0

the English Midlands and those in the Highlands of Scotland although both may correctly be identified as tills.

Traditionally two types of till have been recognized, lodgement and ablation. The former is laid down subglacially when debris is released directly from the sole of the ice, the latter accumulates initially in a supraglacial position and is later lowered to the ground surface by undermelting. Lodgement till tends to be a compact deposit consolidated by the weight of overlying ice whereas ablation till is more open-textured owing to removal of the finer particles by meltwater.

As a result of detailed investigations along the glacier margins in Spitsbergen Boulton has argued that the simple two-fold division into lodgement and ablation till is inadequate. In the first instance it is necessary to consider what happens to englacial debris. Theoretically this may be added to either the subglacial or the supraglacial load. If it is in the lower layers of the ice, basal melting may be sufficiently rapid to ensure its subglacial release. However, if it is slightly higher in the ice, the faster rate of surface ablation will probably mean that it joins the supraglacial material before being deposited. Boulton maintains that the traditional two-fold division fails to take cognizance of the remarkable range of deposits that may accumulate on the surface of a wasting ice margin (Fig. 12.13). He distinguishes two essential types of supraglacial till. The first is flow till which consists of debris that has built up on the ice and after saturation with meltwater becomes so unstable that it flows or slumps into nearby hollows; the depressions in which it comes to rest may be either on the ice or else on earlier sediments. The second is melt-out till which is the direct product of ablation continuing beneath a detrital cover. Often the cover is a flow till or a sheet of material laid down by meltwater, and the final result will be a succession in which the lower members have been released from the ice at a later period than the upper. Undermelting of this type will frequently disrupt any earlier bedding.

One way in which tills of diverse origin may be distinguished is on the basis of their internal structure or fabric. It was realized before 1900 that large erratics in a lodgement till are not arranged in a random fashion but tend to have their long axes preferentially aligned. Little use was made of this information until 1941 when Holmes was able to demonstrate an intimate relationship in certain New England tills between fragment shape and orientation. His primary conclusion was that the long axes of most stones lie parallel to the presumed direction of ice movement, a general property of lodgement till confirmed by many later studies. The alignment applies not only to the larger stones but also to the finer particles in the matrix. It is the former, however, that have been the more widely studied. The orientation of an individual fragment is most conveniently specified in terms of the azimuth and dip of its long axis. These two measurements can be incorporated in a single diagram by plotting the points on an equal-area stereographic net. The array of points may be treated statistically to determine whether it deviates significantly from a random distribution; if it does, the mean vector may then be calculated. In a lodgement till long axes are preferentially aligned parallel to the ice movement and with a gentle up-glacier dip. In a melt-out till the particles appear to retain the disposition they had when being transported englacially. Boulton has shown this can result in two contrasting fabrics. In recently deglaciated regions he has recorded some sections in which the preferred orientation is parallel to the observed ice movement and others in which it is at right angles. The difference is apparently related to the nature of ice-flow, with strong compressional movement leading to a normal orientation and tensional movement to a parallel orientation. In flow tills fragments tend to be aligned parallel to the slope down which the material has slumped rather than to the original direction of ice movement. Ultimately observations of the type made by Boulton should permit a much fuller and more satisfactory interpretation of complex Pleistocene successions than is possible at the present time.

Till plains One of the most distinctive products of a continental ice-sheet is the relatively featureless plain underlain by a thick sheet of glacially-laid debris. Most of the deposit consists of lodgement till, but immediately after ice withdrawal there is often a mantle of supraglacial detritus giving irregular but subdued topography. This superficial layer is the first to be removed by later erosion and so is absent from old till plains. Where well preserved, as in the Mid-West of the United States, such plains can be remarkably smooth and cover many thousands of square kilometres. However, stream erosion will gradually dissect the original surface and eventually erode down to the underlying bedrock. In this way the once continuous cover is reduced to more or less isolated interfluve cappings, such as can be seen today in the glaciated lowlands of midland and eastern England (Fig. 12.10).

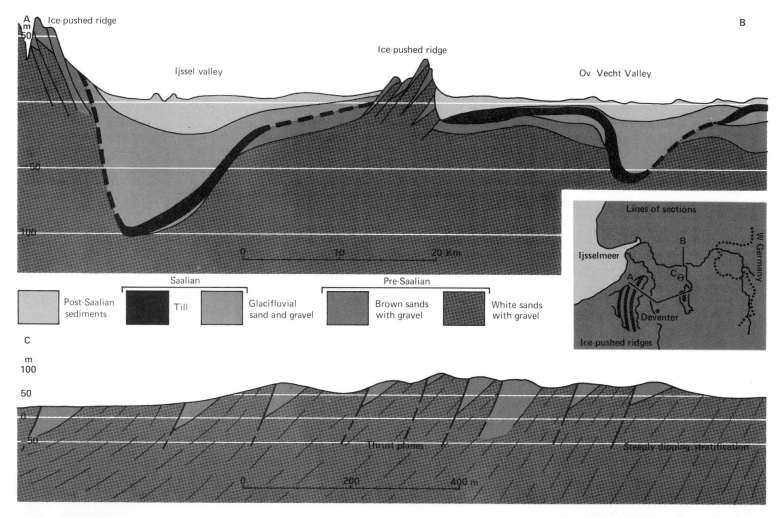

Fig 12.12 Sections through the Pleistocene deposits of the eastern Netherlands (after de Jong). Section A-B portrays two deep ice-scoured basins with intervening ice-pushed ridges. Section C illustrates typical imbricate structures associated with the ice-pushed ridges.

A
m
50
Ice-pushed ridge

B

Ijssel valley

Ice-pushed ridge

Ov. Vecht Valley

50

100

0   10   20 Km.

Saalian   Pre-Saalian

Post-Saalian sediments   Till   Glacifluvial sand and gravel   Brown sands with gravel   White sands with gravel

Lines of sections

Ijsselmeer

B

C

W. Germany

Deventer

Ice-pushed ridges

C

m
100

50

0

50

Thrust planes

Steeply dipping stratification

0   200   400 m

Fig 12.13 Schematic diagram of sedimentation along the margin of a wasting ice-mass.

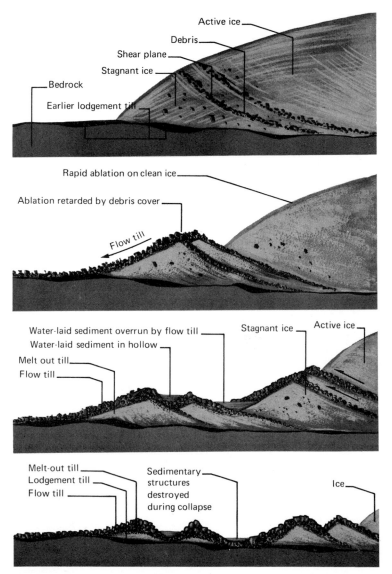

The deposits underlying a till plain often reveal a complex **263** history of accumulation. Beds of contrasting lithology and provenance may succeed each other with almost bewildering frequency. Moreover, since the surface relief is generally less than that of the buried topography the total thickness tends to be greatest over the former valleys and least over the interfluves. Summits in the bedrock landscape may even protrude through the glacial deposits to form local eminences rising above the till plain. Individual beds when traced from valley to interfluve commonly thin out or disappear entirely and this may present formidable problems of correlation. It should also be emphasized that, although by definition the dominant surface material is till, the underlying deposits often include thick sequences of water-laid sediment, particularly along the lines of former valleys. It appears that the long history of deposition has usually involved a progressive reduction in relief and that the smooth till plain is merely a culmination of this trend.

Given the very small quantity of subglacial detritus beneath most modern ice-sheets, the requisite volume of lodgement till could only accumulate after prolonged deposition. The converse argument is equally valid; total stagnation of an ice-sheet would add a relatively insignificant layer of debris once melting was complete. Certain conditions must therefore conduce to the release of the subglacial load from an actively moving ice-mass. The most important is likely to be substantial basal melting which results in the transported fragments coming more and more into contact with the underlying surface. The frictional drag progressively increases until it surpasses the tractive force of the ice; deposition is then accomplished. A minor variant may involve repeated stagnation of the heavily laden basal ice. It is possible that enhanced friction with the floor leads to shear planes developing in the more easily deformed clear ice above the debris-charged sole; continued melting will later release the detritus from the immobilized basal layer. If the general thesis is correct, thermal conditions in the subglacial zone are obviously crucial. During the history of an ice-sheet, conditions favouring deposition of lodgement till are likely to migrate over wide areas of the glaciated terrain; in general till plains are observed to be best developed on low ground some distance from the maximum limit of ice advance.

264	Difficult problems of interpretation are presented by the stratigraphy of till sheets. A single sheet is often divisible into layers, each characterized by a distinctive assemblage of erratics. There has sometimes been a tendency to assume that each layer represents a separate glaciation, but further supporting evidence is needed to justify such an assumption. A single glacial episode has often witnessed major shifts in the direction of ice movement with resultant changes in till composition. If sufficient sections are available it may be possible to demonstrate the interdigitation of till types in a way that leaves no doubt about their broad contemporaneity. This is true for instance in the English Midlands where, during one glaciation, initial ice from the north-west was directly replaced by ice from the north-east.

The earlier advance laid down a reddish till with a predominance of Triassic erratics preferentially aligned in a north-west–south-east direction. Occasional lenses within this red till contain abundant chalk and flint indicating that, early in the depositional sequence, there was occasional transport of erratics from the north-east. Within a vertical interval of a few metres but with much interleaving of the deposits, the red till finally passes up into a greyish till bearing an immense quantity of chalk and other erratics with a preferred north-east–south-west orientation.

Analogous variations in many other areas and in tills of many different ages can only be satisfactorily explained as the result of shifts in the direction of ice movement. Such shifts probably derive from changes in the pattern of ice accumulation, with

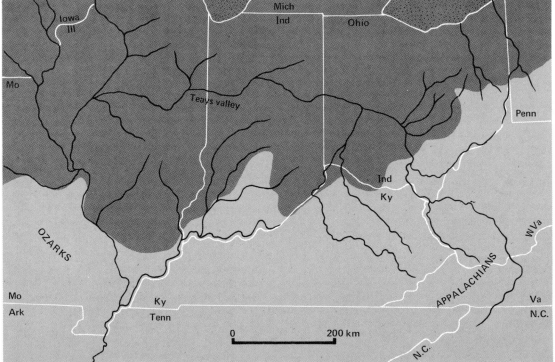

Fig 12.14 The pre-glacial Teays drainage system prior to its destruction by sedimentation at and beneath the margin of successive Laurentide ice sheets (note that lakes Michigan and Erie are inserted solely for purposes of location). The inset map shows the modern drainage pattern of the same region.

Glaciated area

different centres of dispersal achieving temporary dominance. This is most apparent around the margins of upland areas situated beyond the main centres of ice accumulation. At the onset of a glaciation, valley glaciers from these upland regions debouch freely on to adjacent lowlands. However, when the main ice-sheet arrives it constricts the locally generated ice. In consequence the till around the flanks of the upland will tend to contain abundant local material at the base, replaced upwards by more far-travelled erratics marking the arrival of the main ice-sheet. The situation may conceivably be reversed during the declining phase of a glaciation with the production of an overlying till once more rich in local erratics.

With the potential addition of supraglacial tills and outwash during temporary melt phases, it is easy to see how an extremely complex succession can be built up in the course of only one major glaciation. Nevertheless, the thick sequences beneath certain till plains doubtless represent the product of several independent glaciations. The finest examples are probably afforded by the till plains of the American Mid-West. Over vast areas these are underlain by deposits averaging about 30 m in thickness but with the infill along old buried valleys reaching over 150 m. The sediments belong to at least three separate glaciations. This is indicated by interglacial materials and fossil soil profiles at several horizons within the sequence. Such huge accumulations are fully capable of obliterating earlier drainage systems; boreholes supplemented by geophysical investigations have disclosed beneath Ohio, Indiana and Illinois a buried valley pattern totally different from that of today (Fig. 12.14). The major element in this ancient system was the Teays valley which ran westwards across the centre of the three States to the Mississippi. During an early ice advance its lower course was completely blocked and more than 120 m of glacial deposits laid down. A new valley was excavated during the succeeding warm period but when the ice readvanced this in turn was filled with glacial debris. The whole cycle was repeated a third time, but on this last occasion a new marginal drainage route was initiated along the present Ohio valley leaving the Teays system deeply buried beneath the till plain.

Drumlin fields Drumlin fields resemble till plains in being composed primarily of glacially deposited materials, but differ from them in that the extremely subdued surface relief is replaced by distinctive patterns of low, streamlined hills. An individual drumlin may vary in size from a swell only a few metres high to a hill that rises some 50 m above adjacent hollows. Its length is commonly between 1 and 2 km and its width about 0·5 km. A small isolated drumlin might prove difficult to identify as such, but the great majority occur in swarms and in this form constitute a prominent glacial landform. One of the greatest concentrations in the world is in New York State south of Lake Ontario where some 10 000 drumlins occur within an area 250 km by 60 km (Fig. 12.11). Other major swarms are found in Wisconsin and New England, while in the British Isles rather smaller groups are found in northern Ireland and the Vale of Eden.

Drumlins generally lie with their long axes parallel to the inferred direction of ice movement. Combined with an approximately ellipsoid plan, this orientation has encouraged several workers to enquire how far they conform to an ideal streamlined form. It has been pointed out that, as in the cross-section of many aircraft wings, the maximum width is situated about three-tenths of the way from the front to the rear giving a relatively blunt onset side and a longer tapering tail. If the analogy is correct it suggests that drumlins will tend to be more elongated where moulded by rapidly moving ice, although many additional factors may tend to obscure the relationship.

Another technique that can profitably be employed in the study of drumlins is till fabric analysis. In Minnesota Wright found that erratics tend to lie with their long axes parallel to the alignment of the drumlin. The dip of the axes, however, was noticeably steeper than that recorded from most lodgement tills; the mean angle was 23° with the majority of stones dipping up-glacier. Wright suggested that the high dip may be related to steep shear planes within unusually fast-moving ice, and this seems to afford one possible explanation for the development of a drumlin field rather than a featureless till plain.

It has sometimes been maintained that drumlins develop where the basal ice is very unevenly laden with debris. Clean ice will deform more readily than that which is heavily encumbered with sediment. The latter will be retarded by friction with the bedrock floor and will be the first part of the sole to be completely stabilized. The nearby clean ice will continue to move past and so mould the accumulating till into streamlined forms. An interesting variant of this hypothesis attaches particular importance to the physical properties of the subglacial load, suggesting that it may be regarded as a dilatant clay-water system. When dilated it flows readily with the overlying ice, but if the local stresses fall below a certain critical level, as they may do in the lee of an obstruction, the material will collapse into a more stable state. No longer mobile, it will be by-passed by that section of the load in which local stresses have been either maintained or even increased. The effect of the enlarged obstruction will be cumulative so that a bigger streamlined form may evolve by accretion.

Finally it is worth noting that drumlins might conceivably be viewed as the end member of a series having, at its other extreme, delicate flutings less than 1 m high. These are quite common on recently deglaciated till surfaces where they may be seen trending parallel to the direction of ice movement, often in the lee of an obstacle such as a large boulder. They are, however, too fragile to persist for any lengthy period after ice withdrawal. Intermediate forms include long narrow ridges of till which, in North Dakota, are recorded as attaining over 1 km in length but only 50 m in width and 5 m in height; like normal drumlins and many flutings they are highest near their onset ends and taper slowly in the lee direction.

End moraines End moraines may be defined as ridges of glacially deposited material accumulating along the terminal margins of active ice-masses. They can vary in height from less than 1 m to well over 100 m, the larger examples generally being associated with valley glaciers rather than ice-sheets. Across lowland regions end moraines may be traceable for hundreds of kilometres (e.g. Fig. 13.7) but their continuity is usually broken by gaps through which meltwater flowed either at the time of accumulation or during later ice withdrawal. Many different processes may be involved in their formation. Often considerable quantities of water-laid debris contribute to the total volume of an end moraine, but if stratified sediments of that type are in the majority the feature is better termed a kame complex. From a detailed study of Icelandic moraines Okko concluded that two major depositing mechanisms might be recognized which he termed dumping and pushing. Recent investigations have suggested that a third, known as squeezing, needs to be added. In many instances these are complementary, with large moraines the result of the three mechanisms acting in concert.

For a large feature to be formed by dumping there must be prolonged stability of the ice front constantly bringing forward new material. The exceptionally high moraines associated with certain valley glaciers are probably to be explained by the very large loads carried by the ice, and this in turn may be linked with the erosion of Zungenbecken a short distance higher upvalley. Several workers have described one way in which dumping takes place. This involves upward movement of debris along shear planes in the marginal ice. Differential ablation then produces surface irregularities down which detrital material slumps to accumulate as a wedge at the foot of the slope. Although melt-out till may be added from dirty ice beneath the debris cover, the ice-cored moraine formed in this way is unlikely to contain any large volume of sediment. Moreover, a trough developing in clear ice immediately upvalley from the outcrop of the shear planes tends to isolate the glacier terminus from any further supply of detrital material (as for example in Fig. 12.13). To build up a substantial end moraine requires the whole process to be repeated many times, but this is not at all unlikely in view of the constant minor fluctuations that characterize almost any active glacier snout.

Even at a virtually stable ice-front, short-term pulses of advance and retreat frequently disturb previously laid sediments. The action has been likened to that of a bulldozer, the ice pushing debris before it into a prominent ridge. The internal structure of moraines often reveals that such disturbance has taken place. Faulting, thrusting and imbrication may all be present, while in Iceland recent moraines produced by glacial pushing even contain contorted peat beds.

The terminal margins of many temperate glaciers are underlain by detritus saturated with meltwater. Even just beyond the ice edge the debris may contain so much water that it is impossible to walk on it without sinking in to a depth of 0.5 m or more. Such material deforms readily under stress and the over-burden of ice may squeeze it up into crevasses and tunnels; an occasional reticulate pattern of till ridges left by melting of stagnant ice is believed to reflect this process. Development of end moraines may in part be due to extrusion of similarly saturated debris from beneath the snouts of active ice-masses. Price has described individual ridges betwen 1 and 4 m high that are believed to have formed in this way. He analysed the orientation of pebbles in the till and was able to show that their long axes are preferentially aligned at right-angles to the ridge crests. On both sides of the crest he found that the pebbles tend to dip up-glacier, suggesting that the ridges constitute a single dynamic unit and are not the product of slumping. In some localities the ridges appear to be annual features, possibly attributable to summer meltwater reducing the shear strength of the subglacial debris well below its normal winter value. With a stable ice front it is not difficult to envisage a relatively thick morainic accumulation building up in this way.

Whilst most end moraines have developed along the margins of ice-masses terminating on dry land, others have formed where the ice ends in either ponded water or the sea. Termed cross-valley moraines, these asymmetric ridges have been the subject of detailed study by Andrews in Baffin Island where he invokes a particular form of squeezing as the dominant process. On the gentler proximal slopes the stones in the till are preferentially aligned normal to the ridge crest and dipping up-glacier. On the steeper distal side the preferential alignment is much weaker. Andrews suggests that the moraines evolved at the foot of a high ice cliff where seepage of water caused basal melting. Semiliquid debris produced in this environment was squeezed by the pressure of ice into the lake. On the proximal side its movement was closely controlled by the ice and a marked orientation of the stones was imposed, whereas on the distal side greater freedom of movement resulted in significantly weaker orientation. Andrews estimated that the cross-valley moraines he studied in Baffin Island formed beneath water more than 100 m deep.

It may finally be noted that certain moraine-like features appear to be due to ice-pushing on an exceptionally large scale. Termed ice-pushed ridges to distinguish them from conventional moraines, the material of which they are composed was never fully incorporated within the ice. Very good examples are afforded by the hilly regions of the Netherlands, many of which are formed of sediment laid down by the Rhine in early Pleistocene times (Fig. 12.12). Accumulating originally as horizontal beds of fluvial or estuarine sands close to presentday sea-level, they are now found as steeply dipping, imbricated sheets building hill ridges that locally exceed 100 m in height. As the disruption of the original bedding is known to extend a further 100 m below sea-level the scale of the displacement is huge. In general the sands retain their minor depositional structures and are wholly local in origin; however, in places igneous and metamorphic erratics from Scandinavia have been found resting on the surface, suggesting a partial burial beneath the ice edge. Although the evidence for large-scale thrusting is inescapable the precise mechanisms involved are still shrouded in uncertainty. Several early Dutch workers invoked a bulldozing action as the ice thrust its way southwards into a valley system freshly rejuvenated by the glacio-eustatic fall in sea-level. A possible contributory factor was freezing of the surface layers of the sands, concentrating the shearing motion at the base of the permafrost. However, severe doubt must exist regarding the adequacy of this explanation, and recently increasing attention has been paid to the possible role of porewater pressure in facilitating ice pushing. As indicated in Chapter 7, water pressure is a major determinant of the shear strength of porous materials. Several factors can generate exceptionally high pressures beneath the margin of an ice-sheet. Advance of the ice may block previous outlets for groundwater discharge, and by imposing reversed directions of flow raise the pressure of interstitial water; permafrost beyond the ice edge may similarly act as a barrier to the escape of groundwater with comparable results. Where ice overrides a succession of poorly lithified materials, consolidation will tend to expel water from the fine-grained sediments and raise the pressure in the coarse-grained, more permeable strata. Yet the most potent factor of all, accentuating those already mentioned, is the production of meltwater at the base of the ice. Measurements at the sole of several glaciers have revealed pressures in excess of 10 kgf cm^{-2} and these can certainly induce a very significant reduction in the shear strength of underlying permeable sediments. The most favourable conditions are likely to be those where permeable strata are confined both laterally and vertically by impermeable beds.

Moran has argued that large-scale block movement is probably a more important factor in the construction of ordinary end moraines than has formerly been recognized. This, however, is difficult to prove in the absence of detailed stratigraphic sections. Although the ice-deformed ridges in the Netherlands are clearly distinguishable from the more traditional type of end moraine, intermediate forms also appear to exist. Some thick accumulations of glacial debris include huge masses of translocated bedrock. At Cromer on the north Norfolk coast blocks of chalk exposed in coastal sections may be as much as 500 m in length. Other well-known occurrences are found in Denmark and along the southern coast of New England. In each of these instances the bedrock masses are enveloped in normal till and the term moraine may appropriately be used for the topographic form. The fact that the best authenticated examples have a coastal location almost certainly reflects the relatively full sections provided by marine cliffs. On the basis of deep boreholes, several workers have suggested that inland moraines in both North America and Europe may include large displaced slices of bedrock or even of earlier glacial deposits.

Lateral moraines Although lateral moraines are very prominent along the margins of many modern Alpine glaciers, they rarely form ridges of comparable size in formerly glaciated areas. One factor is undoubtedly the ice core that constitutes the bulk of so many presentday lateral moraines. A second is the rapid erosion to which a lateral moraine is subject once it has been deposited on a steep mountain slope. This is evident in the deep gullying that often characterizes moraines stranded a short distance above the present ice surface. Most of the debris in a lateral moraine comes from rock fall rather than glacial erosion. Accumulating on the margin of the ice it reduces the local rate of ablation, and if the glacier is shrinking the clear ice near the centre will downwaste at a faster rate than that at the edges. Inversion of the relief will ensue, an ice-cored ridge being left in place of the original glacier surface sloping towards the valley wall. When the ice core finally melts, a line of debris will be left marking the earlier longitudinal edge of the glacier. In this way lateral moraines may be particularly helpful in the reconstruction of former glacier gradients.

ANDREWS, J. T. and B. B. SMITHSON, 1966. 'Till fabrics of the cross-valley moraines of north-central Baffin Island, N.W.T., Canada', *Bull. geol. Soc. Am.*, 77, 271–90.

BATTLE, W. R. B., 1960. 'Temperature observations in bergschrunds and their relationship to frost shattering' in *Norwegian Cirque Glaciers* (Ed. W. V. Lewis), R. Geogr. Soc. Res. Ser. No. 4.

BOULTON, G. S., 1972. 'The role of thermal regime in glacial sedimentation' in *Polar Geomorphology*, Inst. Brit. Geogr., Spec. Pub. No. 4.

BOULTON, G. S., 1972. 'Modern Arctic glaciers as depositional models for former ice sheets', *J. geol. Soc. Lond.*, 128, 361–93.

BOULTON, G. S., 1974. 'Processes and patterns of glacial erosion' in *Glacial Geomorphology* (Ed. D. R. Coates), State Univ. New York.

BOULTON, G. S. and D. L. DENT, 1974. 'The nature and rates of post-depositional changes in recently deposited till from south-east Iceland', *Geogr. Annlr.*, 56A, 121–34.

CHORLEY, R. J., 1959. 'The shape of drumlins', *J. Glaciol.*, 3, 339–44.

CLAYTON, K. M., 1974. 'Zones of glacial erosion' in *Progress in Geomorphology*, Inst. Brit. Geogr. Spec. Pub. No. 7.

FLINT, R. F., 1971. *Glacial and Quaternary Geology*, Wiley.

GUTENBERG, B., et al., 1956. 'Seismic explorations on the floor of Yosemite valley, California', *Bull. geol. Soc. Am.*, 67, 1051–78

HAYNES, V., 1968. 'The influence of glacial erosion and rock structure on corries in Scotland', *Geogr. Annlr.*, 50A, 221–34.

KAMB, B. and E. LACHAPELLE, 1964. 'Direct observation of the mechanism of glacier sliding over bedrock', *J. Glaciol.*, 5, 159–72.

LEWIS, W. V., 1960. *Norwegian Cirque Glaciers*, R. geogr. Soc. Res. Ser. No. 4.

LINTON, D. L., 1949. 'Watershed breaching by ice in Scotland', *Trans. Inst. Brit. Geogr.*, 17, 1–16.

LINTON, D. L., 1959. 'Morphological contrasts of eastern and western Scotland' in *Geographical Essays in Memory of Alan G. Ogilvie* (Ed. R. Miller and J. W. Watson), 1959.

LINTON, D. L., 1963. 'The forms of glacial erosion', *Trans. Inst. Brit. Geogr.*, 33, 1–28.

LINTON, D. L. and H. A. MOISLEY, 1960. 'The origin of Loch Lomond', *Scot. geogr. Mag.*, 76, 26–37.

MANLEY, G., 1959. 'The late-glacial climate of North-west England', *Lpool and Manchr geol. J.*, 2, 188–215.

MATTHES, F. E., 1966. *The incomparable valley*, Univ. Calif. Press.

McCALL, J. G., 1960. 'The flow characteristics of a cirque glacier and their effect on glacial structure and cirque formation' in *Norwegian Cirque Glaciers*, R. geogr. Soc. Res. Ser. No. 4.

MORAN, S. R., 1971. 'Glaciotectonic structures in drift' in *Till: A Symposium* (Ed. R. P. Goldthwait), Ohio State Univ. Press.

NYE, J. F., 1952. 'The mechanics of glacier flow' *J. Glaciol.*, 2, 82–93.

OKKO, V., 1955. 'Glacial drift in Iceland. Its origin and morphology', *Comm. geol. de Finlande, Bull.*, No. 170.

PRICE, R. J., 1970. 'Moraines at Fjallsjokull, Iceland', *J. Arct. Alp. Res.*, 2, 27–42.

RAY, L. L., 1974. 'Geomorphology and Quaternary geology of the glaciated Ohio river valley – a reconnaissance study', *Prof. Pap. U.S. geol. Surv.*, 826.

SMALLEY, I. J. and D. J. UNWIN, 1968. 'The formation and shape of drumlins and their distribution and orientation in drumlin fields', *J. Glaciol.*, 7, 377–90.

STRAW, A., 1968. 'Late Pleistocene glacial erosion along the Niagara escarpment of southern Ontario', *Bull. geol. Soc. Am.*, 79, 889–910.

References

SUGDEN, D. E., 1968. 'The selectivity of glacial erosion in the Cairngorm Mountains, Scotland', *Trans. Inst. Brit. Geogr.*, 45, 79–92.

SUGDEN, D. E., 1974, 'Landscapes of glacial erosion in Greenland and their relationships to ice, topographic and bedrock conditions' in *Progress in Geomorphology*, Inst. Brit. Geogr. Spec. Pub. No. 7.

SVENSSON, H., 1959. 'Is the cross-section of a glacial valley a parabola ?', *J. Glaciol.*, 3, 362–3.

UNWIN, D. J., 1973. 'The distribution and orientation of corries in northern Snowdonia, Wales', *Trans. Inst. Brit. Geogr.*, 58, 85–98.

WEST, R. G. and J. J. DONNER, 1956. 'The glaciations of East Anglia and the East Midlands : a differentiation based on stone-orientation measurements of the tills', *Q. J. geol. Soc. Lond.*, 112, 69–91.

WRIGHT, H. E., 1957. 'Stone orientation in the Wadena drumlin field, Minnesota', *Geogr. Annlr.*, 39, 19–31.

Selected bibliography

Two modern texts with very extensive bibliographies are C. Embleton and C. A. M. King, 'Glacial geomorphology', Arnold, 1975 and R. J. Price, 'Glacial and fluvioglacial landforms', Oliver and Boyd, 1973. An even more recent survey of the same field is provided by D. E. Sugden and B. S. John, 'Glaciers and landscape', Arnold, 1976.

The comments at the opening of Chapter 12 about the difficulty of observing glacial erosion and deposition in action apply with almost equal force to glacifluvial processes. Whilst drainage around the margins of modern ice-sheets has proved a fruitful source of study, subglacial patterns of water movement are still very imperfectly known. In general investigations of modern ice-sheets have tended to underline the potential importance of basal meltwater as an eroding and depositing agent. Similar conclusions may be drawn from a study of the forms and deposits in previously glaciated areas, and it is with these that the present chapter is primarily concerned.

Glacifluvial erosion

The term meltwater commonly includes not only drainage resulting directly from ice melting but also ordinary stream drainage reaching the ice-mass from surrounding bedrock areas. The routes it follows may be supraglacial, englacial, subglacial or marginal. In all these situations the dominant characteristic of the flow pattern is likely to be its changeability. This arises in the first instance from the mobility of the ice itself. As the ice moves crevasse systems are modified and their relationship with the adjacent rock surface slowly altered. Original routeways over and through the ice are closed while new ones become available. A second cause of change is fluctuating discharge with large seasonal oscillations and, at least in summer, smaller diurnal variations. One effect of altered pathways and contrasting discharge is very uneven flow velocities. At times the water is accelerated to exceptionally high speeds and this may be the cause of the beautifully smoothed and rounded depressions, often several metres deep, which occur in considerable numbers in Scandinavia. Known as p-forms these have been attributed to cavitation erosion (p. 168) at points where water in contact with the rock attains unusually high velocities. Whatever the cause of p-forms, and there have been several conflicting views, there is little doubt regarding the general effectiveness of meltwater as an erosive agent. The channels to be described below are eloquent testimony since they were almost certainly fashioned in a relatively short period of time.

In early studies of formerly glaciated areas most meltwater erosion was assumed to have taken place beyond the actual edge of the ice. The ice was regarded as an impenetrable barrier with water either channelled along its margins or else impounded in ice-dammed lakes. The most influential of these early investigations was that undertaken by Kendall in the North York Moors. In addition to simple 'in-and-out' channels excavated by marginal streams, he identified a series of channels believed to have been cut by water overflowing from ice-dammed lakes. Inherent in this conception was the impenetrability of the ice dam with lake overspill directed across the lowest point in the surrounding watershed. Often this point was where the ice abutted against the valley wall, in which case a 'marginal overflow' was initiated; occasionally it was a separate bedrock col, in which case a so-called 'direct overflow' was initiated. This simple model formed the basis for dozens of further studies in the next 50 years or so. It is strange to recall that, even before Kendall wrote his paper, doubts had been expressed regarding the ability of ice to act as a total barrier to the passage of water. For instance, Prestwich in discussing the parallel roads of Glen Roy had pointed out that the Marjelensee in Switzerland regularly drains through the ice and not by a surface overflow. Such reservations went largely unheeded as thousands of spillways from supposed ice-dammed lakes were identified. The diagnostic criteria most commonly quoted were steep sides, flat ill-drained floors, and apparently underfit streams. However, by themselves, such features do not demonstrate an overspill origin. In 1949 Peel showed that the morphology of two channels in Northumberland, earlier claimed as of the direct overflow type, is inconsistent with that interpretation. The anomalous aspect is the long profile which in both cases is conspicuously hump-backed, in one instance rising over 10 m above the supposed intake level (Fig. 13.1). Various explanations were offered in terms of the Kendall hypothesis, notably a reversal in the direction of flow at different phases of the Pleistocene glaciation. Whilst this might conceivably have happened in a few cases, it could scarcely be of general application and when up-and-down channels were reported from many localities it became apparent that an alternative explanation must be sought.

Meanwhile in Scandinavia studies of meltwater erosion were

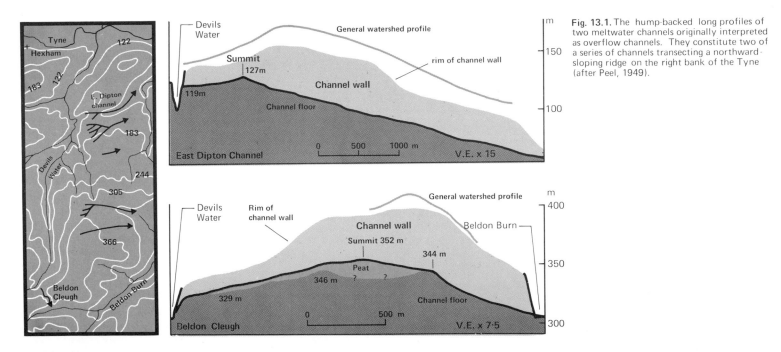

Fig. 13.1. The hump-backed long profiles of two meltwater channels originally interpreted as overflow channels. They constitute two of a series of channels transecting a northward-sloping ridge on the right bank of the Tyne (after Peel, 1949).

following a rather different line. As early as 1945 Mannerfelt was relating channels in Sweden to the patterns of meltwater flow recorded on modern glaciers in Alaska and Spitsbergen. He placed much greater emphasis on marginal and sub-glacial routeways than on ice-dammed lakes and overflow channels. One implication of his work was the inadequacy of the criteria then employed in Britain for identifying lakes and associated spillways. Sissons recognized this shortcoming and largely at his instigation a radical revision of much of the British evidence was undertaken. It is now generally acknowledged that many features earlier interpreted as overflow channels did not in practice originate as spillways from ice-dammed lakes. For present purposes a basic four-fold division of meltwater channels will be employed: marginal and sub-marginal channels, subglacial channels, spillways and coulees.

Marginal and submarginal channels Whilst these two positions of meltwater flow can be readily distinguished along the edge of a modern glacier, it may be difficult to decide in which situation a relict channel was formed once the ice has melted. In theory a submarginal channel may contain debris which collapsed into it during final wastage of the ice, but the evidence is rarely very convincing. It has been argued that truly marginal channels are likely to have lower gradients than those that are submarginal, but the distinction is often difficult to draw and in any case the reasoning is open to question. Another possible criterion is the presence of abrupt changes in direction. Unlikely in marginal streams, these may be very characteristic of submarginal streams in which flow is guided by crevasses penetrating to the base of the ice. Water will alternately be forced along a hillside and then allowed to plunge rapidly downslope, producing an oblique zig-zag course. In recent years the likelihood of a submarginal origin for meltwater features has been increasingly recognized and swarms of shallow channels, often only a few metres deep, scored into the slopes of Scottish mountains have been so interpreted (Fig. 13.2).

Small channels can be excavated extremely rapidly. In some regions it has even been suggested that fresh channels may be eroded annually, each member of a series being the result of summer melt from a rapidly retreating ice front. However, doubts remain about such an interpretation since it appears that several parallel channels can function simultaneously, each occupying a trough between ice-cored ridges that result from differential ablation of dirty ice bands. Moreover, from descriptions of modern glaciers there seems little doubt that submarginal streams occasionally form an anastomosing network; it consequently becomes very difficult to infer details of the former ice edge from a suite of abandoned channels.

A rather different type of marginal drainage occurs along the frontal edge of a glacier or ice-sheet. The simplest case is that of a valley glacier in which small streams emerging from tunnels near the valley wall flow along the ice edge to join the major meltwater stream issuing centrally from beneath the snout. Although such watercourses sometimes cut shallow channels they generally leave little trace of their existence once the glacier has withdrawn. Around the lobate margin of an ice-sheet a similar pattern of drainage converging on a single master stream may be found. This is particularly true on lowlands where earlier arcuate end-moraines often guide the meltwater towards one outlet. The detail of the stream courses is much influenced by the dominant wastage processes. For example, the water may be channelled between ice-cored moraines, occasionally shifting its course as continued undermelting leads to collapse of individual ridges. It is characteristic of meltwater streams in such an environment that they are constantly changing paths, abandoning old channels for new as lower routes become available. Fresh channels can be cut with remarkable rapidity in the unconsolidated sediments, and in an area deglaciated for only a decade, it may be extremely difficult to plot all the various routes followed by meltwater in those few years. A further complication arises from the occasional meltwater stream that, after issuing from a tunnel and flowing along the ice margin, actually turns back under the ice edge.

Much larger marginal channels are found where sizeable rivers, nourished by precipitation in extra-glacial areas, have been diverted along the edge of Pleistocene ice-sheets. Examples occur in both Europe and North America and have been termed Urstromtäler from their frequent occurrence on the north European plain. In some instances the routes initiated along the ice margin are still occupied by large post-glacial

274 rivers; the mouth of the Rhine was diverted westwards to its present outlet by the Scandinavian ice-sheet, while the modern course of the Ohio is believed to have been initiated around the edge of the Laurentide ice-sheet (Fig. 12.14).

Subglacial channels Submarginal meltwater courses are obviously subglacial in the literal sense of that word, but are commonly differentiated because they owe many of their characteristics to proximity to the ice edge. When water penetrates deeper beneath the ice it may appropriately be termed subglacial and the linear depressions that it excavates in that position may be referred to as subglacial channels. Opportunities for direct observations of streams beneath an ice-mass are extremely rare. Water has been recorded emerging with great violence from beneath many ice margins, and in a number of cases abandoned subglacial tunnels have later been explored. The water issuing from such a tunnel system may be derived from a variety of sources. Much of it is likely to be marginal and submarginal drainage that has penetrated to greater depths within the ice. Another part will be supraglacial water disappearing directly down moulins and crevasses on the ice surface. A third contributory source may be ice-dammed lakes

Fig 13.2. Part of a swarm of shallow meltwater channels mapped by Sissons in the central Grampians of Scotland. Most of the channels are only 1–4 m deep but can often be traced for a kilometre or more. They are believed to have been formed in a brief interlude during the decay of a small ice-cap that existed in the period 10 800-10 300 years B P (after Sissons, 1974)

draining periodically either through or beneath the ice. A final part will come from melting either at the sole of the glacier or along the tunnel walls themselves.

So much can be inferred from seeing water disappear into the ice and emerge at the snout. The extent to which the course between sink and resurgence is englacial or subglacial is difficult to establish. In formerly glaciated areas meltwater channels can sometimes be seen running directly down a hillslope. Known as subglacial chutes, these are believed to have been scored by marginal drainage plunging beneath the ice and remaining in contact with the valley wall over a vertical interval of up to 100 m. If such a long chute were formed in a single episode it would imply that water may flow in subglacial channels beneath ice of at least that thickness. Several writers have quoted evidence supporting an even higher figure, but the mechanism by which tunnels are formed at such great depths remains a puzzle. Although water may be readily envisaged flowing through the base of crevasse systems, gradually enlarging them into tunnels, few crevasses have been found extending to depths of over 30 m. Glen has suggested that, as in the case of ice-dammed lakes discussed earlier (p. 240). water in a deep cleft might generate sufficient horizontal stress to cause deformation of the ice. However, the water column would need to be 150 m deep to be really effective so that the argument, although possibly appropriate to the maintenance of very deep openings, does not by itself explain their initiation. It seems feasible that where glacier movement is placing ice under tensile stress, crevasses could be deepened if they were filled with water, but how often this happens is not known. If a figure of 100–150 m is regarded as the normal maximum depth for subglacial stream flow, effective erosion is restricted to a very narrow zone along the edge of an active continental ice-sheet. On the other hand, if a stagnant ice-sheet downwastes, much longer stream courses might be envisaged and it is not unlikely that many features ascribed to subglacial meltwater action were fashioned during such decay phases. Another factor that could well be of great importance, although very little is known about it, is the water released by subglacial melting. To judge from the amount of lodgement till, the total volume of meltwater produced in this position must be very large; whether it is concentrated into channelled flow is unknown.

The hump-backed channels earlier attributed to overflow from lakes are now generally agreed to be subglacial in origin. Many reasons may be adduced for rejecting the overflow hypothesis, but the primary reason for regarding them as subglacial is the inability of water to cut such forms except under hydrostatic pressure. The amplitude of the hump precludes an explanation in terms of normal stream processes and the conclusion must be that the water was capable of flowing uphill. It is not essential, however, for a channel to have a hump-backed long profile to be regarded as fashioned beneath the ice. The disproportionate size of some channels compared with their potential catchment as marginal features indicates subglacial development. A complex anastomosing pattern of channels may defy explanation in terms of shifting ice margins, yet be readily accounted for by large volumes of meltwater flowing subglacially. The relationship to depositional forms that develop beneath the ice can also be significant; in a number of cases channels have eskers running along part of their length. Price has stressed that an englacial drainage system may be superimposed on to bedrock in piecemeal fashion. If the water during a decay phase is constantly seeking a lower route through the ice it may eventually encounter a ridge and begin to cut a channel down the distal slope. Often the meltwater stream will slip laterally until it finds a col through the ridge; the channel carved down the distal slope may then appropriately be termed a 'col gully'. With progressive melting new lower pathways for the meltwater become available. The level of flow may be lowered more rapidly in the ice than across the rock spur and in order to maintain its route the water must begin to flow uphill. By virtue of hydrostatic pressure it will continue to do so until a completely new lower passage becomes available. Such a sequence may explain the tendency for hump-backed channels to consist of short upward and relatively long downward segments.

Analogous in many respects to the subglacial channels just discussed are the remarkable slot-like gorges that sometimes transect rock bars on the floors of glacial troughs. The most famous example is the Aareschlucht in Switzerland where the River Aar cuts through a resistant limestone barrier in a vertical-sided gorge over 200 m deep. The rock bar itself appears to be the result of differential ice erosion, but there is no indication that it ever constituted an unbreached barrier impounding a lake on its upstream side. It seems much more likely that subglacial meltwater initiated the gorge while the whole valley floor still lay beneath the ice.

Of very different form are the much larger features known as tunneldale in Denmark and as Rinnentäler in northern Germany. Excavated in a sheet of glacial drift but locally penetrating the underlying bedrock, these tunnel valleys are commonly 1–2 km wide, up to 100 m deep and in some cases traceable over a distance of 75 km (Fig. 13.3). They tend to have steep sides and fairly flat floors, but in long profile are irregular with enclosed hollows occupied by lakes. Their subglacial origin seems beyond doubt. Cut in tills and outwash of the last glaciation they sometimes have well-formed eskers on their floors. They end very abruptly, being replaced beyond the ice margin by low outwash fans deposited where the emergent meltwater lost its transporting capacity. The nature of the meltwater streams that could erode such broad channels is a puzzle. It is tempting to envisage vast streams completely filling the valleys, yet this would require an ice roof of most unlikely span and one which certainly has no counterpart at the present day. Most workers have preferred the idea that a smaller tunnel has migrated laterally, and this certainly appears more consonant with the late-stage development of eskers. It is strange that such distinctive features occur in considerable numbers in Denmark and north Germany, but have not been reported with corresponding frequency elsewhere. Attention has been drawn to certain valleys in eastern Scotland that display affinities with tunnel valleys while deep elongated pits on the floor of the North Sea have also been likened to the Danish examples.

Another type of subglacial channel is that which Woodland has termed a buried tunnel valley. This is a deep linear depression filled with glacial deposits until it has very little surface expression. Owing to its burial such a feature is difficult to distinguish from a normal stream-eroded valley subsequently filled with glacial drift. Nevertheless, as more borehole evidence becomes available it is increasingly clear that closed depressions quite unlike normal valleys are of widespread occurrence.

Woodland has documented 16 examples in East Anglia where boreholes are believed to have penetrated the fill of subglacially eroded channels (Fig. 12.10). The channels characteristically have very steep sides, even approaching the vertical in places, and extend to depths which preclude normal outfall to the sea. For example, just north of Hitchin far inland from the coast the glacial fill is over 125 m thick and its base lies more than 75 m below present sea-level. The conclusion is inescapable that the boreholes have penetrated a very deep enclosed depression. Further evidence shows that the hollow proved in this way is a particularly deep section of an undulating channel traceable over a distance of at least 20 km. Woodland stresses the morphological similarities between the buried channels in East Anglia and the tunnel valleys in Denmark. However, two striking differences need consideration. The first is the relative narrowness of the East Anglian channels, possibly a consequence of their incision into Chalk rather than drift and poorly consolidated Cenozoic rocks. The second and more important is the infill of the English examples (Fig. 13.4). This consists predominantly of water-laid sediments having obvious affinities with the tills on adjacent interfluves. In general, sand and gravel is found at the distal end of each channel, loams, silts and clays at the proximal end. In the middle reaches the finer material tends to overlie the coarser. Woodland envisages an initial erosional phase during which the deep channel is scoured out and the excavated debris is deposited as outwash beyond the ice margin. As the ice front melts back subglacial erosion continues but the distal segment of the channel is gradually filled with gravel. With further melting and downwasting the subglacial water loses its hydrostatic head and becomes so slow-moving that the proximal end of the channel is filled with silts and clays. No satisfactory explanation has yet been offered for an accumulation phase affecting tunnel valleys in East Anglia but not in Denmark, and until it has it would perhaps be wise not to stress the apparent analogies too strongly. Although East Anglia provides the greatest known concentration of buried tunnel valleys in Britain, they are also found in many other glaciated regions. In addition they have been recorded on the floor of the North Sea.

Fig 13.3 The glacial landforms of Denmark.

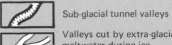	Dunes
	Post-glacial marine sediments
	Weichselian glacifluvial outwash
	Weichselian till with patches of outwash
	Areas of older i.e pre Weichselian glacial deposits
	Sub-glacial tunnel valleys
	Valleys cut by extra-glacial meltwater during ice withdrawal
	Eskers
	Limit of Weichselian ice

0   25   50 Km

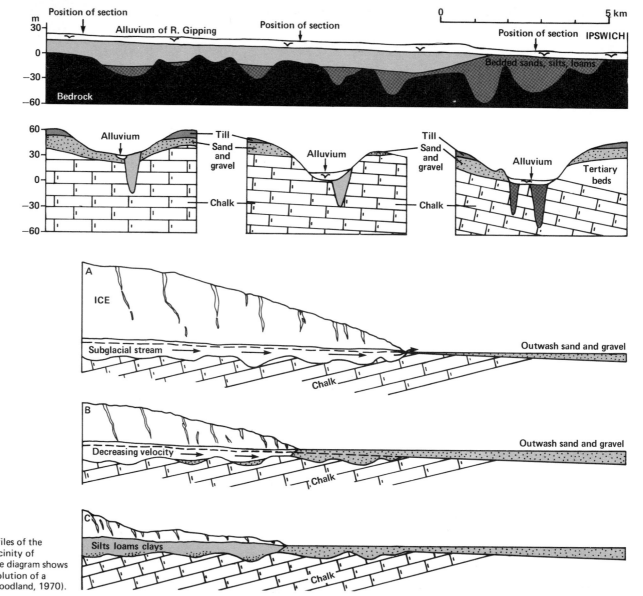

Fig 13.4. Long and cross-profiles of the buried tunnel valley in the vicinity of Ipswich. The lower part of the diagram shows hypothetical stages in the evolution of a buried tunnel valley (after Woodland, 1970).

Spillways Although most proglacial lakes empty through or beneath the impounding ice dam, others conform to the Kendall model described earlier and spill either as marginal streams along the ice edge or as direct overflows across bedrock cols. It is commonly assumed that the overflowing water will have the capacity to modify any col it occupies and to produce a relatively steep-sided flat-floored channel very like those excavated in marginal and subglacial situations. A problem then

arises regarding the criteria by which the various types of channel are to be distinguished. It has been suggested that the term overflow channel should be restricted to those cases where there is independent evidence for the existence of a lake. Such evidence might consist, for example, of either lacustrine sediments or strandline features. Of course, lacustrine sediments can accumulate in a lake that drains past the ice barrier, so that by themselves they do not prove that a nearby channel is a true spillway. More significance generally attaches to the presence of strandline features. For a durable assemblage of shoreline forms to develop the lake surface must remain at the same elevation for a relatively long period, and this seems most likely to occur where the discharge is across a bedrock col. Moreover, an old shoreline permits a relatively precise estimation of the former water-level, capable of being compared with the height and morphology of the putative spillway. The classic illustration in the British Isles is provided by the parallel roads of Glen Roy (Fig. 13.5). It was Jamieson in 1863 who initially interpreted the 'roads' as strandlines of a lake controlled in elevation by three separate overflow routes. During the first stage the lowest available outlet was a direct overflow at 351 m near the head of Glen Roy. Ice withdrawal then uncovered a col on the eastern side of the glen allowing the lake level to fall to 326 m. Further wastage permitted drainage via Glen Spean with the water surface stabilized at 261 m. Not only do the strandline and spillway heights approximately correspond, but the shore features disappear in those areas presumed to have been still occupied by ice. Similarly, a shoreline along the south-eastern edge of glacial Lake Harrison in the English Midlands can be traced for a distance of 55 km before terminating at the presumed position of the ice front (Fig. 13.6). Its elevation of 125 m is virtually identical with that of the gap at Fenny Compton through which the lake water drained, first into the Cherwell valley and thence into the Thames. A striking aspect of the Glen Roy and Lake Harrison overflow routes is that they lack what are often regarded as the diagnostic criteria of spillways. This is especially obvious in the case of the Fenny Compton gap which is broad and gentle-sided and quite unlike the conventional picture of an overflow channel. It is ironic that so many overflow channels have been wrongly identified on the

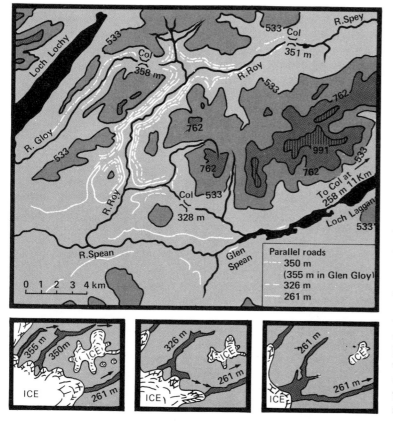

Fig 13.5 The parallel roads of Glen Roy. The diagrammatic sketches below show the three stages in the evolution of the lake system as the main ice-mass withdrew towards the south-west.

basis of their steep sides and flat, ill-drained floors, while some of the best authenticated spillways lack the supposed distinctive morphology.

The frequent assumption that any overflow route is subject to pronounced lowering needs careful qualification. A number of factors will influence the erosion of a spillway. Among the most important are the nature of the material composing the outlet, the volume of water discharging through it, the length of time involved and the original form of the col. Taking the Fenny Compton gap as an example, the present floor is composed mainly of clay, and although buttressed by thin limestone bands, would certainly not be classified as particularly resistant. The volume of water that flowed through it is unknown. Although there is no necessary correlation between lake size and overspill volume, extensive lakes with long ice-cliffs around their margins seem likely to produce large quantities of meltwater; on this basis Lake Harrison would have discharged prolifically through the gap. There is at present no way of assessing precisely how long the overflow functioned, but it cannot have been a particularly short episode since appreciable time is required for the cutting of the associated lake bench. Finally the original col can at most have been 7 m higher than at present since an alternative outlet exists at that elevation. This means that overflow from the lake must have entered a gently sloping valley which it failed to modify to anything like the degree commonly associated with other types of meltwater channel.

It is instructive to consider an example from a recently deglaciated area where the evolution of an ice-dammed lake is recorded in a succession of strandlines. One such example is provided by the southern end of the Lake Michigan basin in North America (Fig. 13.7). Here Lake Chicago, dammed against the Laurentide ice-sheet, discharged via the valley of the Des Plaines River. Successive lake levels are marked by four shorelines at heights of 18, 12, 7 and 3 m above Lake Michigan. Bretz has suggested that the highest level was stabilized because a boulder pavement at the point of outflow protected the underlying weak drift from erosion. The pavement was destroyed by an abrupt increase in discharge when water from the Lake Erie basin suddenly began escaping into Lake Chicago. The outlet was lowered some 6 m before further ice withdrawal led to temporary abandonment. Following a minor re-advance, ponded water again escaped down the Des Plaines River and it was at this stage that the 12 m shoreline received its final shaping. Once more a sudden accession of Erie meltwater lowered the outlet until a resistant bedrock sill stabilized the lake

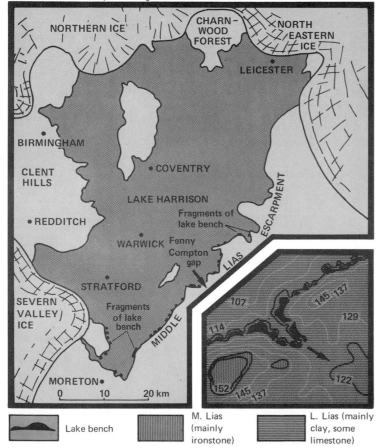

Fig 13.6 Lake Harrison close to its maximum extent as visualized by Shotton. At the time of the overflow through the Fenny Compton gap the lake may have been much reduced in size; the minimum size is presumably indicated by the extent of the lake bench. The inset shows the form and geology of the Fenny Compton gap with a notable absence of any trenching on the floor of the col.

level at the height of the 7 m shoreline. Melting later permitted overflow eastwards along the edge of the main Laurentide ice-sheet and on two separate occasions water-level fell below that of the present Lake Michigan. In time, however, isostatic rebound elevated the more northerly and easterly outlets from the Great Lakes region and the Des Plaines valley briefly functioned as an overflow for the last time. The catchment was a vast lake incorporating both the Superior and Michigan basins of the present day. The sill, however, was lowered very little during this final phase, possibly because other outlets were functioning simultaneously; it was these outlets that apparently controlled the lake level from 3 000 B.P. onwards.

The complete sequence of events at the Chicago outlet is dated to the period between 14 000 and 2 000 years ago, but for about half this time there was no actual overflow. During the spells of active discharge the spillway was lowered some 11 m in two distinct phases of incision. For much longer periods the sill remained stable, permitting strandlines to be etched on the slopes around the southern shores of Lake Michigan.

Coulees In the preceding sections attention has been directed to two common ways in which a glacially impounded lake discharges meltwater. The first involves drainage through en-glacial or subglacial tunnels, the second direct overflow across subaerial cols. A third possibility, although much rarer, is sudden destruction of the ice barrier with almost instantaneous release of the complete contents of a large ice-dammed lake. It has already been emphasized that flow through ice tunnels can produce violent floods, but the scale is small when compared with the cataclysmic events that can accompany a total collapse of the retaining dam. The term 'coulee' is appropriately used to describe the channels scoured by the escaping lake waters since this is the local name for such features in the north-western United States where the most spectacular results of a dam burst have been described. In this instance the Pend Oreille lobe of the cordilleran ice-sheet pushed southwards into Washington and impounded a major lake in the Clark Fork valley (Fig. 13.8). Known as Lake Missoula, this has left prominent strandlines at elevations up to 1 280 m and has been shown at its maximum extent to have had a superficial area of 7 500

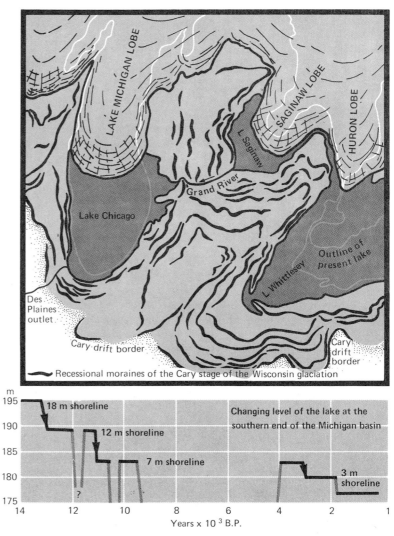

Fig 13.7 The pro-glacial drainage system that formerly drained through the Des Plaines spillway. The stage depicted is that responsible for a sudden increase in discharge as water escaped from the Erie basin via the Grand River outlet (c. 13 000 years B.P.). Below is a plot of changes in water-level around Chicago during the last 14 000 years.

km² and a total volume in excess of 2 000 km³. At the dam the water was at least 700 m deep, resting against ice that cannot have been less than 1 070 m thick. So much can be inferred from the distribution of glacial and lacustrine sediments, but the immediate cause of the ice barrier collapsing is unknown. Within the area of the lake several remarkable features indicate a sudden fall in water-level. Cols that were originally submerged are scored by deep, beautifully streamlined depressions. It is believed that as the dam broke, a steep surface gradient developed across the lake, and water started rushing towards the outlet. At each submerged gap the constriction acted like the throat of a gigantic flume with a fall in pressure sufficient to induce cavitation. By this process hollows up to 30 m deep were rapidly gouged from resistant rock. On the slopes below some of the gaps are gigantic 'ripple marks' 10 m high and with 75 m between successive crests.

However, it is in the area downstream from the ice dam that the most spectacular results are found. So great was the volume of water that sped down the valleys to the west that it spilt across divides and surged up minor tributary valleys, carrying huge boulders with it. A short distance below the dam the floodwaters were over 200 m deep and beyond Spokane they fanned out across the Columbia plateau basalts. Washing earlier loessic materials from the plateau surface, they quarried deep channels into the basalt itself. Steep-walled and irregularly floored, these are the coulees. They owe many details of form to the strong vertical jointing of the basalt, large blocks of which were ripped away by the force of the flood. Dry waterfalls with deep plunge pools characterize many of the coulees. Although the channels are frequently more than 100 m deep, no one by itself could accommodate the immense volume of water and therefore anastomosing networks were formed. Together these constitute the area known as the 'channelled scablands'.

There now seems little doubt that most of the landforms in the channelled scablands result from large-scale meltwater flooding. Yet when Bretz first advanced the idea in 1923 its acceptance was far from unanimous. In the intervening years evidence has accumulated to suggest that there was not a single inundation, but a succession of such events as on several separate occasions the water built up in Lake Missoula. In the most recent major flood, probably just over 20 000 years ago, peak discharge has been estimated at $21 \cdot 3 \times 10^6$ m³sec⁻¹ or 100 times the mean flow of the Amazon. The whole lake seems to have been evacuated in the span of a few weeks so that the fashioning of the coulees and channeled scablands was a remarkably brief episode. This may serve as a salutary reminder of the possible catastrophic origin of certain landforms. There is occasionally a tendency to assume an attitude of extreme uniformitarianism in which the potential significance of the cataclysmic event is denied. There is no fundamental reason why similar devastating floods should not have occurred in other glaciated regions, although the necessary supporting evidence has rarely been identified.

Glacifluvial deposition

Meltwater streams flowing across the surface of an ice-mass vary greatly in the amount of sediment they carry. On an ice-sheet or near the centre of a valley glacier they can be quite clear, but where the ice abuts against a rock wall they often transport considerable quantities of debris picked up from the lateral moraine. Similarly, near the frontal edge of an ice-mass the ablation moraine may supply sediment to the streams before they disappear down moulins or crevasses. The streams that gush out from beneath the ice edge are even more heavily laden. They appear very dirty in colour and it is sometimes possible to hear large cobbles clashing together in the water. Most of the material is believed to be englacial or subglacial debris picked up along the tunnel walls. A further potential source is the bedrock floor if the meltwater is actively eroding submarginal or subglacial channels. The range of particle size being transported is often very wide. The coarsest fraction tends to be deposited quite close to the ice front, but the finer silt and clay is carried long distances and gives many meltwater streams their distinctive milky appearance.

The diversity of meltwater paths provides a great variety of environments in which glacifluvial deposition can take place. A fundamental distinction may be drawn between ice-contact materials laid down on, within or beneath the ice, and pro-glacial materials deposited beyond the contemporary ice limit.

Fig. 13.8. Lake Missoula and the channelled scablands of the north-western United States.

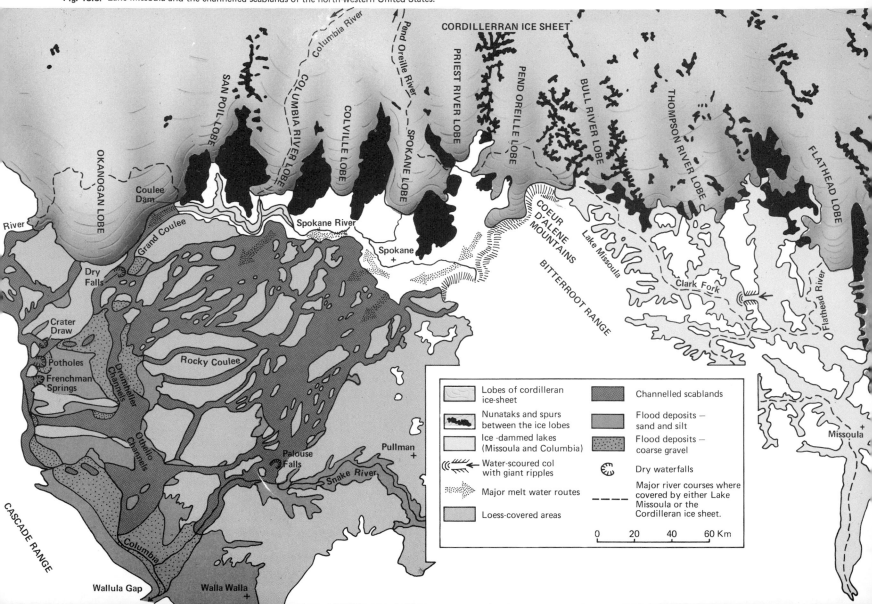

The two types grade into each other near the ice edge, while the proglacial sediments tend to lose their distinctive glacifluvial character the further their point of accumulation from the ice margin.

Glacifluvial deposits naturally display many of the features found in other water-laid sediments. They tend to be reasonably well sorted and at least rudely stratified. Dune and ripple structures are very common. They also display certain distinctive attributes of their own. A fraction of the constituent material commonly comes from sources beyond the immediate catchment but can be matched in local tills. A variable percentage of the pebbles retain traces of glacial faceting and striation. The water-laid sediments are often interleaved with lenses of till. Numerous abrupt changes in particle size reflect the frequent fluctuations in meltwater discharge. Perhaps most distinctive of all are various collapse structures resulting from decay of the ice over or against which many of the materials accumulated.

The terminology that has evolved for the classification of ice-contact glacifluvial deposits is extremely confused. In part this arises from the use of terms taken from different languages and inadequately defined before being adopted for scientific work. More fundamental is the confusion that results from mixing criteria based upon form, constituent material and presumed origin. The problem is well illustrated by the term kame. This was derived from the Scottish word 'kaim' which refers to any steep-sided ridge irrespective of origin: for instance, many sharp-crested ribs of solid rock in Scotland still retain the name. Yet when adopted to refer to a glacifluvial landform, its use was often restricted to circular mounds with ridges specifically excluded. For elongated shapes the term esker, derived from the Irish word 'eiscir', was commonly substituted. To make matters even more confused, certain workers later broadened the term esker to include any hillocky glacifluvial form and introduced the Scandinavian word 'ose' to refer to the elongated shapes. At the moment it is impossible to formulate entirely satisfactory definitions and for the present purposes it is intended to adopt those usages that seem most widely employed among English-speaking workers. The term kame will be used to designate hillocky glacifluvial deposits that owe their current form primarily to removal of the ice against or over which they originally accumulated. Esker will be used to connote an elongated ridge which is essentially the cast of a meltwater channel on, within, or beneath the ice. The Icelandic term sandur will be employed to indicate the relatively smooth surface built up by meltwater streams beyond the ice margin.

Kames Kames may exist either as isolated steep-sided mounds, or as an association of such features forming what has appropriately been called sag-and-swell topography. They range from low hillocks only a few metres high to conspicuous conical hills as much as 50 m high. They characteristically consist of sand and gravel which may be relatively well-bedded in the core of the mounds but often shows evidence of slumping to its angle of repose around the margins. Large striated boulders and till lenses are quite common within the deposit, and many kames have thin superficial cappings of water-washed till.

Kames almost certainly originate in a variety of ways. One potential mechanism is the building of steeply sloping fans or deltas along the ice edge, possibly where heavily laden meltwater streams issue into small proglacial lakes. The proximal side will at first be supported by the ice but will later collapse. Such a feature will tend to have an asymmetric outline with smooth distal slope but irregular ice-contact face. This interpretation for certain kames is supported by the internal bedding which consistently dips parallel to the outer slope of the cone.

A more common origin for kames is believed to be accumulation in an area of stagnant, downwasting ice. The term 'glacier karst' has been employed to describe the complex of sink holes and tunnels that may develop in such an ice-mass. Clayton has described the essential characteristics from his studies of the Martin River Glacier in Alaska. A cover of ablation till is required to prevent rapid melting and to allow evolution of a tunnel system below the surface. Over the outer edge of the Martin River Glacier the till is up to 3 m thick and has been accumulating sufficiently long to support 100-year-old spruce and hemlock trees. The surface of the glacier is pock-marked with funnel-shaped sinks that are typically 100 m across and locally occur at a density of 30 per square kilometre. Many are occupied by lakes and may be as much as 90 m deep. The dominant process is that of melting by water, either around the lake margins or while flowing through passages in the ice. Coalescence of sinks creates compound features over a square kilometre in extent, while tunnel collapse initiates new depressions. Sediment from the surface will either fall or be carried by meltwater into the growing cavities. After the final melting has taken place, the irregularly distributed sediments will produce a chaotic assemblage of materials and forms, closely resembling those found in many Pleistocene kame complexes.

There seems no doubt that many kames have originated under conditions of ice stagnation. Till lenses enclosed within the deposits may be explained by flowage off the adjacent ice surface as continued melting rendered them unstable. The cover of ablation till seen on many kames is probably due to superficial debris being let down from the surface during the final melt phases. Several authors have envisaged a wasting ice-mass so riddled with caves and passages that an englacial water table develops. Above the water table erosion is able to continue, but below it the dominant process is deposition. Sissons has shown that seemingly irregular kame areas sometimes have approximately accordant summits which he believes to mark the level of the englacial water table. In certain cases a succession of accordant levels may be discerned, each apparently related to an intermittently falling water table whose elevation was controlled by the heights of cols unblocked by ice wastage. Intrinsic to this argument is the view that most of the kames accumulated directly on the subglacial floor, since lowering from an englacial or supraglacial position would presumably destroy the accordance. The same reasoning applies with even greater force to the occasional flat-topped form situated in the midst of a kame complex, or even standing as an isolated hill. Termed a crevasse filling, this is usually much larger than that name seems to imply. The remarkably even summit may extend over many square kilometres until it comes to resemble a fragment of sandur. Yet the margins are clearly of ice-contact origin, consisting of glacifluvial sediments that have slipped to their angle of repose above a series of fringing kettle holes. It appears that, penetrating the complete thickness of a stagnant ice slab, there must have been an opening into which meltwater streams poured immense quantities of debris.

A third way in which kames may form is by collapse of an originally smooth sandur laid down over buried ice. As the ice melts the surface becomes pitted with irregular depressions. In some instances the process goes no further than to produce a few kettle holes dimpling the surface of an otherwise even plain. Yet in other cases it is capable of totally destroying the original surface and leaving an irregular, hummocky topography that is virtually indistinguishable from that formed within a stagnant ice-mass. Successive aerial photographs of recently formed sandar have shown conclusively that smooth plains can be converted into a jumbled assemblage of steep-sided mounds and kettles.

Under the general heading of kames, mention must also be made of kame-terraces and kame-moraines. Kame-terraces are the result of marginal deposition where the ice abuts against a sloping rock face. Part of the accumulation takes place on the rock surface and part on the ice. When the ice melts, the outer edge of the terrace collapses but the inner part retains its form largely intact. Kame-moraine is a rather unfortunate term occasionally used to denote a kame complex believed to have accumulated along a stable ice margin. The form may therefore have a similar chronological significance to an end-moraine but a totally different origin. Linear kame complexes are very common. They often appear to consist of coalescent fans or deltas laid down along the ice edge. Many display deltaic structure and appear to be truly subaqueous. A good example is afforded by the ridge of glacifluvial sediments at Galtrim in southern Ireland (Fig. 13.9). Rising 15 m above the adjacent lowland, this ridge is bounded on one side by a steep ice-contact slope, on the other by a more gentle delta front. The channels through which glacifluvial debris was fed to the ice margin are marked by a series of eskers. A thin capping of unsorted debris on the ridge crest probably represents englacial material that slumped to the base of a former ice cliff. On a much larger scale the Salpausselka ridges in Finland are believed to be glacifluvial accumulations bounded by ice-contact slopes. Locally exceeding 100 m in height and 2 km in width they consist almost entirely of stratified materials. Some workers have regarded them as deltaic sediments laid down in a shallow sea when Scandinavia was still glacio-isostatically depressed. However, Virkkala has claimed that ice-contact faces occur on both

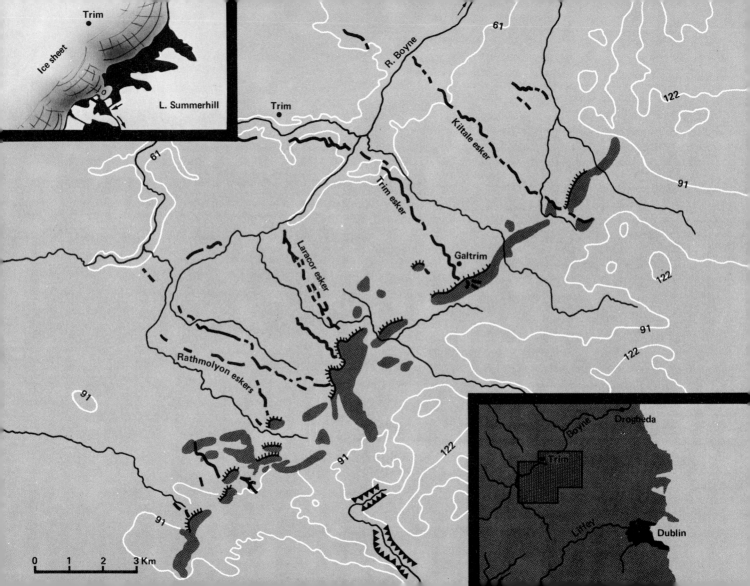

Ice sheet

Trim

L. Summerhill

61

R. Boyne

Trim

Kiltale esker

61

Trim esker

Galtrim

Laracor esker

122

91

122

Rathmolyon eskers

91

91

91

122

91

91

91

122

0   1   2   3 Km

Boyne

Drogheda

Trim

Liffey

Dublin

Fig 13.9 The linear kame-complex at Galtrim in southern Ireland with a series of feeding eskers along its north-western edge. Conditions at the time of accumulation are reconstructed in the inset at the top left (based on Synge, Proc. R. Irish Acad., 1950).

 Sand and gravel deposits of kame complex

 Prominent ice-contact faces.

Eskers feeding into kame complex

 Spillways

proximal and distal sides of the ridges; he contends that they were laid down in deep fracture zones running parallel to, but some distance from, the actual ice margin.

Eskers Eskers are among the most distinctive of all ice-contact forms. They retain their characteristically elongated, steep-sided and narrow-crested form over an immense size range. At one extreme are those that rise as much as 100 m above their surroundings and can be traced for distances of several hundred kilometres (although there are always many gaps in eskers of that length). At the other extreme are ridges less than 2 m high and traceable over distances of only a few hundred metres. In general there appears to be a simple relationship between length and height, with the more massive examples nearly always extending over longer distances. In plan eskers are usually sinuous and occasionally even meandering. They may occur singly, form a confluent pattern resembling a normal drainage net, or constitute part of an anastomosing system. The crestlines are smooth or gently undulating while the bounding slopes may approach the angle of repose of the constituent material. This can range from large cobbles to fine silt and even laminated clay, but the most common component is sand and gravel with cross-bedding indicative of the direction of water flow. The strata sometimes display an arched arrangement, although there may then be doubt whether this is a depositional structure or the result of later slumping. Many studies have been made of the

provenance of pebbles in an esker. This is a particularly interesting approach where a long esker crosses a series of distinctive rock types. Most materials are found to have travelled only a few kilometres from their point of origin, although resistant rocks may sometimes have been carried much further. In Canada esker deposits have been used as a guide to mineralization of the bedrock and a study of the 400-km long Munro esker showed that the maximum concentration of a mineral occurs 3–15 km downstream from the point of outcrop.

There are remarkably few records of really large eskers being observed in the course of formation at presentday ice margins. Many workers have recorded ridges of gravel, some of them as high as 16 m, appearing from beneath the receding snouts of glaciers, but with very few exceptions they have been ice-cored and once melting was complete they were reduced to insignificant proportions. In view of the thorough studies made during recent years along modern ice margins, it is reasonable to infer either that present conditions are not conducive to esker formation and preservation, or that they are developing in locations hidden from direct observation.

It is generally agreed that eskers originate as some form of fill along a meltwater routeway. In 1897 De Geer suggested they develop as deltaic accumulations where subglacial tunnels terminate in ponded water or in the sea. He noted that certain Swedish eskers are divisible into segments, each of which grades downstream from coarse gravel to sand. The gravel reaches tend to be steep and narrow, the sand reaches broader and shaped like deltas. De Geer inferred that each segment represents an annual increment at the edge of a rapidly retreatng ice-mass. Most of the deposition takes place in summer, the coarse debris being laid down in the last few hundred metres of a meltwater tunnel and the finer sediments accumulating in a delta extending just beyond the ice-margin. There seems little doubt that many eskers have originated in this way, although they probably constitute a small fraction of the total number. A currently more favoured hypothesis envisages development entirely within an ice tunnel. This is more consistent with the narrow, steep-sided form of many eskers which lack any sign of deltaic 'beading'. There are two possible locations for the tunnels, subglacial and englacial. The ice core of many esker-like ridges emerging from modern glaciers points quite clearly to their englacial origin. Deposition has first taken place along a channel through the ice and retreat of the glacier margin has progressively exposed the line of debris. The steeply crested form of the ridge in this instance is not an original structure but is due to slumping and differential ablation. One advantage claimed for the englacial hypothesis is that it might explain, by superimposition from within the ice, the observed tendency of some eskers to run short distances uphill. However, englacial tunnels of the dimensions needed to accommodate the sediments of a full-sized esker are almost unknown in modern glaciers; moreover, slump structures seem to be much less widespread than the englacial hypothesis would entail. For these reasons it is usually argued that the tunnels must be subglacial and the occasional uphill trend of eskers must be due to hydrostatic pressure. In many instances the meltwater is assumed to have debouched into either a lake or the sea although this is not an essential condition for esker formation. The fragmentation of many examples into short segments separated by gaps is probably due to erosion by meltwater streams after the ridge was exposed at the ice margin.

A suggestion that some eskers may represent the fill of supraglacial channels receives little support from the study of modern ice-masses where suitable accumulations of water-laid sediment are extremely rare. Another possibility advocated by several Finnish workers is that the materials of certain eskers were laid down in open channels bounded by very deep walls of stagnant ice; this is believed to apply in parts of Lapland where the margin of the Scandinavian ice-sheet ended on dry land rather than in the sea or ponded water. One small group of eskers appears to have been laid down submarginally. Termed subglacially engorged eskers, these originate at points on valley sides where meltwater streams, on passing beneath the glacier edge, locally deposit sufficient material to build glacifluvial ridges. Trending obliquely down the hillslope, such features are usually small in comparison with those formed near the ice front and do not conform to the general rule that eskers trend parallel to the direction of ice movement.

Sandar Beyond the immediate ice margin heavily laden meltwater streams may aggrade their beds to build up extensive accumulations of glacifluvial outwash. The lateral extent of the depositional zone is largely controlled by the local relief; below the snout of an Alpine glacier it is normal to find a relatively narrow valley-sandur, but where an ice margin and associated outwash are not constricted laterally a much broader plain-sandur may develop. Both types are characterized by braided channels, by abundant coarse debris, and by marked variations in seasonal discharge.

The valley-sandur below the Emmons Glacier in Washington has been the subject of a detailed study by Fahnestock. Like the majority of similar features it has a steep long profile declining generally at more than 1 in 100 and locally at more than 1 in 8. The slope is very closely related to particle size and discharge. Along an individual channel with no significant change in discharge, the median diameter of fragments falls at a rate of about 45 mm km.$^{-1}$ Presumably attributable to sorting, this downvalley decrease in particle size is associated with a pronounced concavity in the long profile. Great contrasts exist between winter and summer conditions. During the cold season meltwater discharge is low and a single meandering channel tends to develop; when ablation increases in high summer the much greater volume of water occupies a complex of braided channels. Frequent shifts of watercourse preclude establishment of a continuous vegetation cover and the whole environment is one of almost constant change. It is uncertain how fast a valley-sandur surface can be aggraded; in one reach Fahnestock recorded a net gain of 0·36 m in 2 years, but it is most unlikely that this rate can be sustained over a long interval; in Alaska a mean value of 16 mm yr^{-1} has been calculated for a period covering several millennia. Much probably depends upon the movement of the ice front. During a retreat phase the meltwater streams are likely to degrade the steep surface which had previously been built up close to the ice margin. During an advance, on the other hand, as the glacier overrides earlier outwash the meltwater will rapidly build a new frontal apron at a relatively steep angle. Further downstream the changes are less conspicuous and there often seems to be a slow but continuous aggradation. This may still be much faster than that achieved by unglaciated tributaries and the latter in consequence can suffer damming by the fill across their mouths. The greater the distance from the ice front, the higher the proportion of the discharge that is normal stream flow; in the end, the specifically glacifluvial characteristics are almost totally lost.

Many formerly glaciated areas display fossil valley-sandar, but rarely is one preserved in its original state. The steep long profile encourages rapid dissection by river erosion with the result that the earlier surface remains only as an elevated terrace. Incision often takes place in several stages so that the former accumulation is fashioned into a flight of terraces. Unpaired meander terraces are frequent where the downcutting river encounters buried bedrock spurs.

Plain-sandar are extremely well developed in front of the outlet glaciers descending from Vatnajokull in Iceland. They display many similar properties to those already described for valley-sandar. The constituent materials are very coarse close to the ice edge but diminish in size as the intervening distance increases. Gradients commonly lie between 1 in 25 and 1 in 200, with an appreciable concavity to the overall profile. The meltwater occupies braided channels which cover only one part of the sandur surface at a time. It can take several decades for the channels to migrate backwards and forwards across the full width of the plain. The inactive parts may become partially vegetated but still lack the capacity to stabilize the water courses. Recent glacier recession in Iceland has offered an opportunity to study the contact zone between a plain-sandur and the ice margin. In several cases it can be shown that, prior to retreat, slopes some 20–30 m high had been constructed against the ice front. For example, at one point a narrow proglacial lake impounded against the retreating ice edge and overflowing across the crest of the old sandur attains a depth of 23 m; the lake-floor elevation is only reached again on the sandur surface at a distance of 3 km. The implication that a minimum of 23 m of sediment had accumulated at the crest of the sandur seems to demand that meltwater streams can issue from the ice front at progressively higher levels. Low eskers feeding into the proximal side of many sandar also indicate that meltwater is forced upwards, presumably under hydrostatic pressure, before debouching over the surface.

Fossil plain-sandar are well represented around the lowland margins of many former ice-sheets. A particularly fine example extends westwards from the limit of the last glaciation in Jutland (Fig. 13.3). The outwash buries most of an older landscape composed of earlier till, although the highest parts remain uncovered and protrude as low hills. Coarse gravel in the east grades away from the ice edge into fine gravel, sand and silt. The tunnel valleys of eastern Jutland come to an abrupt end at the ice margin and the meltwater flowing through them must have ascended steeply before spreading out over the sandur. Less well-preserved but of great importance in the history of Pleistocene studies are the sandar along the northern fringe of the German Alps. It was here in 1909 that Penck and Brückner identified four separate glaciations which they termed Günz, Mindel, Riss and Würm from the name of four right-bank tributaries of the Danube. The evidence for each glaciation consists of a separate sandur observed to pass at one locality or another into a till. The two earlier sandar were once extensive plains extending at different levels over the Alpine foreland, but now so thoroughly dissected that only small remnants survive. According to Penck and Brückner the erosional episode following the Mindel glaciation was such as to confine the later Riss and Würm outwash within newly formed lowlands as valley sandar.

Sedimentation in ice-marginal lakes

Characteristics of materials Much of the coarse debris discharged into a lake by meltwater streams is normally trapped in marginal deltas. Owing to their heavy sediment load such streams can construct large deltas with remarkable rapidity. In Canada the Assiniboine delta, built out into Lake Agassiz which existed for only five millennia, covers an area of 5 000 km^2 and has a front which rises 100 m above the lake floor. Fine sediment, however, is normally carried to the central parts of a lake to accumulate as an extensive sheet of clay and silt. This may also contain a varied assortment of pebbles and even boulders rafted into the deeper parts of the lake by icebergs; the distorted bedding caused by the impact of 'dropstones' may be clearly visible in suitable sections.

It is the lamination of fine floor deposits in many marginal lakes that has attracted most attention. The tendency to divide into laminae is due to easy parting along silt horizons while the intervening cohesive clays resist splitting. This is readily seen in hand specimens where it may also be noted that the light silt layers and dark clay layers give the material a banded appearance. Each pair of silt and clay layers is termed a couplet, and may vary in thickness from less than 1 mm to over 100 mm. The deposit as a whole is best termed a rhythmite. As a simple description of the material this name is to be preferred to that of 'varved clay' since the word varve strictly connotes an annual increment which is still unproven for many deposits. The idea that each couplet represents an annual accumulation is a very old one, but it was only with the detailed investigations of De Geer in Sweden starting in the late nineteenth century that its validity was widely established. De Geer contended that the silt and clay layers generally represent summer and winter deposits respectively. During summertime meltwater introduces abundant fresh sediment into the ice-free lake. The coarser fraction of the new input settles rapidly to the floor, while relatively unstable water conditions induce turbidity currents from the marginal slopes. During wintertime, with meltwater flow curtailed and the lake surface frozen, sedimentation is restricted to the very fine particles settling from suspension. Onset of the next spring will see a relatively abrupt reversion to accumulation of coarser material. Within both the summer and winter horizons there is often a tendency towards particle-size grading. It is the coarser fraction that is most frequently graded, sometimes passing up progressively into the overlying clay. Such couplets are termed diatactic by Scandinavian workers to distinguish them from the symmict in which the grading is much less complete. Diatactic couplets are believed to indicate accumulation in fresh water, whereas the symmict type are associated with salt or brackish water in which the presence of electrolytes flocculates the clay particles and causes them to settle rapidly.

The belief that the majority of couplets in glacial rhythmites are annual increments may be justified on several grounds. Where climatic conditions are not too severe the pollen contained at successive horizons within a couplet has been shown to display a seasonal cycle that accords with the period of plant flowering. The abundance of diatoms may also vary systematically within a couplet, being much greater per unit volume of sediment in the silt layers than in the clay. For certain lakes the aggregate quantity of material in each recent couplet has been shown to equal the current annual input by meltwater streams. Finally, since the advent of radiocarbon dating it has become possible to compare assumed ages with those indicated by C^{14} assays; in general, agreement has been close and it has even been suggested that ages calculated from rhythmite studies should be used as a correction factor for C^{14} dates.

Chronological calculations Working on the assumption of annual couplets, De Geer attempted to establish a chronology for the deglaciation of Sweden. At many hundreds of localities he measured the thickness of individual varves, plotting the data for each site as a curve of relative thickness against time. Curves from adjacent localities, generally less than 1 km apart, were then placed side by side and moved up and down until a visual match was obtained. The theoretical basis for this method is that climate acts as an overall control of varve thickness, primarily by determining the amount of summer ablation. Varves thicker than usual for their respective lakes will form during a warm summer when there is exceptional meltwater flow; conversely, unusually thin varves will develop over a wide area during a cool summer. On this basis correlation may be established from one lake basin to another. Assuming that the northern limit of an individual varve marks the contemporary ice edge, De Geer was able to trace the withdrawal of the Scandinavian ice-sheet over a distance of 1 000 km covering a period of about 12 500 years (Fig. 13.10). A critical point in De Geer's time-scale was the sudden drainage of a large ice-dammed lake in central Jämtland. He believed that this took place when the Scandinavian ice-sheet finally split into two smaller remnants. Known as the bipartition, this event produced a particularly thick varve which De Geer employed as his zero varve in working backwards in time across central and southern Sweden. Liden subsequently claimed that the zero varve corresponds to 6839 B.C., although recent investigations have revealed bipartition as a more complicated process than previously supposed. It is now

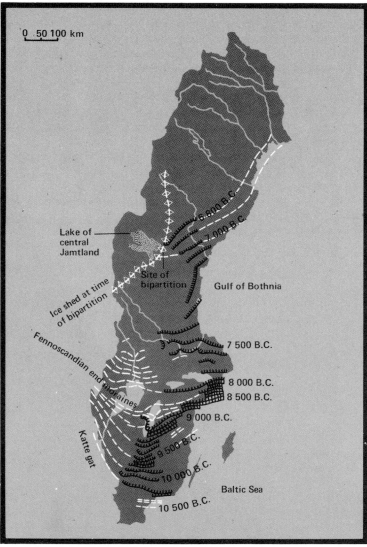

Fig 13.10 Dating by varve chronology of the retreat of the ice front across Sweden. The dashed lines show presumed ice fronts but are not firmly connected with the lines of the main time-scale; the latter are at 100-year intervals and extend from 10 500 to 6 800 B.C. (after Lundquist, 1965).

Map labels:
0 50 100 km
Lake of central Jamtland
Ice shed at time of bipartition
Site of bipartition
Gulf of Bothnia
Fennoscandian end moraines
Kattegat
Baltic Sea
6 800 B.C.
7 000 B.C.
7 500 B.C.
8 000 B.C.
8 500 B.C.
9 000 B.C.
9 500 B.C.
10 000 B.C.
10 500 B.C.

thought that the so-called zero varve actually consists of two varves separated by a time interval of as much as 80 years. Despite unresolved problems of this type, De Geer's chronology has stood up remarkably well to critical examination by later workers. There seems no doubt regarding the annual duration of most of the couplets he employed, nor the existence of a single overriding determinant of varve thickness, presumably the climate.

It is in Denmark that most severe criticism of De Geer's time-scale has been voiced. Several workers have contended that couplets which he used to extend his chronology across the Danish islands of Zealand and Funen are not annual layers but represent shorter-term fluctuations in meltwater discharge. These could be due either to warm and cold spells of weather or possibly to periods of storminess. It has even been inferred that some couplets may be no more than daily layers. It is worth recording that in North America certain discrepancies between radiocarbon dates and varve chronology might also be explained by a proportion of the couplets not being annual layers.

Analysis of lake-floor rhythmites provides a valuable guide not only to the age of the materials themselves but also to the speed with which various geomorphological changes take place. For example, in Sweden it is calculated that during the 400 years immediately prior to bipartition the ice-front receded a distance of 130 km, a mean rate of 325 m yr^{-1}; in Finland for a period of more than 1 000 years after deposition of the Salpausselka ridges the rate of retreat averaged about 260 m yr^{-1}.

BAKER, V. R., 1973. 'Paleohydrology and sedimentology of Lake Missoula flooding in eastern Washington', *Spec. Pap. geol. Soc. Am.*, 144.

BOULTON, G. S., 1974. 'Processes and patterns of glacial erosion' in *Glacial Geomorphology* (Ed. D. R. Coates), State Univ. New York.

BRETZ, J. H., 1966. 'Correlation of glacial lake stages in the Huron, Erie and Michigan basins', *J. Geol.*, 74, 78–9.

BRETZ, J. H., 1969. 'The Lake Missoula floods and the channeled scabland', *J. Geol.*, 77, 504–43.

CLAYTON, L., 1964. 'Karst topography on stagnant glaciers', *J. Glaciol.*, 5, 107–12.

DAHL, R., 1965. 'Plastically sculptured detailed forms on rock surfaces in northern Nordland, Norway', *Geogr. Annlr.*, 47, 83–140.

DURY, G. H., 1951. 'A 400-foot bench in south-eastern Warwickshire', *Proc. Geol. Ass.*, 62, 167–73.

FAHNESTOCK, R. K., 1963. 'Morphology and hydrology of a glacial stream – White River, Mount Rainier, Washington', *Prof. Pap. U.S. geol. Surv.*, 422A.

FROMM, E., 1970. 'An estimation of errors in the Swedish varve chronology' in *Radiocarbon Variations and Absolute Chronology*, Nobel Symp. 12, Almqvist and Wiksell.

GLEN, J. W., 1954. 'The stability of ice-dammed lakes and other water-filled holes in glaciers', *J. Glaciol.*, 2, 316–18.

GOLDTHWAIT, R. P., 1974. 'Rates of formation of glacial features in Glacier Bay, Alaska', in *Glacial Geomorphology* (Ed. D. R. Coates), State Univ. New York.

LEE, H. A., 1965. 'Investigations of eskers for mineral exploration', *Pap. geol. Surv. Can.*, 65–14, 1–17.

LUNDQVIST, J., 1965. 'The Quaternary of Sweden' in *The Geologic Systems: The Quaternary I* (Ed. K. Rankama), Wiley.

MANNERFELT, C. M., 1945. 'Some glaciomorphological forms and their evidence as to the down-wasting of the inland ice in Swedish and Norwegian mountain terrain', *Geogr. Annlr.*, 27, 1–239.

PEEL, R. F., 1949. 'A study of two Northumbrian spillways', *Trans. Inst. Brit. Geogr.*, 15, 75–89.

SHOTTON, F. W., 1953. 'The Pleistocene deposits of the area between Coventry, Rugby and Leamington and their bearing upon the topographic development of the Midlands', *Phil. Trans. R. Soc.*, B 237, 209–60.

SISSONS, J. B., 1958. 'Supposed ice-dammed lakes in Britain, with particular reference to the Eddleston valley, southern Scotland'. *Geogr. Annlr.*, 40, 159–87.

SISSONS, J. B., 1974. 'A Late-glacial ice-cap in the central Grampians, Scotland', *Trans. Inst. Brit. Geogr.*, 62, 95–114.

SYNGE, F. M., 1950. 'The glacial deposits around Trim, Co. Meath', *Proc. R. Irish Acad.*, 53, 99–110.

VIRKKALA, K., 1963. 'On ice-marginal features in south-western Finland', *Comm. Geol. de Finlande, Bull., No.* 210.

WILLMAN, H. B., 1971. 'Summary of the geology of the Chicago area', *Ill. State geol. Surv. Circ.*, No. 460.

WOODLAND, A. W., 1970. 'The buried tunnel valleys of East Anglia', *Proc. Yorks. geol. Soc.*, 37, 521–78.

Selected bibliography

Two modern texts with very extensive bibliographies are C. Embleton and C. A. M. King, 'Glacial geomorphology', Arnold, 1975 and R. J. Price, 'Glacial and fluvioglacial landforms', Oliver and Boyd, 1973. An even more recent survey of the same field is provided by D. E. Sugden and B. S. John, 'Glaciers and landscape', Arnold, 1976.

The term periglacial was introduced by Lozinski in 1909 to refer to the area bordering a continental ice-sheet and characterized by the severity of its climate. There is no necessary identity between the two ideas contained in this statement. Gradually usage has changed until the word no longer demands an ice-marginal location but denotes instead an environment experiencing an intensely cold climatic regime. Nearly all recent workers have accepted this interpretation, but have failed to agree upon a precise definition. It is acknowledged that the periglacial zone is distinguished from the cool temperate regions by the dominance of frost action and prolonged surface freezing. However, frost action and surface freezing are related but not identical concepts and in attempts at definition two schools of thought clearly emerge. On the one hand are those workers who confine the term to areas underlain by permafrost, that is, by perennially frozen ground. On the other are those who regard this as too restrictive and wish to include regions that experience annually a very large number of freeze–thaw cycles. For present purposes the wider interpretation of the word periglacial will be adopted. Permafrost will be discussed first but attention will then be turned to some of the distinctive surface sediments and structures arising, in part at least, from repeated freeze–thaw action.

It must be emphasized that much remains to be discovered about periglacial environments since, until recently, they have been sadly neglected. Early impetus for intensive study came from strategic considerations when it was realized during the Second World War that correct engineering practice in such areas demands a proper understanding of periglacial processes. Further stimulus has since come from the desire of commercial firms to tap the mineral resources of arctic regions. An outstanding example is the search for oil on the north slope of Alaska, where drilling teams had to cope with some of the most extreme conditions on the North American continent. The success of British Petroleum in proving large oil reserves highlighted another argument for intensified research. This is the conservationist argument. In order to move the oil to the consuming areas it must be transported by specially designed pipeline across Alaska to the ice-free Pacific coast. Yet very little is known of the potential long-term effects of the pipeline on what is generally agreed to be a fragile environment that recovers extremely slowly after any disturbance. It seems likely, therefore, that the pace of research will continue to quicken.

Research in modern periglacial areas may assist in defining the environmental parameters that control the development of features now found in fossil form in more temperate zones. For example, Pewe has stressed that

3|14

Periglacial forms and processes

permafrost can form where the mean annual temperature is about −1°C, but that ice wedges can only form where it is less than −6°C. On the other hand, active pingos which were formerly thought to be restricted to high arctic regions have now been found developing in the subarctic forests of interior Alaska where the mean annual temperature is no lower than −2°C. Such information may ultimately be useful in defining climatic zones as they existed during critical periods of the Pleistocene epoch, although it is doubtful whether it will ever be more than a supplement to the biological evidence that is often available for the same purpose. It needs to be stressed, moreover, that climate is not the sole determinant of periglacial processes. The texture and moisture content of the soil, the character of the vegetation, the duration of the snow cover and the aspect of the slope can all have an important influence.

Permafrost

Distribution Excluding Antarctica and Greenland, permafrost has been estimated to underlie over 10 per cent of the world's land surface at the present day (Fig. 14.1). The more extreme climatic zones are underlain by continuous permafrost where the only unfrozen areas are those lying beneath the largest lakes and rivers. Such conditions exist in Canada north of a line from Hudson Bay to the Mackenzie delta, in Alaska along the north slope, and in Siberia throughout a broad belt extending a thousand kilometres southwards from the shores of the Arctic Ocean. Within this area the thickness of the permafrost layer varies considerably. In general it is greatest in the most northerly latitudes, with figures of over 600 m recorded at a number of localities near the arctic coast of Siberia. In northern Canada and Alaska thicknesses of over 500 m have been found in a number of boreholes, while on Spitsbergen colliery workings are said to encounter frozen ground at depths of 320 m. Near the southern margin of the continuous permafrost zone a more typical thickness would be about 50 m. Beyond this margin the permafrost becomes discontinuous. Local geological, topographical and botanical factors combine to limit the development of frozen ground so that its distribution becomes patchy. There is a gradation from regions where all but a small propor-

tion of the area is underlain by permafrost to those where perennially frozen layers occur as isolated pockets. In Canada the zone of discontinuous permafrost extends from the southern end of James Bay eastwards to Labrador and westwards to northern Alberta. It reaches virtually to the Pacific coast of Alaska and in central Asia has been recorded as far south as 45°N. Sporadic outliers are found in such elevated areas as the Scandinavian mountains, the Urals and the Rockies of British Columbia.

Relationship to climate The primary control of permafrost is obviously climate. An equilibrium may be established when the perennially frozen layer attains a depth determined by the atmospheric temperature and the geothermal flux. Since on average the flow of internal heat is less than 0·1 per cent of that from the sun, there is a strong tendency towards seasonal equalization of the atmospheric and surface soil temperatures (Fig. 14.2). Below the surface the amplitude of the seasonal fluctuations gradually diminishes until, at a depth of about 15 m, the temperature of the ground remains stable throughout the year. This is known as the level of 'zero annual amplitude' and the temperature at this horizon can be anywhere between just below freezing point and about −16°C. In practice a close approximation exists between the mean annual air temperature and that at the level of zero annual amplitude. Proceeding to greater depths it is found that the temperature steadily rises, normally at a rate of 1°C per 30–40 m, until freezing point is reached at the level where permafrost ceases.

Many attempts have been made to examine the relationship between permafrost distribution and various climatic parameters. In theory, perennial freezing may occur wherever the mean annual temperature is below zero. In practice, however, the mean temperature is usually a degree or so below freezing point before permafrost actively develops. Studies in Canada have shown that perennially frozen layers are rarely encountered south of the −1°C isotherm, and that they are largely restricted to peatlands between the −1 and −4°C isotherms. It is only north of the −4°C line that permafrost becomes widespread, while it is virtually continuous beyond the −6·5°C line. These figures are surprisingly at variance with those recorded in Asia.

In the eastern USSR permafrost is widespread south of the $-2\,°C$ isotherm and is by no means rare where the mean annual air temperature is at freezing point. The reason for this discrepancy is believed to lie in the fact that much of the Siberian permafrost is a relic from Pleistocene times. Russian workers think that the extent of permafrost is currently decreasing along the southern margin of the discontinuous zone, although it may still be deepening in parts of northern Siberia. Implicit in this interpretation is the idea of a long time lag between a change in climate and restoration of equilibrium conditions.

As climate deteriorates at the onset of a periglacial phase, the top few metres will first be subject to annual freeze–thaw cycles. With further refrigeration the winter freezing cycle will penetrate deeper than the summer thawing cycle so that a layer at depth will be left perennially frozen. Each year this layer will thicken as new permafrost is added at the base. The temperatures of the perennially frozen ground will vary seasonally until the level of zero annual amplitude is reached. Thereafter, as the temperature

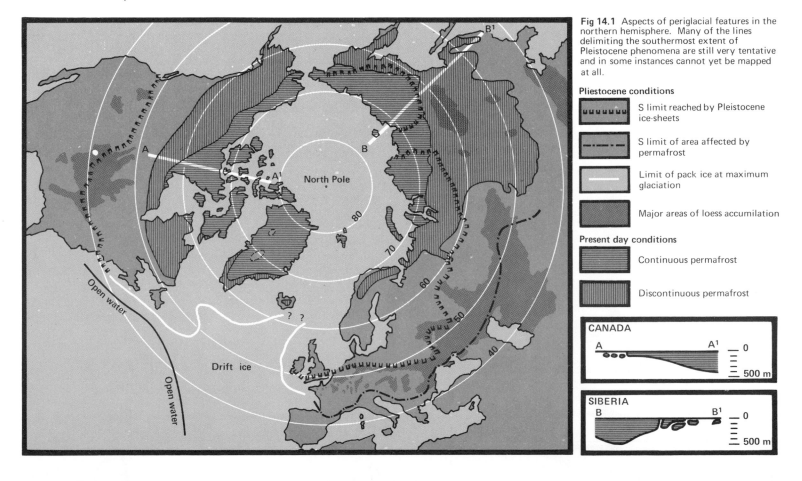

Fig 14.1 Aspects of periglacial features in the northern hemisphere. Many of the lines delimiting the southermost extent of Pleistocene phenomena are still very tentative and in some instances cannot yet be mapped at all.

Pliestocene conditions

S limit reached by Pleistocene ice-sheets

S limit of area affected by permafrost

Limit of pack ice at maximum glaciation

Major areas of loess accumilation

Present day conditions

Continuous permafrost

Discontinuous permafrost

North Pole

Open water

Open water

Drift ice

CANADA   A   A¹   0   500 m

SIBERIA   B   B¹   0   500 m

at the level of zero annual amplitude falls, more permafrost will be added until ultimately it may attain a total thickness of several hundred metres. With a climatic amelioration the process is reversed with the temperature of the frozen layers slowly rising and the depth of permafrost declining. Complications of this simple model may be readily visualized, particularly with rapid climatic oscillations, and in practice observed relationships tend to be very intricate. For example, many boreholes have encountered unfrozen layers known as taliks within other-

Fig 14.2 Thermal relations in an area of permafrost.

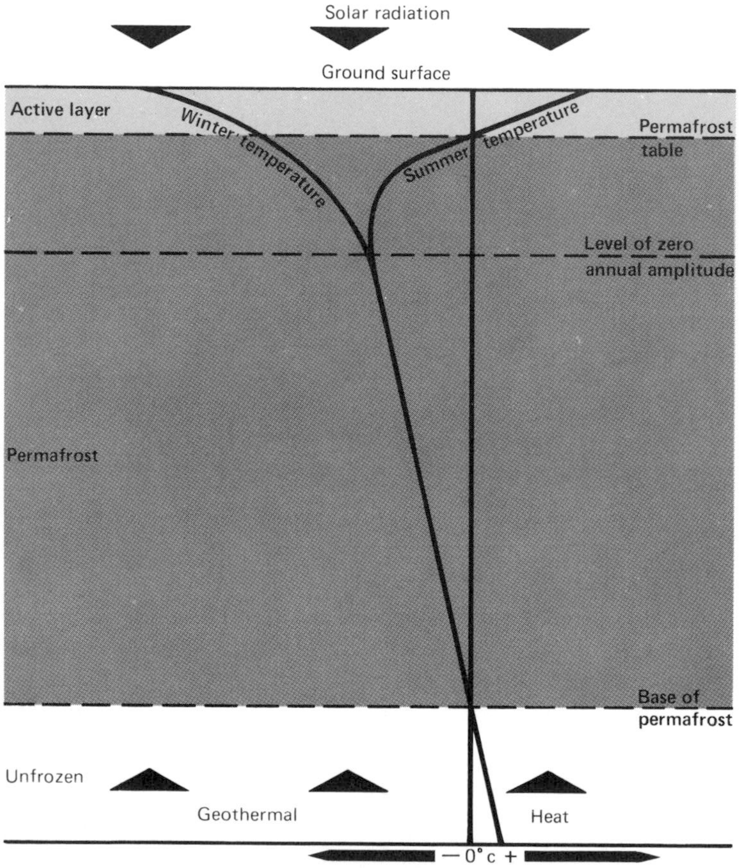

wise continuous permafrost. Equally common is the situation in which the temperatures below the level of zero annual amplitude seem excessively cold, probably due to an overall climatic warming in many permafrost areas during the last few centuries.

The active layer and associated freezing processes A period of perennial freezing has little lasting effect on deep-seated rocks. It is the shallow surface layer subject to summer thawing that suffers the most severe disturbance. Known as the active layer its thickness varies from a few centimetres to more than 3 m. The base of the active layer is known as the permafrost table; whilst in general parallel to the ground surface it tends to show undulations related to purely local factors. In spring and summer the active layer thaws from the surface downwards so that the resulting water cannot percolate freely to greater depths. With the possibility of additional liquid from snowmelt it is scarcely surprising that waterlogged surface horizons are one of the most characteristic features of permafrost areas in summer. In autumn when the temperatures begin to fall it is the surface horizons that freeze first. A layer of water is then trapped between the permafrost table and a freezing front advancing downwards from the surface. Before the whole active layer is refrozen in winter the remnants of the trapped water may be placed under considerable 'cryostatic' pressure.

The freezing process in a porous medium involves much more than simple transformation of interstitial water into ice. It is now many decades since Taber demonstrated that the freezing of saturated sediments can lead to the segregation of large pockets of clear ice. The results are seen in frost-heaved soils underlain by layered bodies of ice, and in boreholes that have encountered in the permafrost layer patches of pure ice 1 m or more in thickness. It is obvious that there were never cavernous voids filled with liquid, but that interstitial water must have migrated towards the growing crystalline body. The processes involved are highly complex and will not be pursued here in detail. There are really two related problems. The first concerns the reason for water moving through the porous medium to feed the expanding ice crystals. Like all crystals, ice grows by virtue of the electrochemical attraction exerted by one stable ion for another in the surrounding medium. On account of its structure, the solid phase exerts a far greater attraction than that linking the liquid molecules together, with the result that water is drawn very strongly towards the growing crystal. Schenk has argued that the force by which water molecules are attracted to an ice crystal must exactly equal the force that the crystal itself can exert. With a value of over 2 000 kgf cm^{-2} this exceeds the force by which most water films are held to sedimentary particles, including even the hygroscopic water adsorbed on to clay minerals. The second problem concerns the reason for part of the water remaining in a liquid state while the rest is frozen solid. The progressive freezing of water in a porous substance is affected by several variables. In general these combine to ensure that the liquid in large pores will freeze before that in small pores. Water may be supercooled to $-0.4°C$ in very fine interstices and still be available for addition to ice crystals that have already formed in the larger pore spaces. A related effect is the desiccation of those zones that are the last to freeze.

Taber showed by a series of experiments that the tendency towards segregation is much greater in some sediments than in others. In general ice segregates most readily in silts. This has long been recognized by engineers who need to identify what they term 'frost-susceptible' materials prone to particularly severe upheaval as a result of freezing; as a guideline they usually reckon that debris containing more than a few per cent silt-sized particles will be especially frost-susceptible. In coarser materials little supercooling takes place and the whole mass freezes as a body when the temperature falls to 0°C. In finer materials segregation is partially inhibited by the bonding of the water to the clay minerals. These variations, combined with more obvious contrasts in the distribution of interstitial water, can lead to extremely irregular development of ground ice. During freezing surface hummocks form, during thawing hollows subside unevenly. Eventually marshy depressions and lakes will evolve with poorly integrated drainage between them. Relief originating in this way is commonly termed thermokarst. Over areas of fossil permafrost in southern Siberia, new thermokarst is currently being created by the clearance of natural vegetation for purposes of cultivation. This permits the deeper penetration of summer thawing cycles and accelerates melting of the ground ice.

Coarse slope deposits Mechanical fracture of rock material is one of the most distinctive attributes of the periglacial environment. It results from the pressure exerted by ice crystallization, and is particularly effective in permeable rocks saturated with water (p. 118). The debris produced in this way ranges from large boulders to fine angular chips depending on the fissility of the parent material. On flat surfaces the detritus may accumulate as a bouldery or rubbly mantle, but more often it will proceed downslope as a sheet of mobile debris.

A wide variety of materials and forms develops as a direct result of frost shattering. Where the debris is exceptionally coarse, blockfields and rock glaciers may be produced. Blockfields, also known by the German term Felsenmeer, consist of angular and poorly rounded boulders, many of them 1 m or more across. Large voids promote the washing down of any finer debris so that the impression of an uneven boulder pavement is created. The arrangement of the materials is seemingly haphazard and there may be no evidence of individual fragments having moved any great distance. On progressively increasing gradients, blockfields give way to sheets of rubble in which the individual fragments tend to lie with their long axes aligned downslope. In modern periglacial areas blockfields are frequently seen on flat summit surfaces, and they are known in fossil form from such regions as Scandinavia and the Appalachians. The lack of fines retards plant colonization so that they often appear as small, poorly vegetated areas in the midst of more thickly forested regions. Rock glaciers consist of lobate, slow-moving bodies of frost-riven debris. They tend to originate at the foot of rocky precipices from which material is constantly being wedged free. A common source is the headwall of an old cirque from which the rock glacier moves downvalley in a form very reminiscent of a true glacier. The debris is often 30 m thick and shows a rough grading from coarse boulders at the surface to a much thicker layer of mixed boulders, sand and silt at the base. The margins are very steep but the upper slopes generally lie at angles between 5° and 20°. Both active and fossil forms are known. Current movement may be indicated by concentric wrinkling of the surface, and superficial debris has been measured travelling at a rate of about 1.5 m yr^{-1} in Switzerland and 0.5–0.7 m yr^{-1} in Alaska. There is no doubt that in the latter region the lower parts of several active rock glaciers are perennially frozen, and it has been inferred that the current motion is almost entirely due to the presence of interstitial ice. When the climate ameliorates a rock glacier is therefore likely to be stabilized and slowly colonized by vegetation. Such fossil forms are known from many parts of the world, although other possible origins for tongues of coarse angular debris need to be considered. It has been suggested that some spreads may be no more than exceptionally thick ablation moraine let down from a stagnant glacier.

Another prominent feature which results directly from freeze–thaw processes is the talus or scree slope accumulating at the foot of a steep rock face. The surface of a scree in a periglacial environment is rarely stabilized. Frost-riving continues to comminute the debris, while the growth of interstitial ice promotes relatively rapid downhill movement. Sections occasionally reveal a distinct stratification composed of alternating beds of fine and coarse particles. Known by the French term grèzes litées, these are most common where the source rock is highly fissile and the average size of fragments correspondingly small. The stratification presumably results from some form of environmental fluctuation, but its exact nature remains uncertain. Some workers have argued in favour of seasonal accumulation layers, while others have invoked longer-term variations in the thickness and duration of the snow cover. Another form that is certainly affected by the snow cover is the protalus rampart. Material wedged by ice from the face of a precipice can skip and slide across a basal snowbank to build a ridge of coarse debris that may locally reach as much as 10 m high. If the climate becomes warmer and the snowbank melts, the ridge or rampart appears to be isolated from its debris source and can easily be mistaken for a small cirque moraine. However, its composition and fabric are different and the materials will show no sign of glacial transport.

Solifluction sheets Of great extent in many periglacial regions is a mantle of ill-sorted, water-saturated debris with the constituent particles ranging in size from clay to large boulders.

In 1906 Andersson proposed the term 'solifluction' for the movement of such material which he observed to be taking place very actively in the Falkland Islands. Indeed, he regarded the removal of waste in that environment as occurring at a rate unsurpassed in other parts of the world. It is important to recognize that in solifluction two processes, flowage and creep, normally operate simultaneously. The former is induced by the tendency towards saturation. This occurs when the surface layers melt in the spring but the underlying horizons remain frozen, either perennially or at least until much later in the summer. Eventually the moisture content may be raised to the Atterberg liquid limit, drastically reducing the strength of the thawed material. High porewater pressures may be generated with similar results. It has also been suggested that during thaw the soil colloids are flocculated and lose their cohesion. The reason appears to lie in a concentration near the surface of free ions which, during the melt phase, coagulate the clay minerals to produce a saturated crumb structure which is mechanically very weak. Moreover, the material becomes thixotropic so that any form of disturbance tends to accentuate the mobility of the debris. All these factors combine to promote rapid soliflual flowage down quite gentle slopes. Solifluction creep, on the other hand, is induced by freeze–thaw cycles. Debris is lifted normal to the slope but tends to drop back vertically under the pull of gravity. Surface layers will generally be displaced more than the lower, while further differential movement may result from irregular freezing patterns and ground-ice segregation. A particularly effective form of freezing that can affect the top few centimetres of the regolith is the development of pipkrake or needle ice. As the air temperature drops acicular ice crystals develop beneath stones and patches of soil that are relatively good conductors of heat. In one major cycle such material may be raised over 0.1 m from its original position, sinking back vertically when melting takes place. In New Zealand rock debris on an $11°$ slope has been shown to move over 0.6 m yr^{-1} by this process.

An active solifluction sheet tends to develop a number of distinctive surface forms. Uneven movement frequently produces conspicuous lobes of rather thicker debris. On steeper slopes many small terrace features may be seen, sometimes extending hundreds of metres parallel to the contours, but more commonly dividing and uniting so as to form a series of irregular crescent-shaped steps known as garlands. They can be divided into two main types depending on the material in the riser. In some instances this consists of large stones, in others of a binding turf layer between more sparsely vegetated flats. The precise mechanism responsible for stone-banked and turf-banked terraces remains uncertain. They are of the same general dimensions with the riser commonly 1–5 m high and the treads 5–25 m across. However, work so far completed suggests that rather different processes operate in their formation. The turf probably acts as a restraining factor with more mobile debris collecting on the upslope side as a terrace. This is indicated by the occasional rupture of the turf allowing detritus to spill across the next lower tread. Yet the riser itself is not usually stationary but moving slowly downhill. This is apparent where sections have been dug through terraces to reveal old, overridden soil profiles. It is possible that the boulders in stone-banked terraces also act as barriers to rapid movement with finer debris piling up behind them. However, several workers have suggested that a sorting process is operative within a solifluction sheet and that larger fragments tend to move to the frontal margin of an active lobe. At a later stage rocky ledges may arrest the movement of the lobes so that they develop into stone-banked garlands, or possibly the larger fragments, once they have been sorted, lose their capacity for fast movement. The latter hypothesis could explain the continued downhill migration of stone-banked garlands attested by the overrunning of old humic layers. From radiocarbon dates the rate of downhill migration in the Colorado Rockies has been estimated to range from less than 1·5 to over 20 mm yr^{-1}.

As the temperate weathered mantle normally moves much more slowly than the corresponding solifluction sheet, refrigeration of the climate tends to increase the rate at which material is conveyed to the valley floor. The transporting capacity of the river may not be proportionately increased since, for long periods in winter, there will be no discharge at all. In consequence, soliflual sediments can accumulate to considerable thicknesses and in suitable sections individual debris sheets may be seen piled one on top of another. A further attribute of solifluction

is its ability to move sizeable boulders down quite gentle slopes. Incorporated as part of the solifluction sheet they are carried towards the valley floor and, if it is beyond the competence of the streams to move them, the larger blocks will be left as an exceptionally coarse residual deposit.

When the climate ameliorates at the close of a periglacial episode, the immobilized solifluction sheets and the thick valley-floor accumulations may remain as witness to the operation of a distinctive set of slope processes. On gentle clay hillsides in eastern England slickensided shear planes at a depth of about 1 m below the surface have been widely identified. These planes are now totally stable, and permafrost preventing downward percolation of water was almost certainly essential for the original movement. As long ago as 1839 de la Beche drew attention to poorly-sorted bouldery deposits at the foot of valleysides in south-western England. He called the material 'head', a term still widely employed to designate the products of former periglacial slope processes. The poor sorting of head can make it difficult to distinguish from till. One of the more obvious diagnostic features of a soliflual sheet is its purely local derivation, but even this criterion may be difficult to apply where there is potential contamination from earlier glacial deposits. It is generally claimed that soliflual debris is more angular, the texture is less compact and the particles are aligned with their long axes pointing downslope. Nevertheless, there are many cases where the distinction between till and head remains a matter of dispute. For example, in Wales detrital sediments enveloping the cliffs and hills of the Cardigan Bay coast have been interpreted by most workers as till; however, Watson has maintained that they could equally well be periglacial deposits and has drawn attention to similarities with deposits in north-western France beyond the normally accepted limit of ice advance.

Fluvial sediments Fluvial sediments accumulating in modern periglacial areas often contrast sharply with those being laid down under more temperate conditions. One reason is that river regimes in arctic areas differ substantially from those of lower latitudes. In winter all but the largest rivers are totally frozen and for extensive areas stream flow is restricted to less than half the year. When the thaw arrives it briefly releases huge volumes of water causing widespread inundation. However, in most cold climates the precipitation totals are so low that the river levels subside well before the onset of the next winter freeze. Conditions are particularly conducive to braiding of the stream courses. Violent fluctuations in discharge, much coarse debris resulting from accelerated movement on hillslopes, weak soil development and poor binding by the vegetation cover, all combine to promote braiding and frequent shifts of stream course. Where large rivers flow polewards into areas of more severe climate, the upper courses may thaw weeks before the lower courses. This results in masses of floating ice becoming jammed against the frozen downstream reaches. Water is ponded back and forced to spill across the floodplain while large ice-floes continue to grind against the channel banks. When the blockage is released a huge surge of water proceeds down the original channel, capable of carrying with it exceptionally large boulders. The overall effect is to build up an alluvial cover of generally coarse grade, moderately well-sorted but with occasional boulder beds and lenses of silt and clay where abandoned distributaries are being infilled. The latter deposits in their fossil form frequently enclose organic horizons which, from the nature of their faunal and floral remains, serve to confirm the extreme climatic regime under which the aggradation took place.

Wind-blown materials Wind-blown sediments are widely distributed in modern periglacial environments. The ability of wind to redistribute silt- and sand-sized particles under periglacial conditions is attributable to several factors. The prevalence of strong winds is a first requirement that is fully met in arctic areas. Moreover, the sparse vegetation cover permits the wind to operate to full effect. Secondly, there must be an abundant supply of suitable fine detritus. This is commonly furnished by broad alluvial flats on which the discharge at certain times of the year is so low as to allow the surface to dry out completely. In addition the growth of ground ice tends to desiccate the uppermost layers of the regolith so that during the cold season a dusty surface is exposed to aeolian action. Winter snowfall can act as a protective blanket, but often the snow itself is blown about in blizzards and carries much loose sediment with it.

Periglacial wind-blown materials vary considerably in grain size, but for convenience may be divided into two major classes, loess and aeolian sand. Loess is normally an unstratified deposit composed mainly of silt-sized quartz grains. Most particles have a diameter between 0·015 and 0·05 mm with only a small proportion falling outside those limits. Péwé has shown that such material is currently being blown each summer from the dried-up distributaries of braided rivers in Alaska, but most information about loess has come from the study of the equivalent Pleistocene deposits.

Pleistocene loess locally attains thicknesses in excess of 50 m although a figure of under 5 m is more typical. Its distribution (Fig. 14.1) points very strongly to glacial outwash as a primary source of the sediment. It tends to be thickest immediately east of those rivers which during Pleistocene times carried prolific meltwater, either from a major continental ice-sheet or to a lesser extent from large valley glaciers. Examples are afforded by the Bug, Don and Volga in the USSR, by the Rhine and Danube in central Europe and by the Mississippi in the USA. However, not all loess comes from such sources. In northern France, for instance, much of the wind-blown material appears to be of purely local origin. Its composition varies according to the calcareous or non-calcareous nature of the underlying bedrock; the sharp boundaries that are found would not be expected if the wind were simply deflating alluvial deposits which are much more mixed in their general composition. One region where loess tends to be relatively sparse is along the cool temperate oceanic margins. In part this may be due to the moist climate having prevented total desiccation of the surface. Another factor may have been the relatively rapid colonization of outwash by plant communities. In Great Britain, for example, true loess in any quantity is rare. The brickearth of south-eastern England is the nearest equivalent, but much of this probably originated as wash from local hillslopes mantled with wind-blown silt. In parts of upland Britain soil analyses have disclosed significant amounts of silt falling within the normal limits of loessic grain size. The belief that this material is of wind-blown origin is strengthened by heavy-mineral studies which show that some of it could not be derived from the local bedrock and must therefore have been introduced from outside.

From research in the Mississippi valley and elsewhere a number of inferences may be drawn about the conditions under which loess accumulation has taken place. The wind-blown silt often contains immense numbers of molluscan shells. They are dominantly land snails and are much too fragile to have been transported long distances. They must have been part of the local fauna gradually interred as deposition progressed. Occasionally so many woodland species are found that it is difficult to avoid the conclusion that some loess was laid down beneath a forest cover. Where radiocarbon dating is available it seems to indicate that the sedimentation rate was normally very slow, the build-up of each metre taking many centuries. Thick loess sequences often contain several fossil soil horizons, implying that conditions favouring accumulation were repeated on many separate occasions. Most of the paleosols have been attributed to interglacial or interstadial episodes when loess accumulation virtually ceased; they constitute a most important line of evidence in the formulation of Pleistocene chronologies for both central Europe and the Mid-West of the United States (Fig. 2.4, p. 27).

Aeolian sand in periglacial regions may assume two forms. In some areas it is blown into dunes, in others spread out as a relatively smooth plain. Constructional relief implies an abundant supply of sand which in most cases has been provided by heavily laden meltwater streams. In northern Germany, for example, prominent dune groups fringe the Urstromtäler, while in Alaska dunes of various ages and degrees of stability are conspicuous elements of the landscape. On poorly cemented sandstones periglacial conditions may be able to initiate deflation. It is believed that the Sand Hills of Nebraska covering some 35 000 km² were active during the last glaciation, although how far destruction of the binding turf is attributable to aridity rather than low temperatures is an open question. Featureless spreads of blown sand are particularly characteristic of the Low Countries in north-western Europe. Known as coversands, these deposits are usually only a few metres thick. The materials are less well-sorted than either dune sand or loess and show a crude stratification. They are believed to have been deposited during winter blizzards when both snow and fine sediments were blown by the wind. When the snow melted the particulate matter was left behind and slow accretion took place. The resulting deposits are sometimes termed niveo-aeolian sands.

Superficial structures due to frost action

Involutions The term involution refers to a variety of structures that may develop in both bedrock (Fig. 14.3) and in unconsolidated sediments subject to repeated freeze–thaw cycles. They range from relatively regular festoons, in which originally horizontal strata are deformed into cup-shaped structures, to highly irregular contortions in which individual beds may be twisted and deformed beyond all recognition (Fig. 14.4). The diversity of forms probably indicates different causes, but these have never been satisfactorily separated and defined. Some involutions are almost certainly connected with the development of earth hummocks or 'thufurs'. In many modern periglacial areas flat surfaces are covered with numerous small mounds generally about 0·5 m high and 1–2 m across. Sections show a core of sediment beneath a more or less continuous mantle of soil. The sediment has clearly been forced upwards but the precise mechanism remains obscure. Most workers have invoked differential freezing, but have clashed over the way this operates. Some have argued that frost first penetrates along the incipient waterlogged depressions, and that the resultant pressure eventually squeezes the intervening material into mounds. Others have contended that the elevation of the hummock arises from frost heave repeatedly affecting the mounds before the encircling depressions. Thufurs could originate in either of these ways without permafrost. Where permafrost is present, however, another mechanism may be responsible for producing surface mounds. As the active layer cools during autumn, water trapped above the permafrost table is under such pressure that it sometimes forces its way to the surface and exudes mud which thereupon freezes into a mound. When the climate ameliorates all earth hummocks will tend to subside but the surface horizons may remain contorted to a depth of a metre or two.

Other involutions probably have no direct surface expression. They apparently result from stresses that develop when layers of

different composition are subjected to freezing temperatures. As already indicated, freezing in such circumstances is an irregular process. Frozen zones expand and exert cryostatic pressures on surrounding material. Segregation leads to lenses and patches of pure ice which on melting collapse and aid contortion. It needs to be remembered that in most cases the materials are so saturated that they deform quite readily under small shear stresses. The frequency with which periglacial involutions are observed leaves no doubt about the effectiveness of differential freezing, although in individual cases it may be difficult to determine the exact mechanisms that have operated. Involutions may be seen in weathered bedrock as well as in stratified superficial deposits. Some of the most spectacular results are found where terrace gravels rest upon clay. Here masses of the clay may be thrust upwards several metres through the overlying sediments, wrapping around pebbles in the gravel. At the present time few climatic inferences can be drawn from the occurrence of involutions since they are obviously capable of forming in several different ways.

Ice wedges In many parts of the periglacial zone thermal contraction of frozen ground engenders a network of frost cracks. The coefficient of linear expansion of ice is relatively high, about 50×10^{-7} per °C, so that a fall in temperature of 20°C can theoretically generate fissures some 10 mm across around a polygon 10 m in diameter. Cracks of this type are often seen in intensely cold regions but they need to develop much further before they attain the dimensions of a typical ice wedge. This consists of a tapering mass of relatively pure ice 1 m or more across at the surface and extending to a depth that may locally reach 10 m (Fig. 14.4). The evolution of a wedge of this size is believed to be due to gradual widening of the original fissure. During winter hoarfrost accumulates in the crack, while during spring water from the active layer percolates down to freeze below the permafrost table. As the temperature of the permafrost

rises in summer, expansion may contort the bedding adjacent to the enlarged wedge. Renewed thermal contraction during the next winter re-opens the crack and the whole process is repeated. Each year a film of fresh ice 1–20 mm thick is added, endowing the wedge with a characteristic vertical foliation. It is believed that a thick ice wedge normally takes several centuries to form.

Modern ice wedges have been studied in many arctic areas, but there is often doubt whether they are currently growing or decaying. Péwé has argued that active development is confined to regions where the mean annual temperature is −6°C or less; in slightly warmer regions they are actually declining but can persist for many centuries without completely disappearing. Ice-wedge networks are best developed on horizontal surfaces composed of fine alluvial sediments. They are therefore particularly conspicuous on river terraces and floodplains. A distinction may be drawn between epigenetic and syngenetic types. The former are those that have developed after fluvial aggradation has ceased. The latter grow concurrently with sedimentation and extend upwards as the level of the ground surface is raised. In general syngenetic wedges are likely to be narrower and deeper than their epigenetic counterparts.

Fossil ice-wedges, or ice-wedge casts as they may appropriately be termed, are often seen in old terrace deposits. They take the form of narrow V-shaped fillings cutting across the normal horizontal bedding. The filling consists of material that has either slumped from the walls of the wedge or been introduced from the surface. During climatic amelioration, an ice wedge melts mainly from the surface downwards. As the permafrost table is lowered, the active layer collapses into the vacant space and a shallow trench develops. Water channelled into the trench percolates downwards past the diminishing ice-mass, carrying with it fine sediment which accumulates in the interstices of the collapsed debris and gives the final infill a different texture from the surrounding beds.

Aerial photography has disclosed very extensive ice-wedge networks (Fig. 14.5). They form large-scale polygonal patterns extending across wide areas of tundra in Alaska, Canada and Siberia. Analogous fossil patterns are often seen in the modern temperate zones of Europe and North America. Soil contrasts

Fig 14.4 Examples of three types of superficial structure due to frost action.

Involutions

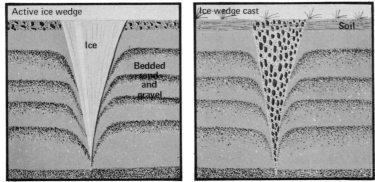

Active ice wedge

Ice

Bedded sand and gravel

Ice-wedge cast

Soil

Stone circles

give rise to crop marks which delineate very clearly the network of ancient wedges. Excellent examples have been described from the Midlands and East Anglia in southern Britain, while similar features can be traced across a broad zone of the north European plain into the USSR. The features can often be dated by reference to the age of the deposits in which they are formed. For example, ice-wedge casts in Britain are particularly common in terrace sediments dated by radiocarbon to the Weichselian glaciation; the ice-wedges themselves must have developed in a permafrost phase no earlier than the second half of the last glaciation. Stratigraphical studies have disclosed a number of earlier periods during which the wedges were formed. Dates in these cases are usually inferred from the age of any overlying deposits that remain unaffected. In East Anglia, for instance, ice-wedge casts have been observed beneath the Cromer Forest Bed. This middle Pleistocene interglacial horizon is older than any known till in the British Isles. The underlying wedge must obviously have formed at an even earlier date and is evidence of intense cold despite the lack of identifiable glacial deposits of similar antiquity.

Before leaving the subject of ice-wedges it is worth noting that not all wedge-shaped forms originate in this way. Even the decay of large tree roots can produce features resembling fossil wedges. In Antarctica Péwé has described sand wedges that are developing by the continual infilling of frost cracks with wind-blown sand. Each year more sand is added to the growing structure until eventually it may reach a width of over 1 m. Nevertheless the great majority of fossil wedges appear to have originated as ice-filled rather than sand-filled forms.

Patterned ground attributable to sorting processes Unlike the ice-wedge networks discussed above, surface patterns due to sorting typically consist of units no more than 3 or 4 m across and often much smaller. Among the many patterns that can result from debris being separated by periglacial processes into different grain sizes, two major groupings are discernible, namely, circles and stripes. Stone circles occur both singly and in groups. The larger fragments form the periphery of the circle with the fine debris in the centre (Fig. 14.4). The diameter is commonly between 0·5 and 3 m. They may be observed in many

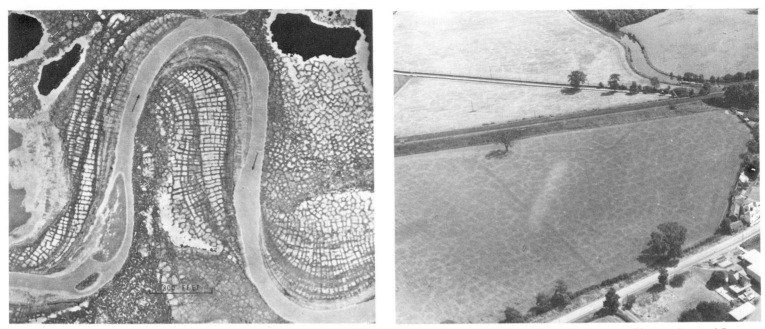

Fig 14.5 Two examples of patterned ground produced by ice wedges. (A) Active ice-wedge polygons in northern Alaska, some 50m southeast of Barrow. (B) Crop-mark patterns attesting to a system of fossil ice wedges beneath the fields of the English Midlands. The ice wedges are believed to have formed during the Weichselian glacial period.

high-latitude areas at the present day and are by no means confined to regions of permafrost. They develop best on flat or very gentle sloping surfaces. On steeper gradients the circles become deformed, gradually assuming an elliptical shape with the long axes trending downslope. Ultimately on slopes of more than about 5° they grade into stone stripes. These consist of parallel ribbons of debris, divided according to fragment size and running directly downhill. The distance between coarse stripes varies widely, but in the majority of cases is under 2 metres. In well-developed examples the stripes can be traced continuously downslope for distances of 100 m or more. They are often rendered conspicuous by the tendency of vegetation to colonize the finer material but to leave the intervening coarse debris totally bare.

Much uncertainty still surrounds the processes involved in development of circles and stripes. It is tempting to assume that the same basic mechanism is responsible for both, but few of the hypotheses advanced in explanation of stone circles seem immediately appropriate for stripes. At the present time it appears likely that several different processes, often acting in concert, are capable of producing circles, but that only a limited number of these can also be responsible for stripes.

Both field and laboratory investigations have been pursued in efforts to determine the cause of the sorting. One of the foremost workers, Corte, has stressed the need for examining not only the surface configuration but also the relationships as seen in section; to this end he regarded a bulldozer as an essential part of the geomorphologist's equipment. He found sorted polygons to be underlain by amorphous masses of ground ice which were composed of large crystals where the debris was coarse and of small crystals where the debris was fine; sorting appeared to be restricted to areas where the superficial layers contained an appreciable proportion of fine sediment. Under experimental conditions he showed that repeated freeze–thaw cycles can induce differential movement within mixed materials. Two situations may be envisaged. In the first the freezing front moves vertically downwards. Under such circumstances, the larger fragments tend to be elevated towards the surface while the fine particles are displaced downwards. Two possible mechanisms have been suggested. The first is that all the material rises together during the freezing phase but that the fines sink back more rapidly during thawing, thus preventing the coarse fragments from resuming their original positions. The alternative view is that freezing begins beneath the larger stones because of their greater thermal conductivity. This leads to differential heaving during the freezing phase, with the new relative positions maintained during the thawing phase by uneven collapse. A sorting process by either or both of these mechanisms appears well established, but of course it does not explain the surface pattern. It is here that the second situation may need to be envisaged. This involves the freezing front migrating laterally through the soil. Under such circumstances it can be shown experimentally that the fine material moves in the same direction as the cooling front and leaves behind a zone enriched in coarse debris. This mechanism might be expected to operate if the initial frost penetration were extremely irregular, but it is doubtful whether it affords the full explanation of sorted stone circles and many supplementary processes have been invoked.

The movement of large fragments to the periphery of the circles has been ascribed to upward bulging of the centres when subjected to intense freezing; that the material of the centres may be particularly frost-susceptible is indicated by measurements showing them as relatively elevated in winter but depressed in summer. A variant of this hypothesis is the idea that needle ice develops preferentially in the core of a stone circle and it is this which displaces the larger stones to the margins. A third theory that has attracted considerable support is due to Philberth. He envisages freezing starting in the coarse debris and affecting any concentration of finer debris at a slightly later stage. The fine material is then placed under considerable cryostatic pressure and may be bulged upwards. Moreover, migration of small particles before the freezing front may lead to accentuation and enlargement of the initial core areas.

The sorting exhibited by stone stripes is a purely surface phenomenon, dying out at very shallow depths. Although the coarser material has occasionally been reported to stand higher than the fine, the reverse is generally the case in actively developing examples. The size of the constituent fragments varies widely and appears to influence the distance between the stripes. Where the fragments are exceptionally large with many of the stones over 0·5 m in diameter, the distance between successive coarse stripes may be as much as 10 m; where the material is appreciably finer with few particles over 0·1 m in diameter the distance is reduced to 1 m or less. These delicate forms are the more common and in them the stones are often arranged so that they are standing on edge. In the large-scale form the cobbles and boulders tend to lie with their long axes preferentially aligned downslope. Differential frost-heaving is generally regarded as the primary mechanism for stripe development. Caine has shown that during prolonged freezing in the English Lake District the surface of the fine debris is raised some 30 mm above that of the adjacent coarse debris. This is a significant superelevation where the coarse stripes are only 0·25 m apart. It is due to the growth in the fine material of thin ice layers, individually only 2 mm thick but in aggregate attaining several centimetres. Of 80 marked stones placed on the fine debris Caine found that, in one winter, 10 had migrated into the neighbouring coarse stripes. Movement was mainly by sliding although there was also evidence of some debris having been overturned. Needle ice has sometimes been invoked as the agent causing displacement, but none was observed in the Lake District. Although frost-heaving appears adequate to explain the maintenance of the stripes, it does not by itself account for their initiation. It may be that minor variations in the original debris cover are progressively accentuated by differential heaving, but it is then necessary to explain the regular spacing. The latter property is probably due to contrasts in the rate of freezing. The coarse debris freezes first and when the cooling front later penetrates the fine materials it places them under considerable pressure. The effect is to equalize the width of the stripes at any one locality, although the actual value depends upon the nature of the debris and the rate of cooling.

Patterned ground due to sorting can develop under many different periglacial regimes. It is currently found in Greenland underlain by thick permafrost, but is also evolving on recent outwash in Iceland under much less rigorous conditions. Stone stripes have been recorded in Arctic Canada with a mean annual temperature below −10°C and on hillslopes in Wales and Scotland where even the coldest month is still above freezing point. Moreover, within even quite small areas active and fossil forms can be found side by side. In Upland Britain, for instance, delicate stripes destroyed to a depth of 0·3 m have been shown to reform within a few years; yet much larger fossil stripes can also be seen to pass beneath a recent peat cover. Great care is obviously needed in using fossil stone circles and stripes for environmental reconstruction.

Pingos The term pingo is used by the Eskimo of the Mackenzie delta in Canada to denote scattered, dome-shaped hills rising sharply above the alluvial plain. Within the area of the delta there are estimated to be some 1 500 such hills. They range from minor hillocks only 2 m high to large examples over 50 m high. Diameters vary from 10 m to over 200 m, but the sides are almost always steep and often exceed 20° in angle. The smaller examples may be perfectly dome-shaped, but many of the larger are breached by crater-like depressions 5 m or more in depth. The cores vary from enormous lenses of pure ice to beds of silt and fine sand interstratified with layers of ice that all dip outwards from the centre. Each pingo on the Mackenzie delta is surrounded either by a shallow lake or by a dried-out lake bed and Mackay has suggested that the insulating effect of the lake plays a critical role in pingo evolution (Fig. 14.6). At first the water is assumed to be adequate to prevent local permafrost, but if the climate deteriorates, or the basin fills with sediment to the point where its insulating effect is lost, the lake floor may become perennially frozen. Water trapped in the underlying sands is then placed under intense pressure as the permafrost

continues to advance. To relieve the pressure the overlying layers are forced up and finally a large mass of ice freezes in the core of the dome-shaped hill that has evolved. This view of pingo growth is sometimes referred to as the 'closed system' hypothesis.

A contrasting 'open system' hypothesis has been advanced by other workers, particularly by those who have studied pingos outside the Mackenzie delta region. They contend the necessary water originates either in a talik or in the active layer trapped by autumnal freezing. Under pressure it migrates considerable distances through unfrozen horizons and passages. Eventually it bulges up the surface at a point of weakness and freezes to form the massive ice lens that characterizes the interior of the normal pingo. That water round a pingo can be under pressure is demonstrated by occasional ruptures in which it gushes out as a minor fountain. For most pingos this 'open system' hypothesis has been favoured. However, the clear association with lake deposits on the Mackenzie delta is then difficult to explain, and Bostrom has suggested that the Mackenzie pingos may be exceptional in owing their formation to continued tectonic downwarping. As the perennially frozen deltaic sediments are carried to greater depths they thaw at the base. The result is that a permafrost layer some 90 m thick rests upon freshly thawed and highly saturated debris. If the water can gain access to the surface, it will be forced up under artesian pressure to form a shallow lake. As sediment in the lake freezes it may be bulged up into a pingo by continued water pressure from beneath.

There appears to be considerable diversity in the rates at which pingos grow. Some of the very small examples on the Mackenzie delta are known to have formed since 1950, and growth rates as high as $0·5$ m yr^{-1} have been recorded. On the other hand the larger pingos appear to be old stable parts of the landscape that, if they are growing at all, yield long-term averages of no more than $0·5$ mm yr^{-1}. A tendency towards a common evolutionary cycle can be discerned. When a pingo grows to a certain size, or when the pressure from beneath exceeds a critical value, the surface cracks and exposes the ice core. This eventually leads to melting and subsidence of the top of the dome into a crater. It can persist in this cratered form for long periods, but any climatic amelioration must lead to further decay. Ultimately all

308 that remains is a shallow pool surrounded by a rampart composed of sediment that was either pushed or slumped to the margin of the ice core. The pool may later be infilled by accumulation of organic materials.

Fossil pingos consisting of a circular rampart around a peaty hollow have now been identified in many former periglacial areas. In the Ardennes over a hundred examples have been recognized, and of those so far dated the vast majority seem to have decayed about 10 000 years ago. Many comparable features have now been mapped in central Wales, while in eastern England a number of depressions formerly attributed to limestone solution almost certainly originated in a similar fashion. As mentioned earlier, the belief that pingos only develop under the most extreme climatic conditions has had to

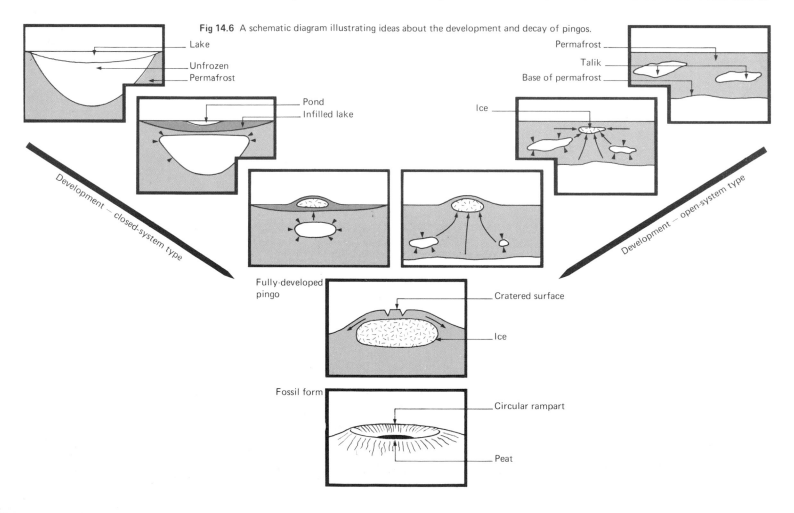

Fig 14.6 A schematic diagram illustrating ideas about the development and decay of pingos.

Lake

Unfrozen
Permafrost

Pond
Infilled lake

Permafrost

Talik

Base of permafrost

Ice

Development – closed-system type

Development – open-system type

Fully-developed pingo

Cratered surface

Ice

Fossil form

Circular rampart

Peat

be revised in view of the discovery that examples of the open-system type are actively growing in central Alaska at the present day. Although clearly indicative of permafrost, fossil pingos in our present state of knowledge permit few precise inferences about Pleistocene climates.

Cambering, gulling and valley bulging In the English Midlands the Jurassic succession of thick clays and much thinner beds of ferruginous limestone and sandstone has long been an important source of iron ore. In consequence the structures affecting these rocks have been studied in great detail. It was early recognized that, despite a simple regional dip of less than 1° towards the south-east, the ironstone beds are draped over the underlying clays in such a fashion that they consistently dip towards the local valley floors (Fig. 14.7). The beds conform so closely to topography that the structures are clearly not tectonic in origin but must be the result of superficial movements that have taken place in the course of denudation. The term cambering aptly describes the dip from each interfluve crest towards the adjacent valleys. In places the beds are cambered down more than 30 m below their regional level. The cambered beds also show clear signs of extension. The most obvious indication is the widening of joints trending parallel to the contours. Such enlarged joints may be several metres across and are known by the quarryman's term of gull. They can often be seen in ironstone pits near the feather edge of the outcrop and are normally filled with either rubbly debris or till. In areas of cambering and gulling

a third type of disturbance, known as valley bulging, is frequently **309** present. This involves disruption of the original stratification at and below the height of the valley floors. Structures in this situation are rarely well exposed, but where available for study they generally show severely contorted clays elevated many metres above their regional level. Beds of more competent rock within the clays can be totally shattered. As with cambering and gulling, there is a clear association with topography since the disturbances are often traceable for many kilometres along winding valley floors.

There is still much uncertainty regarding the origin of the structures. It is generally agreed that excess stress along the margins of each flat-topped interfluve has depressed the edges of the competent beds and extruded clay along the line of the valley. The problem is the cause of the vertical stress which the clays were unable to sustain. An early suggestion was that instability occasioned by rapid stream incision might have produced the structures, but there are many reasons for rejecting this hypothesis. Kellaway and Taylor later argued that the critical factor was not so much an increase in stress as a temporary weakening of the clay. They were able to show that most of the superficial structures developed in a relatively brief interval spanning the extremely cold periods on either side of a major glacial advance. This naturally led to the suggestion that periglacial conditions might have favoured large-scale structural disturbance, particularly during the melt phase when permafrost was gradually thawing. The melting of ground ice would then

Fig 14.7 Diagrammatic section through a hillslope that has been subject to cambering.

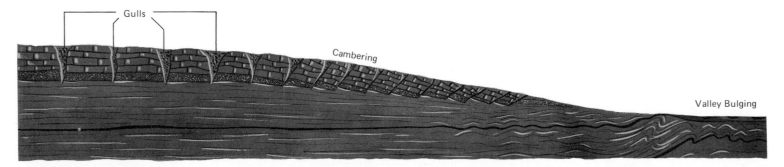

Gulls

Cambering

Valley Bulging

produce a saturated mass of clay lacking the strength to sustain the load it had previously borne. It might therefore deform under the weight of the plateau cap-rock, with much of the contortion concentrated along the valley floors where the ground ice was thickest. A third suggestion that has been offered, is that the instability was brought about by burial of the whole area beneath a major ice-sheet. Against this interpretation, however, is the fact that many of the scarps in the Weald of south-eastern England are prominently cambered and these lie beyond the generally acknowledged limit of glaciation.

Aspects of landscape evolution

The concept of a periglacial cycle Little is known about the Pleistocene climates of Arctic Canada and the USSR, but it is reasonable to assume that they have experienced lengthy intervals during which the climate was at least as severe as to-day. The prolonged operation of periglacial processes might be expected to produce a unique assemblage of landforms, and this belief has encouraged several workers to envisage a distinctive periglacial cycle of erosion. One of the foremost exponents of this idea has been Peltier, who visualized an initial landscape composed of rounded hill of moderate relief with an almost continuous cover of regolith and soil. During the youthful stage, solifluction results in bedrock being exposed on the upper slopes where the debris mantle is removed more rapidly than it is replaced by weathering. The exposed rock faces are subject to frost-sapping so they recede rapidly and soon consume the whole of each hilltop. In a mature periglacial landscape the bedrock faces have all been eliminated and the morphology is dominated by concave slopes mantled with solifluction debris. Further stages in the cycle see a gradual reduction of relief amplitude and mean slope angle but without major changes in geometrical form or dominant process. However, if the periglacial conditions are sufficiently persistent, the debris cover may be so comminuted that aeolian deflation assumes an important role.

Although the Peltier cycle may truly represent a sequence of landform changes, its practical application is fraught with difficulty. It assumes a moderately hilly relief to start with, but clearly many other initial forms could be envisaged which would follow rather different patterns of landform evolution. It pays little attention to geological structure although the relative susceptibility of rocks to frost-riving, and of the weathered mantle to frost-heaving, is of fundamental concern in modern periglacial regions. In an area of shale or clay, for instance, solifluction is likely to dominate all other processes. It is scarcely conceivable that free faces will develop because the accelerated movement of the debris cover will not exceed the rate at which the clay is mobilized to form part of the solifluction mantle. Moreover, no distinction is drawn between regions with and without permafrost, although it can be argued that this factor will be of great significance in the long-term evolution of the landforms. In general the cyclical concept has proved of very limited value in the study of periglacial landscapes.

Tors and altiplanation terraces The tendency in suitable circumstances for free rock faces to develop has two important consequences. The first is the fashioning of bedrock residuals into tors. The origin of tors has been a matter of heated controversy (p.126), but there now seems little doubt that in periglacial areas many owe their form to the stripping of earlier regolith and active frost-riving of the exposed bedrock. Two situations may be visualized. In the first the initial regolith is deep enough to conceal upstanding conical masses of little altered bedrock so that removal of the cover yields tors that are, in their main essentials, already fashioned and simply require exhumation. On the other hand, if the base of the regolith is originally smooth the craggy outline of the tor must be produced either during or after exhumation. In such circumstances varying susceptibility to freeze—thaw activity is likely to be a significant factor. Frost-riving and solifluction will preferentially remove well-jointed rock but leave zones of more massive rock relatively untouched. Both sequences thus lead to tors sited where the rock is poorly jointed. Particular uncertainty has centred around the contribution of freeze—thaw action to the shaping of tors now found in temperate areas. In some localities the amount of coarse debris incorporated in a solifluction apron leaves no doubt that substantial modification has taken place under periglacial conditions. It is difficult to obtain accurate measurements, but around some quite small tors it has been estimated that the total volume of boulders exceeds 250 000 m³.

A second consequence of the active recession of rock faces is the development of altiplanation or goletz terraces. The term altiplanation was used by Eakin in 1916 to denote a group of processes responsible for flattened hill summits and valley-side benches in the extreme climate of Alaska. In the northern Urals and Siberia similar features are known as goletz terraces. In all these arctic areas hillslopes are commonly observed to be divided into a series of almost horizontal steps. The treads vary from a few metres to several kilometres in width and the risers from 5 to 25 m in height. There is sometimes a cover of coarse debris over both the tread and the riser so that the bedrock surface is partially concealed. However, it is known to have a stepped form and many workers have ascribed altiplanation terraces to nivation processes where snowbanks persist in the concavity at the back of each tread. Solifluction and snowmelt remove the frost-shattered debris as fast as it accumulates so that there is a steady recession of the steep face into the hillside. As with many other periglacial processes, the mechanism by which the form is maintained seems relatively clear but the factors controlling its initial development and siting are still obscure. It has been suggested that in a number of instances altiplanation terraces have been mistaken for old erosion surfaces ; it is an important aspect of the terraces or benches that they have no cyclic significance.

Asymmetrical valleys In a periglacial environment where freeze–thaw processes are so dominant, aspect is obviously a potential factor of great importance. There have been many descriptions of asymmetrical valleys believed to owe their form to processes operating with different efficiencies on the sunny and shaded slopes. However, there is still remarkably little agreement about the precise nature of the controls. Insolation is not the only climatic factor that needs to be considered since the prevailing wind direction, particularly as it affects snowdrifting, could conceivably be important. Given that the valley sides at the onset of a periglacial episode are of equal declivity, asymmetry can be produced in a variety of ways. One side may be steepened more than the other, one side may be

flattened more than the other, or one side may be steepened while the other is flattened. The majority of workers in the northern hemisphere have recorded south-facing slopes as steeper than north-facing. On slopes facing south and south-west the greater insolation seems likely to produce greater temperature fluctuations, more freeze–thaw cycles and therefore a much more active debris mantle. It has been maintained that this will lead to steepening of the slope angle, although the arguments sometimes proferred in support of this contention seem rather fragile. It would appear necessary to proceed one stage further and to argue that accelerated movement of regolith provides greater opportunity for frost-sapping and the development of free rock-faces. The slope profile would then be much more irregular, with certain segments appreciably steeper than the corresponding segments on the opposite side of the valley. It has also been suggested that the gentler slopes facing north-east are due to shelter from westerly winds permitting a thicker snow cover to accumulate. The slow release of meltwater would then provide abundant moisture to promote the downhill movement of the debris mantle. By contrast, in an unusually detailed study in north-west Alaska Currey found that, of over 200 cross-sections, approximately one-third displayed no asymmetry, in over half the north-facing slope was steeper, and in only one-eighth was the south-facing slope steeper. He attributed the dominant asymmetry to more intense weathering on the south-facing slopes with the production of finer debris which moves more rapidly as a soliflual mass. The mobile sheet accumulates at the foot of the slope and pushes the stream against the opposite valley wall which is thereby undercut and steepened. The arguments adduced by Currey seem intrinsically sound, and yet valley asymmetry is often reputed to be the reverse of that which he describes from Alaska. This underlines the complexity of the valley-side systems and reinforces the need for much further research into the way they operate in a periglacial environment.

BALL, D. F. and R. GOODIER, 1970. 'Morphology and distribution of features resulting from frost action in Snowdonia', *Field Studies*, 3, 193–217.

BENEDICT, J. B., 1970. 'Downslope soil movement in a Colorado alpine region ; rates, processes and climatic significance', *J. Arct. Alp. Res.*, 2, 165–226.

BOSTROM, R. C., 1967. 'Water expulsion and pingo formation in a region affected by subsidence', *J. Glaciol.*, 6, 568–72.

CAINE, T. N., 1963. 'The origin of sorted stripes in the Lake District, northern England', *Geogr. Annlr.*, 45, 172–9.

CAINE, T. N., 1972. 'The distribution of sorted patterned ground in the English Lake District', *Revue Geomorph. dyn.*, 21, 49–56.

CHANDLER, R. J., 1970. 'Solifluction on low-angled slopes in Northamptonshire', *Q. Jnl. Eng. Geol.*, 3, 65–9.

CORTE, A. E., 1962. 'Vertical migration of particles in front of a moving freezing plane', *J. geophys. Res.*, 67, 1085–90.

CORTE, A. E., 1963. 'Particle sorting by repeated freezing and thawing', *Science N.Y.*, 142, 499–501.

CURREY, D. R., 1964. 'A preliminary study of valley asymmetry in the Ogotoruk Creek area, north-western Alaska', *Arctic*, 17, 84–98.

KELLAWAY, G. A. and J. H. TAYLOR, 1953. 'Early stages in the physiographic evolution of a portion of the East Midlands', *Q. Jl geol. Soc. Lond.*, 108, 343–75.

MACKAY, J. R., 1962. 'Pingos of the Pleistocene Mackenzie River delta area', *Geogr. Bull.*, 18, 21–63.

MACKAY, J. R., 1972. 'The world of underground ice', *Ann. Ass. Am. Geogr.*, 62, 1–22.

MILLER, R., et al., 1954. 'Stone stripes and other surface features of Tinto Hill', *Geogr. J.*, 120, 216–19.

MULLER, F., 1968. 'Pingos, modern' in *Encyclopaedia of Geomorphology*, Reinhold Book Corp.

PELTIER, L. C., 1950. 'The geographic cycle in periglacial regions as it is related to climatic geomorphology', *Ann Ass. Am. Geogr.*, 40, 214–36.

PEWE. T. L., 1959. 'Sand-wedge polygons in the McMurdo Sound region, Antarctica', *Am. J. Sci.*, 257, 545–52.

PEWE, T. L., 1966. 'Paleoclimatic significance of fossil ice wedges', *Biul Peryglacjalny*, 15, 65–73.

PEWE, T. L., et al., 1969. 'Origin and paleoclimatic significance of large-scale patterned ground in the Donnelly Dome area, Alaska' *Spec. Pap. geol. Soc. Am.*, 103.

PHILBERTH, K., 1964. 'Recherches sur les sols polygonaux et stries', *Biul. Peryglacjalny*, 13, 99–198.

PIGGOTT, C. D., 1965. 'The structure of limestone surfaces in Derbyshire', *Geogr. J.*, 131, 41–4.

PISSART, A., 1963. 'Les traces de "pingos" du Pays de Galles (Grande Bretagne) et du plateau des Hautes Fagnes (Belgique)', *Z. Geomorph. N.F.*, 2, 147–65.

SCHENK, E., 1955. 'Die periglazialen Strukturboden-bildungen als Folgen der Hydratations-vorgänge im Boden', *Eiszeitalter und Gegenwart*, 6, 170–84.

SOONS, J. M., 1971. 'Factors involved in soil erosion in the Southern Alps, New Zealand', *Z. Geomorph.*, 15, 460–70.

SOONS, J. M. and D. E. GREENLAND, 1970. 'Observations on the growth of needle ice', *Wat. Resour. Res.*, 6, 579–93.

SPARKS, B. W., et al., 1972. 'Presumed ground ice depressions in East Anglia', *Proc. R. Soc.*, A 327, 329–43.

WAHRHAFTIG, C. and A. Cox, 1959. 'Rock glaciers in the Alaska Range', *Bull. geol. Soc. Am.*, 70, 383–436.

WATSON, E., 1972. 'Pingos of Cardiganshire and the latest ice limit', *Nature Lond.*, 236, 343–4.

WATSON, E. and S. Watson, 1972. 'The coastal periglacial slope deposits of the Cotentin peninsula', *Trans. Inst. Brit. Geogr.*, 49, 125–44.

Two admirable surveys of periglacial features, together with extensive bibliographies, are provided by A. L. Washburn, *Periglacial Processes and Environments*, Arnold, 1973 and C. Embleton and C. A. M. King, *Periglacial Geomorphology*, Arnold, 1975.

4|15

Coastal processes

In certain respects the coast provides the geomorphologist with exceptionally fine opportunities for studying contemporary processes. A beach, for example, is one of the most changeable of all landforms, and even cliffs are subject to more rapid modification than the majority of physiographic features. Yet the coast also presents particular difficulties for the geomorphologist interested in the measurement of processes. Many of the changes take place in the zone of wave action where direct observation is by no means easy. Moreover there is an increasing awareness of the intricate exchange of sediment between the foreshore and offshore zones. At one time coastal geomorphology tended to concentrate almost exclusively on two basic subjects, the erosion of cliffs and wave-cut platforms and the redistribution of sediment by longshore drifting. It is now recognized, however, that the narrow ribbon between high- and low-water marks cannot be treated in isolation but must be regarded as part of a much broader zone of erosion and accretion. Study of sediment movement across the full width of that zone poses obvious problems, but fortunately a long tradition of investigation for engineering purposes has evolved sophisticated instrumentation.

The study of water movement is clearly fundamental to an understanding of coastal processes. For a long time tides were regarded as the dominant influence in shaping the edge of the land and virtually all marine features have, at one time or another, been attributed to their action. With inadequate observational data to form a sound judgement, a reaction ensued by which almost every coastal form was ascribed to wave activity. Recent investigations indicate that the pendulum swung too far and that in certain circumstances tides can play a significant role in the fashioning of the coast. Their role is two-fold: they raise and lower the level of wave attack, and by mass·transfer of water generate currents capable of transporting large quantities of fine sediment. The way in which water movement arising from both tides and waves is translated into processes of erosion and sediment transport forms the major theme of the second part of this chapter.

Tides

Tide-generating forces Tides differ from other types of energy input in geomorphology by being attributable to neither solar nor terrestrial heat. Admittedly their amplitude may be affected by temporary meteorological conditions, but basically they are the outcome of gravitational attraction between the earth, moon and sun. The whole subject of tide generation is one of immense complexity, but for present purposes is most easily approached by considering first a highly idealized situation (Fig. 15.1). One may imagine the earth to be entirely covered with water and subject only to attraction by the moon moving in a circular orbit which it completes once every 28 days. For the whole system to remain in equilibrium the earth and moon must move round a common centre of gravity since the gravitational and centrifugal forces must be exactly balanced. The common centre is found to lie within the earth at a depth of about 1 700 km. The simplest type of motion about this point capable of maintaining the equilibrium is the revolution of the earth. The centrifugal force in this case is everywhere the same. The lunar gravitational attraction, on the other hand, being inversely proportional to square of the distance from the centre of the moon, must vary from place to place on the globe. It has a mean value equal to the centrifugal force at the centre of the earth, rises to a maximum at the sublunar point closest to the moon and falls to a minimum at the antipodal position. It follows that the tide-generating force at any point is essentially equal to the deviation of lunar gravity from its mean value, and that there is a

maximum residual force acting towards the moon on one side **315** and away from it on the other. Strictly speaking, the tide-generating force needs to be resolved into a horizontal tractive force to explain the actual movement of water. Tractive values are zero at the sublunar point and its antipodes, and also along the great circle at an angular distance of 90°; they reach a maximum value at an angular distance of 45°. The tractive forces are all directed towards the points closest to the moon or furthest therefrom. Water being drawn from the great circle towards these two locations transforms the spherical shape of the water surface into a prolate spheroid with its major axis directed towards the moon. The actual distortion from sphericity will be minute since the extreme deviation of lunar gravity from its mean figure is equivalent to no more than one nine-millionth part of the earth's own gravity. In the hypothetical circumstances at present being visualized, antipodal tidal bulges will each proceed round the globe in 28 days.

It is now possible to relax some of the artificial constraints so far imposed. The moon moves in an elliptical rather than a

Fig 15.1. Schematic protrayal of the tide-generating forces.

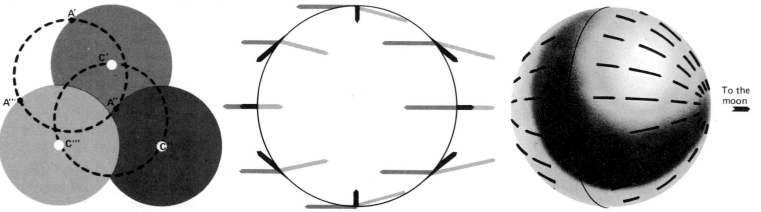

A. Revolution of the earth without rotation.

Note that all points on and within the sphere describe a similar circle (e.g. C′, C″, C‴ and A′, A″, A‴), indicating that under these conditions centrifugal force is everywhere the same.

B. The tide - producing forces.

Note that these vary with latitude, being the resultant of the uniform centrifugal force and the variable attractive force of the moon.

C. The tractive force.

These reach a maximum at 45° from the line to the moon and decline to zero at 0° and 90°.

circular orbit so that its tide-generating force varies with a periodicity of approximately 28 days. The tidal range will consequently show an oscillatory cycle depending on the distance between the moon and earth. A fluctuation of even greater amplitude is introduced by the sun. The solar tide-generating force is also capable of deforming the ocean surface into a prolate spheroid, although the potential deviation from the sphere in this case is just under half that due to the moon. In practice, of course, the tidal effects of sun and moon interact. When all three bodies lie in a straight line the lunar and solar tides reinforce each other and it is under these circumstances that the greatest tidal ranges occur. Conversely, when the bodies lie in quadrature the gravitational pulls tend to counteract each other and the tidal ranges are at a minimum. This means in effect that the amplitude of the tides tends to be greatest during periods of full and new moon, and least at time of the first and third quarter. The former are known as spring tides and the latter as neap tides. A minor complication is introduced by the fact that the lunar orbit lies at an angle of about 5° to the plane of the ecliptic; one consequence is that spring and neap tides themselves show systematic oscillatory patterns.

Observed oceanic tides In the preceding section attention was confined to the tide-generating forces that arise from relative movements of the earth, moon and sun. The tide produced under these idealized conditions is known as the equilibrium tide, and is obviously very different from the changes in sea-level observed at coastal gauging stations. Three major factors explain the difference between the equilibrium and observed tides: the earth's rotation, the configuration of the ocean basins and temporary meteorological conditions. The first two are systematic and can be allowed for in tidal predictions. The third is a random factor which can cause significant deviation from the predicted levels.

Rotation of the earth has been ignored so far since it is not a tide-generating force. The globe is simply spinning within the gravitational fields of the sun and moon. With respect to the sun each full rotation takes 24 hours. However, owing to passage of the moon along its orbital path a full rotation with respect to that body takes about 24 hours 50 minutes. Because lunar gravitation is the dominant influence high tides are commonly observed at intervals of 12 hours 25 minutes; these are known as semi-diurnal tides. If the moon's orbit and the equator lay in the same plane successive high tides would reach approximately the same level. However, the rotational axis of the earth is tilted with respect to the lunar orbit. This means that only twice in one orbit does the sublunar point move along the equator. At other times it moves along parallels of latitude as far north and south as the two tropics. When it moves along the Tropic of Cancer the antipodal point will be moving along the Tropic of Capricorn and vice versa. At any location close to the tropics there will be a marked inequality of successive semidiurnal tides. Another important consequence of rotation is the Coriolis effect which becomes particularly significant when the division of the earth's surface into separate ocean basins is considered.

The configuration of any water-filled basin is of fundamental importance in determining its response to the tide-generating forces. It is simplest to consider first a hypothetical basin of great length but small width. When the water in such a unit is disturbed, it moves backwards and forwards heaping up first at one end and then at the other. The form is known as a standing wave, with a central nodal line at which the water elevation remains constant. The period of oscillation depends upon the rate of travel of a normal progressive wave along the trough. Since a wave travels at a speed equal to $\sqrt{gD}$ where g is gravitational acceleration and D is depth, the period may be defined as $2L/\sqrt{gD}$ where L is the length of the basin. In other words it is the time taken by a wave to move from one end of the basin to the other and back again. If it is approximately equal to the interval between successive high tides a resonance effect will increase the amplitude of the tides. If the period is either much shorter or much longer, the oscillations will be out of phase and the tidal range will be correspondingly smaller.

In this way the natural period of oscillation has a profound effect upon tidal range. However, in a normal ocean basin the water does not simply move backwards and forwards in a narrow trough but is capable of moving laterally. In such movement the Coriolis effect deflects the water to the right in the northern hemisphere and to the left in the southern. The result is to produce a high tide that sweeps around the margins of a basin in an anticlockwise direction north of the equator and in a clockwise direction south of the equator. Instead of a nodal line there will be a central point of zero tidal range known as an amphidromic point. Most of the ocean and sea basins act as one or more large amphidromic systems. The exact configuration of some of these systems is still imperfectly known since it is very difficult to measure tidal changes away from the coast. However, in the North Sea basin three separate systems have been identified giving the complex pattern of tidal ranges and co-tidal lines shown in Fig. 15.2.

Tidal currents Tidal rise and fall obviously requires large-scale horizontal transfer of water. This motion is known as the tidal current and in the open ocean assumes a slow rotatory form. For the geomorphologist much more interest attaches to the currents engendered in shallow water closer to the shore. Here the rotatory movement tends to be replaced by an alternating one in which the water moves for just over 6 hours in one direction and then approximately reverses that direction for the next 6 hours. Two aspects of such tidal currents are of vital importance : the maximum velocity at the sea-bed which will determine the size of any sediment that can be entrained, and the circulatory pattern which will control the net transport direction. It must certainly not be assumed that there is an exact reversal of the tidal currents every 6 hours or so, since field investigations nearly always reveal a much more complex circulatory system. Several techniques have been devised by which measurements can be made. The simplest involves the use of sea-bed drifters. These consist of plastic discs carefully weighted to travel close to the sea-floor and capable of moving under the influence of low-velocity currents. Released at predetermined points they drift under the influence of tidal and other currents until washed up on the coast. They often have an attached card which the finder is asked to return to the investigator with details of place and time of recovery. Their value lies in the guidance they provide regarding overall directions of bottom movement; unfortunately the precise route followed between the points of release and recovery remains unknown, although it may often be inferred from other evidence. To obtain fuller information more elaborate methods must be employed. One technique involves submerging to known depths small bottles partly filled with hot liquid jelly and enclosing a compass card. The alignment of the bottle is recorded by the setting of the jelly, and from its tilt and orientation the current velocity can be assessed. In water 7 m deep off the East Anglian coast tidal currents have been measured by this method at velocities up to 5 km hr^{-1}. At a depth of over 150 m in the Bay of Biscay a tidal stream with a velocity of 1 km hr^{-1} has been recorded. It is not practicable with jelly bottles to attempt continuous monitoring, and for this much more complex instrumentation is required. By installing arrays of specially designed current meters a continuous record of water flow over a complete tidal cycle may be obtained. When plotted in the form of a vector diagram, the resulting data will often disclose significant residual net movements (Fig. 15.3).

The most elaborate studies have been undertaken in estuaries where patterns of tidal flow are often extremely intricate. Movement may be concentrated in relatively narrow threads of faster-flowing water, with the positions of maximum-velocity threads on the flood tide differing from those on the ebb tide. In this way flood and ebb currents may be spatially differentiated, as is well exemplified in Fig. 15.8. Many factors impinge upon details of the circulatory system. The shape and depth of the estuary will have an important bearing on the velocity attained by the inflowing and outflowing water. In all cases there must be a net outflow as a result of freshwater discharge into the head of the estuary; however, the ratio of river discharge to the volume of the tidal prism entering and leaving the estuary varies greatly from one example to another. The differing densities of river and sea water can have a profound effect on the circulation. In some estuaries a very pronounced stratification is found with the less dense fresh water passing seawards over a bottom layer of salt water; owing to the Coriolis effect the interface, as viewed from the land, often slopes down to the right in the northern hemisphere. In other cases mixing is so complete that the estuary waters can be regarded as virtually homogeneous. A final factor worthy of mention is the meteorological situation which, among other things, influences the volume of tidal prism entering and leaving the estuary. It is evident that many of the influences on the circulatory system vary from time to time and that, for a full understanding, observations need to be made over a wide range

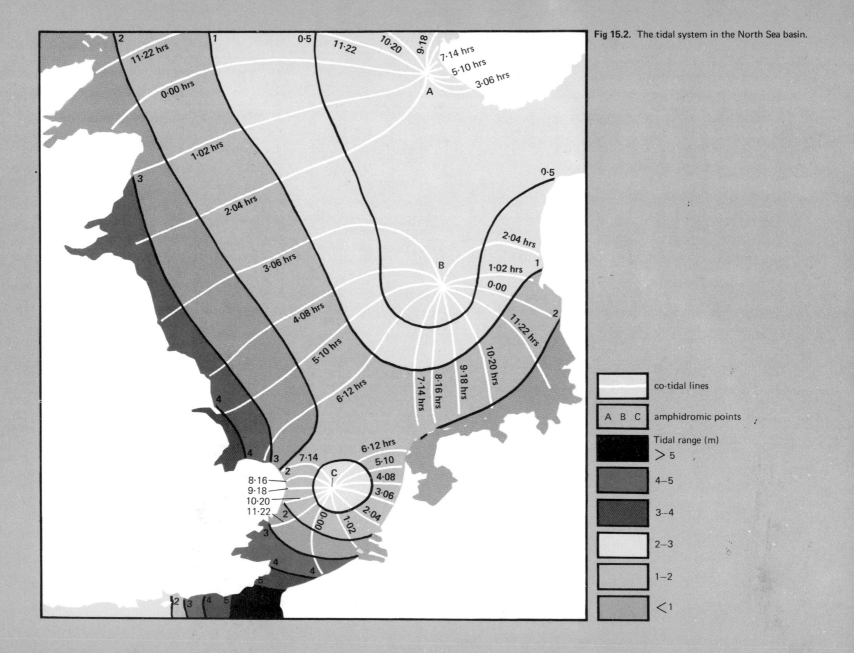

Fig 15.2. The tidal system in the North Sea basin.

co-tidal lines

A B C   amphidromic points

Tidal range (m)

> 5

4—5

3—4

2—3

1—2

< 1

of conditions.

Many of the ideas derived from estuarine studies are also applicable to the open coast and, to a lesser extent, to the continental shelf as a whole. The most fundamental finding as far as the geomorphologist is concerned is that tidal movement does not consist of a uniform, low-velocity drift regularly reversing in direction, but involves instead a concentration into distinct tidal streams which can have a profound effect on sediment transport and sea-bed morphology.

Waves

Wind-generated waves in deep water The vast majority of sea waves are generated by wind blowing over the water. They represent a transference of the kinetic energy of the atmosphere to the hydrosphere, with most of this energy ultimately expended where the waves break along the coast. It is important to distinguish between waves in deep water which are oscil-

Fig 15.3. An example of a tidal cycle plotted as a vector diagram from measurements made at a single point in the Elbe estuary. The plot is of water movement; sediment movement is likely to be significantly different since only at certain periods will the water reach the critical velocity required to entrain bed material

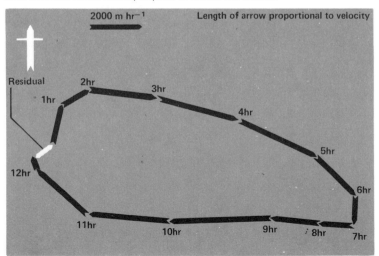

latory in character and those that break along the shoreline which are translatory in character. In the former the motion of the individual water particle is approximately circular so that it returns very nearly to its original position after the passage of the wave. In a wave of translation, on the other hand, the water particles move forward at approximately the same velocity as the wave form.

A standard terminology is applied in the description of waves (Fig. 15.4). It is found that the profile of a wave with a very low steepness ratio approximates to a sine curve, but that steeper waves are more closely trochoidal in form. A trochoid is the curve traced out by a point within a circle when that circle is rolled along a straight line. It has a flatter trough and sharper crest than the sine curve. It is a matter of common observation that the orbits described by the water during the passage of a wave are not exactly circular but involve a slow forward movement. This gradual creep in the direction of wave advance is known as the mass transport velocity. It increases rapidly with wave steepness and can be extremely important near the coast where it leads to the piling up of water and the raising of the tide above its predicted level.

The precise mechanism by which waves are generated remains uncertain. When the wind blows across a smooth water surface pressure differences due to turbulence will cause random stress variations. These will be adequate to generate irregularities of different shape and size; it is the way in which certain of these irregularities grow into fully-fledged waves while others decline that is not entirely understood. Phillips has suggested that there may be oscillatory pressure fluctuations of a frequency coinciding with that of certain of the incipient waves which would then be preferentially developed by a form of resonance. Clearly, an important factor is the increasing effect of the disturbed water surface upon the pattern of air flow. Many workers have argued that there must be significant pressure differences between the windward and leeward sides of a developing wave, complemented by the greater drag stress on the windward side.

In the absence of an entirely satisfactory theory of wave generation, much attention has been devoted to the collection and analysis of empirical data. It is generally agreed that there

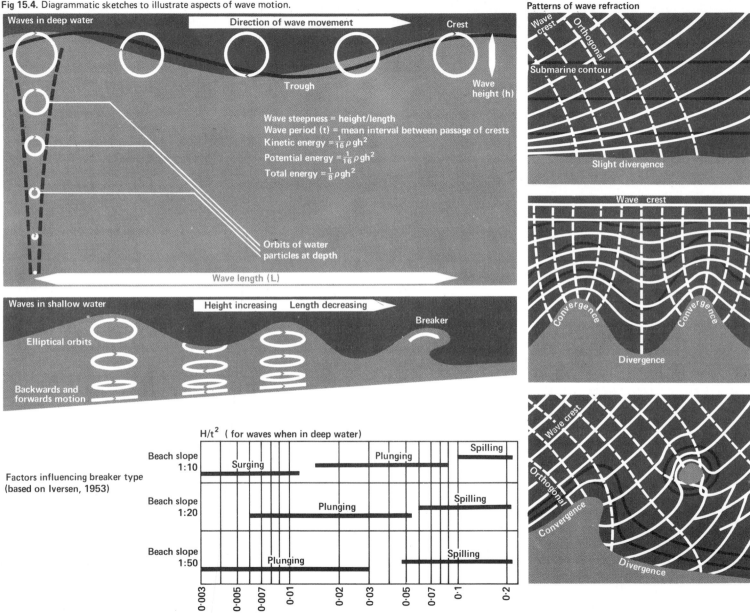

320 Fig 15.4. Diagrammatic sketches to illustrate aspects of wave motion.

Patterns of wave refraction

Waves in deep water

Direction of wave movement

Crest

Trough

Wave height (h)

Wave steepness = height/length
Wave period (t) = mean interval between passage of crests
Kinetic energy $= \frac{1}{16}\rho gh^2$
Potential energy $= \frac{1}{16}\rho gh^2$
Total energy $= \frac{1}{8}\rho gh^2$

Orbits of water particles at depth

Wave length (L)

Waves in shallow water

Height increasing   Length decreasing

Breaker

Elliptical orbits

Backwards and forwards motion

H/t^2 (for waves when in deep water)

Beach slope 1:10

Surging

Plunging

Spilling

Factors influencing breaker type (based on Iversen, 1953)

Beach slope 1:20

Plunging

Spilling

Beach slope 1:50

Plunging

Spilling

0·003 0·005 0·007 0·01 0·02 0·03 0·05 0·07 0·1 0·2

Wave crest

Orthogonal

Submarine contour

Slight divergence

Wave crest

Convergence   Convergence

Divergence

Wave crest

Orthogonal

Convergence

Divergence

are three basic controls of oscillatory deep-sea waves; wind velocity, wind duration and fetch. Any one of these can act as a limiting factor. For example, fetch constitutes an important limiting control in the development of waves in both the North Sea and the Irish Sea, whereas in the Atlantic and Pacific Oceans it is wind duration and velocity that impose the ultimate limit on wave size. Tables have been prepared relating the three variables to observed wave height. For example, the fetch of an easterly wind blowing across the North Sea to the English coast is about 600 km. With a wind blowing at 50 km an hour for a period of 30 hours, waves with a height of just over 4 m might be expected. In the Irish Sea with a fetch restricted to 200 km the same wind would generate waves under 3 m high; in the Atlantic it would produce waves up to 6 m high.

In practice such figures need to be treated with caution. Rarely is a single train of uniform waves produced. Interacting trains result in complex wave patterns with locally exaggerated highs and lows where two crests or two troughs coincide. The waves may then steepen dramatically until the crests overturn and 'whitecaps' are formed. Moreover, the figures are restricted to true 'sea', that is, forced waves produced directly by wind action. They do not apply to swell once the waves have moved outside their immediate area of generation. Many wave trains have now been traced thousands of kilometres from their source regions, and the changes that they undergo have been analysed. The height gradually declines but the velocity remains fairly constant. There is also a small increase in the wave period indicative of growing wave length. Storm waves initially 10 m high diminish to swell little more than 2 m high after travelling 2 000 km, while the typical period extends from under 10 seconds to about 15 seconds. These changes are important for the geomorphologist since they gradually alter the characteristics of the ocean waves reaching the coastal zone.

Wind-generated waves in shallow water In the open sea the orbital motion associated with the passage of a wave diminishes rapidly from the surface downwards. At a depth equal to one wavelength the orbital movement is virtually imperceptible, theoretically being about 1/500 of that experienced at the surface. At a depth equal to one-half of the wavelength the orbital diameter is about 1/20 of that at the surface. It is not surprising, therefore, that waves only begin to 'feel bottom' and suffer rapid modification of form when they enter water of a depth equivalent to less than one-half of the wavelength. Although the period remains constant, nearly all other properties change. Velocity diminishes until, in very shallow water, it conforms to the equation for other wave forms of $V = \sqrt{gD}$. Since the period is unchanging, there must be a proportionate decrease in length and the waves appear to be piling up along the shoreline. Wave height increases only slowly at first but with increasing rapidity as very shallow water is entered. The opposed changes in height and length mean that the steepness increases dramatically before the waves actually break. An associated change affects the orbital paths of the water. In the open sea these are approximately circular, but in shallow water they become elliptical with the long axis horizontal. Close to the sea-floor the orbits are flattened until there is simply a backwards and forwards motion of the water. During the passage of a crest the water moves rapidly landwards, while during the ensuing trough it recedes slowly seawards.

There are two related explanations for waves breaking on entry into very shallow water. The first stresses that the enhanced wave height, accompanied by the change to elliptical orbits, implies an increase in orbital velocity. The water is therefore actually accelerating at the very time the rate of travel of the wave form is diminishing. In effect, the water overtakes the wave form and plunges forward unsupported. The second explanation points to the diminishing volume of water within the wave despite the enlarged orbital movement. Ultimately a point is reached when the water is inadequate to complete the orbit and a cavity forms on the landward side of the wave. Once again the crest of the wave is unsupported and plunges forward towards the beach. It is a matter of observation that waves normally break when the depth of water is about 1·3 times the wave height.

Three different types of breaker may be distinguished. The first is the spilling breaker in which the water never falls freely but constantly spills down the front of the wave to provide a prominent foaming crest. The wave form is preserved but gradually declines in height as the water advances up the beach. Several spilling breakers can often be seen advancing simultaneously one behind the other. The second type is the plunging breaker. In this case the water at the crest falls vertically and traps a pocket of air within the wave. Following the plunge the wave form disintegrates and the water rushes forward as a turbulent foaming mass. The third type is the surging breaker in which the water lunges forward with a rapidly collapsing crest but with no clearly defined plunge. Although waves can often be assigned to one or other type with confidence, it is also possible to encounter intermediate forms. For instance, some waves first break by spilling but later plunge when they pass into much shallower water. The factors determining whether a wave spills, plunges or surges are not fully understood. It has been maintained that a plunging breaker occurs where the velocity of water in the wave crest significantly exceeds that of the wave body, while a spilling breaker occurs where the two velocities are approximately equal. Empirically it has been shown that the two dominant factors are wave steepness and beach gradient. Spilling breakers occur with steep waves, plunging breakers with intermediate waves, and surging breakers with gentle waves. Beaches of low gradient are more commonly associated with spilling than plunging or surging breakers.

The shoaling of the water also has an important effect on the angle at which the waves strike a coast. There is a pronounced tendency for waves, irrespective of their direction of approach, to swing round until more nearly parallel to the shoreline. Known as refraction, this process begins as soon as the waves feel bottom and are thereby retarded. It is seen in its simplest form where a train of waves approaches a straight and uniformly sloping coastline at an oblique angle. The wave crests curve round until they approach the shore almost at right angles. It is unusual for refraction to be complete so that there is normally some residual obliquity. In the case of an irregular, indented coastline the situation is more complex. It must be remembered that the basic control is water depth so that the pattern of refraction depends ultimately upon the submarine contours. However, in most cases the waves first feel bottom opposite the headlands while the advance into the bays continues relatively unimpeded. A very important consequence of refraction in such circumstances is the concentration of wave energy on the headlands. The breakers here will be very much larger than those in the bays.

A final aspect of wave action in shallow water is the tendency for mass transport in a landward direction to induce circulatory current systems. These consist of two major components, longshore currents parallel to the coast and rip currents normal to the coast. The simplest case is again the straight uniform coast with obliquely approaching waves. Longshore currents with a velocity of up to 3 km hr^{-1} may then be initiated, with the accumulation of water at the coast relieved by occasional rip currents as exemplified in Fig. 15.5. The strength of the longshore and rip currents will vary with the rate of mass transport, the result being that diverse circulatory systems may be observed at different times on the same section of coast. For example, it has been found on some sandy beaches that storm waves produce a few, very powerful rips whereas smaller waves induce numerous minor rips. On indented coastlines a longshore current is often found moving from the headlands into the bays; this results from wave refraction causing an uneven mass transport pattern with a slight raising of water-level on the headlands. Compensating rip currents flow seawards near the centres of the embayments.

Other coastal waves Of all coastal waves, infinitely the most common are the normal wind-generated type described above. However, brief reference must also be made to occasional waves of other origin since some of these are capable of modifying coastal landforms out of all proportion to their frequency. Two types are of particular importance, tsunamis and storm surges. As wave forms both are characterized by their immense length and extended period.

A tsunami is generated by some form of submarine disturbance, most commonly a displacement of the sea-floor during seismic activity. The resulting oscillatory waves can travel thousands of kilometres with little loss of energy. The wavelength often extends to several hundred kilometres and the period to over an hour. Velocity may be inferred from the time it takes a tsunami to cross an ocean basin; mean speeds of about 700 km hr^{-1} have been recorded in this way. By contrast the heights of tsunamis in the open ocean are very small. With a vertical amplitude of less than 1 m they are normally imperceptible at sea where their passage is obscured by the much higher wind-generated waves. However, as they approach the coast dramatic changes occur. The shoaling of the water leads to marked deceleration accompanied by a rapid growth in height. A ten-fold change in height is not unusual and the waves that reach the coast can attain very formidable proportions. They do not normally break in the fashion of ordinary waves, but owing to the great concentration of energy can wreak immense damage on low-lying coasts. They have been responsible for a number of major disasters with great loss of life, and this has encouraged international cooperation in their study. A particularly thorough investigation has been made of the waves generated by the Alaskan earthquake of 1964. Those that spread across the Pacific had a period of about 108 minutes, but as they entered shallower water they often gave rise to secondary waves of higher frequency. The study underlined the role played by the configuration of the continental shelf in controlling the nature of waves that actually strike a coast. In this way local factors can profoundly affect liability to attack by tsunamis.

A storm surge may be regarded as a form of long wave initiated by exceptional weather conditions. It normally requires a cyclonic centre of unusually low pressure, together with the high-velocity winds capable of inducing mass transport on a very large scale. By themselves reduced atmospheric pressure and on-shore mass transport can raise sea-level by about 2·5 m. A further contributory factor in the largest storm surges is some degree of resonance. This occurs, for instance, where the induced wave moves at the same velocity as the storm centre itself. Many tropical hurricanes produce surges of exceptional magnitude where they pass over wide, shallow continental shelves. Under such circumstances sea-level may be elevated by at least 5 m and very extensive areas of low-lying coast inundated. Surges of this type are well exemplified by those that afflict the Ganges delta when a tropical hurricane develops over the Bay of Bengal. In November 1970 an estimated 600 000 people were killed in what has been described as the greatest natural disaster ever experienced. Certain other areas of the world are subject to storm surges which owe their severity mainly to the configuration of the coast. The North Sea falls into this category. Since it opens northwards into the Atlantic Ocean periods of exceptionally strong northerly winds drive huge volumes of water into the basin. During the great storm surge of 1953, for instance, it is estimated that $42·5 \times 10^{10}$ m³ of water, enough to raise the mean level of the North Sea by over 0·6 m, entered the basin in a 15-hour period. The passage of this water could be traced as a great wave moving southwards along the east coast of England and gradually rising in height. By the time it reached the Rhine estuary it had attained a height of about 3 m and in flooding across 1 800 km² of reclaimed land was responsible for 1 800 deaths.

Marine erosion

Mechanical processes Waves approaching the coast possess kinetic energy by virtue of the orbital motion of the water, together with potential energy arising from the elevation of the wave crest above mean water-level. When the waves break the potential energy is converted into kinetic energy, greatly enhancing the capacity of the water to produce such features as cliffs and shore platforms. At least two mechanical processes are commonly involved, hydraulic action and corrasion.

Hydraulic action refers simply to the impact of the moving water. Anyone who has walked along a coastal promenade subject to battery by large storm waves can testify to the veritable hammer-blows that it receives. Dynamometers installed in a number of coastal protection structures have recorded momentary pressures amounting to well over 50 kgf cm⁻². Such pressures are not sustained for more than a fraction of a second, and in any case are much less than the crushing strength of most natural rocks. Nevertheless, repeated battering of this ferocity is certainly capable of enlarging incipient fracture patterns and helping to divide an apparently massive rock into smaller joint-bounded blocks. These blocks need not be small before they can be dislodged, since the momentum of the waves is capable of displacing fragments weighing many tonnes. The process is sometimes described as quarrying, and the significance of rock structure is often clearly visible in the detailed form of the cliff face. Major joint and fault planes are preferentially attacked, and

324 deep narrow inlets excavated along lines of structural weakness. A factor that is often emphasized is the abrupt compression and decompression of air trapped in crevices and hollows within the rock. The real effectiveness of this process has yet to be established, although it undoubtedly transmits some of the pressure changes associated with breaking waves deep into the rock face. Blow holes far to the landward of the cliff can expel large volumes of air during periods of storm waves, although it should not automatically be assumed that air compression is responsible for initiating the holes.

The efficiency of hydraulic action depends very much upon the precise location of the wave break-point. If a wave breaks more than a few metres in front of a cliff face much of its energy is dissipated in the turbulent motion of the water. If, on the other hand, it fails to break before reaching the cliff the wave form will be reflected seawards and the energy expended on the rock face will be minimal. The interference of the advancing and reflected waves produces the form known as clapotis. It is when a breaker makes its initial plunge directly against a cliff face that the greatest pressures are generated. Since it is shoaling that causes a wave to break, vertical cliffs descending directly into deep water will be much less prone to hydraulic action than those which rise from a moderately wide basal platform.

Corrasion is the action of rock fragments regularly set in motion by the sea water. The particles may range in size from sand grains to large cobbles. It is difficult to obtain quantitative data, but the effect of abrasion is readily apparent in the prominent rounding of marine shingle. In part this is doubtless due to attrition as the fragments hit against each other, but it is also the result of impact with exposed rock surfaces. Smoothed and polished faces bear testimony to this action, while swirling water armed with small pebbles and sand frequently drills out circular pot-holes.

Both hydraulic action and corrasion are most effective under conditions of high-energy storm waves. Exposure is therefore an important factor in determining variations in the rate of mechanical erosion. On a regional scale, a long fetch tends to favour the development of powerful storm waves, while more local variations may be associated with refraction concentrating attack on prominent headlands. At an even more detailed scale, such factors as the growth of seaweed can influence the effectiveness of the waves; a thick cover of seaweed will act as a protective agent by sharply reducing the velocity of the water in contact with the rock.

Chemical processes The chemical alteration of rocks in contact with sea water is generally referred to as corrosion. It is not restricted to the immediate vicinity of the breakers, but may occur where spray drifts high up above the surf zone. Investigations have tended to concentrate on limestone coasts although even here much remains to be learnt about the solutional processes involved. The surface waters of the oceans are generally close to saturation with calcium carbonate so that the water is not particularly aggressive towards limestone. Minor solutional forms are frequently seen in the intertidal zone, and their formation may depend either on precipitation or on organic processes that increase the acidity of the water in rock pools during low tide. Biochemical activity is certainly significant on many limestone coasts. Both plants and animals are capable of boring into calcareous rocks. Blue-green algae are especially important since they form the food source of many marine creatures which consequently pull away the weakened surface of the rock. Other borers include various species of sponge, mollusc and echinoid, all dissolving limestone and reducing the overall strength of the rock.

There is as yet little detailed evidence regarding the relative effectiveness of hydraulic action, corrasion and corrosion along most coasts and many other simultaneous processes certainly contribute to the overall erosion rate. For instance, rocks between high and low tide are subject to very regular wetting and drying. This can promote a wide range of chemical processes and in cold climates favours the disintegration of rock on the foreshore by freeze—thaw action. Some of the fastest rates of marine erosion are found where the cliffs are composed of till. In such cases wetting reduces the cohesion of the material to the point where it constantly slumps to the cliff base, is disaggregated by the waves, and finally transported to the seaward side of the surf zone. In this and all other cases the ability of the sea to remove the products of earlier erosion is an important factor in the continued effectiveness of the marine processes.

Movement of sediment

Material in suspension Waves and currents are responsible for transporting vast quantities of sediment, in part as a suspension load and in part as a mobile bed layer. In estuaries such as the Wash, on coasts backed by clay cliffs undergoing rapid erosion, and at the mouths of large rivers huge quantities of suspended mud are redistributed by the sea. There is, moreover, an almost constant rain of fine terrestrial sediment over the ocean floors. Yet in these areas far from the continents the sediment concentration by weight often amounts to no more than 1 part in 40 million. Around the open coast, and across extensive areas of the continental shelves, much of the total transport takes place by sand particles sliding, rolling and occasionally skipping over each other. It is only in the accelerated water movements associated with breakers that any appreciable quantity of this coarser material is raised into temporary suspension. The sediment concentration in breaking waves can be measured by collecting water samples from different depths below the surface with special suction equipment. The results generally show an almost uniform vertical distribution close to the break point, presumably as a result of the fast orbital motion carrying material upwards to the very crest of the wave. Representative figures range from 1 to 20 parts per mille by weight, with the value tending to rise with increasing wave height. Seaward from the break point the concentration drops sharply; for 100 m or more a suspension load may still be detectable, but is confined to a shallow layer next to the sea-bed. In the swash the turbulent motion of the water keeps the sand grains evenly distributed, but on the backwash there is a marked concentration close to the bottom.

King has described experiments designed to elucidate the maximum depth to which sand is disturbed within the breaker zone. At low tide a column of sand is removed from the beach, dyed a distinctive colour and replaced. If the position of the column is marked, the dyed sand can be excavated after one tidal cycle and the depth of disturbance ascertained. As might be expected. the results display a close relationship with wave height. In few instances during King's experiments did the depth exceed 40 mm, and she concluded that even under very high waves it is unlikely that sand would be disturbed to depths in excess of 0·2 m. The results suggest that a relatively thin sheet of sediment, far from acting as an abrasive agent, may form a protective cover for a rock platform. However, it should be noted that the observations only apply where the beach profile is approximately the same at the beginning and end of the tidal cycle; under storm conditions many beaches suffer very substantial accretion or depletion, as the repeated profile surveys to be discussed below so clearly demonstrate.

Tracer experiments The use of tracers for studying the movement of marine sediments has a long history. Early experimenters were content to use a wide variety of materials despite the obvious desirability of matching the physical properties of the tracer to the natural sediments. At the present time widespread use is made of two types of tracer, the radioactive and the fluorescent, although valuable results may still be obtained by marking natural shingle with marine paint. The advantages of fluorescent and radioactive tracers have been outlined in the discussion of sediment movement in rivers (p. 172) and need not be repeated here. However, the marine environment presents special problems which do require further comment.

The most common use of marine tracers is to provide data relating to longshore movement. Different techniques need to be adopted for the study of sand and shingle. Sand may be marked either by a fluorescent dye or a radioactive isotope with a half-life of no more than a few weeks. In early work the sand was placed on the beach between high- and low-tide marks and its subsequent dispersal examined at frequent intervals, in the case of fluorescent dyes with an ultra-violet lamp after dark, and in the case of radioactive tracers with a scintillation counter. Such simple procedures provided little quantitative data other than the maximum observed distance of travel. Various more elaborate techniques have subsequently been developed. Sampling has been improved by the use of adhesive cards which, when pressed on to the surface of the beach, retain a large number of sand grains which can be examined at leisure to determine the proportion of marked material. This is particularly valuable in the case of work with fluorescent tracers. In addition to providing much greater flexibility, it permits the construction of isopleth diagrams indicating the changing concentration of marked material away from the point of initial insertion (Fig. 15.5). Even more important is the opportunity it affords to extend observations below low-tide level. This facility was already available with radioactive sand where a scintillation counter could be dragged across the sea-floor on a sled. However, its application to fluorescent tracers was particularly significant at a time when it was becoming increasingly clear that considerable transfer of material between beach and off-shore zone might be taking place. So long as observations were confined to the inter-tidal zone, only longshore movement could be traced and the potential loss of marked material seawards was an unknown factor. Moreover, with a little ingenuity, the method could be adapted to study sediment movement in much deeper water. Scuba-divers may release sand and then sample with greased cards at depths of up to 40 m, while special adhesive weights sinking to the sea-floor are effective at more than twice that depth.

Shingle movement demands such high water velocities that only in special circumstances is it likely to be important much beyond the wave break-point. Tracers have therefore been employed mainly to determine rates and patterns of movement within the surf zone. Considerable attention has been focused upon the rates at which materials of different size move. For instance, on beaches at Deal and Winchelsea in south-eastern England, Jolliffe inserted selected natural pebbles marked with marine paint, together with artificial pebbles made of concrete chips and a fluorescent constituent. Observations lasted for periods of from 4 to 17 days. The most mobile material travelled over 150 m and proved to be, not the smallest of the particles, but those of intermediate size. This important result demonstrated that longshore movement is not simply a winnowing

Fig 15.5. An example of seaward transport by a rip current, identified by the use of fluorescent sand tracers. The isopleths represent conditions 40 minutes after release of the tracer and refer to the number of fluorescent grains per 6.45 cm^2 (after Ingle, 1966).

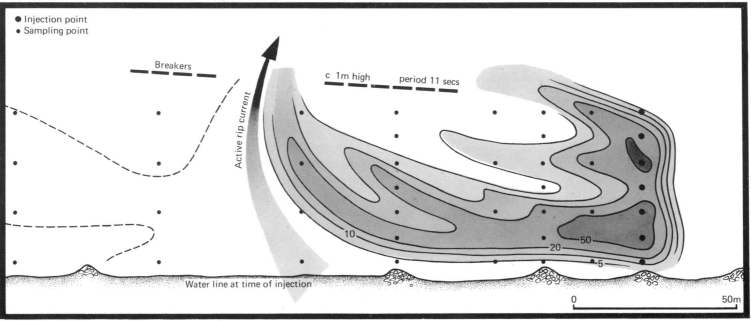

action by which fine debris is sorted from coarse but involves much more complex relationships. Work in deeper water is well exemplified by investigations on the East Anglian coast close to the shingle spit of Orfordness. Here 600 radioactive pebbles were carefully located in water where the depth varied between 6 and 8·5 m according to the state of the tide. Their position was surveyed at intervals over a period of almost 1½ months but no movement was detected. By contrast 2 000 pebbles were placed in much shallower water 11 m seaward of the low-tide mark. Some of these travelled northwards more than 1·6 km during a prolonged spell of light southerly winds, but when the winds reversed an even more rapid southward migration occurred. This study clearly demonstrated the need for maintaining observations over a sufficient period of time to cover a wide range of weather conditions.

sediment movement are also seen in short-term changes of such features as beach profiles, the distal ends of spits, and the bottom morphology of estuaries. These are all coastal forms that can be profitably studied either by the technique of repeated surveys or by recourse to the evidence of old charts and maps.

Beach profiles constitute one of the most mobile of all coastal features. They often display systematic patterns of erosion and accretion related to such controls as tidal cycles and weather sequences. A foreshore profile at right angles to the water-line may be conveniently surveyed during low tide. It is extending the profile into the offshore zone that presents the greatest difficulty. On a gently shelving beach normal survey methods can continue to be used with the staff-holder wading into shallow water. On steeper beaches it is necessary to use

Fig 15.6. An example of superimposed beach profiles and the 'sweep zone' that they define (after King, 1972).

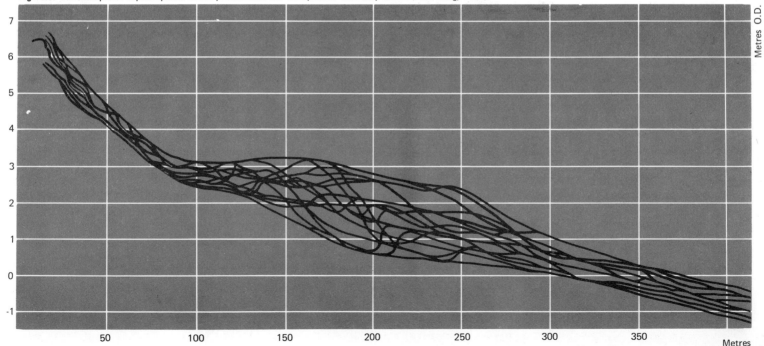

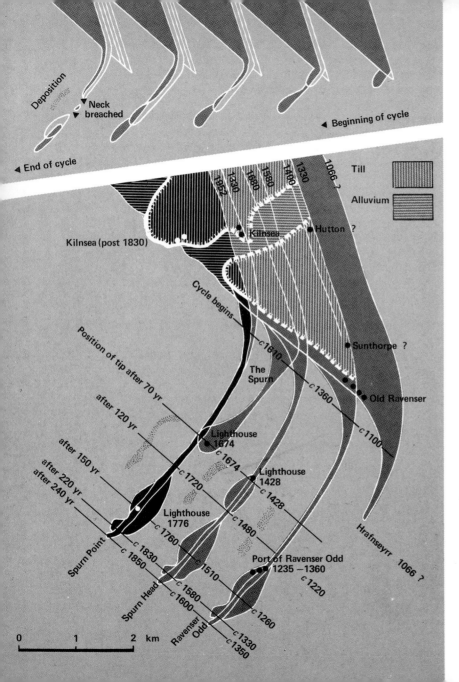

sounding procedures and the accuracy of the results usually suffers in consequence. Once selected profiles have been surveyed several times, the results can be superimposed so as to allow a visual inspection of the changes that have occurred (Fig. 15.6). The area between the highest and lowest lines is known as the sweep zone and is a measure of the maximum changes that have taken place within the period of study. Volumetric computations may be made by assuming that individual profiles are representative of the areas between them.

Rather longer intervals are usually treated when considering such constructional forms as spits. Nevertheless, significant changes can often be detected within a period of a few years and repeated surveys may provide valuable evidence regarding the processes involved. Yet care needs to be exercized in the interpretation of the results since it has been shown in a number of cases that the sequence of changes is cyclical; in such circumstances the continuation of the observed trends cannot be assumed. The longer the time interval to be considered, the greater the reliance that needs to be placed upon historical records. Old charts and documents can prove extremely informative if interpreted with due caution, remembering in general the older they are the greater the doubt regarding their reliability. The value of documentary evidence is well illustrated by the historical study of Spurn Point by de Boer. He showed that this spit at the mouth of the Humber estuary undergoes a phase of elongation followed by narrowing of its neck as the coast to which it is attached recedes under the influence of storm waves (Fig. 15.7). Eventually the neck is breached, leaving the distal end of the spit as an island. This persists briefly but is ultimately destroyed by further wave action. A new spit is then initiated and grows rapidly southwards. Each cycle is believed to last about 250 years, with new spits starting to develop

Fig 15.7 (left). Evolution of Spurn Head since 1100 A.D. envisaged by de Boer. Above schematic diagram of a single cycle of growth and destruction. Below, application of the cycle concept to the present Spurn Point and its immediate predecessor, Spurn Head and Ravenser Odd (after de Boer 1964).

Fig 15.8 (right). An illustration of the use of radioactive tracers for studying the nature of sediments movement in an ebb and flood channel system in the Firth of Forth. Both diagrams represent conditions one day after release of the tracer; measurements at the points indicated were made by lowering a scintillation head to the sea floor (after Smith et al. 1965).

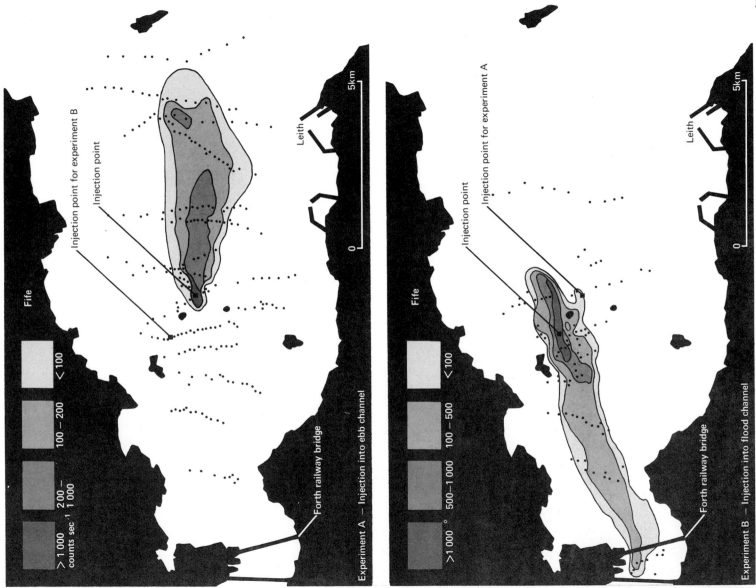

Fife

Leith

> 1 000

200 –
1 000 counts sec⁻¹

100 – 200

< 100

Injection point for experiment B

Injection point

Forth railway bridge

Experiment A — Injection into ebb channel

Fife

Leith

> 1 000

500–1 000

100 – 500

< 100

Injection point for experiment A

Injection point

Forth railway bridge

Experiment B — Injection into flood channel

around the years 600, 850, 1100, 1360 and 1610. In the mid-nineteenth century the present spit was breached by storm waves and a gap of almost 500 m cut across the neck. This was artificially closed and since that date the form has been carefully stabilized as a matter of policy to protect the Humber estuary.

In the hands of an experienced hydrographer charts of different date can provide valuable insights into the processes modifying the bottom topography of near-shore areas. Unchanging features may first be identified and used as reference points. Investigations have generally shown these to be composed of materials that local currents are incapable of moving. They may be contrasted with topographic features that vary in either form or position from one chart to another. These often display a distinctive morphology permitting their origin to be inferred with a good deal of confidence. Among the most common are sand waves and systems of ebb and flood channels. Examples of both have been intensively studied to elucidate the general principles of formation. Sand waves are long undulations that corrugate the surface of many sand banks. They vary widely in size. In large examples individual crests may be over 1 km apart, rise more than 10 m above the intervening troughs, and be traceable laterally for long distances. Smaller waves, on the other hand, have lengths of less than 10 m and heights of under 1 m. The common characteristic shared by all mobile sand waves is asymmetry with the steeper face pointing in the direction of movement. Ascribed to tidal currents, they are believed to be a manifestation of large-scale sediment transport. It has been estimated that 40×10^6 m³ of sand move northwards along the coast of the Netherlands each year, with a compensating return movement southwards past the East Anglian coast.

Ebb and flood channels are a prominent feature of many hydrographic charts close to the coast. They are particularly conspicuous in estuaries where abundant mobile sediment is available for shaping by tidal currents. The ebb and flood components interdigitate, with the mobile sand between the channels moulded into elongated banks. Both channels and banks shift their positions with time, and it is often possible to discern cyclical patterns with the original bottom configuration restored after a matter of a few years or decades. The direction of dominant water movement in the channels can usually be deduced from their morphology. At their upstream end they are relatively deep and narrow but flare in the direction of dominant movement. Velocity measurements have shown that water frequently continues to flow inwards along the flood channel for a short time after the tide has begun to recede. The ebb channel is similarly occupied by outflowing water after the tide has turned. There is a net transport of sediment landwards along the flood channel and seawards along the ebb channel (Fig. 15.8). Individual grains may pursue a gyratory course, the residual movements first carrying them from the flood channel into the ebb and later returning them to the seaward end of the same flood channel. Changes in channel pattern are of obvious economic importance in estuaries widely used by shipping, and may also be highly significant where waste material is discharged into the sea.

Use of scale models Two types of physical model are in common use for the study of coastal processes. The first consists of a simple tank equipped with a means of producing artificial waves. By its use attempts may be made to investigate the relationships between various wave parameters and the movement of beach material. The second type is a scale reproduction of a specific section of coast. This is particularly valuable where engineering works are proposed since it allows the likely changes in coastal morphology to be studied. Considerable difficulties arise in the process of scaling. Certain factors defy direct scaling. For instance, gravity must be the same in both model and natural prototype. Nevertheless, as far as possible the geometrical, kinematic and dynamic properties of the prototype must be retained. This can in theory be achieved by adherence to what is known as Froude's law. This states that time and velocity scaling should be equal to the square root of the linear scaling. This means, for example, that if a model is constructed at a scale of 1 : 1000 the tidal cycle should be reduced to just under 1/30 of its natural duration; for a normal semidiurnal tide this will amount to a period of about 25 minutes. Geometrical scaling is therefore accompanied by an acceleration of coastal change which is an obvious advantage to the user. However, even simple geometrical scaling poses problems.

At a scale of 1 : 1000 water that is 1 m deep will be represented by a layer 1 mm thick. The nature of water movement in such circumstances is totally altered. To overcome this difficulty most coastal models are built with a vertical exaggeration and the motion of the water is regulated by numerous small pins having no natural counterparts. A further problem is occasioned by the fact that the properties of sediments do not vary as a simple function of grain size. It is no use employing clay particles to simulate sand since they have totally different properties; they require higher velocities to set them in motion, but once in suspension have slower settling speeds. To overcome this difficulty special particles lacking the cohesion of clays and of different density from sand are often employed.

These points illustrate a few of the many problems involved in model building. Yet the practical utility of models can scarcely be doubted in view of the time and money spent both on their construction and on the large buildings required to house them. Well-known models include : those of the Mersey, Humber and Thames estuaries in Britain ; the Rhine delta in the Netherlands ; San Francisco Bay, Puget Sound and the Delaware estuary in the United States.

CARR, A. P., 1965. 'Shingle spit and river mouth : short term dynamics', *Trans. Inst. Brit. Geogr.*, 36, 117–29.

CARRUTHERS, J. N., 1962. 'The easy measurement of bottom currents at modest depths', *Civ. Eng.*, 57, 486–8.

CARRUTHERS, J. N., 1963. 'History, sand waves and nearbed currents of La Chapelle Bank', *Nature Lond.*, 197, 942–7.

DE BOER, G., 1964. 'Spurn Head : its history and evolution', *Trans. Inst. Brit. Geogr.*, 34, 71–89.

INGLE, J. C., 1966. 'The movement of beach sand', *Developments in Sedimentology*, 5.

IVERSEN, H. W., 1953. 'Waves and breakers in shoaling water', *Proc. 3rd Conf. Coast. Eng.* (Council on Wave Research), pp. 1–12.

JOLLIFFE, I. P., 1964. 'An experiment designed to compare the relative rates of movement of different sizes of beach pebbles', *Proc. Geol. Ass.*, 75, 67–86.

KIDSON, C. and A. P. CARR, 1959. 'The movement of shingle over the sea bed close inshore', *Geogr. J.*, 125, 380–9.

KIDSON, C. and A. P. CARR, 1962. 'Marking beach materials for tracing experiments', *J. Hydraulics Div. Proc. Am. Soc. Civ. Eng.*, 3189, HY4, 43–60.

KING, C. A. M., 1951. 'Depth of disturbance of sand on beaches by waves', *J. Sed. Pet.*, 21, 131–40.

MACKENZIE, P., 1958. 'Rip current systems', *J. Geol.*, 66, 103–13.

McCAVE, I. N., 1971. 'Sand waves in the North Sea off the coast of Holland', *Marine Geol.*, 10, 199–225.

PHILLIPS, O. M., 1957. 'On the generation of waves by turbulent wind', *J. Fluid Mech.*, 2, 417–45.

PHILLIPS, O. M., 1966. *The Dynamics Of The Upper Ocean*, CUP.

ROBINSON, A. H. W., 1960. 'Ebb-flood channel systems in sandy bays and estuaries', *Geography*, 45, 183–99.

ROSSITER, J. R., 1954. 'The North Sea storm surge of 31 Jan and 1 Feb, 1953', *Phil. Trans. R. Soc.*, A 246, 371–99.

SMITH, D. B., et al., 1965. 'An investigation using radioactive tracers into the silt movement in an ebb channel, Firth of Forth, 1965', UKAEA Research Group Rep., HMSO.

Selected bibliography

Nearly all the topics discussed in this chapter are very thoroughly reviewed in C. A. M. King *Beaches and Coasts*, Arnold, 1972. Another excellent text is F. P. Shepard, *Submarine Geology* (3rd edn.), Harper and Row, 1973.

The processes considered in the preceding chapter are virtually time-independent. The longest time interval to which reference was made amounted to no more than a few centuries. Yet many coastal landforms are clearly the result of a much longer history and a greatly extended period must be taken into account when considering their evolution. As soon as that period exceeds a few millenia it is essential to take cognizance of sea-level changes which undoubtedly constitute one of the major factors in their development.

Relative changes of land and sea-level may arise from a number of different causes. Eustatic changes affecting all coasts equally may be occasioned by variations in the volume of oceanic water, or by fluctuations in the capacity of the ocean basins. Local changes having little effect beyond their immediate area of occurrence arise either from tectonic causes (Fig. 16.1) or from depression of the crust beneath the weight of an ice-sheet. Given these circumstances the ideal procedure would be to construct a eustatic sea-level curve to which local variations might then be related; major residuals between global and local levels would clearly be ascribable to either tectonic or glacio-isostatic causes. This objective is still far from fulfilment since, despite many attempts, no entirely satisfactory method of constructing a eustatic curve has yet been devised. The fundamental problem is identifying a section of coastline that has been stable sufficiently long to provide a reference standard. There is always a grave risk of circular arguments being employed. A coast may first be classified as stable on the basis of apparently undeformed shorelines, and then used as a standard to which shoreline sequences in other areas are referred. Thereafter similarities tend to be stressed, and it is very easy for these to be interpreted as strengthening the claims of the original area to provide a global reference standard. However, such is the complexity of eustatic changes that superficial similarities must be treated with considerable caution. As the discussion below will show, the whole subject of sea-level changes is still fraught with uncertainty.

The nature of the evidence

Marine features located above present sea-level Elevated marine phenomena fall into two major categories, those due to erosion and those due to deposition. The former include abandoned cliffs, shore platforms, fossil stacks and marine caves. Where all are present in close proximity identification is relatively easy, but often the features occur individually and in much degraded form. There may then be considerable doubt regarding

Fig 16.1. One of a suite of raised marine shorelines that characterize the Pacific coast of California between San Francisco and Los Angeles. In this tectonically active region the elevation of the shorelines provides little direct guide to eustatic changes in sea level.

correct identification. For example, gently sloping benches backed by much steeper slopes on the flanks of Exmoor in south-western England have been interpreted by some workers as marine and by others as subaerial in origin. Of possible criteria by which elevated shore platforms can be distinguished from features of similar morphology originating in other ways, the most widely employed is the consistent height of the concave break of slope at the foot of the presumed cliff. Yet even when features of marine erosion have been confidently identified, it may still be difficult to infer the exact height of the sea at the time of their formation. Two factors militate against a precise determination. The first is an initial variation in height relative to any datum such as high water mark of ordinary spring tides. Lithology, exposure and fetch can all influence the height at which a shore platform abuts against a cliff. The second is the progressive degradation that originally clear forms suffer. Even recently abandoned cliffs tend to have their base obscured by slumping and rockfall. With older cliffs the destruction may be so total that the break of slope observed today affords almost no information regarding either original form or even exact position.

Marine deposits sited above the reach of presentday storm waves present some of the most convincing and informative evidence on sea-level changes. They range from old shingle ridges to extensive flats of estuarine clay and silt. Distinguishing unfossiliferous marine shingle from other coarse gravel deposits can prove troublesome, although the fabric and surface textures of the material may aid in identification. Where fossils are present the task is much easier, and a further great advantage of biological evidence is its potential for shedding some light on water temperatures at the time of accumulation. This can be particularly useful in indicating whether the high sea-level occurred during a glacial or interglacial phase. It may also exclude as most improbable those correlations which altitudinal factors alone would suggest as quite reasonable. As with erosional features, it is not always easy to determine the exact height of sea-level at the time deposition was taking place.

Large waves may build storm ridges to a significantly greater height on exposed than on sheltered sections of coast. Fine sediments are subject to much compaction and it would be unwise to assume that the surface level of raised estuarine silts coincides exactly with, say, the old mean tide level. Occasionally isolated pockets of sediment are found that contain the remains of organisms known to live within clearly defined depth limits. It is tempting to use such information to reconstruct the former sea-level though no contemporaneous beach features have survived. Whilst the procedure may be sound in theory, in practice it often gives rise to much dispute, particularly over such questions as whether the shells are still in their original growth positions. It is often safest to adopt the conservative argument that the upper limit of marine sediments indicates the minimum level attained by the sea.

In the preceding paragraphs emphasis has been placed on some of the difficulties of interpreting evidence for former high sea-levels. This stress is intentional for in the past the problematic nature of much of the evidence has received too little notice. Nevertheless, this approach should not be allowed to detract from the prolific evidence in many parts of the world that the sea formerly stood at higher levels relative to the land than at the present day. Both erosional and depositional forms attest quite clearly to this fact.

Terrestrial features located below present sea-level The increasing pace of research on the continental shelves has yielded abundant indications that parts of the present sea-floor were once dry land. Both physical and biological evidence bears witness to one or more phases of low sea-level. Among the most prominent physical signs is the seaward prolongation of many continental valleys in the form of sediment-filled channels. It is important to note that these channels are not restricted to glaciated areas in which some form of subglacial scour might be invoked but are also found in regions far beyond the limits reached by the ice-sheets. They commonly descend to depths· of about −90 m, although there is no complete uniformity. In addition to these submerged valleys, other identifiable features include old soil profiles, deltas, cliffs and even former spits. In more northerly latitudes forms due to glacial deposition can be traced from the land surface across the adjacent sea-floor without observable change; it must be presumed that at the time of accumulation the sea-floor was dry land.

Biological materials recovered from below sea-level tell the same story. Submerged 'forests', generally consisting of tree stumps in position of growth on the foreshore, have long provided an indication of a rise in sea-level. During the nineteenth century when large docks were being dug along the coasts of both Europe and North America, many beds of fresh-water peat were encountered below modern sea-level. As investigations have extended into areas of deeper water signs of terrestrial life have continued to be recovered. Peat has been dredged from the floor of the North Sea, bones of such animals as mammoth, bison, elk and moose have been found on the continental shelf off the Atlantic coast of North America, and shell banks composed of marine species restricted to water less than 10 m deep have been located at depths of over 100 m in the western Pacific. In the majority of cases there is good reason to believe that the material is in situ and has not been disturbed by recent submarine transport. Further important sources of evidence are the great deltas of the world. In many of these boreholes have penetrated organic deposits that indicate fresh or brackish water conditions at depths well below modern sea-level. Complications in such environments may be introduced by tectonic subsidence and sediment compaction, but the extent of these factors can often be computed and an appropriate allowance made. Minimum sea-levels of -100 m or more are often indicated by such procedures.

The evidence for lower sea-levels is usually inadequate to fix the precise height of the sea. Peat and animal remains merely show that an area was once dry land without any indication of its elevation above the contemporaneous shoreline. More significance attaches to the remains of organisms requiring a habitat in either brackish water or very shallow salt water. However, such finds constitute only a small proportion of those that are made. Correlation from one locality to another is extremely difficult when dealing with submarine features, and often all that can be done is to record the minimum depression of the sea consistent with a particular find.

Models of sea-level change

The amplitude of glacio-eustatic fluctuations It was realized soon after the idea of an ice age was propounded in the nineteenth century that the growth of large ice-sheets would have a profound effect upon the volume of water in the oceans and consequently upon global sea-level. Indeed, the effect upon sea-level was suggested as a means by which the whole concept of an ice age could be tested.

When evidence was later adduced in favour of multiple glaciation, the explanation for several large-scale eustatic oscillations was recognized to be at hand. Yet in subsequent years it has proved remarkably difficult to obtain an entirely satisfactory basis for estimating the amplitude of such glacio-eustatic fluctuations.

There are two obvious approaches to the subject. The first is to see whether the evidence for sea-level changes sheds any light on oscillations that occurred simultaneously with major glacial advances. The difficulty with this approach has proved to be gross uncertainty over dating and correlation; in general, it is not that the observations are inconsistent with glacio-eustatic fluctuations, but that without positive dating they are capable of so many interpretations that they cannot be held to support one thesis rather than another. The second approach involves calculating the volume of water locked up in major continental ice-sheets during a specific glaciation. Once the volume is known it is a relatively simple matter to transform the figure into the change of sea-level that it would represent. Many workers have attempted the necessary computations, among the more recent being Flint who employs the concept of a model glacial age in which ice is assumed to cover simultaneously all the territory glaciated at any time during the Pleistocene epoch. In such circumstances the volume in excess of that in the modern ice-sheets is estimated at just over 50×10^6 km^3, corresponding to a water volume of 47×10^6 km^3. This is equivalent to a fall in sea-level of 132 m. A correction factor is then required to convert this theoretical maximum value to that appropriate for any particular glaciation. For the last glaciation the correction is believed to be a reduction of about 10 per cent.

Flint stresses that there are at least three sources of potential error in his calculations. The first is that the ice advances may not be exactly synchronous over the whole globe. The second is the difficulty of estimating the thicknesses of the Pleistocene ice-sheets; the method most commonly adopted involves reconstruction of transverse profiles by analogy with existing ice-sheets. The third is the isostatic consequence of abstracting large volumes of water from the ocean basins. Much attention has been devoted to the isostatic depression of continental areas by ice-sheets but relatively little to analogous effects when a thick layer of water is alternately removed from and returned to the ocean basins. There is no reason to suppose that the ocean floors do not sink and rise in sympathy with the alternate loading and unloading they experience. The magnitude of changes in water level might be reduced by up to one-third owing to this factor. It may be concluded, therefore, that a major glacio-eustatic oscillation has an amplitude of about 100 m with an uncertainty of ±20 m.

The foregoing analysis is based upon the assumption that presentday conditions are typical of an interglacial period. The volume of ice in Antarctica and Greenland is formidable, being equivalent to a water layer thickness over the oceans of almost 65 m. If this were actually returned to the oceans it might induce a rise in water-level of just over 40 m when allowance is made for isostatic effects. This is one possible explanation for sea-level having risen above its present height at various intervals in the past. However, it seems almost certain that neither the Antarctic nor the Greenland ice-sheet disappeared during recent interglacial periods, and in all probability neither deviated greatly from its modern dimensions.

The concept of a secular decline in Pleistocene sea-levels If glacio-eustatic oscillations were the sole control of sea-level changes it would be expected that, after each glaciation, the sea would return to its original elevation. In such circumstances any raised shoreline would inevitably imply disturbance by either tectonic or glacio-isostatic processes. Yet early investigations revealed so many elevated strandlines beyond the limits of glaciation and without obvious signs of deformation that it was widely agreed that some other control must be operating. The most influential pioneer studies were

those undertaken by French workers around the western Mediterranean. De Lamothe in Algeria, Dépéret in southern France and Gignoux in southern Italy distinguished a succession of strandlines to which the following names and heights are conventionally assigned: Calabrian 150–180 m; Sicilian 80–100 m; Milazzian 55–60 m; Tyrrhenian 30–35 m; and Monastirian 15–20 m. It was originally maintained that each of these beaches is undeformed and the sequence must therefore imply a long-term eustatic fall. The observations in the western Mediterranean prompted formulation of a simple model of sea-level change that has dominated thinking on the subject ever since. The model envisages glacio-eustatic oscillations superimposed on a secular decline (Fig. 16.2). As a result the culminating sea-level during successive interglacials becomes significantly lower, defining what has been termed the interglacial geoid. During successive cold periods minimum sea-level shows an almost identical fall, defining the corresponding glacial geoid.

Even general testing of a model of this type has proved remarkably difficult. Profiles drawn through many coasts of the world from, say, 200 m above to 200 m below modern sea-level will disclose clear signs of periods when the sea stood both higher and lower than at the present day. Yet arranging the various levels in a chronological sequence poses immense problems. The fragmented nature of the evidence often precludes use of simple stratigraphic principles. Any organic remains are usually too young to permit application of the normal concepts of biological evolution. Ecological arguments, because of the repeated climatic cycles of the Pleistocene epoch, are of limited value, even for purposes of relative dating. Radiometric methods are locally very important but much of the evidence is not amenable to this approach; moreover, if one excepts the recently de-glaciated regions of the globe, elevated shorelines are generally too old for dating by means of radiocarbon.

Despite these problems of validation, theory continues to demand glacio-eustatic fluctuations and observations to imply a secular decline. Yet certain postulates of the model have not always been substantiated by the field evidence. This is probably more true in the Mediterranean region than most others, and a

338 vast literature has accumulated on a number of controversial issues. An early subject of investigation was the relationship between raised strandlines and the advance of the Alpine glaciers. Preliminary studies sometimes suggested a synchroneity which would have undermined one of the basic tenets of the model. Many of these early studies have since been discredited, but it remains difficult to establish a clear link between glacial advances and sea-level changes. A second way in which the climatic associations of the Mediterranean beaches may be studied is through the faunal assemblages they contain. This approach has also produced results not entirely in accord with predictions of the model. Whereas the Milazzian, Tyrrhenian and Monastirian beaches have yielded assemblages with sub-tropical affinities and therefore acceptable as being interglacial in origin, the Calabrian and Sicilian beaches contain a number of molluscan species with northern affinities that do not accord with a water temperature as high as that of today. This problem is still unresolved, possibly because the Mediterranean coastline presents a singularly difficult area in which to make close faunal comparisons. The temperature and salinity of the water is much influenced by the pattern of currents flowing through the Straits of Gibraltar. At the time the Calabrian and Sicilian beaches were being formed the Straits were both wider and deeper than at the present day, allowing a much more ready exchange of water with the Atlantic. It is difficult to evaluate this effect in precise terms, but it does underline the need for caution in making comparisons with the present day. A further indication of the complex relationships of the Mediterranean basin comes from studies in southern Italy to fix the stratigraphic position of the Plio-Pleistocene boundary. At several localities these have revealed a sudden influx of what are usually regarded as cold species, but without any evidence of a corresponding fall in sea-level. Moreover, analysis of oxygen isotope ratios has disclosed no simultaneous sharp fall in water temperature.

The long chapter of controversies surrounding the Mediterranean beaches has undoubtedly eroded their claim to be regarded as the prototype for Pleistocene eustatic fluctuations. Nevertheless, in the decades following their first scientific description they were widely used as a reference standard and many apparently sound correlations were established. The

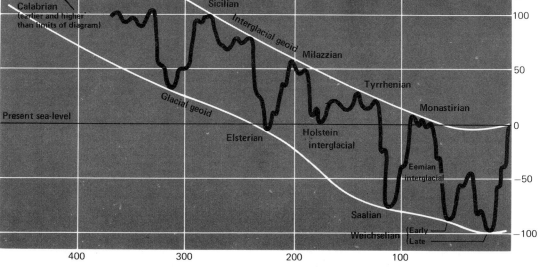

Fig. 16.2. An example of a Pleistocene eustatic curve based upon the traditional Mediterranean sequence. This particular interpretation is that of Fairbridge, 1961.

Years x 10^3 B.P.

classical sequence implies an early Pleistocene sea-level of between 150 and 200 m, and this value has been accepted by many workers throughout north-western Europe. For example, a Calabrian age has frequently been assigned to a prominent bench around the uplands of south-western England and Wales at about 180–200 m (see, for example, Fig. 4.11, p. 69). Although marine sediments are generally lacking the dating is often held to be strengthened by analogy with south-eastern England where shingle at the same elevation has yielded a fauna consistent with an early Pleistocene age. Numerous other examples could be quoted, and although there is a very real risk of circular arguments being employed in the way that was outlined in the introduction to this chapter, it is difficult to ignore the success that has attended what might be termed the Mediterranean model of sea-level change.

As research has proceeded in other parts of the world, this Mediterranean model has come under increasingly critical scrutiny. One region that has been the focus of much study is the Atlantic seaboard of the United States. This has been claimed as much more suitable for establishing eustatic changes than the Mediterranean region because of its greater tectonic stability. Considerable conflict emerges when the findings from the two areas are compared. In the coastal zone from Virginia to Florida Pleistocene marine features in the view of many workers do not extend higher than about 30 m. This is in sharp contrast to the classic sequence of the Old World, although it should be added that where dating is available in North America it indicates a younger age for the highest shorelines than would normally be accorded the Calabrian, or even the Sicilian, beaches in Europe. To that extent it could be contended that the earlier and higher episodes are unrepresented along the American seaboard, although this obviously poses the question why that should be so if the Mediterranean levels are truly eustatic. The enigma is all the greater when it is recalled that the Atlantic coastal plain bears the imprint of several mid-Cenozoic transgressions; it requires special pleading to argue that the early Pleistocene examples have been selectively destroyed by erosion.

On account of its presumed tectonic stability, Australia has been claimed as another area particularly suited to the study of eustatic changes. The height ranges recorded from that continent are generally closer to those of North America than those of the Mediterranean. It is impossible to summarize all the evidence, but it would be fair to conclude that many workers outside the Old World accept a model involving secular decline of sea-level but believe the amplitude of the decline during Pleistocene times has been much less than the 180 m envisaged by early European workers. This would imply that the elevation of the Mediterranean beaches owes much to tectonic uplift. There is no doubt that exaggerated claims have been made regarding the horizontality of those beaches. They are all warped in one area or another and several may even pass below present sea-level for short stretches. Yet there are one or two areas, most notably in northern Libya, where the beaches are claimed to maintain a constant altitude over long distances. These areas of undeformed strandline are held by some to justify retention of the Mediterranean as the standard reference sequence. It should be added that the claims on behalf of other regions to have been tectonically stable throughout the Pleistocene do not all bear close examination. Several investigators, for instance, have maintained that parts of Florida have been significantly displaced during Pleistocene times. There is increasing evidence from geodetic surveys of current instability along the whole seaboard from Florida to Maine (p. 56). It should also be remembered that one of the largest earthquakes ever experienced in North America was located near Charleston in South Carolina.

Doubt regarding the long-term stability of the crust has encouraged a small but apparently increasing number of workers to question the whole concept of a secular decline. They maintain that raised shorelines outside glaciated regions are merely an indication of tectonic uplift. In many ways this is an attractive hypothesis and is certainly difficult to disprove. Majority opinion, however, still seems to support the view that there has been a long-term fall. Assessments of the amount vary from 200 m downwards. The higher figures approach the maximum theoretical value for any eustatic change (p. 92) and now appear rather improbable. A further weakness of all models involving a large-scale secular decline is the lack of any satisfactory mechanism to explain the fall.

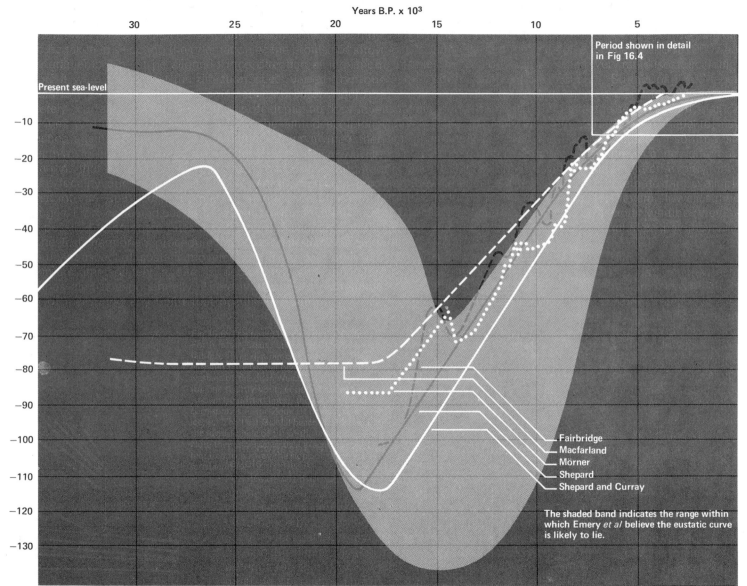

Years B.P. x 10³

Present sea-level

Period shown in detail
in Fig 16.4

Fairbridge
Macfarland
Mörner
Shepard
Shepard and Curray

The shaded band indicates the range within
which Emery *et al* believe the eustatic curve
is likely to lie.

Fig **16.3** A diagram to illustrate some of the views that have been expressed regarding eustatic sea-level changes during the last 35 000 years.

Fig **16.4** Examples of eustatic curves for the last 7 500 years constructed by eight different authorities (based on a compilation by Mörner, 1971).

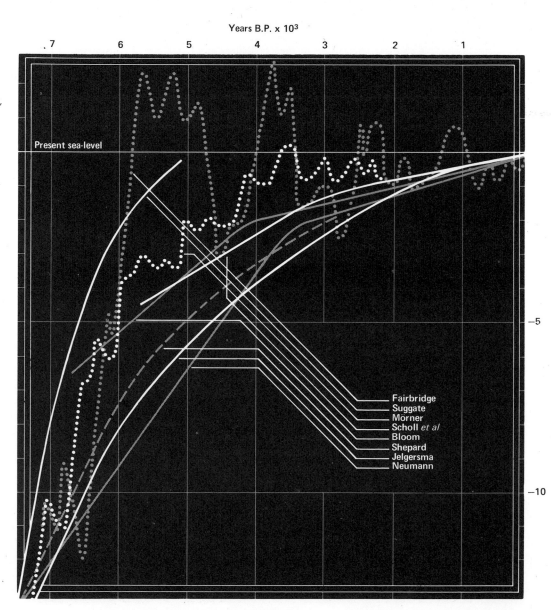

Years B.P. x 10³

.7 6 5 4 3 2 1

Present sea-level

Fairbridge
Suggate
Mörner
Scholl *et al*
Bloom
Shepard
Jelgersma
Neumann

—5

—10

Eustatic changes during the last 35 000 years

The preceding paragraphs have inevitably stressed the diversity of views regarding long-term sea-level variations. It is a relief to turn to a topic on which, despite divergences over points of detail, there is a broad measure of agreement. Selection of the 35 000-year interval for discussion reflects the period spanned by radiocarbon dates since these have contributed immeasurably to our knowledge of recent sea-level changes.

As indicated earlier, elevated beaches outside the glaciated regions are too old to be dated by radiocarbon techniques. It is the material recovered from below sea-level that has proved to lie within the range of radiocarbon dating. This is not surprising inasmuch as we live just after the close of a major continental glaciation and might expect the return of water to the oceans to have resulted in a sharp rise in sea-level. A common procedure adopted by workers in many different parts of the world has been to plot the age of materials against the depth of recovery,

sometimes with the added refinement of distinguishing those finds that are particularly sensitive indicators of sea-level. From a wide variety of coastal environments the results have been reasonably similar (Fig. 16.3 and 16.4). They nearly all imply an extremely rapid rise in sea-level during the period between 15 000 and 7 500 years ago. Most estimates of sea-level for 15 000 years ago lie within the range from −100 m to −60 m, and those for 7 500 years ago within the range from −15 m to −10 m. This gives a sustained rate of rise averaging about 10 mm yr^{-1}. Rapid flooding of the upper parts of the continental shelves took place. Where these are gently sloping very extensive areas of dry land were drowned. This is well exemplified by the recent evolution of the North Sea basin as illustrated in Fig. 16.5.

Differences of opinion about sea-level changes during the last 35 000 years have centred on three major points: movements prior to 15 000 years ago, the steadiness of the rise during the main transgression, and minor fluctuations within the last 7 500

Fig 16.5. Flooding of the North Sea basin in the millennium between 9 300 and 8 300 years ago (after Jelgersma, 1961).

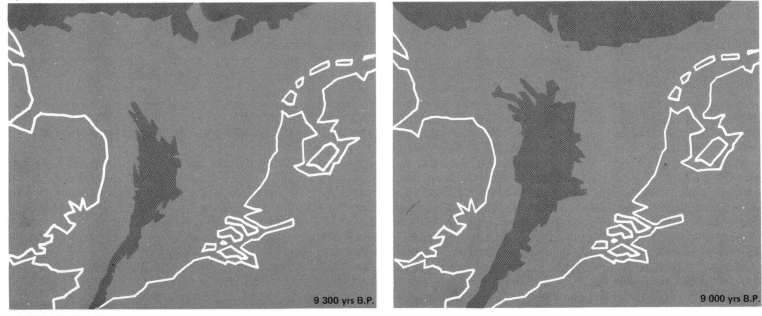

9 300 yrs B.P.

9 000 yrs B.P.

years. It is now generally acknowledged that the ice-sheets of the last glaciation attained their greatest extent in both Europe and North America between 18 000 and 20 000 years ago. The lowest sea-level would be expected to coincide with that maximum advance and many workers have inferred that a slow rise was probably taking place for several millenia prior to 15 000 years ago. The number of radiocarbon dates for these earlier periods becomes so small that increasing latitude must be allowed in the estimates. However, two contrasted schools of thought have arisen regarding the eustatic changes in the period between 35 000 and 20 000 years ago. The first contends that sea-level remained substantially below that of the present day. Support for this view comes from a number of finds indicating the highest shoreline to lie at least 10 m below present water-level. On the other hand, a few radiocarbon assays on marine material at, or even a little above, current sea-level have yielded ages of some 30 000 years. This has encouraged the

belief that, during Mid-Weichselian times, sea-level may even have been slightly higher than at the present day. In some respects this view is difficult to reconcile with the evidence of the contemporaneous climate. Although several Weichselian interstadials have been identified the prevailing temperatures were still well below those of today. They seem to imply at least moderately large continental ice-masses which would correspondingly depress sea-level. The whole subject has been reviewed by Thom, but more data are required before the divergence of views can be finally resolved. In the meantime it is worth noting that the notion of a secular decline in Pleistocene sea-levels does little to resolve the problem; even at the fastest feasible rate of long-term decline the differences between today and 35 000 years ago would be no more than 5 m.

A number of workers have claimed to identify minor regressive phases during the main rise in sea-level between

Fig 16.5 (continued)

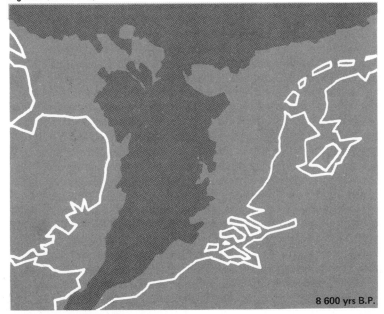

8 600 yrs B.P.

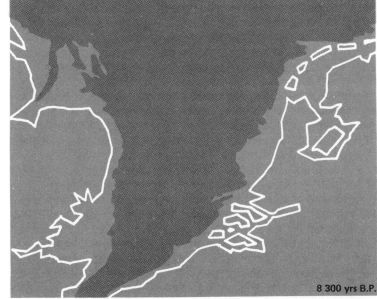

8 300 yrs B.P.

344 15 000 and 7 500 years ago. Fairbridge, for instance, suggested that four separate periods when the sea temporarily receded could be recognized. However, past lower sea-levels can rarely be assessed with any precision. There is a range of uncertainty attaching to both dates and depths. In the case of radiocarbon assays this may be expressed statistically in the form of confidence limits, but it is difficult to apply such an objective scheme to depths. When the data come from widely separated coasts that may well have suffered slight but varying degrees of tectonic displacement, the problem of computing an exact eustatic curve is virtually insurmountable. For this reason most workers have greeted attempts to identify minor cyclical patterns with considerable scepticism.

The rate of sea-level change was clearly beginning to decline quite sharply about 7 000 years ago. During the preceding two millenia the sea had risen over 20 m, but in the next two that figure was to be virtually halved. Thereafter any fluctuation was within relatively narrow limits. Disagreement has centred on whether the sea has ever risen above its current level within the last 5 000 years. Fairbridge has collated observations that seem to demonstrate rises to 1 m or more above modern datum around 5 000 and 3 500 years ago. However, his evidence is drawn from many different parts of the globe and the interpretations upon which he relies have often been the subject of challenge by local workers. Detailed studies restricted to a single segment of coastline frequently lead to rather different conclusions. In the Netherlands where hundreds of radiocarbon dates are available but where there are acknowledged problems of subsidence and compaction, most workers maintain that the global sea-level was below its present height 5 000 years ago and has since shown a very slow rise. A similar conclusion has been drawn from intensive investigations on the Everglades coast of Florida. Here the final transgression was across a broad limestone shelf that is claimed to offer the best site in the United States for measuring recent eustatic changes. The sea 5 000 years ago was some 4 m lower than at present and the current level represents the highest point reached at any subsequent date. An interesting aspect of these findings has been discussed by Bloom. He points out that the major Scandinavian and Laurentide ice-sheets had already disappeared 5 000 years ago, but sea-level has continued to rise since, albeit at a much slower rate. He suggests, as the most likely sources of water, continued melting of those ice-sheets that have survived to the present day and 'decantation' as water is displaced from Hudson Bay and the Baltic Sea by sustained glacio-isostatic uplift. Other oscillations due to climatic fluctuations may be superimposed upon this rise, but their amplitude is likely to be very small.

Sea-level changes in areas of glacio-isostatic uplift

General principles It is over a century since the likelihood of depression of the earth's crust by continental ice-sheets was first mooted. Support for this idea came from the association of recent raised beaches with severely glaciated terrain in Canada and Scandinavia. Not only did the preservation of delicate marine forms testify to the post-glacial age of the beaches, but the limits of the areas over which they could be found also ran roughly parallel to the limits of glaciation. It was soon recognized that the beaches are tilted and that lines drawn through places of equal elevation on the same strandline tend to encircle the areas of major ice accumulation. Gravity measurements disclosed negative anomalies over both Scandinavia and North America, pointing to a lack of crustal equilibrium such as might be expected from the sudden removal of a thick ice load. These and other observations left little doubt about the reality of glacio-isostasy. It was concluded that the growth of any large ice-sheet leads to a basin-like depression of the crust, in general form resembling that beneath Antarctica today. As an ice-sheet grows in size the downwarping must increase in amplitude at the centre of the basin and slowly extend outwards. With melting the sequence is reversed. As the ice-sheet decays the depression becomes shallower and less extensive. However, there is a delay between the melting of the ice and the recovery of the land surface. It is this relaxation time that accounts for the negative gravity anomalies currently observed over both Scandinavia and north-eastern Canada.

In many instances depression of the land surface was sufficient to permit the sea to flood in across the terrain vacated by ice withdrawal. This occurred despite the contemporary lower sea-level, and the invading water fashioned shoreline features that were to be warped during the later uplift. So long as the rate of uplift exceeded that of eustatic rise, successively lower strandlines could be formed. If the rates were temporarily reversed a transgression took place. By careful dating and mapping of strandlines it should be possible to reconstruct in some detail the history of isostatic recovery.

Of the many diagrammatic techniques that have been employed for recording the history of uplift three will be quoted to illustrate the range of possibilities. The first involves the plotting of lines known as 'isobases' (e.g. Fig. 16.7). The term isobase has been used in slightly different ways by different authors. In its simplest form it refers to an isopleth depicting the elevation of an individual strandline above sea-level. By extension it has also been used for an isopleth that indicates the amount of isostatic uplift suffered by an area within a specified interval; the calculations in this case require that an appropriate allowance be made for eustatic changes. The second common diagram is the profile drawn parallel to the presumed direction of tilt (e.g. Fig. 16.6). Although a very convenient method of comparing the heights and gradients of a group of beaches, this does not illustrate directly the amount of uplift that has occurred since it takes no account of eustatic fluctuations. Great care must also be taken in selecting the orientation of the profile. To show true gradients it must be aligned in the direction of maximum tilt; if it were found that the direction of tilt varied from one beach to another its value for comparative purposes would be severely curtailed. A third method of depicting sea-level changes within a small area is to plot the altitudes of local beaches against their ages. An 'emergence curve' may then be drawn to link the plotted points. In effect, this shows the variations that would have been noted by an observer situated at the locality during the post-glacial period. It can be converted into an 'uplift curve' by making a correction for eustatic changes. From corrected curves for a series of different localities it is possible to compare rates and times of maximum uplift.

The last-mentioned procedure has demonstrated one very significant aspect of isostatic movement: the rate of uplift is at a maximum immediately a site is deglaciated and thereafter declines almost exponentially. There is at present no reliable way of measuring any uplift that occurs at a site prior to ice

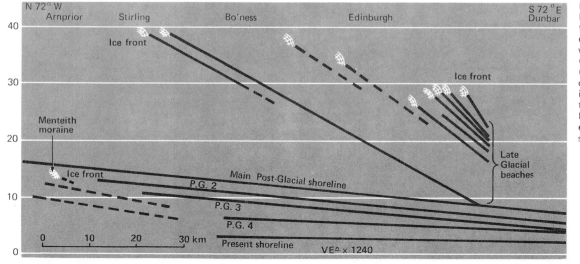

Fig 16.6 The raised beaches of the Firth of Forth (after Sissons et al., 1966). The beaches fall into three distinct categories: (a) The elevated Late-Glacial beaches formed during wastage of the main Weichselian ice-sheet; (b) The slightly younger buried raised beaches (shown by dotted lines), the earliest of which corresponds to the Zone III readvance of the ice; (c) The Post-Glacial beaches which exhibit the lowest angle of tilt of the whole sequence. Note the immense vertical exaggeration required to portray the slope of the beaches satisfactorily.

withdrawal, but it seems almost certain that a large proportion of the total recovery precedes deglaciation. For example, if the Scandinavian ice-sheet were originally 3 000 m thick, as seems likely, the aggregate depression of the crust could well have been 1 000 m. Yet the maximum uplift registered by marine strandlines is some 520 m and, even when allowance is made for the uplift that is undoubtedly still to come, a large component of the overall recovery remains unrecorded. In Canada Andrews has inferred that the total uplift curve is likely to be S-shaped, with maximum velocity at the moment of deglaciation; most of his calculations suggest that at least half of the overall recovery precedes deglaciation.

Having examined some of the basic concepts of glacio-isostasy, attention may now be directed to two specific areas of rather different dimensions that have suffered recent glacial unloading.

Scotland Although the idea of glacio-isostasy was enthusiastically endorsed by many early workers in Scotland, it is ironic that a rigid mode of thought evolved which was reluctant to acknowledge the tilt that affects Scottish raised beaches. In part this was due to an unfortunate terminology, apparently developing almost by accident, in which the beaches were classified by their approximate height above sea-level. It is only since 1960, largely at the instigation of Sissons, that serious attempts have been made to place them in a truly rational framework. The recent age of the beaches can only be reconciled with eustatic changes if they are tilted. It follows that it is totally inappropriate to designate them by their elevation. This has led to abandonment of the classic terminology of '100-foot', '50-foot' and '25-foot' raised beaches, although no wholly satisfactory alternative has yet been proposed. A classification based upon age appears to be the most logical, and a simple division into Late-Glacial and Post-Glacial beaches is in common use.

Early investigators in the glens of western Scotland noted that certain high shorelines, clearly visible at the seaward ends of the glens, terminated abruptly when traced inland. As the points at which the shorelines disappeared were often marked by moraines or thick glacifluvial accumulations, it was inferred that the heads of the glens must still have been occupied by glaciers. A similar conclusion was drawn by Sissons and his collaborators from their study of the Forth valley in south-eastern Scotland. They distinguished ten separate high-level beaches, each rising gently west north-westwards and terminating at its highest point in a mass of outwash believed to mark the position of the contemporaneous ice front (Fig. 16.6). This whole assemblage of Late-Glacial forms was interpreted as recording the spasmodic recession of an ice-sheet, with the area vacated by ice withdrawal being immediately submerged by ingress of the sea. The earlier beaches are steeper than the later, confirming that uplift and ice retreat were taking place simultaneously. The oldest and highest of the ten beaches has a gradient of 1·63 m km^{-1}, the most recent a gradient of 0·55 m km^{-1}. In order to measure the total amount of isostatic recovery represented by these shorelines it is obviously necessary to know their ages. This remains uncertain but the most recent probably dates from about 13 000 years ago. At that time global sea-level was still 60 m below its present height. If this figure is added to the 37·5 m by which the shoreline rises above current sea-level near Stirling, isostatic recovery at that point approaches 100 m in the last 13 000 years.

Around 13 000 years ago the land must have been rising very rapidly. This can be deduced from the fact that although sea-level was rising at a mean rate of about 10 mm yr^{-1} it was still being outstripped by isostatic recovery. After the last of the high-level beaches had been formed in the Forth valley, relative sea-level fell very swiftly and the next event of note was accumulation of the features which Sissons has termed buried raised beaches. These are a group of three beaches that span the boundary between Late-Glacial and Post-Glacial times. The term 'buried' is used because they were all submerged during a later transgression in which a cover of fine estuarine sediment was laid down. The highest is almost contemporaneous with the 10 300-year-old Menteith moraine. The intermediate or main beach, dated to about 9 500 years ago, is approximately 1 m lower and can be traced from the Menteith moraine downstream to beyond Stirling. Sea-level near Stirling was then 8 m above its present height, and since the contemporaneous eustatic level was −27 m the area must have been elevated isostatically some 35 m during the last 9 500 years. When this is compared with the 100 m of uplift since the higher suite of beaches was formed, it is evident that in the preceding 3 500 years isostatic displacement amounted to 65 m. The lowest buried beach indicates a relative fall in sea-level of about 2 m and is dated to some 8 800 years ago.

The next formative episode in south-eastern Scotland was a major transgression. For a brief interlude during the Post-Glacial period the continued eustatic rise outstripped the decelerating uplift of the land. The sea flooded across areas that

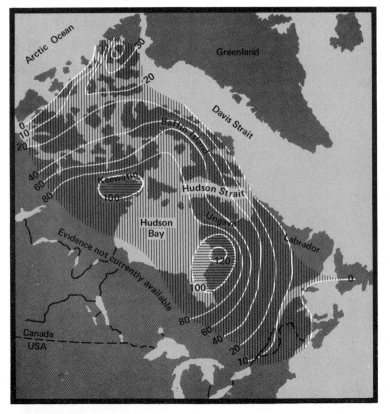

Fig 16.7 Isobases depicting emergence (i.e. discounting eustatic sea-level changes) in north-eastern Canada during the last 6 000 years (after Andrews, 1970)

had already become dry land. Without the plentiful coarse **347** debris previously supplied from nearby glaciers, most of the sediment laid down was fine estuarine silt, known locally as carse clay. The sea reached its culminating height 5 500 years ago. The main Post-Glacial shoreline lies at an elevation of 16 m near the Menteith moraine, but descends eastwards to under 6 m near Dunbar. There has clearly been both overall uplift and differential tilting during the last five millenia. Relative sea-level has fallen spasmodically so that at least three further shorelines have been fashioned, each now sloping more gently than its predecessor.

Most of the recent work on Scottish raised beaches has been concentrated in the south-eastern part of the country. Investigations in other areas have shown consonant strandline patterns, but the detailed sequences still require further study.

Arctic Canada Uplift associated with decay of the Laurentide ice-sheet was first studied by means of the deformed strandlines that encircle the Great Lakes. However, since about 1950 increasing attention has been devoted to the coastal regions of north-eastern Canada where analysis of raised marine shorelines has helped to clarify the later history of isostatic recoil. The earliest major event was a transgression along the St Lawrence valley known as the Champlain Sea. This is dated to about 12 000 years ago when global sea-level was some 50 m lower than at the present day. The Champlain strandlines locally attain heights of over 200 m giving a total uplift since deglaciation of more than 250 m. Melting back very rapidly across southern Canada the ice-sheet finally split into a number of separate remnants around 8 000 years ago. A prime factor in this fragmentation was penetration by sea water into the Hudson Bay depression. For a long time the low eustatic sea-level and the barrier of high ground along the Atlantic seaboard had prevented such penetration. However, about 8 500 years ago the rapidly rising sea-level promoted vigorous calving in the vicinity of the Hudson Strait, eventually permitting the water to leak into the Hudson Bay area. Thereafter the sea advanced rapidly until the former ice-sheet was divided into three separate masses, each sited on higher ground beyond the reach of glacier calving. The mass on Baffin Island has persisted to the present day, but those on the Ungava peninsula and in the Keewatin district both disappeared completely between 5 000 and 6 000 years ago. Divested of its ice cover within the space of about 1000 years the Hudson Bay area was subject to very rapid isostatic uplift.

Studies of glacio-isostasy in Arctic Canada have concentrated on the marine limit, that is, the maximum elevation attained by the sea in a particular locality since deglaciation. Few attempts have been made to map individual strandlines in the fashion attempted in south-eastern Scotland. Under the leadership of Andrews a very different approach has been adopted. Andrews has collated twenty-one curves depicting isostatic uplift at sites where dates are available not only for the marine limit but also for several lower strandlines. After analysing the curves to discover whether they have a common mathematical shape, he concludes that they all approximate to an exponential decay form and indicate a uniform response by the earth's crust to deglaciation. As a consequence realistic uplift curves can be constructed for the large number of sites where information is restricted to the age and elevation of the marine limit. By means of this technique isobase maps for the whole of Arctic Canada can then be drawn (Fig. 16.7). Making certain further assumptions, Andrews is also able to assess the amount of isostatic recovery that is still to come; he concludes that along the eastern shores of Hudson Bay further uplift is likely to surpass 140 m.

The approach employed by Andrews provides an interesting contrast with that normally used in Britain. In part, of course, this is a reflection of the immense area under consideration and the reconnaissance nature of most of the investigations so far. As Andrews is at pains to emphasize, his predictions require checking by detailed field measurement.

ANDREWS, J. T., 1970. *Post-glacial uplift in Arctic Canada*, Inst. Brit. Geogr., Spec. Pub. No. 2.

BLOOM, A. L., 1971. 'Glacial eustatic and isostatic controls of sea level since the last glaciation', in *Late-Cenozoic Glacial Ages* (Ed. K. K. Turekian), Yale.

EMERY, K. O., et al., 1971. 'Post-Pleistocene levels of the East China sea' in *Late-Cenozoic Glacial Ages* (Ed. K. K. Turekian), Yale.

FAIRBRIDGE, R. W., 1961. 'Eustatic changes in sea level', *Physics and Chemistry of the Earth*, vol. 4, 99–185.

FLINT, R. F., 1971, *Glacial and Quaternary Geology*, Wiley.

GUILCHER, A., 1969. 'Pleistocene and Holocene sea-level changes', *Earth Sci. Rev.*, 5, 69–97.

JELGERSMA, S., 1961. 'Holocene sea-level changes in the Netherlands', *Med. Geol. Sticht.*, Ser. C, vol. 6.

MORNER, N-A., 1969. 'Eustatic and climatic changes during the last 15 000 years', *Geol. en Mijn.*, 48, 389–99.

MORNER, N-A., 1971. 'The Holocene eustatic sea-level problem', *Geol. en Mijn.*, 50, 699–702.

SISSONS, J. B., 1967. *Evolution of Scotland's Scenery*, Oliver and Boyd.

THOM, B. G., 1973. 'The dilemma of high interstadial sea levels during the last glaciation', *Prog. in Geogr.*, 5, 167–246.

WARD, W. T., et al., 1971. 'Interglacial high sea levels – an absolute chronology derived from shoreline elevations', *Palaeogeog. Palaeoclim. Palaeoecol.*, 9, 77–99.

Because of the ambiguous nature of so much of the evidence, there is no single authoritative source on Pleistocene sea-level changes. Useful reviews with extensive bibliographies are to be found in A. Guilcher, 'Pleistocene and Holocene sea-level changes', *Earth Science Reviews*, 5, 1969, and P. Evans *Towards a Pleistocene Time-Scale*, Special Publication No. 5 of the Geological Society of London, 1971.

Valuable accounts of raised beaches in areas subject to glacio-isostatic recovery are provided by the works of J. B. Sissons and J. T. Andrews listed above.

The sea-level changes discussed in the preceding chapter have two important implications for the study of coastal landforms. In the first place on a stable coast the sea attained its present level less than 5 000 years ago, so that neither erosional nor depositional features can properly be attributed to marine processes acting unchanged over a protracted period. Secondly, all models of eustatic change require that the sea has stood at its present level at a number of periods in the past. This is clearly the case if there has been no secular fall in sea-level since the water surface would be expected to return to approximately the same position during each interglacial. If there has been a long-term decline, each glacio-eustatic oscillation should have carried sea-level below its present height once the interglacial geoid had fallen to about 100 m. The date at which this stage was reached is still a matter of dispute, but in virtually all interpretations the sea must have stood at its current height on at least five separate occasions, and probably several more. A significant aspect of any model involving secular decline is its prediction that sea-level will have coincided with its present position only on the falling and rising legs of each glacio-eustatic fluctuation. There seems to be no intrinsic reason why it should in aggregate have stood any longer at its present elevation than at many others.

The notion of sea-level change has formed the basis for many attempts at coastal classification. Since these often incorporate important ideas about coastal evolution, they will be briefly reviewed before attention is turned to more specific aspects of landform history.

Schemes of coastal classification

For many years ideas of coastal classification were dominated by the work of the American geomorphologist, Johnson. In his classic work *Shore processes and shoreline development* he placed immense stress on the dual concepts of submergence and emergence, although admitting the need to recognize two further classes which he termed neutral and compound coasts. Neutral coasts according to Johnson are those in which the form is due, not to sea-level change, but to essentially non-marine processes such as faulting and vulcanicity. Compound coasts are those which involve some combination of the other three. In many ways the most unfortunate aspect of Johnson's classification has proved to be its emphasis on sea-level change. He proposed two major models of coastal evolution, one to represent the sequence of changes after a rise in sea-level, the other after a fall. The former involves the drowning of earlier river valleys, the concentration of wave attack upon headlands, and the filling of initial inlets with bay-head beaches. Gradually the promontories are cut back as a series of cliffs until the plan of the coast is eventually converted from crenulate to straight. Emergence, on the other hand, was held to lead typically to uncovering of a low coastal plain. The gradient of the submarine profile is reduced and in consequence an off-shore bar created. The resultant tidal lagoon is slowly silted up with both marine and fluvial sediments. Wave attack concurrently pushes the bar landwards until it ultimately reaches the original coastline. Any further erosion induces cliffing but this only happens after a very protracted period of stable sea-level.

There is no doubt that Johnson recognized these models of coastal development as illustrative of two commonly occurring situations. He also realized that submergence in given circumstances can produce features resembling those he had described as typical of emergence. He wrote 'A shoreline of submergence (emergence) is one of which the dominant, not necessarily the latest, features reflect submergence (emergence) – or in which the features resemble those which would be produced by submergence (emergence).' Although this sentence might be construed as implying a morphological rather than a genetic classification, the very terms emergence and submergence clearly refer to coastal evolution, and many of Johnson's disciples have treated them as indicating specific sea-level changes. The results have often been totally paradoxical. The Atlantic seaboard of the United States has been quoted as the prototype of an emerged coast, whereas all the evidence now available identifies it as a classic example of a gently shelving area drowned by the post-glacial marine transgression. Fjords are conventionally described as features of submergence, but are often sited on coasts that have experienced substantial isostatic rebound and a net fall in sea-level since deglaciation.

It is clear that if the terms submergence and emergence are to be retained, they must be closely defined with respect to the time interval under consideration. In the short term nearly all coasts have experienced inundation, the only exceptions being those where either isostatic or tectonic uplift has outstripped the post-glacial rise in sea-level; in the longer term many must have experienced relative uplift, particularly if the model of sea-level change envisaging a secular decline is correct. In this limited respect the Atlantic seaboard of the United States could

still be regarded as an emerged coast, although the value of such long-term changes as a basis for coastal classification seems rather dubious.

Although the ideas of Johnson have continued to exercize a profound influence, many alternative classificatory schemes have been suggested. Shepard has proposed a division into coasts fashioned mainly by terrestrial agencies and those shaped mainly by marine processes. The former are subdivided into land erosion coasts, subaerial deposition coasts, volcanic coasts and coasts shaped by diastrophic movements; land erosion coasts include both drowned river valleys and fjords, while subaerial deposition equally embraces both fluvial and glacial activity. Coasts shaped by marine processes are divided into three major classes, those due to erosion, deposition and the activity of organisms. Shepard's classification has much to commend it; its major weakness in application is the subjectivity of deciding when a coastline has been sufficiently modified to justify a designation as shaped mainly by marine agencies. A second interesting variation on the traditional approach to coastal classification has been proposed by Valentin. He suggests that the primary division should be between those coasts that are advancing and those that are retreating. The changes on the former may be due either to emergence or to deposition, on the latter either to submergence or to erosion. Various combinations can be readily visualized, and in certain instances a balance may be struck between the forces leading to advance and those leading to retreat so that the position of the coastline remains constant.

Inman and Nordstrom have proposed a classificatory scheme based upon the ideas of plate tectonics (Fig. 17.1). They recognize three primary classes, collision coasts at the destructive margins of plates, trailing-edge coasts where continental blocks are moving away from spreading centres, and marginal-sea coasts. The trailing edge coasts are divided into three types depending on whether continental separation is geologically recent (neo- type), both margins of the continent are trailing edge (afro- type), or one margin of the continent is trailing edge and the other collision (amero- type).

Finally, mention may be made of the approach to coastal classification advocated by Davies. Owing to the significance of wave action in shaping coastal forms he maintains that it is vital to distinguish contrasting wave environments. Four major types are recognized (Fig. 17.1). The first is the storm-wave environment in which a significant proportion of the waves are generated by local gale-force winds. These winds are most common in the temperate storm belts so that such an environment occurs with greatest frequency in middle-to-high latitudes where waves of high energy produce prominent cliffs and shore platforms.

Fig. 17.1. Two possible methods of classifying the coasts of the world: (A) according to tectonic type. (B) according to wave environment.

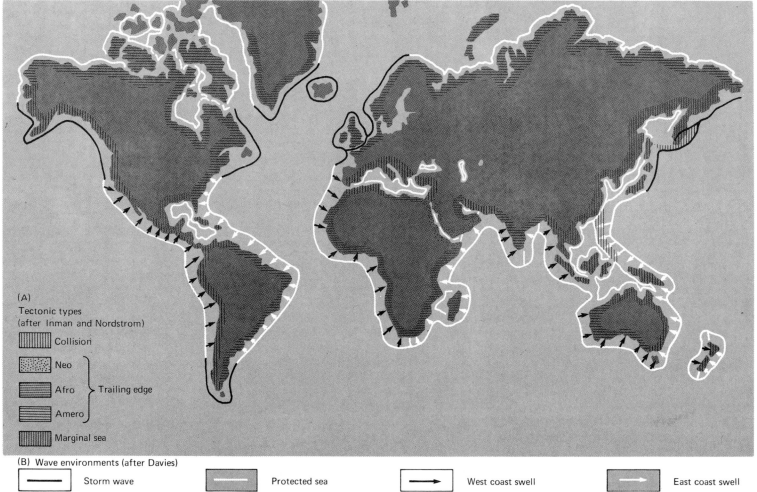

(A)
Tectonic types
(after Inman and Nordstrom)

▥	Collision
▦	Neo
▨	Afro } Trailing edge
▤	Amero
▥	Marginal sea

(B) Wave environments (after Davies)

| ── | Storm wave | | Protected sea | → | West coast swell | ⇨ | East coast swell |

Constructional features are very abundant and tend to be composed of shingle. The second environment is that of west-coast swell. This occurs primarily within the tropics where gale-force winds are rare and most of the large waves are produced by swell originating from temperate storms. The mean wave energy is moderately high around the tropics but tends to decrease towards the equator. Cliffing is less active and constructional forms more commonly consist of sand than of shingle. The third wave environment is that of east-coast swell. Owing to the strong westerly component in most temperate storms the swell reaching east coasts is on average lower than that on west coasts. Mean energy levels may be characterized as low to moderate, although the tropical cyclones which occur on these coasts can have an effect out of all proportion to their frequency and duration. The final wave environment is that of the protected sea. Here either coastal configuration prevents the full penetration of oceanic swell, or an ice cover damps the effects of waves approaching the coast. The typical features of such a low-energy environment ·are intricate constructional forms related to the complex patterns of fetch and wind direction.

No single classificatory scheme commands overwhelming support at the present day, much depending upon the purpose for which the classification is to be used. Morphology, evolution and contemporary processes have all been suggested as primary criteria, sometimes with unfounded assumptions being made about the relationships between them. This has led Russell, one of the foremost students of coastal landforms, to argue that attempts at coastal classification are premature until we possess a much better factual knowledge.

Forms due to wave erosion

Cliffs Although coastal cliffs can be among the most spectacular of all landforms, their origin and development has attracted surprisingly little detailed investigation. On all active cliffs two groups of processes operate simultaneously. At the base the sea performs the dual functions of erosion and transportation. Higher on the cliff face a variety of subaerial processes induces collapse and supply of debris to the waves below. The nature and balance of these activities together determine the form of

the cliff. In many respects the evolution of marine cliffs can be regarded as a particular case of slope development. The full range of subaerial processes as outlined in Chapter 8 can be observed to operate on different sections of coast. The main distinguishing feature is the constant backwearing at the base. As long as sea-level remains stable this can only act horizontally, and it is easy to visualize a simple equilibrium form being conserved during continuous recession. A slope is first steepened by wave attack until the subaerial processes are able to keep pace with the basal retreat. On some materials this situation may be reached when the cliff angle is only 20° or so, but in the case of massive bedrock the profile may be steepened to the vertical before the whole face begins to retreat uniformly. A particular form can only be maintained so long as the balance between marine and subaerial processes remains unchanged. In theory this is unlikely to occur over a protracted period since wave action will gradually be modified by the increasing width of the shore platform.

One way of classifying cliffs is by reference to the dominant subaerial processes acting on them. In weakly coherent materials movement may take place by means of flowage, gullying, or sliding. Often all three processes can be observed in operation within a short distance of each other. Flowage is particularly characteristic of cliffs formed in clay where interstitial water substantially reduces the mechanical strength of the material; the toe of each earthflow may be regularly trimmed back by wave action at high spring tide. On certain cliffs weakly coherent deposits are rapidly degraded by surface run-off. This is especially true of poorly cemented sands and silts which are sometimes deeply gullied to form veritable badland areas. Detrital fans built out on the beach during periods of heavy rain are obliterated when waves next wash the cliff base. It is, however, sliding that constitutes the dominant mechanism of many cliffed coasts, particularly those composed of thick clay successions. Relatively shallow slab slides are extremely common along cliffs cut in till, whereas deep-seated arcuate slides tend to develop most fully on well-bedded sedimentary rocks. Especially favourable circumstances occur where massive strata overlie an impermeable clay. By attacking the clay the waves steepen the cliff until the basal material is unable to sustain the weight of overlying rock. The strength of the clay may be reduced by percolating water after heavy rain, while a more gradual reduction often arises from long-term chemical alterations. The precise shape of the plane along which movement occurs is much influenced by structural details, but in many instances rotational displacement can be demonstrated. This results in back-tilting of the surface strata, and elevation at the toe of the slip. The latter occasionally produces a prominent ridge on the foreshore which is subject to later destruction by wave action. The large slumped mass itself is rarely removed by wave action before the next segment of cliff is dislodged ; the result is an under-cliff of highly irregular relief and jumbled structure. Amid the displaced slices of rock, flows of saturated clay often help to carry part of the debris further seawards. Well-known examples of rotational slipping are found at two localities on the coast of the English Channel. At the Warren, east of Folkestone, Chalk and Upper Greensand rest on impermeable Gault clay .The water percolating through the Chalk is heavily charged with calcium carbonate, and on reaching the glauconitic Gault clay the calcium ions replace the original potassium ions. This reduces the strength of the clay and is a contributory factor in the frequent movements that occur on this section of coast. Ten major slips were recorded in the period between 1765 and 1915, since when there has been rather greater stability. Further west near Lyme Regis the cliffs also consist of Chalk, Upper Greensand and a rather sandy facies of the Gault clay, the whole succession here resting unconformably on Liassic clays, marls and limestones. Deep-seated rotational movement is very common. The largest recorded displacement is that of a mass known as Goat Island, some 6 hectares in surface area and over 8 million tonnes in weight. The movement took place in 1839, leaving a chasm at the foot of the main cliff 60 m deep and 100 m wide. In this particular instance there is still some uncertainty about the precise form of the slip plane, but many smaller slides undoubtedly occur along arcuate surfaces. An interesting aspect of many of these is their accompaniment by earthflows indicating that the clay is temporarily converted to a liquid state in which it is capable of relatively rapid flowage. A few kilometres east of Lyme Regis Brunsden has monitored the movement of a large multiple slide and has shown that on certain arcuate shear-planes sporadic displacements occasionally total as much as 80 m in a year.

On less deformable materials the primary mechanism of cliff recession is rockfall. This is the process associated with the traditional concept of marine cliffs as high rock faces, often approaching the vertical and sometimes overhanging due to the undercutting action of the waves. Structure and lithology are the dominant controls of form. Where massive, horizontally bedded strata are involved, collapse of huge vertical slabs and pinnacles is probably the most significant means of recession. Isolated by erosion along prominent vertical joints, they are eventually rendered unstable by continual undercutting at the base. Sea stacks are a characteristic feature of cliffs retreating in this way. Where similar massive strata are tilted, the cliff profile depends on the disposition of the planes of potential movement (see Fig. 8.3, p.145). Where the joints are more closely spaced, numerous minor collapses tend to bring blocks of various sizes crashing down. Many limestone and chalk cliffs recede in this way. An excellent example is afforded by the high chalk cliffs of the Seven Sisters in Sussex.

As on all slopes, it needs to be remembered that many different processes operate simultaneously. Although the basic cause of rockfall in marine cliffs is instability due to wave action, the actual triggering will often be a period of exceptionally heavy rain or intense freeze–thaw activity. In cold climates rockfall may be so rapid that it prevents the cliffs attaining the angle of slope they would reach in other climatic zones. Decomposition and disaggregation of rocks has been little studied on the face of marine cliffs, but there seems little doubt that in such an exposed position individual mineral grains are particularly prone to loosening. One possible factor worth recalling is the great volume of salt spray that can be carried high up on a cliff face during a severe storm.

Structure and lithology affect not only the profiles of cliffs but also their development in plan. This can be seen at several different scales. On the broad scale lithology is particularly influential. Where the geological strike is normal to the trend of the coast, rapid erosion of weaker rocks produces indented bays between promontories of more resistant material ; where the

Fig 17.2 The relationship between geology and coastal landforms on the Dorset coast. The lower diagram shows the area around Lulworth cove in greater detail.

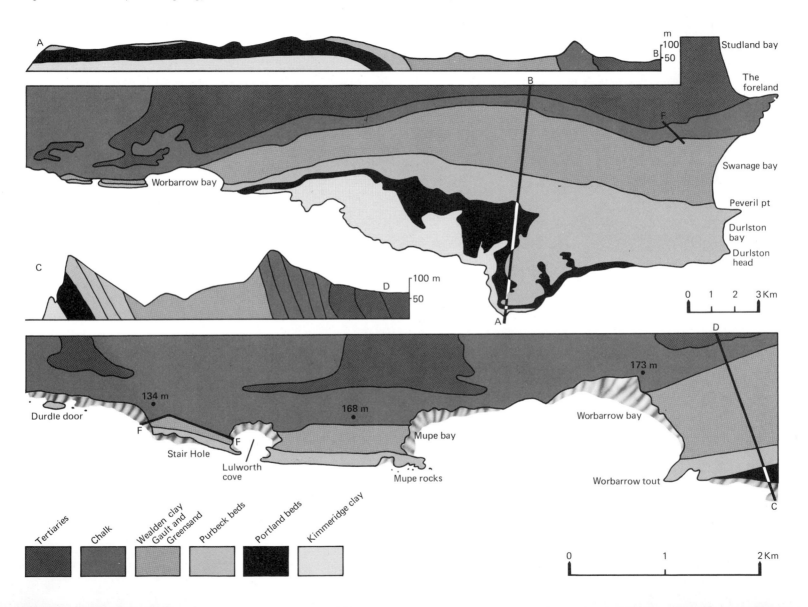

A

B

m
100
50

Studland bay

The foreland

F

Worbarrow bay

Swanage bay

Peveril pt

Durlston bay

Durlston head

C

D

100 m

50

A

0 1 2 3 Km

B

D

134 m

173 m

Durdle door

Worbarrow bay

F

168 m

Stair Hole

F

Mupe bay

Lulworth cove

Mupe rocks

Worbarrow tout

C

Tertiaries

Chalk

Wealden clay Gault and Greensand

Purbeck beds

Portland beds

Kimmeridge clay

0 1 2 Km

Fig 17.3. The coastal landforms at Tintagel in north Cornwall. The cliffs commonly exhibit a slope-over-wall form, well exemplified in the headland at the top of the photograph. The effectiveness of wave erosion at the cliff base is much influenced by rock structure, producing an intricate pattern of minor bays and headlands.

strike runs parallel to the coast, alternating weak and resistant outcrops can engender a highly irregular outline. Both situations are beautifully exemplified on the coast of the English Channel in Dorset (Fig. 17.2). At the more detailed scale structure often appears to be the prime influence. The crenulate cliffs of north Cornwall illustrate very well the way in which bedding planes, joints and faults are all subject to preferential erosion (Fig. 17.3). It is not always clear why one set of fractures has been eroded more successfully than another. Other things being equal fault planes may suffer particularly rapid enlargement owing to the shattered nature of the adjacent rock. However, it is the orientation of the fractures relative to the direction of wave approach that is probably most significant. On the coast of north Cornwall features aligned parallel to the approach of the largest storm waves are said to be the most rapidly attacked. On the coast of northern Oregon, on the other hand, the fractures suffering most rapid erosion are those facing north-west, despite the fact that the most powerful waves come from the south-west; the explanation appears to lie in the prevalence or greater frequency of the slightly smaller storm waves generated by the north-westerlies.

So far it has been assumed that cliff forms are to be explained in terms of processes operative at the present day. However, this is an assumption that needs to be treated with caution. It has become increasingly clear during the last few years that many lines of cliffs have had a much more complex evolution than was formerly supposed. On a stable coast wave erosion at the cliff base can have been going on continuously for 5 000 years at most. Prior to that there must have been a minimum of 25 000 years during which the sea was too low to have had any effect. Two specific lines of evidence point to the sea having recently re-occupied cliffed coastlines that were abandoned earlier in the Pleistocene epoch. In a number of localities old cliffs buried by till have been identified. One of the most striking instances occurs along the Welsh coast in Cardigan Bay. Here

the apparently modern cliffs pass laterally behind a plug of drift with very little change in form. It is evident that a major function of wave action during the last 5 000 years has been removal of a drift cover plastered against the ancient cliffline. The current cycle of erosion may have trimmed and freshened the fossil forms but has certainly not fashioned them ab initio. South of the limit attained by Weichselian ice it is not uncommon to find an old beach lying only a metre or two above present sea-level and backed by ancient cliffs beyond the reach of contemporary wave action. In many instances this beach can be shown to pre-date the last glaciation. In these localities the sea has clearly returned to a position near, but not exactly coincident with, one that it occupied earlier in the Pleistocene epoch. Where recent wave erosion has enabled the sea to destroy the old beach, it is reasonable to assume that it has frequently reoccupied an ancient cliffline.

To a certain extent recognition of this two-stage evolution removes a predicament for the geomorphologist, since it is often difficult to conceive that major clifflines have been produced within the space of 5 000 years. Yet it also raises the question of the modifications suffered by the abandoned coastline at the height of the last glaciation. In much of south-western Britain, for instance, it appears that periglacial processes rapidly modified the vertical cliff faces. Freeze–thaw activity reduced the angle of slope and led to accumulation of a wedge of frost-shattered debris on the former beach or shore platform. The results of this interlude are most obvious where the deposit of head is still preserved at the base of the old cliff. However, in many localities the post-glacial marine attack has removed the head and the waves are once more undercutting the bedrock. Nevertheless, the upper part of the cliff retains the sloping form imposed during the periglacial episode, often thinly mantled with frost-shattered debris. The resultant profile is described as 'slope-over-wall', with the height of the wall largely determined by current exposure to storm waves. Although found on many high-latitude coastlines, slope-over-wall profiles are particularly well developed in south-western England, probably because the powerful North Atlantic breakers eroded very high interglacial cliffs which were then subject to unusually intense freeze–thaw activity.

On coasts affected by substantial glacio-isostatic rebound, rather different considerations must apply. The present cycle of marine cliffing can only have commenced after local deglaciation. The period available for periglacial activity is comparatively short. The dominant change in sea-level has generally been an emergence continuing right up to the present day. This means that the idea of 5 000 years of stable sea-level no longer applies. As long as glacio-isostatic recovery remains incomplete, old interglacial clifflines will be tilted and so cannot be exhumed simultaneously over long distances by present wave action. The evolution of cliffs in such glaciated areas is obviously complex and little detailed work has yet been done. It seems inconceivable that some of the high cliffs in Scotland, for instance, could have been fashioned entirely in the short post-glacial interval, especially when the local fluctuations of sea-level are taken into account. Exhumation of buried features seems probable, and this is confirmed by observations in north-eastern Scotland where apparently modern cliffs, complete with sea-stacks, disappear beneath a cover of glacial drift; elsewhere the long, structurally guided inlets known as geos contain pockets of till, implying that they too must be older than the last local glaciation.

In the foregoing paragraphs it has been suggested that certain clifflines appear too large to have developed wholly within the post-glacial period. This implies some appreciation of the rate of cliff evolution, and it is necessary to examine in more detail what is known of the rate of cliff recession. An immediate problem arises from the tendency to confine measurements to coasts undergoing obvious retreat. Even within this category, measured values vary by a factor of 50 or more. The fastest rates are generally found on coasts of weakly coherent material. The Holderness coast of Humberside is particularly famous in this respect. Many former villages have been lost to inroads of the sea (Fig. 15.7) and it has been estimated that since Roman times the coast has retreated some 4 or 5 km. Yet it is only since the advent of accurate surveys that precise measurements have been possible. Valentin has examined the cartographic evidence for the period between 1852 and 1952. He estimates the average rate of erosion to be about $1 \cdot 8$ m yr^{-1}, with the figure tending to increase from north to south, and in the vicinity of Easington attaining a local value of $2 \cdot 75$ m yr^{-1}. The use of mean values hides considerable variations in both space and time. Some sections of the coast remain stable for many decades and are then subject to brief but very severe erosion. Around Withernsea, for instance, some 130 m were lost in the relatively brief period between 1852 and 1876. Another area renowned for the rapidity of cliff erosion is the East Anglian coast near Dunwich. Here for more than a century the rate averaged about 4 m yr^{-1}, while at Covehithe 10 km to the north it exceeded 5 m yr^{-1} between 1925 and 1950. During the 1953 storm surge cliffs at Covehithe were driven back a full 12 m in a single day. If continued for several millenia, all the above rates would involve major changes in coastal position. However, it must be emphasized that they are quite exceptional and found only on cliffs composed of weakly coherent materials. On rocky coasts the values are invariably lower and often so small as to be extremely difficult to measure. Relatively fast bedrock rates have been recorded on Miocene sandstones along part of the Oregon coast where cliffs since 1880 have been receding at about $0 \cdot 6$ m yr^{-1}; chalk cliffs are currently retreating at about $0 \cdot 3$ m yr^{-1} on the Isle of Thanet; at $0 \cdot 5$ m yr^{-1} in parts of Sussex; and at almost $0 \cdot 25$ m yr^{-1} in northern France. Interesting measurements have been made in North Yorkshire where the cliffs are composed of both till and bedrock. Since 1892 those consisting of Lias shales have retreated at an average rate of $0 \cdot 09$ m yr^{-1}, while intervening reaches consisting of glacial drift have retreated at $0 \cdot 3$ m yr^{-1}. Yet on many coasts composed of more resistant bedrock, cartographic evidence reveals no detectable change within the period of a century or more. Ancient structures presumed to have been built near the cliff edge remain untouched. A problem on such coasts is the erratic nature of much cliff destruction. This is well illustrated in north Cornwall. Long sections of that coast show negligible alteration within the last few centuries but at Tintagel erosion has severely modified the island on which a castle was built about 1145 (Fig. 17.3). Part of the castle originally stood on an isthmus linking the island with the mainland. This collapsed and in the thirteenth century a bridge had to be built in its place. By the sixteenth century the gap was too large to be bridged and the remains of the castle were isolated from the mainland. Yet the foundations of some of the original castle walls, almost certainly built at the cliff edge, remain intact today. Measurement of cliff recession in such conditions poses obvious problems, but there seems no doubt that many segments of the coastline are retreating at less than 0.01 m yr^{-1}.

Extrapolating current rates of erosion backwards in time is always a perilous procedure. On some sections of coast it is evident that Man is having a considerable effect. In part this is intentional since attempts have often been made to stabilize cliffs undergoing rapid retreat. There are also instances where erosion has been accidentally exacerbated by Man's activities. In south Devon removal of shingle from the foreshore during the 1890s led to accelerated erosion and destruction of the old fishing village of Hallsands. Groynes intended to diminish long-shore drifting have frequently had the effect of accentuating nearby erosion because a protective beach has been starved of its former supply of shingle. Such examples serve as a reminder of the varied factors that can influence cliff recession. It may also be recalled that few workers have regarded world sea-level as totally unchanging during the last 5 000 years, and even a slight rise can have a significant effect in maintaining the efficacy of wave erosion. Despite the many uncertainties, it is reasonable to conclude that on resistant coasts the current cycle is often rejuvenating older, degraded cliffs and that the modern forms can only be fully understood in the context of this two-stage evolution.

Shore platforms Many of the factors discussed in the preceding section with respect to cliffs are also apposite in considering the development of shore platforms. For instance, it is clear that certain shore platforms at present swept by the sea are exhumed features and that the current cycle of wave action has done little more than strip off a superficial cover of till or head. Stripping may occur over a relatively wide altitudinal range since waves are capable of washing away weak material high above the level at which they would normally be effective as abrasive agents. This is one reason why shore platforms do not bear any consistent relationship to such a datum as the high-water mark of ordinary spring tides. However, an even more important reason is the complexity of processes involved in the develop-

ment of platforms.

The classical approach to shore platforms maintained that they are almost exclusively the product of abrasion, first of all by material carried backwards and forwards in the swash zone and later by debris being transported seawards in the off-shore zone. Continued cliff recession requires that the wave-cut bench should be progressively lowered so that wave energy is not entirely dissipated in a wide area of very shallow water. In the inter-tidal zone there is often field evidence of abrasion in the presence of potholes, smoothed surfaces and well-rounded pebbles. Below low-tide level the evidence is more ambiguous. At one time abrasion due to wave agitation was held to be effective at depths up to about 180 m, but modern studies suggest the true figure is much closer to 10 m. Below that depth large storm waves may stir fine sediment into movement but its erosive capacity is negligible. Assuming a typical minimum gradient for a platform of 1 in 100, the greatest width ascribable to abrasion would be 1 000 m in areas of low tidal range, rising to 1 500 m in areas where the tidal range reaches 5 m. Yet, with a stable sea level, these values would only be attained after very protracted erosion; greater values would only be likely if there were a steady rise.

Several workers have measured the form of the platforms that front the cliffs of southern Britain. Those at the foot of actively receding cliffs have attracted particular attention since they are clearly being fashioned at the present day and not simply inherited from older forms. The altitude of the notch at the base of the chalk cliffs on the Isle of Thanet has been surveyed by both Wood and Wright who find that it lies some 3 m higher in the embayments than on the headlands. This means that the notch has a seaward slope that commonly amounts to 1 in 15. Since this is due neither to a slope in the water surface nor to recent tectonic tilting, Wood contends that it must represent deeper excavation by waves breaking on the headland. On the other hand, So, who made an independent study of the Thanet platforms, argues that they tend to extend to a higher level where the exposure is greatest, pointing to contrasts between the two sides of a single headland. There is some conflict of evidence here which still has to be resolved. Both Wood and So agree that the seaward slope of the platforms is much greater in the bays than on the headlands. This can only be due to variations in the intensity of vertical erosion once the notch has been formed and pushed further inland. Both authors dismiss the explanation that the rocks in the bays are less resistant than those on the headlands, and maintain that there must be some process by which the platform in the bays is steepened to a ramp-like form. Wood suggests this may be scouring as a result of the concentration of sediment in the bays. A final aspect of the Thanet platforms is that, around the headlands, they have a convex-upward profile which may well be linked either to eustatic changes during the last few millennia or to continuing depression of the coasts around the North Sea basin.

Wright has compared the shore platforms along the whole southern coast of England (Fig. 17.4). He finds that eastwards from Torquay the form of the platforms is relatively simple, with little convexity or concavity and gradients ranging from 1 in 14 to 1 in 90. The maximum width is about 270 m, and the mean about 100 m. The junction between platform and cliff is below the uppermost level of marine action, often corresponding with the mean high-water neap position. Occurrences above this datum can generally be ascribed to a structural control in which the upper surface of the platform coincides with a particularly resistant bed; occurrences below it Wright attributes to the effectiveness of locally abundant abrasive materials. Westwards from Torquay platforms are much more complex in form. Two or even three distinct levels can be detected in a single profile,

with the upper one often rising above the mean high-water spring level. There is little doubt that many of these multiple platforms arise from rejuvenation of formerly abandoned shorelines.

Although workers who have examined the shore platforms along the south coast of England have disagreed about the nature of the factors determining altitude and morphology, they have generally concurred in the belief that the basic erosional process is wave abrasion. Yet on other coasts, particularly around the Pacific and Indian oceans, multiple platforms have been described which are very difficult to explain solely in terms of corrasive action. Consisting of nearly horizontal benches separated by short segments of much steeper slope, even miniature cliffs in some cases, they may be divided into two major groups, the high-tide and low-tide platforms. The former are found at or just above mean high-tide level. They may be flooded by the sea at spring tide, but during the rest of the tidal cycle remain dry unless there are large storm waves. They have in fact been ascribed to the action of storm waves, although it is difficult to see why one particular level should be preferentially developed to the exclusion of others. A number of alternative explanations have been offered. It has been suggested that they are the product of wave action at a slightly earlier Holocene period, possibly kept fresh by modern surf washing over them. If it is accepted that sea-level has not significantly exceeded its present height since the last glaci-

Fig 17.4 The elevation of shore platforms along the south coast of England in relation to various tide levels (after Wright, 1970).

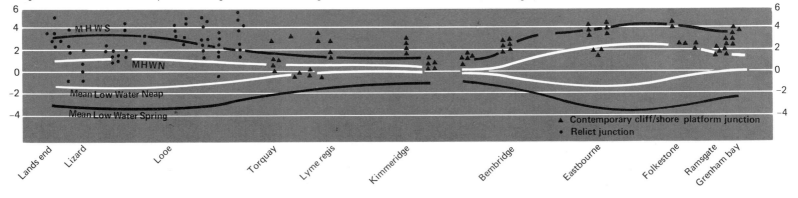

ation, high-tide platforms of this origin would be restricted to recently uplifted coasts. An alternative hypothesis ascribes the high-tide platform to water-layer weathering. This process involves the gradual enlargement of pools near and just above high-tide level where there is regular replenishment of water by spray. The constant wetting and drying around the margins of the pools leads to their progressive extension. For the process to be fully effective occasional washing over by storm waves is necessary to remove the loosened debris. It is unlikely that a platform entirely due to water-layer weathering can exist where rapid cliff recession is taking place, since such recession implies large storm waves actively breaking against the cliff face. This may explain why high-tide platforms are less common around the stormy North Atlantic than on coasts affected mainly by moderate swell.

Low-tide platforms are near-horizontal surfaces exposed only when the sea falls below mid-tide level. Being characteristic of limestone coasts, they are usually attributed to solutional processes. As already pointed out, sea water is normally saturated with respect to calcium carbonate so that special conditions are required for active solution to take place. Circumstances favouring solution occur in rock pools temporarily isolated from the sea, and these are most frequent in the inter-tidal zone. The downward limit to corrosion in this way obviously approximates to low-water mark and explains why the platforms should be found at that height. On the other hand, it is equally true that many limestone coasts do not show low-tide platforms. In general they seem to be less well-developed in stormy temperate areas than along tropical coasts. It appears that active abrasion can mask the effects of solution, as happens in the case of the sloping platforms that front the chalk cliffs around the British Isles. Here the presence of nodular flint may be a significant factor since it arms the waves with a powerful abrasive tool; in tropical latitudes there are long reaches of coastline composed exclusively of calcium carbonate.

It seems clear that shore platforms can originate in many different ways and at a variety of levels. Many problems remain to be solved. Compared with measurements of cliff recession there are few assessments of the rate of platform erosion. In part this is due to the slower pace of change and the corresponding difficulty in making short-term measurements. On one section of the foreshore around the Isle of Thanet it has been estimated that lowering of the chalk platforms averaged about 25 mm yr^{-1} between 1904 and 1961. Although such a figure could not be sustained over a long period of time since it would be inconsistent with local values of cliff recession and platform gradient, it does indicate the potential efficacy of downwearing by wave action. The seaward termination of platforms is a further subject requiring investigation. Many show an abrupt steepening of the gradient, often approaching the dimensions of a low cliff. This has been referred to as the low-tide cliff and has been shown locally to migrate landwards in the same fashion as the high-tide cliff. In such circumstances it is obviously an important factor in determining the width of the shore platform.

Finally, reference may be made to the widest shore platform of all, the strandflat, which still defies satisfactory explanation. Found off the coasts of Norway, Spitsbergen, Iceland and Greenland, it is clearly associated by both distribution and morphology with glaciation. It consists of a platform up to 60 km wide terminating abruptly against coastal mountain ranges. Although lying close to sea-level and undergoing slow alteration by wave action, it still preserves clear signs of glacial moulding. Opinions differ as to whether it is essentially a glacial landform modified by the sea, or a coastal landform modified by the passage of ice. Some workers have suggested that ice and sea may have acted simultaneously, ascribing the strandflat to erosive processes beneath an ice-shelf that rose and fell with the tide.

Forms due to marine deposition

Beaches Beaches may be regarded as accumulations of marine sediments attached to the coast along their full length. The composition of the materials varies from very coarse cobbles to fine sand, but a useful distinction is that between shingle and sand since intermediate sizes are relatively rare. Davies has attempted to map the distribution of pebbly beaches around the coastlines of the world and has shown that they are only common in high latitudes. The popular picture of tropical sandy beaches thus has some justification in fact. Several factors help to explain the distribution pattern, but it is first necessary to identify the four major sources of beach material. The first is the cliff face and adjacent shore platform subject to active wave erosion. The second is the sediment transported to the coast by agencies involved in the denudation of the land. The third is sediment moved laterally along the coast by longshore drifting. The final possible provenance is loose sediment lying in the off-zone and fed to the beach as a result of wave disturbance. All the primary sources tend to contribute finer material in low latitudes than in high latitudes. In tropical areas the swell generally lacks the erosive capacity of the steep waves encountered in the temperate storm belt, while the rivers entering the tropical oceans are laden with fine sediment rather than coarse. A further factor in northern latitudes is the widespread cover of glacial, glacifluvial and periglacial deposits which all constitute a ready source of coarse detritus.

The provenance of shingle can be ascertained by examining the lithology of the pebbles. The majority are generally of local derivation. On the other hand, many beaches also contain a small proportion of far-travelled constituents that have followed long and intricate routes since first being removed from their point of origin. For example, shingle on the coast of north Cornwall is found to contain material from Wales. This is thought to have been transported first by an ice-sheet from the Welsh upland and deposited on the floor of the Bristol Channel. Later, during the post-glacial marine transgression, it was pushed shorewards by the advancing sea and eventually incorporated as part of the present coastal shingle. On sand beaches the provenance of the sediment may be indicated by the nature of the heavy minerals. Along the Dutch coast most of the coastal sand was formerly assumed to be derived by longshore drifting from the Rhine estuary. However, heavy mineral analysis has demonstrated a very significant admixture of material coming from the off-shore zone on the floor of the North Sea. It should also be noted that by no means all beaches are siliceous in composition. This applies most obviously to the coral sands of the tropical beach but even in temperate latitudes a surprisingly large proportion of beaches are highly calcareous. Many sands on the exposed coasts of western Ireland and Scotland consist of comminuted shells derived from abundant organisms living in the off-shore zone. In these cases the powerful storm waves are an important factor in driving the shell fragments landwards.

Virtually all sections of a normal coastline have potential sources of beach material in one direction or another. Absence of beaches may in part be a function of time. For instance, rocky coasts subject to continuing glacio-isostatic uplift have often experienced a stable sea-level for too short an interval to permit much beach accumulation. Yet the single most important factor explaining the absence of a beach is longshore drifting. The effect of this transfer from one section of coastline to another is most readily apparent on a 'cape-and-bay' coast where the rocky headlands remain free of sediment while the re-entrants have thick bay-head beaches. Davies has distinguished coasts of free transport from those of impeded transport. On the former, material may be distributed great distances by lateral movement from a single source; the beach is virtually continuous. In the case of impeded transport there is little or no exchange of sediment between adjacent re-entrants; each bay-head beach is an isolated unit fed from a purely local source.

In plan, beaches are usually smoothly curved with their concave side facing seawards. This is most obvious on small bay-head beaches but may also be seen on a much larger scale along coasts of free transport. There is no doubt that the outline is controlled mainly by wave action. As a general rule it is the powerful waves capable of moving both the greatest quantity and largest size of material that exert most influence; the beach tends to assume an alignment parallel to the crest of these dominant waves. The factors determining their direction of approach seem to vary from one area to another. Refraction plays the crucial role on coasts affected mainly by oceanic swell. Where there is free transport the beaches often exhibit beautifully regular alignments in accord with the wave pattern as refracted on entering shallow water. On crenulate coasts bay-head beaches may show a more diverse orientation but again are aligned parallel to the refracted waves. Material is first swept by longshore drift into sharp, rocky re-entrants. Initially the beach will be sharply curved but continued drift towards the centre will eventually reduce the curvature until the trend accords with that of the approaching wave crests. On coasts affected mainly by storm waves refraction is still significant but at least equal importance attaches to the frequency and direction of gale-force winds. Schou has advocated calculation of a so-called 'wind resultant' from anemometer data. He suggests that winds of less than Beaufort Scale 4 (21 km hr⁻¹) should be ignored as ineffective. Vectors for those of greater velocity are obtained by multiplying the frequency of winds for each Beaufort class by the cube of the mean velocity for that class, and then summing the products for each wind direction. For any section of coastline attention is often restricted to the three vectors representing on-shore winds, permitting calculation of the so-called onshore resultant. After appropriate allowance has been made for refraction, beaches might be expected to lie parallel to the waves associated with the onshore resultant. Although this is often found to be true, where there are marked variations in fetch these tend to be of overriding importance in determining beach orientation; the coasts of the North Sea and Baltic Sea illustrate this situation very clearly.

Beach profiles vary according to the nature of the constituent materials. Three common beach types may be distinguished, those composed of sand, those composed of shingle and those in which a shingle ridge forms the landward margin to a broad apron of sand. Shingle beaches are usually much steeper than those formed of sand, while on composite profiles a sharp break of slope often divides shingle ridge from sand apron. Beach form is obviously determined by wave action and, to a lesser extent, by tidal range. In general terms the maximum height of a beach is set by the upper limit of swash action. However, there is a marked difference between shingle and sand in this respect. During major storms large pebbles can be thrown high above the reach of normal waves, building shingle ridges 10 m or more above mean high-tide level; the prominent features thus produced are among the most enduring of beach forms, persisting unaltered while many profile changes occur along their seaward margin. Sand, on the other hand, does not normally form prominent storm ridges. Storm waves tend to be destructive on sand beaches and the higher parts of the profile are usually constructed during periods of moderate swell. The feature built up in this way is termed a berm. It typically has a flat upper surface separated by a sharp break of slope from the face regularly washed by the waves.

The seaward face of a beach profile varies according to the nature of recent wave action. As indicated in Chapter 15, repeated surveys can be used to record the changes in form, which may in turn be related to wave and tide observations. In their effect upon the profile waves are conventionally divided into two major groups, destructive and constructive. The former are characterized by vertically plunging crests and relatively weak swash, the latter by elliptical orbits and a powerful rush of water up the beach. Destructive breakers comb material down the beach, steepening the gradient near the limit of swash action; constructive breakers return material to the beach, building up the surface and generally reducing the seaward slope. In an early influential paper on wave types Lewis recorded that he had found destructive waves to have a frequency of 13–15 per minute and constructive waves a frequency of 6–8 per minute. By counting the number of waves breaking per unit-time Lewis was, in effect, measuring wave length. However, further investigations suggest that the primary factor is the related parameter of wave steepness. Studies have been made to determine the steepness at which wave action changes from constructive to destructive. For shingle beaches the critical value appears to lie between 0·016 and 0·020, for sand beaches between 0·010 and 0·014; this means that waves which are constructive on shingle can be destructive on sand.

Repeated surveying of beach profiles reveals many different types of cyclical pattern. Some cycles are annual, with conspicuous differences between winter and summer forms. On many beaches, for instance, winter storms remove a cover of sand to leave exposed a deposit of coarse shingle or even boulders. The summer visitor would often be very surprised if he could see his favourite beach after a severe winter storm. Fortunately by early summer spells of constructive waves have normally replaced the blanket of finer sediment. Of much shorter duration are cyclical changes that occur between individual storms. For example, if a beach has been steepened by destructive breakers at the time of high spring tide, later constructive action often tends to build a series of minor prograding ridges, each marking the level reached by swash during successive high tides as the sea regresses from its spring to its neap stage.

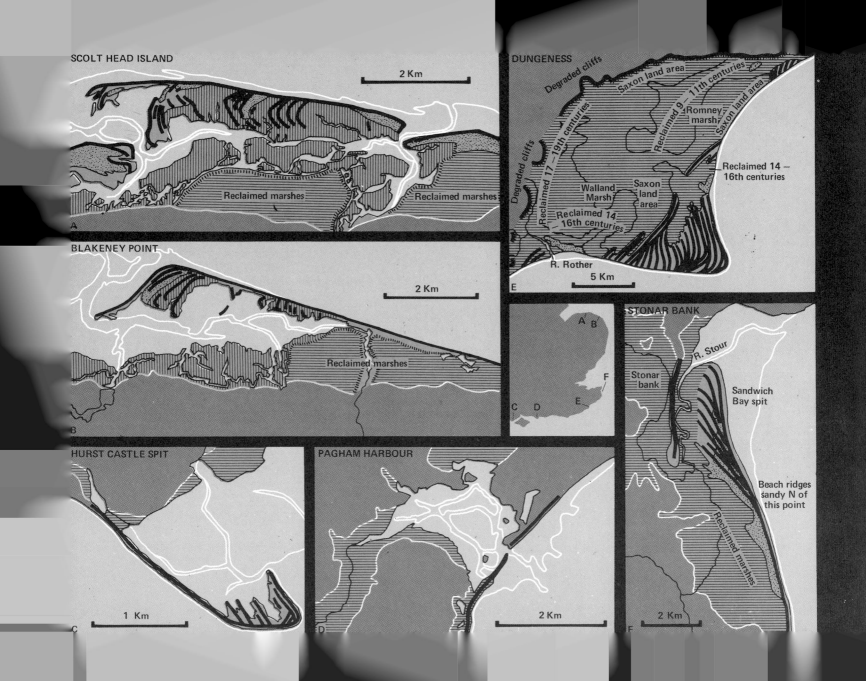

SCOLT HEAD ISLAND

2 Km

Reclaimed marshes

Reclaimed marshes

A

BLAKENEY POINT

2 Km

Reclaimed marshes

B

HURST CASTLE SPIT

1 Km

C

PAGHAM HARBOUR

2 Km

D

DUNGENESS

Degraded cliffs

Saxon land area

Reclaimed 9 – 11th centuries

Romney marsh

Saxon land area

Degraded cliffs

Reclaimed 17 – 19th centuries

Reclaimed 14 – 16th centuries

Walland Marsh

Saxon land area

Reclaimed 14 – 16th centuries

R. Rother

5 Km

E

A B

F

C D E

STONAR BANK

R. Stour

Stonar bank

Sandwich Bay spit

Beach ridges sandy N of this point

Reclaimed marshes

2 Km

F

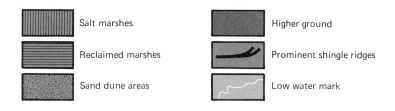

Fig 17.5 Sketch maps to illustrate the varied morphology of six constructional features located along the coasts of southern and eastern England.

Salt marshes		Higher ground	
Reclaimed marshes		Prominent shingle ridges	
Sand dune areas		Low water mark	

Sand beaches can show many deviations from a smooth seaward slope, but probably the most distinctive is that known as ridge-and-runnel. This occurs on gently shelving beaches protected from powerful storm waves. The ridges commonly have a height of about 1 m above the adjacent runnel. They tend to run parallel to the shoreline, although examples are known where they run obliquely across the foreshore. As the tide ebbs, water left in the runnels normally escapes by way of small transverse channels. The ridge-and-runnel pattern is a persistent feature which can survive many tidal cycles. The number of ridges varies. It is not uncommon to find a single large ridge just above low-water mark, but in other cases three separate ridges may be present. They appear to originate as small bars built up by the swash in positions where it has the longest opportunity for constructive action, usually the high- and low-water marks of the neap and spring tides. The small berms built in this way can survive submergence by the rising tide, and although the pattern they form may be destroyed by storm waves, it is then re-established during succeeding spells of calm weather.

Spits Spits are linear depositional forms attached at one end to the coast and most commonly sited where there is an abrupt change in coastal alignment. Material is supplied by longshore drifting and shaped into the distinctive form by wave action. General requirements are an abundant supply of sediment and reasonably shallow water into which the spit can grow. Many spits built across river mouths divert the river parallel to the coast. In the case of Orfordness on the East Anglian coast the diversion of the Alde amounts to over 16 km but when the spit

becomes elongated to this extent there is always the risk of storm waves breaching its neck. As mentioned earlier, this is believed on historical evidence to have happened at regular intervals to the spit at Spurn Point.

The alignment of a spit is generally determined by the direction from which the largest storm waves approach. Given an adequate supply of shingle these tend to build high ridges parallel to the crest of the advancing breakers. In many sheltered seas it can be shown that spits have developed normal to the maximum fetch since it is this which controls the size of the breakers. At their distal end spits often develop lateral hooks or recurves. Smoothly rounded hooks are due to wave refraction as the end of a spit advances into deeper water. Angular hooks, on the other hand, occur where material swept round the end of a spit is shaped into ridges by waves approaching from a second direction. A well-known example is the fashioning of the distinctive recurves at the end of Hurst Castle spit by waves coming down the Solent (Fig. 17.5). The hooks are here aligned at an angle of 150° to the trend of the main stem.

The pattern of recurves is of great value in deciphering the growth stages of a spit. Development is often spasmodic with phases of rapid development alternating with others of relative stability. At certain periods large recurves may evolve with little lengthwise extension of the feature. At others the dominant change is lengthwise growth with few lateral hooks. At yet others there may even be erosion trimming back the tip of the spit. The determinants of such spasmodic evolution are still poorly understood. Many studies have shown elongation near the tip occurring simultaneously with recession of the main stem; the point at which erosion ceases and gives place to aggradation is known as the fulcrum. If the point of attachment to the mainland is receding, the spit itself must retreat in order to maintain its orientation. This retreat means that the main storm beach, instead of meeting the old hooks in a smooth curve, intersects them at an angle. The same landward retreat can also lead to fine sediments originally deposited in the lee of the spit being exposed on the seaward face. It is worth noting that, with very abundant sediment, progradation of a spit can occasionally take place with successive storm beaches being constructed one in front of the other.

Pairs of spits occur at the seaward end of certain estuaries. These pose a problem in that they appear to grow towards each other and imply longshore drifting from opposite directions. In some instances there may well be convergent drifting towards the head of an estuary. Where this occurs the ebb and flow of the tide will be an important factor in fashioning the ends of the spits. It should not always be assumed, however, that this is the full explanation of 'double spits'. At Pagham on the English Channel coast Robinson has described two spits, one trending north-eastwards and the other south-westwards (Fig. 17.5). He has presented historical evidence to suggest that the two features derive from a single spit that was breached near its centre by exceptional storm waves. Further east the same author has described two shingle accumulations superficially resembling spits that have grown in opposite directions across the mouth of the River Stour. These are the features known as the Stonar Bank and the Sandwich Bay spit. The former extends southwards from the Isle of Thanet, the latter northwards from the cliffs near Deal. The outer feature has all the characteristics of a true spit, but the inner Bank is much less typical. Its morphology is not that of an active spit and its composition is not such as can be derived by longshore drifting from the adjacent cliffs. It probably originated as an offshore bank that migrated landwards during the post-glacial rise in sea-level and has subsequently been shut in behind a conventional spit growing from the south.

Nesses and cuspate forelands The terms ness and cuspate foreland have both been applied to prograding shorelines in which the sedimentary accumulation assumes a triangular shape with the broad base attached to the mainland and the apex facing seawards. It now seems likely that several different processes can be involved in the development of features of this general shape. On the East Anglian coast Robinson has shown that each of five nesses is associated with a distinctive pattern of ebb and flood channels a short distance offshore. A flood channel delivers sediment to the northern side of the ness, while an ebb channel does likewise on the southern side. The abundant supply of material thus depends on conditions off-shore and is not related to the pattern of longshore drifting. Indeed, at least one of the nesses has been shown to move north-wards while the direction of drifting is southwards. The movement has been accompanied by a corresponding northward migration of the ebb and flood channel system. This origin may be contrasted with that of other nesses in which the main supply of sediment is by longshore drifting. More closely related to spits, these features might be distinguished as cuspate forelands. The best-known example is that of Dungeness on the coast of the English Channel. It has been suggested that this began as a normal spit aligned almost parallel to the trend of the coast. At a later stage it swung round to face the direction from which the dominant storm waves were coming. The early evolution of the foreland may be deduced from historical evidence dating back to Roman times, more recent evolution from an exceptionally fine suite of shingle ridges. The section facing south-south-west is currently undergoing erosion while accretion continues near and round the point. Recession has taken place under the attack of destructive storm waves, accretion under the influence of constructive waves during the periods of high spring tide. The rate of progradation averaged just over 5 m yr^{-1} between 1600 and 1800. It appears that a very important factor in maintaining the sharpness of a cuspate foreland like Dungeness is the narrowness of the marine strait in which it is situated. At Dungeness powerful storm waves come either from the south-south-west with a fetch of 200 km to the coast of Normandy, or from the east-north-east through the Straits of Dover with a slightly shorter fetch. The waves approaching from directly across the Channel are relatively impotent and unable to blunt the point of the foreland.

Barrier bars and islands Offshore accumulations of marine sediment enclosing a lagoon on their landward side are referred to as barriers. They may assume many forms but two of the most common are long narrow bars trending approximately parallel to the coast, and shorter islands that are slightly arcuate in plan with their convex side pointing seaward. Although barriers of this type are very common, being well exemplified on the Atlantic seaboard of the United States, along the coast of the Gulf of Mexico, and on the southern shores of the North and Baltic Seas, their origin is still a matter of dispute. They are characteristic of very gently shelving coasts that have a low tidal range. In general, the more continuous bars are found where the tidal range is least, and the barrier becomes increasingly fragmented as the tidal amplitude rises. As early as the mid-nineteenth century it was suggested that most of the material in a barrier is formed from loose sea-floor sediments. This origin has been confirmed by many later investigations, but it leaves unexplained how the features are built above water-level. It is well known that waves can construct a small submarine bar close to their breakpoint. Requirements are a plentiful supply of debris and steeply plunging waves. However, there seems no simple way in which this process can continue to build the bar above water-level, and in any case the position of the breakpoint will fluctuate with the changing tide. Several hypotheses have been proposed to overcome this problem. The first envisages a fall in sea-level which exposes the submarine bar so that it becomes the beach and can thereafter be built up by the swash. The difficulty with this explanation is a lack of convincing evidence for a post-glacial decline in sea-level. Moreover, many of the best examples of barrier coasts are found in areas subject to long-term tectonic depression. A second hypothesis suggests that very temporary rises in sea-level may form small bars which become the nucleus for later development by swash action. For instance, a surge associated with a hurricane might initiate a breakpoint bar capable of enlargement in this way. There is, however, little direct evidence to support this view, and many workers have sought alternative explanations. When the nature of Holocene sea-level changes is borne in mind, barriers seem more likely to be associated with a transgression than a regression. Hoyt has argued that barriers off the south-eastern United States are old beach ridge systems, topped by sand dunes, that have been partially submerged by the continually rising water level. Others have maintained that no change in level is necessary for the production of a barrier. Otvos has claimed that most of such features in the Gulf of Mexico are no more than 5 000 years old so that they began to form after the post-glacial rise was virtually complete. He contends that submerged shoal areas can be built up so close to sea-level that eventually the action of swash takes over and constructs them into an island or bar. He quotes examples of islands which have evolved in this way within historical times.

Although sea-level changes cannot be ignored it seems unlikely that they are essential for the formation of barriers. The major need is a gently shelving coast with a copious supply of debris. Material may come from the sea-floor after inundation of a coastal plain strewn with unconsolidated sediments. Alternatively material may be supplied by rivers, as off the Mississippi delta, or by fluvioglacial streams, as along parts of the Icelandic coast. There is no doubt that longshore movement is an important mechanism by which sediment is added to the developing barrier. Many islands move laterally along the coast, being eroded at one end and simultaneously extended at the other. Other barriers may prograde seawards or retreat landwards. Seaward movement of an actual bar is unusual and progradation is normally by addition of an extra ridge on the oceanic margin. More common is a landward migration by which bars and islands originating offshore eventually connect with coastal headlands. A barrier attached to the mainland in this way may be difficult to distinguish from a true spit, and it has become increasingly clear that features once described as spits could have had a rather different origin. The problem is well exemplified by Scolt Head Island and Blakeney Point on the northern coast of East Anglia (Fig. 17.5). Early workers tended to assume that both these features were spits, but more recent studies have disclosed many details inconsistent with that origin. Steers, for instance, has concluded that Scolt Head Island was probably initiated as an offshore feature which migrated landwards until finally lodged in its present position at the coast. This interpretation was suggested in part by a gap which superficially resembles a recent storm breach but which detailed study shows to be of considerable antiquity. In other words the insular form is not a late modification but is an integral part of the structure. At first sight Blakeney Point appears a much more characteristic spit with no break in its continuity. Nevertheless, there are certain puzzling aspects which imply that even this shingle feature is more complex than might be supposed. Observations reveal that the beach material currently moves both eastwards and westwards according to the direction of wave approach; if anything, easterly movement predominates over westerly at the present time. There is no evidence for presentday replenishment by longshore drifting, so that the feature is virtually a fossilized shingle ridge being constantly reshaped by wave action.

Chesil Beach on the coast of the English Channel is another shingle structure which fits uncomfortably into most classificatory schemes. It has been the subject of much research owing to the unusually clear particle-size grading that it exhibits. Whether properly classified as a barrier is still in doubt, but it is worth fuller description for some of the principles it illustrates. Chesil Beach extends some 28 km westwards from Chesilton on the Isle of Portland to an arbitrary limit near Bridport on the mainland (Fig. 17.6). Although attached at both ends, for a distance of about 18 km it is backed by the tidal lagoon known as the Fleet with a maximum depth of 3 m below mean sea-level. The beach is over 150 m wide where it shelters the Fleet but rather narrower at both extremities. Its crest tends to rise eastwards, reaching a maximum elevation of over 14 m above mean sea-level near Chesilton. At Bridport the constituent pebbles are some 8 mm in diameter, but towards the Chesilton end they reach a mean diameter of about 60 mm. The size grading has been studied in great detail by Carr who collected samples totalling 89 000 pebbles from twenty-three sections across the beach. All the samples near high-water mark were well sorted and unimodal, but those at low-water mark tended to show bimodality and a rather greater spread of values. Skin-divers have demonstrated that size grading scarcely exists immediately offshore, where much of the material is coarser than that found anywhere on the beach, and the mean size of recovered samples is about the same as that occurring only 4 km from Chesilton. From these observations the grading appears to be the product of waves breaking on a bank of poorly sorted debris; in essence, longshore transport in the surf zone carries material to the point on the beach appropriate to its size. Many

Fig 17.6 A map and section of Chesil Beach. Figures at intervals along the beach indicate average size of material (in cm) as given by Neats (1967). Section and dotted contour of bedrock surface from Carr (1973).

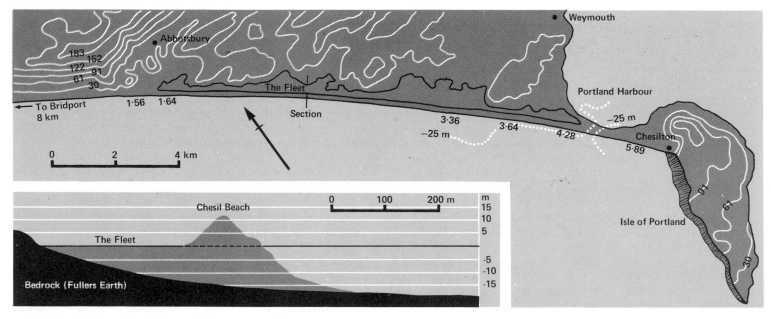

experiments have been undertaken to monitor the movement of specially prepared pebbles. It has been shown that coarser debris tends to move eastwards at a variable rate. Seventeen thousand pebbles of quartz granulite were injected on the beach by Carr. Taken from a beach in Scotland these consisted of sizes at least as large as those occurring naturally at the point of injection. After one day the mean distance travelled was 87·6 m and after 165 days just over 1 km. It appears that sediments move to the point where they are in equilibrium with prevailing wave conditions. Chesil Beach is aligned almost normal to the approach of the largest storm waves and although many smaller waves approach obliquely it seems improbable that they are the fundamental cause of the size grading. It is more likely that the grading is related to wave energy variations along the length of the beach. Wave energy is greater at the eastern than at the western end owing to the steeper offshore profile near Chesilton.

In the course of studying the size grading Carr also analysed the composition of the pebbles on Chesil Beach. About 98·5 per cent of the material exposed at the surface consists of flint and chert. A significant proportion of the remainder is composed of quartz, quartzite, sandstone and limestone, together with rarer pebbles of such rocks as granite, serpentine and greywacke. Although it is believed that virtually all the material can be matched somewhere in south-western England, owing to impeded transport it is difficult to conceive of some of the far-travelled fragments being supplied solely by longshore drifting; moreover, the idea of material currently being added to the western end of the beach is inconsistent with such perfect size grading. The observed degree of sorting is only possible where one is dealing with a virtually closed system. This, of course, does not eliminate the feasibility of longshore drifting as a significant factor in the earlier growth of the beach.

Evidence on the development of the beach is provided by several deep boreholes. The marine sediments rest on a gentle bedrock slope composed of a series of separate benches. The lowest lies at about −15 m, and the sequence continues upwards until eventually encompassing several that lie above current sea-level on the landward side of the Fleet; the whole series must presumably have been cut during Pleistocene interglacial episodes. Beneath the current shingle beach boreholes have proved a layer of sand that is in turn underlain by a pebbly deposit containing up to 45 per cent limestone. This limestone shingle occurs at a height of −13 to −15 m and is thought to be a locally derived shoreline deposit that accumulated at a time when the precursor of Chesil Beach was still some distance out in the English Channel. On the basis of eustatic curves this stage is likely to have been reached 7 500 years ago. Shortly afterwards the shingle accumulation as we know it today began to affect conditions near the present coast. This is shown by a bed of peat which underlies much of the Fleet at between −2 and −5 m. Radiocarbon assays have proved the peat to date from between 6 000 and 4 000 years ago, and by this time there must presumably have been a barrier protecting the lagoon from the sea. The shingle ridge may subsequently have pushed slightly further landwards, but since the mid-nineteenth century its location has been relatively constant. A survey of the beach by Sir John Coode in 1852 has been compared with one completed in the 1960s. This reveals that the crest of the ridge maintained its position almost unchanged for over 100 years, but in that time was raised along much of its length by about 1·5 m; on the other hand, near Chesilton its crest fell by 2·5 m in the same period.

Chesil Beach illustrates clearly that modern depositional forms, despite the rapid changes they sometimes exhibit, can only be fully understood in the context of a long and complex history. Two points merit particular emphasis. Firstly, the sea along the Dorset coast has recently re-occupied a position that it held in the past and the relict landforms from that earlier period significantly affect the coastal features observable today. Secondly, during the Holocene transgression constructional forms initiated beyond the limits of the present coastline have been pushed landwards as the sea continued to rise. Chesil Beach is one of a number of major shingle accumulations that appear to be fossilized in the sense that they are no longer being actively nourished by longshore drifting.

Tidal flats In localities sheltered from wave action accretion takes the form of mud flats more or less flooded during each tidal cycle. Such mud flats are especially common in estuaries and in the lee of spits and barriers. A very significant factor in their development is growth of vegetation adapted to periodic inundation by salt water. The pioneer plant communities vary according to the local climate, but two of the most common are those associated with salt marshes in temperate regions and mangrove swamps in tropical regions.

Salt marsh plants display a delicate adaptation to the varied environment on a tidal flat. This commonly involves a zonation related to altitude and therefore to the proportion of time during which the flat is submerged. The first colonizer is often *Zostera* spp. or *Spartina* spp. The latter has been a particularly effective pioneer in north-western Europe since the vigorous hybrid *Spartina townsendii* originated in Southampton Water as a cross between native British and American species in the 1870s. Another early colonizer on many tidal flats is *Salicornia* spp. which is unusually well adapted to conditions of high salinity. These pioneers assist in the trapping of fine sediment brought in by the tide, thereby accelerating the rate of surface accretion. This in turn permits other plant communities, less tolerant of frequent submergence, to develop. Ultimately the mature marsh is inundated only at the highest spring tides and is characterized by such species as *Juncus maritimus* and *Artemisia maritima*. A very slow change to fresh-water conditions may occur thereafter. This is promoted by invasion of reed-swamp communities dominated by species like *Phragmites communis* that can withstand brackish water.

The plant succession varies considerably from one locality to another, depending on such factors as exposure, climate, tidal range and nature of the substrate. One of the most interesting morphological features of tidal flats is the meandering system of creeks along which discharge on the ebbing tide is concentrated. These are kept clear by scouring while the adjacent flatter areas are raised by accretion. Miniature levees may be constructed by sedimentation around such plants as *Halimione portulacoides* which tends to grow well in that situation. Attempts at measurement have generally shown the rate of accretion to vary significantly over short distances according to the nature of the vegetation cover and other factors. It is usually highest on areas subject to regular tidal inundation and where the plant cover is most dense. On flats covered with *Spartina townsendii* upward growth not infrequently attains values between 10 and 50 mm yr^{-1} and it is clear that natural reclamation of land from the sea can take place extremely rapidly; a fully mature salt marsh may evolve in a matter of a century or two.

Mangrove swamps are almost totally confined to the tropics. All mangrove species are readily killed by frost and well-developed mangrove is restricted to coasts where the coldest month has an average temperature of 20°C or more. Growth is confined to low energy coasts where there is an abundance of fine mud. Two different types of tree are common, those with stilt roots and those that throw up vertical pneumatophores. Both have the effect of reducing water velocity and thereby promoting settlement of any fine suspended load. There has been considerable dispute as to their effectiveness, but most workers agree that they do substantially increase the rate of accretion. As on a salt marsh, drainage on the ebb tide takes place along well defined channels, in some instances forming a reticulate pattern, in others a dendritic pattern.

Coasts dominated by organic activity: coral reefs

In the preceding section attention was directed to coastal forms in which sedimentation is greatly affected by organic growth. Many further instances might be quoted of interaction between physical and biological processes. However, the outstanding example of a shoreline dominated by organic activity is that constructed by the coral polyp.

Reef-building corals have demanding environmental requirements that strictly limit their distribution. Although the main structure of a coral reef is built by the polyps, they live in symbiosis with minute algae, and it appears to be the life requirements of the algae that are the limiting factor in a number of vital respects. Global distribution is primarily determined by temperature. The extreme limits are generally taken to be the 18° and 36°C surface-water isotherms for the coldest and hottest months respectively. However, optimum growth conditions occur within a much narrower band, usually defined as between 25° and 30°C. Given these temperature requirements, the polyps need salinity values between 27 and 40 parts per mille, water that is reasonably free of sediment, and a rapid circulation that constantly replenishes the supply of nutrients and oxygen. Such conditions are met along many tropical coasts with low to moderate swell. Opposite large river mouths dilution with fresh water and increased turbidity may provide sites that are locally inimical. The final necessity for active reef development is light for algal photosynthesis. This is of crucial importance since it limits the depth of prolific coral growth. Although living corals have been recorded at −100 m they are very rare at these depths and vigorous reef formation is restricted to within 25 m of the surface. The significance of this last figure lies in the fact that ancient reefs often extend into much deeper water, but before discussing the problem that this poses it is necessary to outline the characteristic reef forms.

In describing coral reefs in the mid-nineteenth century Darwin employed a simple classification which is still in use today. It distinguishes three main types: fringing reefs that directly border the coast; barrier reefs that are separated from the coast by a deep channel; and atolls that form islands encircling a lagoon. Although difficulties occasionally arise in applying this classification, and there are certainly isolated platform reefs that do not fit readily into any of the three categories, it still constitutes a valuable basis for discussing concepts of reef development.

Darwin propounded a hypothesis which neatly linked together the three types of reef which he had distinguished and the known fact that corals could not grow in deep water. He argued that the only feasible explanation of atolls, all completely accordant in level yet arising from the deep ocean floors, is subsidence of the foundations on which they rest. Such subsidence would account for fringing reefs around an island first becoming barrier reefs by continued upward growth; eventually, if the process were to continue, the barrier reef would be converted into an atoll by disappearance of the central peak.

Many of the characteristics which led Darwin to propose the theory of subsidence can be explained equally well by a rising sea-level and it was natural that, as the idea of glacio-eustatic fluctuations gained currency, its implications for the growth of coral reefs should be considered. In 1910 Daly proposed a theory of glacial control. He argued that, during a glacial period, continued reef growth outside a narrow equatorial zone would have been precluded by the drop in water temperature. At the same time any existing reef would have been subject to rapid marine erosion, with potential truncation of oceanic islands yielding platforms from which atolls could later grow upwards as sea-level rose during amelioration of the climate. In support of this thesis he claimed that atoll lagoons show a general accordance of level, rarely being deeper than about 75 m.

In order to discriminate between the subsidence and glacial control hypotheses it is essential to have information regarding the structure and age of the foundations on which modern coral reefs are built. The earliest deep boring on an atoll was at Funafuti in the Ellice Islands where coral was penetrated to a depth of 339 m without the bottom being reached; more recent seismic explorations suggest that the base of the reef extends to a depth of about 900 m. In 1936 a borehole through the uplifted atoll of Kita Daito Jima just south of Japan continued to a depth of 432 m without striking basement rocks. A later drilling on Bikini atoll in the Marshall Islands proved coral to a depth of 640 m, at which level the limestone is believed to be of Oligocene age. On the nearby Eniwetok atoll a borehole encountered the bedrock foundation at 1 267 m immediately underlying shallow-water Eocene limestones. All these findings leave little doubt regarding the validity of Darwin's original hypothesis, although the sequence of events appears to have been rather more complex than he envisaged. Breaks in the continuity of reef growth indicate that some of the islands were periodically elevated above sea-level and subject to erosion. There is still inadequate evidence to show whether the periods of erosion were contemporaneous over wide areas, but it is generally presumed they were due to local tectonic movements rather than eustatic changes.

Confirmation of the subsidence hypothesis does not itself refute the glacial control hypothesis. Glacio-eustatic fluctuations are potentially a factor of great significance in the Pleistocene evolution of all reefs. How far an existing reef can be modified during a single glacio-eustatic oscillation obviously depends upon the rates of reef erosion and growth. Daly contended that, during a glacial period of low sea-level, extensive planation might occur. More recent assessments have cast doubt upon the validity of this idea. Current marine solution along notches at mean sea-level has been found to average about 1 mm yr⁻¹, and if this figure is only approximately correct it appears to preclude extensive marine planation during a short-lived glacial stillstand. Recourse cannot be had to sub-aerial denudation as the agent responsible for reef erosion. If the final glaciation is regarded as representative, a glacio-eustatic cycle lasts some 100 000 years. Adopting this value and an annual runoff of 1 m, it can be calculated that, even if fully saturated, the water would remove a layer only some 10 m thick. No great accuracy can be claimed for such figures, but even if only of the right order of magnitude they demonstrate the inability of erosional processes to destroy an exposed reef within the brief time available. The logical conclusion is that a glacio-eustatic fall in sea-level can modify a reef but not destroy it. As sea-level rises renewed growth may induce further change. The rate at which reef growth can take place varies with different species but commonly lies within the range 10 to 40 mm yr⁻¹. An allowance must be made for the frequency with which coral material is broken from a reef and added to submarine talus slopes, but upwards extension at a rate of 15 mm yr⁻¹ appears representative of reasonably favourable conditions. In such circumstances coral growth would be able to keep pace with the post-glacial eustatic rise in sea-level which took place at about 10 mm yr⁻¹. The result is presumably a veneer of recent reef adhering to a much older core. As yet there is little direct evidence to confirm this hypothetical evolution. Nevertheless, material from close to sea-level on a number of reefs has yielded radio-carbon dates in excess of 30 000 years. On others, dates of only a few thousand years have been obtained so that there is limited support for the provisional history outlined above. An interesting aspect of the very young dates is that they have sometimes come from material above present sea-level and so have helped to keep alive the controversy concerning the date at which the sea reached its culminating post-glacial level.

ARBER, M. A., 1973. 'Landslips near Lyme Regis', *Proc. Geol. Asso.*, 84, 121–33.

BLOOM, A. L., 1965. 'The explanatory description of coasts', *Z. Geomorph. NF*, 9, 422–36.

BRUNSDEN, D., 1974. 'The degradation of a coastal slope, Dorset, England', in *Progress in Geomorphology*, Inst. Brit. Geogr., Spec. Pub. No. 7.

BYRNE, J. V., 1963. 'Coastal erosion, northern Oregon' in *Essays in Marine Geology in Honor of K. O Emery*, Univ. S. Calif. Press.

CARR, A. P., 1969. 'Size grading along a pebble beach : Chesil Beach, England', *J. Sed. Pet.*, 39, 297–311.

CARR, A. P., 1971. 'Experiments on longshore transport and sorting of pebbles : Chesil Beach, England', *J. Sed. Pet.*, 41, 1084–104.

CARR, A. P. and M. W. L. BLACKLEY, 1973. 'Investigations bearing on the age and development of Chesil Beach, Dorset, and the associated area', *Trans. Inst. Brit. Geogr.*, 58, 99–111.

CARR, A. P. and R. GLEASON, 1971. 'Chesil Beach, Dorset and the cartographic evidence of Sir John Coode', *Proc. Dorset Nat. Hist. and Arch. Soc.*, 93, 125–31.

DAVIES, J. L., 1972. *Geographical Variation in Coastal Development*, Oliver and Boyd.

HOYT, J. H., 1967. 'Barrier island formation', *Bull. geol. Soc. Am.*, 78, 1125–36.

INMAN, D. L. and C. E. NORDSTROM, 1971. 'On the tectonic and morphologic classification of coasts', *J. Geol.*, 79, 1–21.

MAY, V. J., 1971. 'The retreat of chalk cliffs', *Geogrl. J.*, 137, 203–5.

NEATE, D. J. M., 1967. 'Underwater pebble grading of Chesil bank', *Proc. Geol. Ass.*, 78, 419–26.

OTVOS, E. G., 1970. 'Development and migration of barrier islands, northern Gulf of Mexico', *Bull. geol. Soc. Am.*, 81, 241–6.

ROBINSON, A. H. W., 1955. 'The harbour entrances of Poole, Christchurch and Pagham', *Geogrl J.*, 121, 33–50.

ROBINSON, A. H. W., 1966. 'Residual currents in relation to shoreline evolution of the East Anglian coast', *Marine Geol.*, 4, 57–84.

ROBINSON, A. H. W. and R. L. CLOET, 1953. 'Coastal evolution of Sandwich Bay', *Proc. Geol. Ass.*, 64, 69–82.

SCHOU, A., 1945. 'Det Marine Forland', *Folia Geogr. Danica*, 4, 1–236.

SO, C. L., 1965. 'Coastal platforms of the Isle of Thanet', *Trans. Inst. Brit. Geogr.*, 37, 147–56.

STEERS, J. A., 1960. *Scolt Head Island*, Heffer.

STODDART, D. R., 1969. 'Ecology and morphology of Recent coral reefs', *Biol. Rev.*, 44, 433–98.

TERWINDT, J. H. J., 1973. 'Sand movement in the in- and offshore tidal area of the S.W. part of the Netherlands', *Geol. en Mijn.*, 52, 69–77.

TRENHAILE, A. S., 1974. 'The geometry of shore platforms in England and Wales', *Trans. Inst. Brit. Geogr.*, 62, 129–42.

VALENTIN, H., 1952. *Die Kusten der Erde*, Petermanns Geogr. Mitt.

VALENTIN, H., 1954. 'Der Landverlust in Holderness, Ostengland von 1852 bis 1952', *Die Erde*, 3, 296–315.

WILSON, G., 1952. 'The influence of rock structures on coastline and cliff development around Tintagel, north Cornwall', *Proc. Geol. Ass.*, 63, 20–48.

WOOD, A., 1959. 'The erosional history of the cliffs around Aberystwyth', *Lpool Manchr geol. J.*, 2, 271–9.

WOOD, A., 1968. 'Beach platforms in the Chalk of Kent, England', *Z. Geomorph.*, 12, 107–13.

WRIGHT, L. W., 1970. 'Variation in the level of the cliff/shore platform junction along the south coast of Great Britain', *Marine Geol.*, 9, 347–53.

Selected bibliography

Among recent texts dealing with the evolution of coastal landforms in general are : J. L. Davies, *Geographical Variation in Coastal Development*, Oliver and Boyd, 1972 ; and E. C. F. Bird, *Coasts*, M.I.T. Press, 1969.

The standard works on the coastal landforms of the British Isles are J. A. Steers, *The Coastline of England and Wales*, CUP, 1964 and *The Coastline of Scotland*, CUP, 1973.

Throughout the text measurements are quoted in familiar metric units which have the advantage of being relatively easily visualized. However, for scientific work increasing use is being made of the Système International d'Unités which is an extension and refinement of the traditional metric system. As this gains currency, the SI units will doubtless themselves become familiar, and below is given a table for converting the units quoted in this book into the corresponding SI units.

Physical quantity	Metric unit used in text	SI unit	Conversion equivalent
Mass	kilogramme (kg)	kilogramme (kg)	
Density	gramme per cubic centimetre ($g\ cm^{-3}$)	kilogramme per cubic metre ($kg\ m^{-3}$)	$1\ g\ cm^{-3} = 10^3\ kg\ m^{-3}$
Force	kilogramme force (kgf)	newton (N)	$1\ kgf = 9 \cdot 81\ N$
Pressure	kilogramme force per square centimetre ($kgf\ cm^{-2}$)	newton per square metre ($N\ m^{-2}$)	$1\ kgf\ cm^{-2} = 9 \cdot 81 \times 10^4\ N\ m^{-2}$
Viscosity	poise (P)	newton second per square metre ($N\ s\ m^{-2}$)	$1\ P = 10^{-1}\ N\ s\ m^{-2}$
Energy potential	kilogramme force elevation (kgf m)	joule (J)	$1\ kgf\ m = 9 \cdot 81\ J$
Heat	gramme-calorie (cal)	joule (J)	$1\ cal = 4 \cdot 1868\ J$

Index